Dynamics of Polyatomic Van der Waals Complexes

NATO ASI Series

Advanced Science Institutes Series

A series presenting the results of activities sponsored by the NATO Science Committee, which aims at the dissemination of advanced scientific and technological knowledge, with a view to strengthening links between scientific communities.

The series is published by an international board of publishers in conjunction with the NATO Scientific Affairs Division

A	**Life Sciences**	Plenum Publishing Corporation
B	**Physics**	New York and London
C	**Mathematical**	Kluwer Academic Publishers
	and Physical Sciences	Dordrecht, Boston, and London
D	**Behavioral and Social Sciences**	
E	**Applied Sciences**	
F	**Computer and Systems Sciences**	Springer-Verlag
G	**Ecological Sciences**	Berlin, Heidelberg, New York, London,
H	**Cell Biology**	Paris, and Tokyo

Recent Volumes in this Series

Series B: Physics

Dynamics of Polyatomic Van der Waals Complexes

Edited by

Nadine Halberstadt

CNRS, Université de Paris-Sud
Orsay, France

and

Kenneth C. Janda

University of Pittsburgh
Pittsburgh, Pennsylvania

Plenum Press
New York and London
Published in cooperation with NATO Scientific Affairs Division

Proceedings of a NATO Advanced Research Workshop on
Dynamics of Polyatomic Van der Waals Complexes,
held August 21-26, 1989,
in Château de Bonas, Castéra-Verduzan, France

Library of Congress Cataloging-in-Publication Data

NATO Advanced Research Workshop on Dynamics of Polyatomic Van der
  Waals Complexes (1989 : Castéra-Verduzan, France)
   Dynamics of polyatomic Van der Waals complexes / edited by Nadine
Halberstadt and Kenneth C. Janda.
      p. cm. -- (NATO ASI series. Series B, Physics ; v. 227)
   "Proceedings of a NATO Advanced Research Workshop on Dynamics of
Polyatomic Van der Waals Complexes, held August 21-26, 1989 in
Château de Bonas, Castéra-Verduzan, France"--T.p. verso.
   "Published in cooperation with NATO Scientific Affairs Division."
   Includes bibliographical references and index.
   ISBN 978-1-4684-8011-5   ISBN 978-1-4684-8009-2 (eBook)
   DOI 10.1007/978-1-4684-8009-2
   1. Molecular association--Congresses. 2. Van der Waals forces-
-Congresses.   I. Halberstadt, Nadine. II. Janda, Kenneth C.
III. North Atlantic Treaty Organization. Scientific Affairs
Division. IV. Title. V. Series.
QD461.N333 1989
541.2'2--dc20   90-44241
                                                          CIP

© 1990 Plenum Press, New York
Softcover reprint of the hardcover 1st edition 1990
A Division of Plenum Publishing Corporation
233 Spring Street, New York, N.Y. 10013

PREFACE

This publication is the Proceedings of the NATO Advanced Research Workshop (ARW) on the Dynamics of Polyatomic Van der Waals Molecules held at the Château de Bonas, Castéra-Verduzan, France, from August 21 through August 26, 1989.

Van der Waals complexes provide important model problems for understanding energy transfer and dissipation. These processes can be described in great detail for Van der Waals complexes, and the insight gained from such studies can be applied to more complicated chemical problems that are not amenable to detailed study. The workshop concentrated on the current questions and future prospects for extending our highly detailed knowledge of triatomic Van der Waals molecule dynamics to polyatomic molecules and clusters (one molecule surrounded by several, or up to several tens of, atoms). Both experimental and theoretical studies were discussed, with particular emphasis on the dynamical behavior of dissociation as observed in the distributions of quantum states of the dissociation product molecules. The discussion of theoretical approaches covered the range from complete *ab initio* studies with a rigorous quantum mechanical treatment of the dynamics to the empirical determination of potential energy surfaces and a classical mechanical treatment of the dynamics. Time independent, time dependent and statistical approaches were considered.

The workshop brought together experts from different fields which, we hope, benefited from their mutual interaction around the central theme of the Dynamics of Van der Waals complexes. It was especially fruitful for experimentalists and theoreticians to interact which each other. New experiments raised new challenges to the theory and new theoretical developments provided answers to some old questions and raised new ones for the experimentalist to study.

We would like to offer at this point our deep appreciation to the lecturers, who devoted so much of their time and talents to make this workshop successful. As the reader of these proceedings can see, they succeeded in being both pedagogical and prospective, which was necessary for being understood from such a diverse community and for showing at the same time the perspectives of their own techniques.

The meeting was enhanced by fifty poster presentations which served as a focus for stimulating informal discussions. The posters remained on display for the entire week, and one evening was devoted to them. The discussion that evening was amongst the most animated of the conference as participants were able to show their latest results, often preliminary, and discuss future directions for the work. Several poster results have been selected to be included in this volume in a separate section entitled POSTERS.

We are especially grateful to the NATO Scientific Affairs Division, which provided not only the financial support for the Workshop, but equally important organizational and moral support. Also, the contribution from the french Atomic Energy Commission (CEA) and from the QUANTEL FRANCE company is gratefully acknowledged.

We would like to thank Mr and Mrs Simon for allowing the meeting in their beautiful property, the Château de Bonas. The location, the organization and the fact that the place is dedicated to educational or cultural activities, make Bonas an ideal site for a NATO ARW. The efficiency of Mr Stockmann, who was in charge of accomodations, food service and meeting facilities, was greatly appreciated.

Finally, most of the organizational work relied on Odile Dubost. We warmly acknowledge her attention to detail and professional skills, which were a key factor in creating an atmosphere that allowed the rest of us to concentrate on the scientific purpose of the meeting. We also thank the local committee: J. Alberto Beswick, Octavio Roncero and Arne Keller for their kind assistance.

Nadine Halberstadt,   and   Ken C. Janda,
Orsay,   Pittsburgh,

December, 1989

CONTENTS

INTRODUCTION

DYNAMICS OF GROUND STATE MOLECULES; MOSTLY EXPERIMENTS

SMALL MOLECULE THEORY

EXPERIMENTS: SMALL MOLECULES, UNPAIRED ELECTRONS

DIATOMIC DIMERS: THEORY

DYNAMICS: LARGER MOLECULES; MOSTLY EXPERIMENTS

MORE DYNAMICS OF LARGER MOLECULES

APPROXIMATE TREATMENTS FOR POLYATOMIC DIMERS

DYNAMICS OF LARGER CLUSTERS

POSTERS

Current Problems and Future Prospects for Polyatomic Van der Waals Molecules and Small Clusters: Experiments

William Klemperer and David Yaron

Harvard University

12 Oxford St., Cambridge MA 02138

The topic of van der Waals molecules has been reviewed repeatedly for a number of years and is likely to continue to attract widespread attention.[1, 2, 3, 4] There are a variety of reasons for this interest, although clearly the role of intermolecular forces in the aggregation of matter is the major motivating force. The characterization of the bound states arising from an intermolecular potential is made possible by a wonderful variety of tools. Introspectively, some of the enjoyment of this subject, the spectroscopy of weakly bonded complexes, has been that much of the early work in this area was intrinsically detailed high resolution spectroscopy which provided relatively definitive experimental data. Thus from the early work of Harry Welsh and his students[5], it was clear that the spectroscopic studies of bound states could indeed provide us with a reliable data base upon which to discuss intermolecular forces. It is fair to point out that the relative complexity of the molecular dynamics that occur for non-rigid systems requires a great amount of high resolution data to provide a sense of assurance in one's structural conclusions. This type of data is in hand for a number of binary complexes, thus much progress has been made toward determining and understanding the detailed nature of anisotropic intermolecular forces.

This report is intended to be primarily a qualitative reflection on the characterization of weakly bound molecular complexes— van der Waals molecules. There has been a veritable explosion of spectroscopic studies on these systems, mostly under cryogenic conditions in supersonic jets. The spectrometric methods available range in frequency from the radio region through the ultraviolet, effectively dc to daylight. The present knowledge of the structure of gas phase molecular complexes is extensive. A great advantage of measurements at low temperatures is that they direct our attention to what is, at least initially, perceived as most important. In the case of a molecular complex consisting of polyatomic subsystems it is difficult, with a finite data base, to summarize the totality of bound states. Thus it is appealing, certainly as the first goal of a structural study, to determine the geometry of the region of minimum energy and it is, of course, this region that cryogenic methods clarify spectrally. The most primitive characterization of minimum energy geometry consists of at least two parts; average structure and an indication of the dispersion or zero point oscillation amplitudes. A second clear

Dynamics of Polyatomic Van der Waals Complexes
Edited by N. Halberstadt and K. C. Janda
Plenum Press, New York, 1990

advantage of the supersonic jet is that it leaves the molecule in isolation during spectral measurement, and thus amenable to relatively complete theoretical treatment. For example, the species of interest is perfectly characterized by having angular momentum as a good quantum number (until the investigator shifts spectral lines with the application of electric and magnetic fields). It is remarkable how reliable the high resolution methodologies made possible by the free molecule condition have been. Certainly, at least for rigid molecules, the characterization of molecular structure by rotational spectroscopy has been relatively error free given a modicum of patience in the analysis and interpretation.

The jet has been called a new state of matter; which is ultimately due to the extremely low temperatures that are obtained in a jet. As with matrix isolation techniques, the apparent metastability of species in a jet produces this effective "new state of matter". It is this metastability that can also produce ambiguity in a predictive understanding of the species and states existing in jets and also those likely to exist in matrices. The concept of an all encompassing Boltzmann distribution with one temperature characterizing all degrees of freedom of the system is clearly not applicable. The problem of systematizing our knowledge of the relaxation of different degrees of freedom is one that, with present techniques, is certainly capable of solution. By far the most important concern is that of isomeric forms. Initially one of us assumed that isomers never occurred. In our own work, always utilizing argon as expansion gas, the jet appeared characterized by a 10K temperature, with respect to soft degrees of freedom such as rotation. It is clear that there is a very low probability for structural degeneracy at the 10K level. However, it is clear from numerous studies that isomeric forms frequently occur and that they appear more frequently when He is used as the carrier.[6]

We have believed that the reason isomeric forms did not occur was in the kinetics of dimer formation. The mechanism of formation of the generic form XY under conditions where the species X and Y were mixed at about 1% in Ar was thought to be:

$$3Ar = Ar_2 + Ar \tag{1}$$

$$Ar_2 + X = ArX + Ar \tag{2}$$

$$Ar_2 + Y = ArY + Ar \tag{3}$$

$$ArX + Y = \{ArXY\} = XY + Ar \tag{4}$$

$$ArY + X = \{ArXY\} = XY + Ar \tag{5}$$

The decomposition of the intermediate complex $\{ArXY\}$ to the lowest energy form of XY was simply the consequence of the number of accessible translational states for the argon. With helium, He_2 is in all likelihood insignificant as a reagent. The species HeX and HeY are also less likely as significant reagents. The production of XY will then occur by a) cooling of the XY existing under stagnation conditions, and b) the three body process $X+Y+He = XY + He$. Again due to the small binding energy of He complexes, the likely path of the three body process is formation of metastable XY followed by collisional stabilization with He. It appears obvious that the freezing in of high energy isomers is more likely in this case than for Ar.

As the emphasis of this report is on experimental structural characterizations of van der Waals complexes, perhaps the most fundamental question to ask is whether there is any need for experimental structural characterizations or whether indeed theory has evolved to the point where it can provide a more accurate overall representation of the potential energy surface of any molecular system than can experiment. This is certainly a legitimate question to ask and probably should be fully explored. While for the diatomic molecule it is well known that the rotation-vibration energy levels may be inverted through the RKR procedure to provide *the* potential energy curve to totally adequate accuracy, there really is little reason at present to believe that a similar procedure is available for the polyatomic system. [Parenthetically, recent studies by our colleagues Alice Smith and Kevin Lehmann[7, 8] on the vibrational energy levels and band absorption intensities in hydrogen cyanide lead one to suspect that the accurate ab initio electron structure calculations of Peter Bottschwina[9, 10] on this system reflect the real potential energy surface and vibrational wave functions more reliably in the whole than do potential energy surfaces derived by fitting the observed vibrational energy levels. It is found that while band origins may be well fit with an empirical potential (about thirty bands with energies up to 16000cm^{-1} are fit with an rms error of 3cm^{-1}), the band intensities, which display a dynamic range of greater than eight orders of magnitude, are relatively poorly fit when the wave functions from the empirical potential are combined with the dipole moment function from electron structure calculations. A much better intensity fit on the whole is achieved by Bottschwina using completely theoretical methods, i.e. using both the potential energy function and dipole moment function from electron structure theory. The error in band origin fit is of course poorer than in the empirically fit potential, however, the band intensities are fit to within a factor of two for all bands but one (105 shows an anomalously high intensity). It therefore appears likely that the potential energy function based purely on electron structure calculations is a better representation of the real molecular potential than is the empirical potential.] We have in the polyatomic systems of weakly bound complexes, a clear example of the uncertainty about what methods will produce reliable potential energy surfaces. Presently, we probably do not have a fully theoretically justifiable answer to this question; thus this subject is one indeed for both formal theoretical investigation and also for steady tests of experimental and theoretical observations.

The simplest comparison between theory and experiment is that of the structure or minimum of the potential energy surface. The question arises as to whether there are virtually engineering recipes by which structures, and what is meant here is the equilibrium geometry of a complex, can be obtained in a reliable straightforward manner. There is little doubt that the present majority view is that the structure of a complex can be accurately predicted by considering the electrostatic interaction between the unperturbed subunits, along with repulsion between the hard sphere atom cores and perhaps a modest admixture of well determined polarizabilities. (It is truly important, at this time, to recall that approximately 15 years ago, Roy Gordon suggested that one could model interactions by treating the electrostatic interaction explicitly and modeling the exchange repulsion terms with an electron gas density functional theory[11]. Initially this appears unsatisfactory due to the overlap of charge distributions and the inherent distortions which result. However, in view of the wide spread success of electrostatic modelling, it may be that the repulsive cores, which have been so hard to treat explicitly, may be approachable through these approximate density functional theories.) The electrostatic models appear to provide an excellent description of hydrogen bonded systems. However, failures of the models can occur in situations where hydrogen bonds could but do not occur. The ammonia dimer is certainly an example of such a failure. It is also, interestingly enough, a difficult system for which to obtain the correct potential energy function by electron structure theory. We have used the model of Buckingham and Fowler,[12] and of Stone [13, 14] to examine the complexes of water and ammonia with CO, CO_2 and N_2O. Of the experimental structures for these complexes, only H_2O–CO is hydrogen bonded, and the model correctly predicts the geometric arrangement of this complex. The model fails to predict the correct structures of the remaining complexes, all of which have the lone pairs of the water or ammonia directed toward the binding partner. (We find that there is not even a local minimum at the experimental structure, except in NH_3–CO_2; in this case however, there is another nearby minimum, the effects of which would certainly have been observed in the experiments.) It appears that the model underestimates the tendency of water and ammonia to act as electron pair donors.

It is in the area of comparing experiment and theory beyond just the structure of the complex that a considerable chasm exists. The complexity of executing solutions to the multidimensional Schrodinger equation with accuracy adequate for the prediction of rotation–vibration transitions is probably too well appreciated (in the sense that each group hopes someone else will do it). Furthermore, the number of weakly bonded molecular complexes for which a surplus of rotation-vibration energy levels exists, is quite low. In this sense, the polyatomic van der Waals molecule is quite similar to the polyatomic chemically bonded molecule. It is probably useful to reexamine experience painfully gained in the study of stable molecules to see what is likely to be important in weakly bonded complexes. It is clear that matrix elements of operators other than energy have been extremely useful in providing independent checks of wave functions deduced from fitting energy levels. In particular, electric dipole matrix elements have been virtually the only independent test of the reliability of rotation-vibration wave functions. This point has been emphasized by Bob Leroy[15], in studies of inert gas-hydrogen complexes. While it is of course necessary to have a dipole moment function to execute this strategy, there are both theoretical and empirical procedures which are probably of adequate accuracy for this purpose. With the steady increase in the number of observed vibrational spectra, this point will in all likelihood become increasingly important.

It is fortunate that argon-hydrogen chlorine is truly, by the standards of the field, well characterized and thus provides a test of the adequacy of dynamical modelings. The far infrared studies, primarily by Saykally's group[16, 17], have given a very detailed data base upon which to model potential energy surfaces. Detailed calculations by Jeremy Hutson and Brian Howard[18, 19] seem to be consistent with experiment. Hutson has now produced a potential function that fits all present observations.[20] [We had thought that only rotation-vibration energy levels were used to fit the potential. Thus the many hyperfine and stark coefficients that have been measured to obtain essentially angular distributions of the HCl unit could be used to test the wavefunctions from the potential. The remarks by Jeremy Hutson at this meeting are important. He points out that all available data has been used in determining the parameters of the potential. It will therefore be important to determine new, unobvious data to test the adequacy of this sophisticated empirical potential.] At present, there appears to be no major problems since further studies in the 3000 cm^{-1} region by Brian Howard and Alan Pine[21] also show a fairly self consistent model of the problem. In some respects, however, and this has been pointed out to us by Fraser, Ar–HCl is a relatively simple system in that adiabatic separation techniques work well. Thus averaging over angle in the ground state produces an effective potential for radial motion. This reduction to a one dimensional effective radial problem allows the use of arguments such as those used in diatomic systems. Certainly, this is not likely to be the case for more typical systems where strong couplings between modes occur.

For the more complex coupled system, where perhaps Ar–HCN is the simplest example, what is required is less clear. The rotational spectrum of the Ar–HCN complex has some peculiar features which most likely arise from the presence of two potential minima: a global minimum which has a linear, hydrogen bonded geometry and a secondary minimum at the T-shaped configuration[22, 23]. The distortion constants of the complex are unusually large, and the isotopic scaling of $\cos\theta$ and $\cos2\theta$ is not as is expected for a reasonable single minimum potential. A considerable success has been obtained by Helen Leung and Mark Marshall[24] using Jeremy Hutson's full 2 dimensional approaches, the more recent technique being close coupling calculations. The calculations were computationally intensive and the number of parameters in the potential was large. Part of the motivation for the full 2 dimensional approach was that the minima in the potential differ in both R and θ (the linear hydrogen bonded form has a center of mass separation which is about an Angstrom longer than the T shaped form.) Therefore, it seemed that separation of variables was not feasible. Part of this perceived problem was do to the order in which variable separations are done in the BOARS method (Born Oppenheimer Angular Radial Separation). In BOARS calculations, an effective potential in R is generated by solving the angular problem for fixed distances. This technique misses an important part of the physics in the Ar–HCN problem. Namely, since the separation distance, R, is not the same in the two minima, the reduced mass for the angular motion is also different for the two wells. This effect can be included by reversing the order of the separation so that an effective potential and also an effective reduced mass are generated as functions of the polar angle, as in the rigid bender model. The wavefunction generated by the reversed BOARS method was checked using numerical relaxation techniques. This involves setting up the approximate wavefunction on a 2 dimensional numerical grid and letting it relax into agreement with the differential equation. The reversed BOARS' wavefunctions are reasonably accurate and if higher accuracy is needed reversed BOARS followed by relaxation appears to be a very time efficient approach. The initial calculations did not include the Coriolis terms, however, it should not be very difficult to generate higher vibrational states and then include the Coriolis term perturbatively. This is probably not a disadvantage of the technique since even in close coupling calculations it is usually more efficient, and less susceptible to numerical instabilities, to calculate the distortion constants perturbatively than by direct numerical methods. (The latter requires extremely high accuracy be achieved for the energy of each rotational state.)

The reduction from two to one dimensions is a great advantage for a qualitative understanding of the Ar–HCN system. The effective one dimensional potential has significantly fewer parameters and given a limited amount of experimental data, allows something concrete to be said about the potential. We find that the potential minimum at the T shaped configuration is not classically allowed, but the potential is low and flat enough that significant penetration into the region does occur (see figure 1). The first excited bending states should be interesting as they will be classically allowed in both regions and may even approach a free rotor. The infrared spectroscopic observation by Fraser[25] of the Π bending state

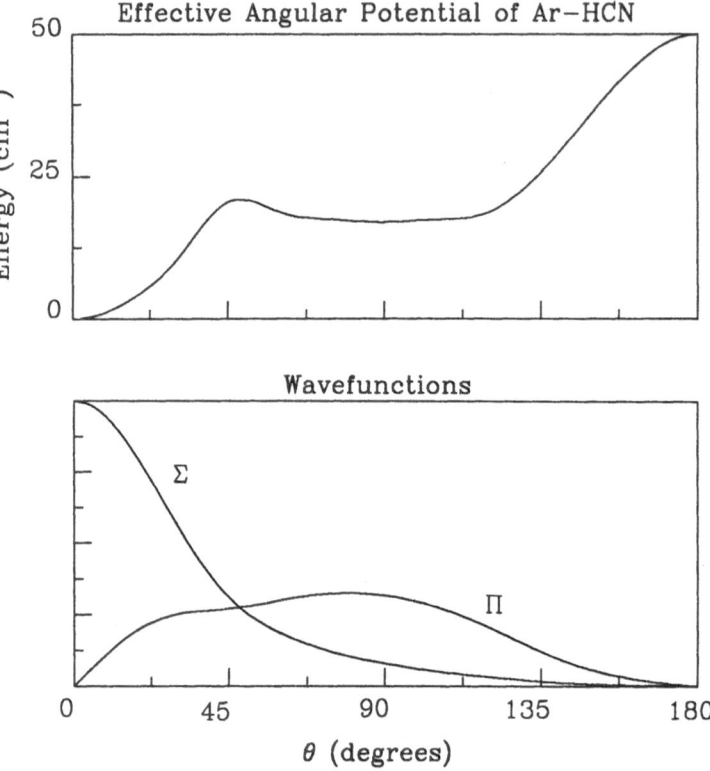

Figure 1. Effective one dimensional bending potential for Ar-HCN from the reversed BOARS procedure. $\theta=0$ corresponds to the linear hydrogen bonded geometry and $\theta=180^{\circ}$ to the linear nitrogen bonded geometry. Wavefunctions for the lowest Σ and Π bending states are shown. The center of mass separation was fixed at 4.52A for $\theta=0$ and at 3.40A for $\theta=90^{\circ}$. The fit is insensitive to the value of the potential at 180° and this was arbitrarily held at 50cm^{-1}.

at ≈ 8 cm^{-1} in a combination band of the CH valence stretching vibration is in good agreement with the picture presented here and also with predictions of Leung and Marshall's full 2 dimensional close coupling calculations. (The fully ab initio potential developed and reported at this conference by David Clary[26] certainly appears to be in good agreement with experimental observations. It also appears to us to be similar to the empirical potentials that we have used to fit properties of the ground vibrational state.)

There is little doubt that striking advances have been made in the area of polyatomic van der Waals molecules and small clusters. The increasing application of high resolution infrared spectroscopy has provided solutions to a number of problems. An intrinsic requirement of pure rotational spectroscopy, the existence of a (permanent) dipole moment, is removed allowing in particular the study of the non-polar complex. This is now firmly established by a number of well known examples[27].

A second dramatic change that infrared spectroscopy has brought about is a determination of the vibrational frequencies of the soft or van der Waals modes. In the previously discussed example of Ar-HCN a strong indication of the frequency of the bending mode is obtained from fitting of hyperfine structure constants in the ground vibrational state. These indeed indicated a small force constant for bending. The pathologically large and isotopically sensitive centrifugal distortion if naively (and incorrectly) interpreted as a stretching of the van der Waals bond produced an unbelievably soft bond. The demonstration by Fraser of the low bending frequency then is important in clarifying this problem. It seems likely that expectation values of positional operators are simply interpretable while those of energy perturbation effects are fraught with peril.

References

[1] van der waals interactions. *Chemical Reviews*, 88(6):815–988, 1988.

[2] A. Weber, editor. *Structure and Dynamics of Weakly Bound Molecular Complexes*. Reidel, Dordrecht, 1987.

[3] William Klemperer, David Yaron, and David D. Nelson Jr. *Faraday Discuss. Chem. Soc.*, 86:261, 1988.

[4] E.R. Bernstein, editor. *Atomic and Molecular Clusters*. Elsevier, 1989.

[5] see for example A.R.W. Mckellar pg.141-147 in ref. 2.

[6] G. Fraser has pointed out to us that higher energy isomers are usually observed in helium expansions.

[7] A.M. Smith and K.K. Lehmann. (to be published).

[8] A.M. Smith. PhD thesis, Harvard University, 1988.

[9] P. Bottschwina. *Faraday Discuss. Chem. Soc.*, 81:73, 1986.

[10] P. Bottschwina. *Faraday Discuss. Chem. Soc.*, 84:1263, 1988.

[11] R.G. Gordon and Y.S. Kim. *J. Chem Phys.*, 56:3122, 1972.

[12] A.D. Buckingham and P.W. Fowler. *Can. J. Chem.*, 63:2018, 1985.

[13] A.J. Stone. *Chem. Phys. Lett.*, 83:233, 1981.

[14] A.J. Stone and M. Alderton. *Molec. Phys.*, 56:1047, 1985.

[15] R.J. Leroy and J.S. Carley. *Adv. Chem. Phys.*, 42:353, 1980. (see early discussions).

[16] R.L. Robinson, D. Ray, D.H. Gwo, and R.J. Saykaly. *J. Chem Phys.*, 86:5211, 1987.

[17] R.L. Robinson, D. Ray, D.H. Gwo, and R.J. Saykaly. *J. Chem Phys.*, 87:5149, 1987.

[18] J.M. Hutson and B.J. Howard. *Mol. Phys.*, 43:493, 1981.

[19] J.M. Hutson and B.J. Howard. *Mol. Phys.*, 45:769, 1982.

[20] J. Hutson. Public Comment, Bonas, France Aug. 21, 1989.

[21] B.J. Howard and A.S. Pine. *Chem. Phys. Lett.*, 122:1, 1985.

[22] K.R. Leopold, G.T. Fraser, F.J. Lin, D.D. Nelson, Jr., and W. Klemperer. *J. Chem Phys.*, 81:4922, 1984.

[23] T.D. Klots, C.E. Dykstra, and H.S. Gutowski. *J. Chem Phys.*, 90, 1989.

[24] M. Marshall, H. Leung, and W. Klemperer. (poster at this conference).

[25] G.T. Fraser and A.S. Pine. *J. Chem Phys.*, 91:3319, 1989.

[26] David Clary. (at this conference).

[27] D. Nesbitt. *Chemical Reviews*, 88:843, 1988.

CURRENT PROBLEMS AND FUTURE PROSPECTS FOR POLYATOMIC VAN DER WAALS

MOLECULES AND SMALL CLUSTERS: THEORY

George E. Ewing

Department of Chemistry
Indiana University
Bloomington, IN 47405

A BRIEF HISTORY

The beginnings of the theory of dynamics of van der Waals molecules and small clusters go back many decades. Over forty years ago, Stepanov[1] made the suggestion that short vibrational predissociation lifetimes might account for the diffuse bands commonly observed in the infrared spectra of hydrogen bonding systems. In 1974 an estimate of this lifetime was provided by Klemperer[2] who scaled the gas phase vibrational relaxation rate constant for HF + HF* collisions by the frequency of vibration against the hydrogen bond in the complex HF$\cdots$HF*. Over the next several years three independent derivations of an identical analytical expression that provided vibrational predissociation lifetimes of van der Waals molecules were published by Coulson and Robertson[3], Beswick and Jortner[4] and Ewing[5]. Within this same period, Child and Ashton[6] performed the first numerical calculation of vibrational predissociation based on a realistic anisotropic intermolecular potential. The decade that has followed the development of these early theoretical approaches has seen a proliferation of models and calculations in response to the excellent experimental measurements that continue to call for explanation.

The supersonic jet molecular beam electronic spectroscopy experiments of Smalley, Levy and Wharton[7,8] on He$\cdots$I$_2$* were the first to reveal, from lineshape measurements, unambiguous vibrational predissociation lifetimes. These lifetimes, dependent on the vibrational quantum number of I$_2$ and of the order of 10^{-10} s, were successfully interpreted by the analytical model of Beswick and Jortner[4,9]. Another pioneering experimental technique developed by Gough, Miller and Scoles[10] used infrared spectroscopy of super-sonic molecular beams to probe vibrational predissociation of clusters. High resolution infrared laser measurements combined with a long path optical cell also provide valuable dynamics information. Pine and Lafferty for example, have studied HF$\cdots$HF*[11]. Their spectra of the bonded hydrogen stretching vibration region revealed a vibrational predissociation lifetime of 10^{-9} s. This time is 11 orders of magnitude shorter than the early prediction of Ewing[12] but 2 orders of magnitude longer than the estimate of

Dynamics of Polyatomic Van der Waals Complexes
Edited by N. Halberstadt and K. C. Janda
Plenum Press, New York, 1990

Klemperer[2]. Calculations by Coulson and Robertson[3] on other hydrogen bonding systems yielded times of $\sim 10^{18}$ s. While these discrepancies between theory and experiment are probably not a record, they give one pause to wonder about the factors which determine the efficiency of the vibrational predissociation process.

A SELECTION RULE AND SOME PROBLEMS

Not only do the experimental vibrational predissociation lifetimes require interpretation, so do the increasingly sophisticated theoretical calculations whose results often fall out of a web of coupled differential equations or the convoluted algebra of quantum mechanics. In order to offer a qualitative overview of dynamical processes in van der Waals molecules, we shall introduce a selection rule which can provide insight into possible relaxation channels of vibrationally excited molecules. This selection rule concerns the change in a quantum number, Δn_T, which is to remain small for efficient vibrational predissociation processes. It bears a close analogy to the selection rules of optical spectroscopy which require small changes in quantum numbers Δv, ΔJ, ΔS, etc. for efficient transitions between molecular states. Let us review the origin of the vibrational predissociation selection rule which has been developed in more detail elsewhere[13].

We begin by describing the vibrational predissociation of a generic van der Waals molecule:

$$A\text{-}B^* \cdots C \xrightarrow{\tau^{-1}} A\text{-}B + C + \Delta E \ . \tag{1}$$

The complex is described by $A\text{-}B^* \cdots C$ where $A\text{-}B^*$ is a vibrationally excited chemical bonded molecule attached by a van der Waals bond to atom (or molecule) C. Energy from the excited chemical bond is transferred to the van der Waals bond at rate, τ^{-1}, causing rupture and fragments A-B and C fly away with translational and rotational energy $\Delta E = E_t + E_r$. These fragments can in general be vibrationally excited as well.

In analytical treatments to obtain the vibrational predissociation rate[3-5], the van der Waals interaction is modeled by a Morse potential

$$V(r) = D_e[e^{-2a(r-r_e)} - 2e^{-a(r-r_e)}] \ . \tag{2}$$

Here D_e is the intermolecular potential well depth, r is the van der Waals bond length and its equilibrium separation is r_e. This radial coordinate is shown in Fig. 1. A measure of the steepness of the van der Waals potential is given by the range parameter a.

The rate is obtained from the Golden Rule expression[3-5,9,12,13]

$$\tau^{-1} = (4/\hbar^2 v_f) |<f|V_c|i>|^2 \tag{3}$$

where

$$v_f = (2E_t/\mu_t)^{1/2} \tag{4}$$

is the final velocity of the fragments A-B + C, in terms of their translational kinetic energy E_t with $\mu_t = m_{AB}m_C/(m_{AB} + m_C)$ the reduced mass of the
complex. The initial state wavefunction, given by $|i>$, defines the vibra-
tional and rotational state of the excited van der Waals molecule A-B*···C,
while $|f>$ gives the translational, vibrational and rotational motions of the
fragments. Couplings of the vibrational and rotational motions through the
van der Waals potential of eq. 2 is given by V_c. Using simple assumptions
about rotational and vibrational motions, analytical expressions for τ^{-1}
have been derived[9,12].

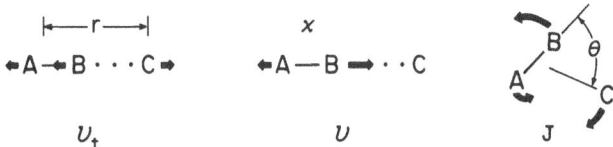

Figure 1.   Coordinates and quantum numbers for an idealized van der
            Waals molecule A-B···C. Adapted from ref. 12 with
            permission.

We can capture the qualitative aspects of this involved analytical
treatment if we write the rate as

$$\tau^{-1} \approx 10^{13} \exp\text{-}\pi(\Delta n_t + \Delta n_r + \Delta n_v) = 10^{13} \exp\text{-}\pi(\Delta n_T) \ . \tag{5}$$

The method of the extraction of eq. 5 from theoretical models is provided
elsewhere[13]. The selection rule expression, which loosely resembles the
early method for estimating vibrational predissociation rates by Klemperer[2],
has a satisfying physical explanation. The pre-exponential factor $\approx 10^{13}$ s^{-1}
gives the typical collision frequency of A-B against C through the van der
Waals bond. The exponential term reflects the probability that the initial
discrete state of the bound A-B*···C complex and final state of the fragments A-B + C will mix during the half-collision and is thus a measure of
the reluctance of the complex to change quantum numbers during vibrational
predissociation.

The exponential expression contains the change in the total <u>effective</u>
quantum number (Δn_T) defined as the sum of the translational (Δn_t),
rotational (Δn_r), and vibrational (Δn_v) quantum changes. The general

selection rule thus states that relaxation will proceed most efficiently by
that particular channel which corresponds to the smallest value of Δn_T, i.e.
the smallest change in the total effective quantum number. We will now
offer an interpretation of these quantum numbers.

The energy flow in van der Waals molecules requires a variety of coordinates and quantum states. Allow A-B to represent a chemically bonded diatomic molecule like I_2, HF or even a polyatomic molecule such as C_2H_4 or p-difluorobenzene. This molecule is attached through a van der Waals bond to atom (or molecule) C. Stretching vibration against the van der Waals bond along this coordinate is identified by the quantum number v_t. Vibration against the chemical bond of A-B is given by the displacement x and the quantum number v as in Fig.1. For polyatomic molecules we can use normal mode displacements and quantum numbers v_1, v_2, v_3, etc. Internal rotatory motions of A-B against C through the van der Waals interaction is given by the relative displacement angle θ and quantum number J. For cases where the barrier to internal rotation is very high, the J quantum number goes over into a vibrational motion of rotatory origin identified by quantum number v_r[12]. For polyatomic van der Waals molecules, additional angles as well as rotational and bending mode quantum numbers may be added.

The quantum numbers for the fragments of vibrational predissociation are correlated through the same van der Waals molecule coordinates. For the radial coordinate r, the v_t quantum number of the bound complex A-B$\cdots$C goes over into q_t, which is proportional to the translational quantum number of the fragments A-B + C. The free rotation of A-B relative to C about angle θ is identified by the J quantum number. The quantum numbers of the vibrations involving chemical bonds have the same definitions in the fragments as in the complex.

We illustrate in Fig. 2 the vibrational predissociation of A-B$^*\cdots$C. (The example we have chosen is actually H-F$\cdots$H-F*[12].) The intermolecular parameters are imagined to remain unchanged for the A-B* and C interaction, and the potential curve is just shifted up by W_{AB}, the vibrational energy in the A-B* chemical bond. The excited van der Waals molecule A-B$^*\cdots$C in our example has its chemical bond vibrationally excited to the $v = 1$ state. The wavefunction $|v_t\rangle$ for the stretching motion against the van der Waals bond is shown in Fig. 2 for the $v_t = 0$ state. Its exact analytical form is obtained by solving the wave equation for the Morse potential function, and it can be seen to resemble a harmonic oscillator wavefunction[12].

Vibrational predissociation of A-B$^*\cdots$C produces A-B + C flying away from each other over the lower potential curve of Fig. 2. The chemical bond vibration of A-B is now relaxed to $v = 0$. Motion of the fragments against the van der Waals bond is described by the translational wavefunction, $|q_t\rangle$, shown in Fig. 2 as a plane wave for large A-B + C separations but damped out rapidly near the repulsive wall of the potential curve. The translational wavefunction is characterized by the dimensionless quantity[3,4,5]

$$q_t = (2\mu_t E_t)^{1/2}/a\hbar. \tag{6}$$

For our example we have $\Delta E = E_t = W_{AB} - D_o$ since the translational kinetic energy is the difference in the vibrational energy, W_{AB}, between A-B* and A-B, and that lost in breaking the van der Waals bond from the $v_t = 0$ zero point level. If we ignore the zero point energy we have $D_o \approx D_e$. We may interpret q_t as proportional to the number of nodes of the (near) plane wave describing the motion of A-B + C. For our purposes, what is important is

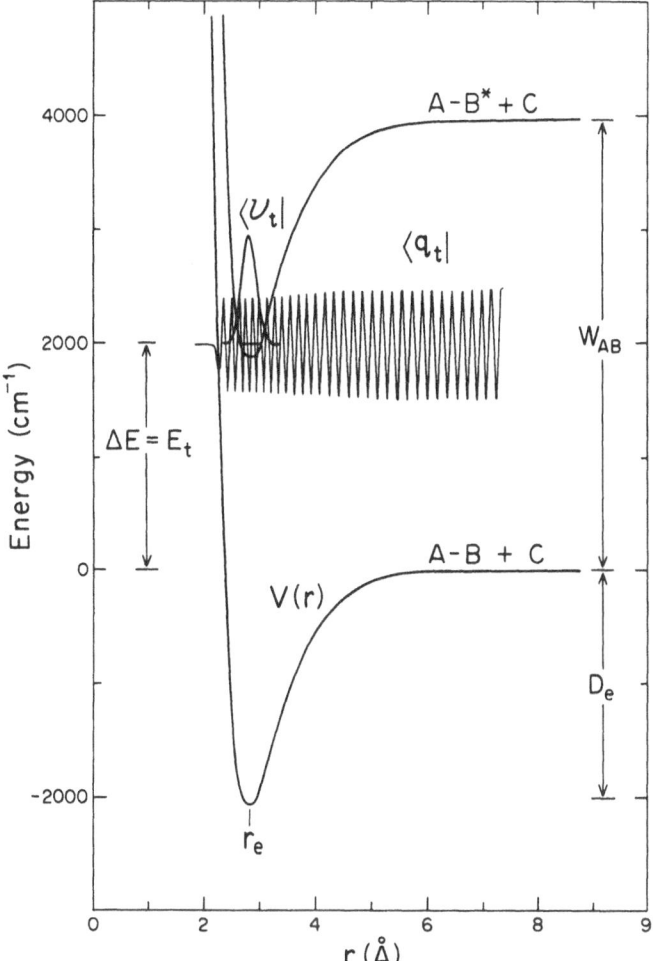

Figure 2. Potential surfaces, energy terms, and translational wave
functions for vibrational predissociation of A-B*···C. The
example chosen is for the complex HF···HF. Adapted from
ref. 12 with permission.

the number of nodes embraced by the intermolecular potential well since that is the region in which vibrational predissociation occurs. As we describe below, this number of nodes is approximately equal to $q_t/2$. Since quantum numbers are closely associated with nodal behavior, we may view $q_t/2$ as the effective translational quantum number of the fragments. In our example of Fig. 2 we have $q_t/2 \approx 11$.

We are now prepared to set down the effective quantum numbers for use of the selection rule expression of eq. 5. Application of the analytical expression for vibrational predissociation rates of A-B$^*\cdots$C for a wide variety of van der Waals molecules bound by Morse intermolecular potential functions like those shown in Fig. 2 reveals[13] the effective translational quantum number change

$$\Delta n_t \approx |q_t/2 - v_t| . \tag{7}$$

This quantum number change is essentially the difference between the effective number of nodes, $q_t/2$, of the translational wavefunction of the predissociation fragments, A-B + C, and the number of nodes, v_t, in the van der Waals stretching vibrational wavefunction of A-B$^*\cdots$C. The exponential dependence on these quantum numbers in eq. 5 is consistent with the poor Franck Condon type overlap expected for wavefunctions with widely differing numbers of nodes such as those shown for example by A-B$^*\cdots$C predissociation in Fig. 2.

The influence of rotational degrees of freedom during the vibrational predissociation process is the most difficult to model simply. In the spirit of this presentation we have used the simplest possible treatment by replacing the reduced mass μ_t for the translational motion in eq. 6 by I/r_v^2 for the rotational motion. This substitution, first used by Moore[14] to relate the transfer of collision pair vibrational relaxation to rotational motions, involves I, the vibrating molecule moment of inertia, and r_v the distance between its center of mass and the vibrating atom. With CH_4^* for example, r_v becomes just the C-H bond length. For a diatomic molecule we have $r_v = \alpha r_{AB}$ where r_{AB} is the A-B bond length and

$$\alpha = m_A/(m_A + m_B) \tag{8}$$

gives the fraction of the vibrational displacement of B toward C in the A-B$\cdots$C complex of Fig. 1. With $I = \mu_r r_{AB}^2$ and $\mu_r = m_A m_B/(m_A + m_B)$, the reduced mass of A-B, Moore's substitution yields the dimensionless quantity

$$q_r = (2\mu_r E_r)^{1/2}/\alpha a \hbar \tag{9}$$

where E_r is the rotational kinetic energy of the fragment A-B . Since $E_r = J(J+1)\hbar^2/2I$, substitution into eq. 9 reveals $q_r \approx J$ since for typical values of the molecular parameters $\alpha r_{AB} a \approx 1$. The rotational quantum number of A-B, J, is then quite close to q_r. By analogy to the effective translational quantum numbers, we shall take

$$\Delta n_r \approx |q_r/2 - v_r| . \tag{10}$$

Overlap between the chemical bond vibrational wavefunctions is also given in terms of their effective quantum numbers. As described elsewhere[13], we write

$$\Delta n_v \approx \gamma |v_f - v_i| = \gamma |\Delta v| \qquad (11)$$

where v_i is the vibrational quantum number of A-B* within the complex and v_f labels the vibrational state of the A-B + C fragments. The factor γ is the measure of the effectiveness of the coupling of vibrational motions of the chemical bonds in A-B* with the intermolecular or van der Waals bond which holds it to C in the complex. Without this specific vibrational coupling with the van der Waals bond there can be no predissociation. The value $\gamma \approx 1$ for diatomic A-B is typical[13]. In cases where the vibrations within polyatomic A-B* do not effectively couple with the van der Waals stretching motions, γ becomes large and $\exp{-\pi\gamma|v_f - v_i|}$ goes to small values. Predissociation through changes involving these vibrational modes is thus unlikely.

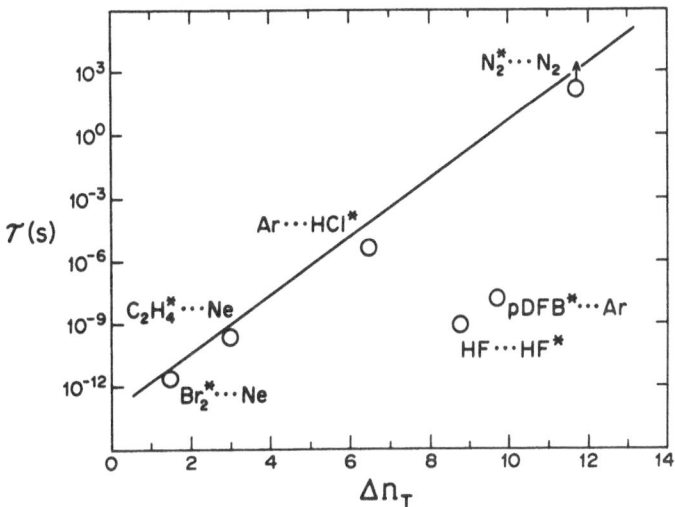

Figure 3. The total quantum number change, Δn_T, and lifetimes, τ, for vibrational predissociation. The line is the selection rule expression of eq. 5. Experimental measurements are indicated by the open circles described in the text.

We may summarize the preceding discussion. In undergoing vibrational predissociation as in eq. 1, energy is transferred to translational and rotational kinetic energy (ΔE) and possibly also deposited into vibrational levels of the fragments. Our selection rule states that the most efficient vibrational channel is the one that results in the smallest total change in effective quantum numbers. Using the crude relations we have just given for Δn_t, Δn_r and Δn_v, we can explore the possible channels and uncover the most efficient. In the examples to follow we shall see that the selection rule enjoys many successes but suffers failures as well.

The straight line plotted in Fig. 3 is the lifetime for vibrational predissociation of eq. 5. The data points are from experimental measurements or close-coupling calculations. The times involved span 14 orders of magnitude. Let us explore the mechanisms of these vibrational predissociation processes.

The shortest lifetime, $\tau = 2 \times 10^{-12}$ s, is from spectroscopic bandwidth measurements of Cline et al.[15] on $Br_2(v = 27)\cdots Ne$ and indicated on Fig. 3. Vibrational predissociation produces $Br_2(v = 26) + Ne$ within the B electronic state of Br_2. With $D_0 \approx 62$ cm^{-1} and $W_{AB} = 64$ cm^{-1} we have $E_t = W_{AB} - D_0 = 2$ cm^{-1} (4×10^{-23} J). Taking $a = 2$Å^{-1} (2×10^{10} m^{-1}) as a typical range parameter[4,5] and $\mu_t = 17.8$ amu (3×10^{-26} kg), we obtain $\Delta n_t = 0.4$ using eqs. 6 and 7. We have $\Delta n_v = 1$ and ignoring rotational excitation so $\Delta n_r = 0$, we obtain $\Delta n_T = 1.4$. This point on Fig. 3 nearly lies on the theoretical curve. If we explore rotational excitation we must contend with the rather large moment of inertia of Br_2. This is consistent with the product distribution that shows little rotational excitation of the Br_2 fragment. The selection rule also explains the observed propensity for $\Delta v = -1$ (or $\Delta n_v = 1$) relaxation on two counts. First, an increase in the Δv change would increase Δn_v and second, it would also increase Δn_t since E_t would become larger.

At the other lifetime extreme, consider vibrational predissociation of $N_2^*\cdots N_2$. This lifetime has not been measured for the isolated van der Waals molecule and its infrared spectrum under low resolution reveals no predissociation effects[16]. However, in a related experiment, the vibrational relaxation of $N_2(v=1)$ by collisions in the neat liquid occurs with $\tau \geq 10^2$ s[17]. If we imagine liquid nitrogen as a collection of van der Waals molecules, then this measurement provides a lower limit of 10^2 s for $N_2^*\cdots N_2$. This lifetime is indicated on Fig. 3 with the arrow to indicate the lower limit of the measurement. This long lifetime for $N_2^*\cdots N_2$ vibrational predissociation is consistent with the calculated total change in quantum number of $\Delta n_T = 11.8$. This large value of Δn_T is a consequence of the large product of reduced mass and vibrational frequency of $N_2^*\cdots N_2$. Here we have used $E_t = W_{AB} - D_0 = 2270$ cm^{-1} and $a = 2$Å^{-1}, a typical range parameter, to obtain $\Delta n_t = 10.8$. After adding $\Delta n_v = 1$ we obtain the value of Δn_T given on Fig. 3. We favor a predominately translational channel over a rotational channel because of the relatively large moment of inertia of N_2. Thus it would require $J = 33$ to place a rotating N_2 into near resonance with the energy released on vibrational predissociation: $E_r = J(J + 1)B = 2244$ cm^{-1} with[18] $B = 2$ cm^{-1}. Using this value of E_r into eqs. 9 and 10 results in $\Delta n_r = 14.9$ and $\Delta n_T = 15.9$. Possibly a more favorable channel would divide angular momentum equally between the two N_2 fragments. In this case μ_r in eq. 9 is replaced by the reduced mass of the two partners, $\mu_r/2$. As a consequence $\Delta n_r = 10.5$ and $\Delta n_T = 11.5$ which becomes competitive with the equally inefficient translational channel. In any event the lifetime of $N_2^*\cdots N_2$ is very long and consistent with our selection rule.

Consider next vibrational predissociation of $Ar \cdots HCl^*$ as explored in a thorough theoretical investigation by Hutson[19]. His close-coupling calculation result appears in Fig. 3 with $\tau = 4 \times 10^{-6}$ s corresponding to a relaxation channel that produces vibrationally relaxed HCl in the $J = 15$ level.

In applying our simple model we first explore the translational channel for which $\Delta E = E_t$. We use $W_{AB} = 2886$ cm^{-1} and $D_o = 116$ cm^{-1} to obtain $E_t = 2770$ cm^{-1}. An exponential fit of the intermolecular potential surface[20] gives $a = 2.6$ Å^{-1}, and with $\mu_t = 18.9$ amu, we obtain $\Delta n_t = 10.7$. Since $\Delta n_v = 1$, the total quantum number change becomes $\Delta n_T = 11.7$. It is this large quantum number change, due both to the high vibrational frequency and the large reduced mass that is responsible for the sluggish vibrational predissociation by the translational channel as found by Hutson[19].

We can understand efficient rotational relaxation by comparing rotational and translational channels if each accept the same kinetic energy. Taking the ratio of eqs. 6 and 9 we find $q_r/q_t = (\mu_r/\mu_t)^{1/2}/\alpha = 0.2$, where $\mu_r = 0.97$ amu, and $\alpha = 0.97$. Thus, rotational kinetic energy requires a much smaller effective quantum number than does translational kinetic energy. Suppose that vibrational predissociation of $Ar \cdots HCl^*$ produces the near resonant HCl fragment in its $J = 15$ level so that it carries away $E_r = J(J+1)B = 2540$ cm^{-1} of rotational kinetic energy ($B = 10.6$ cm^{-1})[18]. Use of eqs. 9 and 10 gives $\Delta n_r = 2.4$. The remaining kinetic energy is translational, $W_{AB} - D_o - E_r = E_t = 230$ cm^{-1}, giving $\Delta n_t = 3.1$ by eqs. 6 and 7. With $\Delta n_v = 1.0$ we have $\Delta n_T = 6.5$ for this channel to give the point lying close to the line in Fig. 3. Thus our simple selection rule shows that the channel producing large amounts of rotational energy into HCl is more efficient than the channel producing only translational energy in accord with the close-coupling calculation.

We turn now to the vibrational predissociation of $C_2H_4^* \cdots Ne$ with a 5×10^{-10}s lifetime reported by Casassa et al.[20]. The excited vibration is an out-of-plane mode $\nu_7 = 950$ cm^{-1}, and the van der Waals bond has $D_o \approx 100$ cm^{-1}. The translational channel puts $E_t = 950-100 = 850$ cm^{-1} into the relatively massive C_2H_4 and Ne fragments. Use of our selection rule expression (with $a = 2$ Å^{-1} and $\gamma = 1$) yields $\Delta n_T = 13$, a value so large as to be incompatible with the observed lifetime. Likewise considerations of rotational channels reveal too large a value of Δn_T. The channel placing vibrational excitation into the predissociation fragment, however, explains the experiment. We assume that a low lying C_2H_4 mode, $\nu_{10} = 826$ cm^{-1}, absorbs energy leaving only 23 cm^{-1} into fragment translational energy corresponding to $\Delta n_t = 1.0$. Now with $\Delta n_v = 2$ (since two vibrational quanta are involved in this channel) we have $\Delta n_T = 3.0$ and with $\tau = 5 \times 10^{-10}$s we position the point on Fig. 3. We see that this point lies near our theoretical curve of eq. 5. Detailed close-coupling calculations of Hutson et al.[21] bear out our simple arguments favoring this channel. Since changes in vibrational level can absorb the greatest amount of energy with the least quantum change, we can expect these channels to be particularly efficient in vibrational predissociation processes.

We have demonstrated more successes of the selection rule elsewhere[13]. However, the failures also can help us better understand van der Waals molecule dynamics. Two such failures are listed on Fig. 3: vibrational predissociation of HF$\cdots$HF[*] and vibrational predissociation of excited p-difluorobenzene complexed with argon (pDFB[*]$\cdots$Ar).

Our model does account for the relative mode selective efficiencies of HF$\cdots$HF[*] versus HF[*]$\cdots$HF vibrational predissociation[11,22] through changes in the coupling parameter of eq. 11[13]. When the donor hydrogen is excited γ is small. For the non-bonded hydrogen γ is large. However the absolute lifetime of $\tau = 10^{-9}$ s for HF$\cdots$HF[*] is not reproduced. We anticipate rotational excitations to be important in the fragments as in the case of Ar$\cdots$HCl[*] relaxation. Using our recipes and constants from the literature we imagine for example an HF fragment in J=10. The total quantum number change becomes $\Delta n_T = 8.8$ and is also indicated on Fig. 3. Relaxation channels producing translational excitation into HF + HF is hopelessly inefficient. Why does the selection rule fail?

The problem lies in the assumption required to derive the selection rule that the $v=1$ and $v=0$ surfaces are the same shape and are merely displaced vertically as we have illustrated in Fig. 2. For HF$\cdots$HF on the contrary, the intermolecular potential is highly anisotropic and rotational excitation of the fragments results in an effective potential which is shallow and may actually cross other surfaces. This has been demonstrated in calculations of Halberstadt et al.[23]. The surfaces taken from their work are shown in Fig. 4. The curve crossing yields relaxation times orders of magnitude more efficient than those calculated by our selection rule. It is a challenge to the theorists to model the predissociation process, consistent with experiment[24], that allows both HF molecules to rotate on fragmentation. Clearly anisotropic effects will play an important role in understanding vibrational predissociation in other systems as well—for example, in the electronically excited state of OH[*]$\cdots$Ar by Lester et al.[25].

The second failure is in trying to explain experiments of Parmenter et al.[26]. Here for example the 5^1 level of the S_1 electronic state of pDFB[*]$\cdots$Ar, $|\bar{5}^1;0,0,0\rangle$, is interrogated to find vibrational predissociation to the ground vibrational state with lifetime $\tau = 5 \times 10^{-9}$ s. (The notation $|\bar{v}_i;v_x,v_y,v_z\rangle$ indicates $\bar{v}_i$ chemical bond vibrational level of the S_1 electronic state of the complex with v_x, v_y and v_z van der Waals mode excitation.) With its large moment of inertia, rotational excitation of the fragments would appear to be inefficient. Assuming the predissociation process deposits its energy into translational motions, $\Delta E = E_t = 420$ cm^{-1} and using[26] $a = 1.7 \times 10^8$ cm^{-1} we find $\Delta n_T = 9.5$ by using eqs. 5-7 and eq. 11 with $\gamma = 1$. Here is another dramatic failure of the selection rule as the datum on Fig. 3 shows.

It has been shown that Fermi resonances between chemical bond vibrational levels and van der Waals modes can dramatically reduce vibrational predissociation lifetimes[27]. The selection rule becomes altered because the definitions of the quantum numbers become blurred by the Fermi resonances. This is illustrated in the recent study of Tiller, Peet and Clary[28] and shown in Fig. 5. Here $|\bar{5}^1;0,0,0\rangle$ mixes with nearby $|\bar{17}^2;0,0,10\rangle$ whose vibrational predissociation lifetime is calculated to be two orders of magnitude shorter than the prepared state.

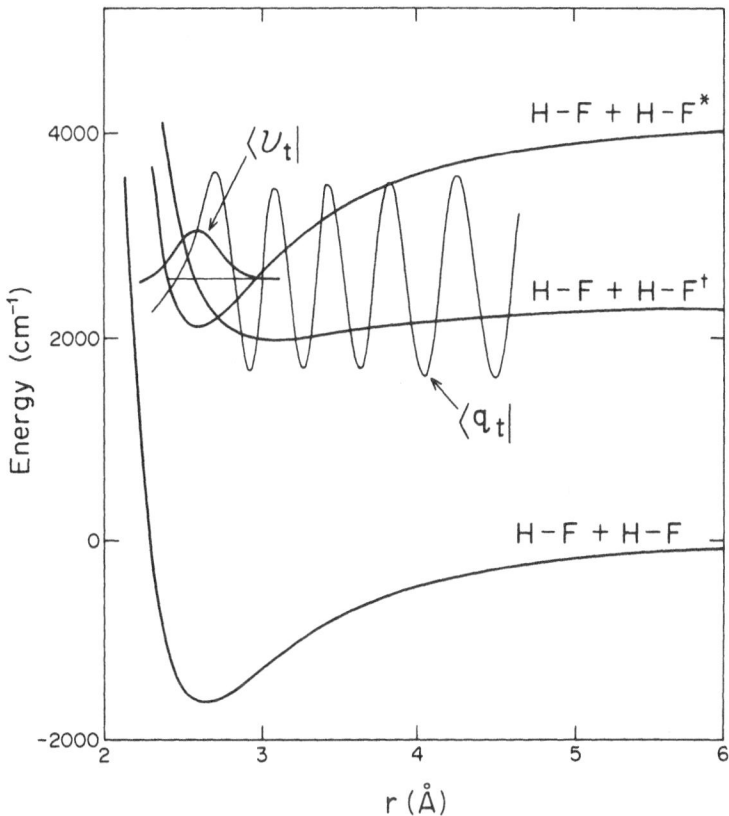

Figure 4.   Potential surfaces for HF + HF*, HF + HF and HF + HF†. (The dagger indicates J = 10.) Translational wave functions of HF$\cdots$HF*, $\langle v_t |$, and HF + HF†, $\langle q_t |$, are also shown. Adapted from ref. 23 with permission.

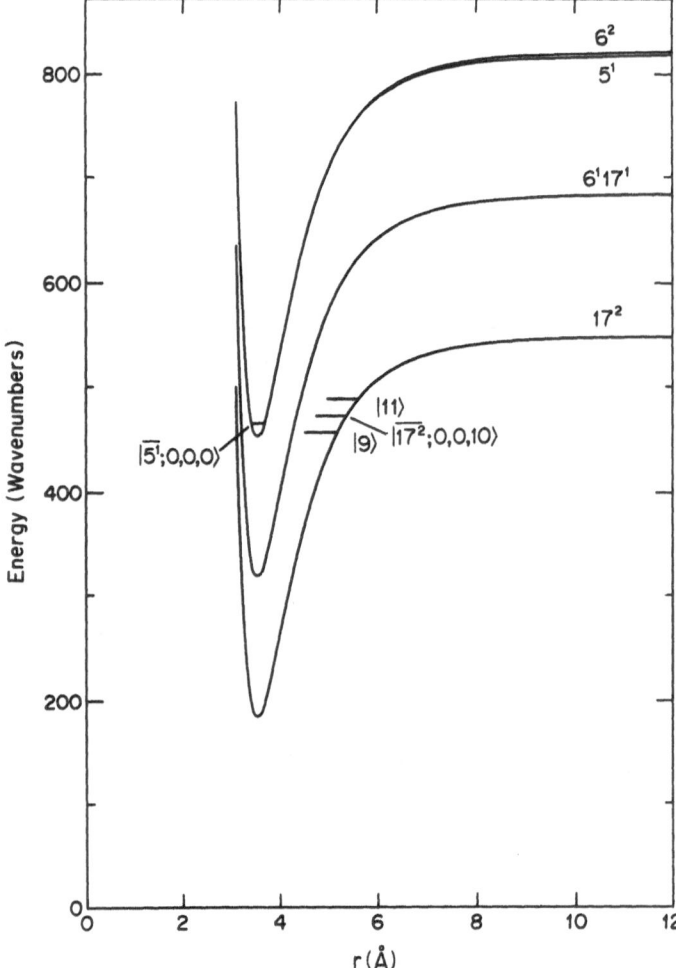

Figure 5. Potential surfaces of pDFB*···Ar in various vibrational states. Taken from ref. 28 with permission.

Unfortunately, the problem has not yet been solved by these close-coupling calculations. The calculated lifetimes still remain orders of magnitude too long. While there is both experimental[29,30] and theoretical evidence[27,28] for these Fermi resonances it is very difficult to model them quantitatively. As we might imagine from Fig. 5 small changes in the potential surfaces could bring the mixing states in or out of resonance. A variety of the many other possible resonances besides $|\overline{17}^2, v_x, v_y, v_z\rangle$ have been ignored. Moreover it is not clear that the surfaces for pDFB$\cdots$Ar used, reminiscent of Fig. 2, should not be replaced by surfaces qualitatively resembling those of Fig. 4 because the relaxed fragments pDFB + Ar may contain large amounts of angular momentum.

The van der Waals molecule pDFB$\cdots$Ar is not particularly large. Considering the number of difficulties experienced in the excellent close-coupling calculations so far performed, it is not clear what the future holds for this approach to van der Waals molecular dynamics. At some stage a practical solution will have to involve statistical methods[13]. We can read more about these approaches in following papers.

NEW SYSTEMS

We have just seen how our limited knowledge of the multidimensional intermolecular potential surface can impede our understanding of energy flow in rather simple van der Waals molecules. What of large systems? To ease into this question let us consider either small molecules (e.g. CO) adsorbed onto a surface or small clusters of small molecules (e.g. CO_2).

Photodesorption

Let us begin by casting a slightly different interpretation on vibrational predissociation of the generic van der Waals molecule and replace eq. 1 by

$$CO^*\cdots NaCl(100) \xrightarrow{\tau^{-1}} CO + NaCl(100) + \Delta E \quad . \qquad (12)$$

Here NaCl(100) is the face of a single crystal of salt and the dots represent the physisorbed bond dominated by electrostatic contributions as in the case of van der Waals molecules. The vibrational energy[32] in CO^* of 2100 cm^{-1} exceeds the physisorbed bond strength[33] of $D_0 \approx 1500$ cm^{-1} so predissociation is possible and relaxed CO flies away with kinetic energy $\Delta E \approx 600$ cm^{-1}. For the same reason vibrational predissociation is inefficient for $N_2^*\cdots N_2$ so it is for the relaxation of eq. 12. Indeed attempts to photodesorb CO from NaCl(100) have failed and the quantum yield for this relaxation channel is[34] $\Phi \leq 10^{-7}$. Theory to account for relaxation channels from excited molecules physisorbed to surfaces is developing[34-40] but the paucity of experimental data prevents a calibration of the calculations. The selection rule of eq. 5 can be easily extended[34] to qualitatively account for photodesorption and becomes

$$\tau^{-1} \approx 10^{13} \exp{-\pi(\Delta n_t + \Delta n_r + \Delta n_v + \Delta n_p)} = 10^{13} \exp{-\pi(\Delta n_T)} \quad . \qquad (13)$$

where the additional term, Δn_p, accounts for excitation of phonon modes which can reduce the total quantum number change Δn_T. Again a thorough knowledge of the multidimensional potential surface is required before reliable photodesorption rates can be expected. In two recent calculations, the same theoretical model gave the wildly different photodesorption life-times $\tau \approx 10^2$ s[34] and $\tau \approx 10^{-13}$ s[38] for eq. 12 because of different choices in the range parameter, a, of the Morse potential used describe the physisorbed bond.

Further theoretical and experimental work will need to deal with poly-atomic molecules stuck to surfaces and also consider vibrational dynamics among neighboring adsorbates. Because of rapid energy transfer between the excited adsorbate and the substrate, the photodesorption channel will be quenched for many systems. Models to account for this behavior will be needed to explore the possibilities of surface photochemical processes[41].

Small Clusters

To begin how small is small? For our purposes we shall take a small cluster to have a comparable number of molecules on the exterior as the interior. Two such examples are shown in Fig. 6. In one case we have presented a cluster of $(CO_2)_{72}$ where the structure is consistent with that for bulk crystalline carbon dioxide[42]. Electron diffraction studies of clusters of this size[43] suggest that this is a reasonable structure. Infrared spectra of small clusters is also consistent with the proposed geometry[42,44,45]. As Miller et al.[46-48] show, the structures of $(CO_2)_2$ and $(CO_2)_3$ bear no clear relationship to that of the bulk crystal.

In another example of a small cluster consider the two-dimensional sheet of monolayer CO_2 on NaCl(100)[49] also shown in Fig. 6. (While the mono-layer is in contact with the substrate, it has been displaced here to demonstrate more clearly the structure.) The forces holding this sheet together and to the substrate are again dominated by electrostatics. The two-dimensional structure resembles a slice of the bulk crystal or equiv-alently $(CO_2)_{72}$.

We have selected the single example of the CO_2 cluster for historical reasons. Van der Waals theoretical studies[50] over a century ago on the PVT properties of CO_2 in a way gave birth to our present interests. Further theoretical modeling of these small clusters to understand the equilibrium structures as well as their vibrational dynamics is in its infancy and no doubt will see development in the near future.

Acknowledgment
We appreciate the financial support by the National Science Foundation for these studies.

a

b

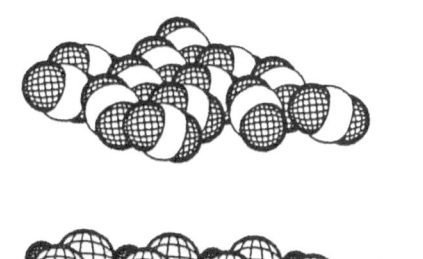

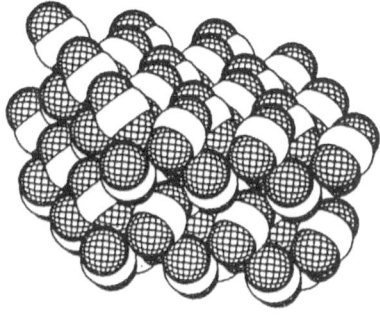

c   d

Figure 6.   Structures of small clusters of CO_2.   a) The dimer[46,47],
b) the trimer[46,48], c) a two-dimensional cluster, displaced
for clarity from the NaCl(100) surface to which it is
bound[49] and d) a three dimensional cluster[42] $(CO_2)_{72}$.

REFERENCES

1. B. I. Stepanov, _Nature (London)_ 157:800 (1946).
2. W. Klemperer, _Ber. Bunsenges. Phys. Chem._ 78:128 (1974).
3. C. A. Coulson and G. N. Robertson, _Proc. R. Soc. London_ A342:289 (1975).
4. J. A. Beswick and J. Jortner, _Chem. Phys. Lett._ 49:13 (1977).
5. G. E. Ewing, _Chem. Phys._ 29:253 (1978).
6. M. S. Child and C. J. Ashton, _Faraday Disc. Chem. Soc._ 62:307 (1976).
7. R. E. Smalley, D. H. Levy, and L. Wharton, _J. Chem. Phys._ 64:3266 (1976).
8. D. H. Levy, _Adv. Chem. Phys._ 47:323 (1981).
9. J. A. Beswick and J. Jortner, _Adv. Chem. Phys._ 43 (Part 1): 263 (1981).
10. T. E. Gough, R. E. Miller and G. Scoles, _Appl. Phys. Lett._ 30:338 (1977).
11. A. S. Pine and W. J. Lafferty, _J. Chem. Phys._ 78:2154 (1983).
12. G. E. Ewing, _J. Chem. Phys._ 72:2096 (1980)
13. G. E. Ewing, _J. Phys. Chem._ 91:4662 (1987).
14. C. B. Moore, _J. Chem. Phys._ 43:2979 (1965)
15. J. I. Cline, D. D. Evard, B. P. Reid, N. Sivakumar, F. Thommen, K. C. Janda, _"Structure and Dynamics of Weakly Bound Molecular Complexes"_, A. Weber ed., Reidel, Dordrecht (1987).
16. C. A. Long, G. Henderson, G. E. Ewing, _Chem. Phys._ 2:485 (1973); A.R.W. McKellar, _J. Chem. Phys._ 88:4190 (1988).
17. D. W. Chandler, and G. E. Ewing, _J. Chem. Phys._ 73:4904 (1980).
18. G. Herzberg, _"Spectra of Diatomic Molecules"_, Van Nostrand, Princeton, 1950.
19. J. M. Hutson, _J. Chem. Phys._, 81:2357 (1984); 81:6413 (1984).
20. M. P. Casassa, D. S. Bomse, and K. C. Janda, _J. Chem. Phys._, 74:5044 (1981
21. J. M. Hutson, D. C. Clary, and J. A. Beswick, _J. Chem. Phys._, 81:7747 (1984).
22. Z. S. Huang, K. W. Jucks and R. E. Miller, _J. Chem. Phys._ 85:3338 (1986).
23. N. Halberstadt, Ph. Brechignac, J. A. Beswick, and M. Shapiro, _J. Chem. Phys._ 84:170 (1986).
24. D. C. Dayton, K. W. Jucks and R. E. Miller, _J. Chem. Phys._ 90:2631 (1989).
25. M. T. Berry, M. R. Brustein and M. Lester _J. Chem. Phys._ 90:5878 (1989) and these proceedings.
26. K. W. Butz, D. L. Catlett, Jr., G. E. Ewing, D. J. Krajnovich and C. S. Parmenter, _J. Phys. Chem._ 90:3533 (1986).
27. G. E. Ewing, _J. Phys. Chem._ 90:1790 (1986).
28. A. R. Tiller, A. C. Peet and D. C. Clary, _Chem. Phys._ 129:125 (1989).
29. J. J. F. Ramaekers, H. K. Dijk, J. Langelaar, and R. P. H. Rettschnick, _Faraday Discuss. Chem. Soc._ 75:183 (1983).
30. D. V. Brumbaugh, J. E. Kenny and D. H. Levy, _J. Chem. Phys._ 78:3415 (1983).
31. D. F. Kelley and E. R. Bernstein, _J. Chem. Phys._ 90:5164 (1986).
32. H. H. Richardson, H. -C. Chang, C. Noda and G. E. Ewing, _Surf. Sci_ 216:43 (1989).
33. H. H. Richardson, C. Baumann and G. E. Ewing, _Surf. Sci._ 185:15 (1987).
34. H. -C. Chang and G. E. Ewing, _Chem. Phys._ (in press).
35. D. Lucas and G. E. Ewing, _Chem. Phys._ 58:385 (1981).
36. Z. W. Gortel, H. J. Kreuzer, P. Piercy and R. Teshima, _Phys. Rev._ B27:5066 (1983).

37. A. Ben Ephram, M. Folman, J. Heidberg and N. Moiseyer, <u>J. Chem. Phys.</u> 89:3840 (1988).
38. J. T. Muckerman and T. Uzer, <u>J. Chem. Phys.</u> 90:1968 (1989).
39. A. Nitzan and J. C. Tully, <u>J. Chem. Phys.</u> 78:3959 (1983).
40. I. Benjamin and W. P. Reinhardt, <u>J. Chem. Phys.</u> 90:7535 (1989).
41. E. J. Hielweil, M. P. Casassa, R. R. Cavanagh and J. C. Stephenson, <u>J. Chem. Phys.</u> 81:2856 (1984).
42. R. Disslekamp and G. Ewing, <u>Disc. of the Faraday Soc.</u> (1989) (in press).
43. G. Torchet, H. Bouchier, J. Farges, M. F. de Feraudy and B. Raoult, <u>J. Chem. Phys.</u> 81:2137 (1984).
44. J. A. Barnes and T. E. Gough, <u>J. Chem. Phys.</u> 86:6012 (1987).
45. De T. Sheng and G. E. Ewing, <u>J. Phys. Chem.</u> 92:4063 (1988).
46. R. E. Miller, <u>Science</u> 240:447 (1988).
47. K. W. Jucks, Z. S. Huang, R. E. Miller, G. T. Fraser, A. S. Pine and W. J. Lafferty, <u>J. Chem. Phys.</u> 88:2185 (1988).
48. G. T. Fraser, A. S. Pine, W. J. Lafferty and R. E. Miller, <u>J. Chem. Phys.</u> 87:1502 (1987).
49. O Berg and G. E. Ewing, <u>Surf. Sci.</u> (in press).
50. J. D. van der Waals, Dissertation, Leiden, The Netherlands, (1973).

DYNAMICS OF SIZE-SELECTED CLUSTERS

J. Reuss

Molecular and Laser Physics, University of Nijmegen

Toernooiveld, 6525 ED Nijmegen, The Netherlands

1. SIZE-SELECTION

From an experimental point of view size selection can be achieved by the Göttingen collisional deflection technique, but sometimes also by purely spectroscopic methods. If the purely spectroscopic approach is possible, one has the big advantage of much stronger signals which renders the use of weak lasers (e.g. FCL, diode laser or cw CO_2 laser) possible, too.

The spectroscopic approach does not necessarily imply highly resolved spectra: more often than not we have to content ourselves with low resolution results because of spectral congestion in larger complexes. For size and isomer selective spectroscopic techniques see contribution of Leutwyler on clusters containing aromatic molecules.

The Göttingen technique of size-selection can be applied before further experiments (e.g. IR laser predissociation) are performed. The energy deposited in the clusters by collisional deflection has then to be reckoned with. This feature can be turned into an advantage since the amount of energy deposited can be determined and controlled - within limits - by the choice of deflection angle (to be looked at) and the collision partner (normally He atoms).

Alternatively, size selection can be performed after the event of e.g. predissociation. In this case the laser attenuation of a size-selected beam signal of "cold" complexes is measured. Both methods have been used successfully, as described in the contribution of U. Buck to this volume.

Again in some cases - discussed below - an energy deposition - before the event to be measured - can be realized by optical means.

2. DYNAMICS THROUGH LINEWIDTH

It is without danger to connect a predissociation lifetime to fully resolved and well assigned ro-vib spectral lines, see e.g. the contribution of Miller to this volume.

As to the opposite extreme, it is often completely impossible to deduce information on lifetimes from broad spectral structures as we are going to demonstrate here for $(NH_3)_4$.

The ammonia trimer and tetramer do not dissociate upon absorption of a single

Dynamics of Polyatomic Van der Waals Complexes
Edited by N. Halberstadt and K. C. Janda
Plenum Press, New York, 1990

27

CO_2 laser photon, see next section. A broad excitation spectrum (FWHM of 4 cm^{-1}) has been observed. In a two-laser experiment no hole burning effects have been obtained, i.e. the broad spectrum has a homogeneous appearance [1]. A straightforward estimate yields a lifetime of about 1 ps, from the homogeneous width of 4 cm^{-1}. Still, the spectrum is due to excitation without ensueing dissociation. Thus it makes no sense to talk of a predissociative lifetime.

In the following we will dwell in the no man's land between the just mentioned extremes. The main example shall be the ethylene dimer. For this system two types of spectra have been observed which have to be explained simultaneously [2,3,4,5].

First, narrow transitions (FWHM of 3 MHz) occur which are still unassigned but there is little doubt that they belong to resolved rotational transitions. The observed lines are broadened with respect to the C_2H_4-monomer transition (FWHM<1 MHz, the instrumental width). Therefore, we can derive a minimum dissociative lifetime of 40 ns. For a Van der Waals combination band about 40 cm^{-1} put into the dimer does not lead to measurable shortening of the lifetime [2,3,4,5].

Second, there is also present a broad-banded dimer predissociation spectrum (FWHM of 12 cm^{-1}) with a much weaker transition strength. Its relative contribution (as compared to the narrow line contribution) depends on the source conditions. For increasing source pressures this contribution (not size selected) goes through a minimum, the high pressure rise being attributed to the appearance of larger clusters. The low pressure drop is attributed to a stronger expansion cooling (with increasing stagnation pressure). Therefore, the picture is that "warm" dimers show a broad spectral contribution. The picture is tested by variation of the stagnation temperature, leading to the same conclusion. Furthermore, with size selection before laser induced dissociation, the extra collisional energy deposited into the dimers led to a still stronger broadening (FWHM of 32 cm^{-1}) in agreement with the qualitative picture sketched above.

To explain this behaviour one has to take into account the ν_{10} mode of C_2H_4, 125 cm^{-1} below the ν_7 mode. Fast internal energy redistribution within a ν_7 excited C_2H_4 dimer leads to an energy of 125 cm^{-1} eventually available for dissociation, which, together with a thermal energy of about 80 cm^{-1}, leads to an excitation just below or above the threshold of dissociation. The high density of coupled states around the threshold is to be held responsible for the appearance of the broad spectral structure with its FWHM of 12 cm^{-1}, in our opinion. A theoretical analysis is needed which confirms or rejects this conjecture and/or comes up with alternative explanations. Hopefully, this analysis will also lead to an assignment of the observed sharp structures.

The system $(CH_3OH)_2$ similarly shows a very strong dependence on the stagnation conditions. Here warm dimers do not only show a significant broadening of spectral structures but the excitation spectrum (without dissociation for cold dimers) even changes to a dissociation spectrum. Together with other systems which do not dissociate upon absorption of a single CO_2 laser photon, $(CH_3OH)_2$ will be discussed below.

In the contribution of Leutwyler to this volume one finds examples of spectra with narrow features on broad absorptions. This relative strength can be varied with stagnation conditions similarly as has been discussed here.

3. THE CASE OF $(SF_6)_n$ AND SIMILAR COMPLEXES

The predissociation spectrum of $(SF_6)_n$ is mainly determined by resonant dipole-dipole forces, which produce clear size-dependent splittings. Therefore, $(SF_6)_n$ possesses a nice fingerprint spectrum which allows to distinguish the contributions from different cluster sizes to the predissociation signal. Even isotopic substitutions have been identified [6,7,8,9].

The linewidth of the various spectral features attributed to dimers, trimers, etc., have been investigated by two-laser techniques. Hole burning has been observed, testifying the at least partially homogeneous character of the observed bands. In the course of these measurements three features emerged which concern the dimer two-peak spectrum. For $(SF_6)_2$ the splitting amounts to 20 cm^{-1}. Both peaks arise from the same dimer structure, one being excited along the dimer axis (red peak) and one perpendicular to it. By the hole burning technique we observed a correlation between the outer flanks of the two peaks, i.e. hole burning in the red flank of the red peak is accompanied by a hole appearing simultaneously in the blue flank of the blue peak. Similarly, the inner flanks of the two peaks are correlated. This is the first feature observed.

The second feature is more connected to the dynamics of the system and concerns the width of the holes burned into the inner and outer flanks, respectively. There is a significant difference; the inner flank holes are about 2 to 3 times broader than the outer flank holes.

The model to explain at least qualitatively these two features is based upon the dominance of the resonant dipole-dipole forces. For dimers a splitting is predicted proportional to $< R^{-3} >$, where R stands for the distance between the two SF_6 molecules forming the dimer. Warm (cold) dimers will have a larger (smaller) average value of R and therefore they will manifest themselves at the inner (outer) flank of the two dimer peaks. Also, warm dimers have a stronger coupling to dissociative and other states and therefore they show a larger linewidth, as observed for $(C_2H_4)_2$ and $(CH_3OH)_2$ (see above). Here calculations including these couplings would be very interesting and helpful.

The third feature has to do with (the absence of) polarization effects in two-laser experiments. The hole-burning strength has been measured for different (parallel and mutually perpendicular) laser polarizations without any observable difference. Also, this third feature points to a strong coupling, this time between internal rotation of the constituents and the end-over-end rotation. In principle this observation is in agreement with recent discussions of anisotropic forces between SF_6 molecules.

The behaviour of $(SiF_4)_2$ and $(SiH_4)_2$ has been investigated both experimentally and theoretically. Especially $(SiH_4)_2$ shows a strong deviation from the simple model applied to $(SF_6)_2$, since here the resonant dipole-dipole forces are much less dominant, see the contribution of J. van Bladel to this volume.

4. CLUSTERS SURVIVING A 1000 cm^{-1} EXCITATION

In this section $(NH_3)_3$, $(NH_3)_4$, $(CH_3OH)_2$ and $(CH_3OH)_3$ will be discussed. To observe the survival upon excitation of these complexes the application of optothermal detection is essential. A bolometer, as molecular beam detector, makes it possible to distinguish between excitation without dissociation (yielding a positive signal since more energy is deposited on the bolometer element) and excitation resulting in dissociation (yielding a negative signal since the fragments leave the molecular beam and less molecules arrive at the detector).

The most fortunate case is formed by $(CH_3OH)_2$, which almost dissociates upon excitation by a 1040 cm^{-1} photon. With a total extra (thermal) energy of 90 cm^{-1} (corresponding to 30 K with He seeding instead of the estimated 5 K with Ne seeding)

Table 1. Estimated relative detection efficiencies for varied beam conditions applying to optothermal detection

	100% CH_3OH	1% CH_3OH in Ne	1% CH_3OH in He
$\frac{E_{ex}}{E_{dis}}\mid_{OT}$	1	(1.6)-1	(7.5)-1

this dimer is found to dissociate (negative signal, FWHM of 40 cm^{-1}), whereas at low T a dominant two-peak excitation spectrum is found (FWHM of 2 to 4 cm^{-1}). This way the dissociation energy (well depth - zero point energy) can be estimated to be 1085±45 cm^{-1} [10]. Theoretical estimates come very near to this experimental value.

Using high intensity (especially pulsed) lasers it is possible to dissociate the methanol dimer by two-photon excitation. Again a two-peak dissociation spectrum about twice as broad as the one that is found for pure excitation is observed without essential shifts. Embarrassing is the fact that also beams seeded in He show this two-peak spectrum (only slightly broadened) [11], in contrast to the observations discussed in section 2, where optothermal detection yielded very broad banded dissociation spectra (FWHM of 40 cm^{-1}) after absorption of a single photon. To explain this discrepancy we have to keep in mind that the detection efficiency E_{dis} for the dissociated dimers - caused by one or two-photon transitions as it may be - is different from that of excited dimers E_{ex}, for optothermal detectors (see Table 1).

In comparison, $E_{ex}/E_{dis} = 0$ for mass spectrometric detectors. Even after correcting for these relative efficiencies there remains a disturbing discrepancy between the two types of measurement. Huisken estimates a one-photon background of only 5%, whereas the measurements of Bizzarri et al show a cold- dimer-excitation to warm-dimer-dissociation signal ratio of 13%, under similar source conditions. This discrepancy has got to be cleared up experimentally. It does not, however, invalidate the above estimate of binding energy.

Methanol trimers were found not to dissociate even after absorption of two CO_2 laser photons. The absorption of the second photon shows a maximum at a frequency 4 cm^{-1} shifted to the red [10].

Methanol trimers were not observed in excitation spectra. Probably the excitation is so sharply peaked that it was missed by the coarse grid of frequency points obtainable from a line tunable CO_2 laser. Also for the trimer excitation only one CO_2 laser frequency yields excitation - a very strong excitation signal, by the way.

The discussion whether $(NH_3)_2$ needs one or two 1000 cm^{-1} photons for dissociation has ended in favour of the one-photon thesis [12]. However, $(NH_3)_3$ was found to dissociate only after absorption of two photons [1]. The excitation spectrum has a FWHM of about 10 cm^{-1}. By two-laser experiments hole burning with very narrow holes (FWHM of 1 MHz) has been observed. The excitation spectrum with its single broad peak must thus be interpreted as a congested inhomogeneous line form.

The trimer excitation is peaked around 1017 cm^{-1}, the subsequent dissociation by absorption of a second photon at 1007 cm^{-1} with a FWHM of 14 cm^{-1}. As discussed above the NH_3 tetramers behave similar to the trimers without showing hole burning effects.

REFERENCES

[1] B. Heijmen, A. Bizzarri, S. Stolte and J. Reuss, Chem. Phys. 126 (1988) 201

[2] B. Heijmen, C. Liedenbaum, S. Stolte and J. Reuss, Z. Phys. D 6 (1987) 199

[3] K.H. Baldwin and R.O. Watts, Chem. Phys. Lett. 129 (1986) 237

[4] U. Buck, Ch. Lauenstein, A. Rudolph, B. Heijmen, S. Stolte and J. Reuss, Chem. Phys. Lett. 144 (1988) 396

[5] M. Snels, R. Fantoni, M. Zen, S. Stolte and J. Reuss, Chem. Phys. Lett. 124 (1986) 1

[6] J. Geraedts, M. Snels, S. Stolte and J. Reuss, Chem. Phys. Lett. 106 (1984) 377

[7] J. Geraedts, S. Stolte and J. Reuss, Z. Phys. A 304 (1982) 167

[8] B. Heijmen, A. Bizzarri, S. Stolte and J. Reuss, Chem. Phys. 132 (1989) 331

[9] F. Huisken and M. Stemmler, Chem. Phys. 132 (1989) 351

[10] A. Bizzarri, J. Reuss, A. v. Duyneveld and B. van Duyneveld, to be published

[11] F. Huisken, private communications; see also the contribution of U. Buck to this volume

[12] M. Snels, R. Fantoni, R. Sanders and W.L. Meerts, Chem. Phys. 115 (1987) 79

THE VIBRATIONAL DYNAMICS OF HYDROGEN BONDED MOLECULAR COMPLEXES

AT THE STATE-TO-STATE LEVEL

Roger E. Miller

Department of Chemistry
University of North Carolina
Chapel Hill, N.C. 27599.

ABSTRACT

In recent years our understanding of the vibrational predissociation dynamics of weakly bound molecular complexes has improved greatly owing to both the experimental and theoretical advances that have been made in this area. We are now in the very fortunate position where both theory and experiment often can be brought to bare on the same system. The developments made in our laboratory involve the use of the opto-thermal detection technique to measure the infrared spectra and state-to-state dissociation rates for a number of these complexes. In the present report we examine the affects of molecular orientation on the rate of vibrational relaxation. This is done by measuring the predissociation lifetimes associated with several vibrational modes of different isomeric forms of a binary complex. The vibrational relaxation rates are found to be highly anisotropic and several interesting correlations can be made between this data and the collisional relaxation results already available in the literature.

In addition to the relaxation data obtained from measuring lifetimes, one also would like to obtain information on the disposal of excess energy in the fragments which result from vibrational predissociation of the parent complex. This clearly involves the measurement of state-to-state dissociation rates for these complexes. Results of this type have recently been obtained from measurements of the angular distributions of the photofragments and will be discussed for systems such as HF and HCN dimers. In favorable cases data of this type can provide relative state-to-state rates which can be used as sensitive tests of the emerging theoretical methods. Comparisons with simple statistical theories show that the dissociation dynamics associated with these systems is highly non-statistical.

1. INTRODUCTION

With the introduction of several new infrared laser-molecular beam methods[1-3] has come a wealth of new experimental data on both the structure and dynamics of weakly bound molecular complexes. Of particular importance to the present discussion is the fact that the

Dynamics of Polyatomic Van der Waals Complexes
Edited by N. Halberstadt and K. C. Janda
Plenum Press, New York, 1990

33

intermolecular bond is often so weak that the energy associated with the excitation of an intramolecular vibration is sufficient to dissociate a binary complex into its constituent monomer units. This type of photochemical dynamics, often referred to as vibrational predissociation, has been the subject of considerable current interest owing to the fact that the modest excitation energy places these systems in regions with very low state densities such that the ensuing dynamics is expected and observed[1,4-6] to be highly non-statistical. As we shall later see, these deviations from statistical behavior provide important clues concerning the nature of the dissociation process.

Information concerning the unimolecular dissociation rate for these systems can be obtained directly from the infrared spectra in those cases where the instrumental resolution is sufficient to uncover the associated homogeneous broadening of the individual ro-vibrational transitions[1]. In addition to this, evidence of intramolecular vibrational dynamics is observed in the form of spectral perturbations, which can give rise to both frequency shifts[7] and the appearance of new transitions in the spectrum[8,9] due to coupling to other discrete vibrational states of the molecule. From this large body of experimental data now available, taken together with the theoretical work being carried out in this field[10-12], it is clear that the rate of vibrational predissociation in these complexes is related directly to the strength of the coupling between the intramolecular vibrational mode originally excited and the intermolecular degrees of freedom. It is interesting to note that the spectral shift (due to formation of the complex) of the intramolecular vibrational mode is dependent also on the strength of this coupling so that the vibrational predissociation rate and spectral shift are correlated[1]. Figure 1 summarizes the data presently available for a wide range of complexes. The correlation proposed by the author[1] is "lifetime $\propto 1/(shift)^2$". As discussed by Le Roy and co-workers[13], this correlation can be justified theoretically.

It is interesting to point out that the magnitude of the lifetime and frequency shift can generally be rationalized by considering the structure of the complex. For example, if the intramolecular vibrational mode initially excited is involved directly in the formation of the weak bond then the frequency shift generally will be large and the lifetime short, while for vibrational modes that are remote from the intermolecular bond, the shifts are very small and the vibrational predissociation lifetimes long. In the linear hydrogen cyanide dimer, for example, excitation of the "hydrogen bonded" C-H stretch yields an excited state lifetime of 6 ns, while the lifetime of the "free" C-H stretch excited state exceeds 140 ns[14]. As we shall see in the present paper, these geometrical effects are particularly noticeable and interesting in cases where more than one isomeric form of a given binary complex is observed.

In addition to the structure and rate information, which is obtained directly from the spectroscopic data, much more can be learned about the photodissociation dynamics of these systems if state-to-state rate constants can be obtained. One approach, which has proven very successful, is the traditional pump-probe technique. For example, Cassassa et al.[5] have made use of pulsed IR laser to pump the NO dimer and a pulsed UV laser to probe the NO fragment. The resulting determination of the rotational product state distributions reveals both the average internal and translational energy released to the fragments. In our laboratory we have measured photofragment angular distributions for several vibrational modes associated with systems such as HF[6] and HCN dimers. In all cases the angular distributions differ substantially from those expected from a statistical dissociation and in the case of the HF dimer the individual rotational channels are at least partially resolved. In the latter case these distributions can be used to determine the relative state-to-state rate constants. The advantage of this method over the traditional pump-probe technique is that the correlations between the rotational states of the two diatomic fragments can be established. In essence, this type of experiment is equivalent to a coincidence experiment involving two neutral fragments. In addition to reviewing our latest results using this method, we will also discuss some of the future prospects of the technique.

2. EXPERIMENTAL

The opto-thermal detection technique, as used in molecular spectroscopy, has been discussed in detail in the literature[1,15,16] so that there is no need to do so here. Nevertheless, its application in the measurement of photofragment angular distributions deserves some attention. Since the bolometer is an energy detector, it can be used to measure directly the energy of the photofragments, as well as the missing energy from the molecular beam. The only difficulty in the former case is that the signal is distributed over the available solid angles, while in the depletion experiments all of the dissociated complexes are detected simultaneously. Nevertheless, the sensitivity of the opto-thermal detection method is sufficient to make this possible for many systems. Figure 2 shows a schematic diagram of the experimental arrangement used in this case. The molecular beam is formed by free jet expansion of the gas mixture needed to form the desired complex, after which it is collimated by a skimmer and a second collimator in order to eliminate direct beam contributions to the signal at the scattering angles of interest. The angular distributions are measured by rotating the molecular beam source about the photolysis point, while the bolometer remains fixed. An F-center laser is used to photolyze the species of interest on the axis of rotation of the instrument. Owing to the fact that the complexes of interest dissociate from a moderately long lived "predissociating" state, the spectrum is highly resolved so that the initial state of the parent molecule can easily be selected. Final state selection then depends upon the ability of the instrument to resolve structure in the angular distributions which can be assigned to individual rotational and\or vibrational channels in the fragments.

3. THE ANISOTROPY OF VIBRATIONAL RELAXATION RATES

As indicated above, the vibrational predissociation rates of hydrogen bonded complexes tend to be strongly dependent upon the nature of the vibrational mode initially excited. Another example of this is in the HCN-HF complex where excitation of the C-H stretch yields a lifetime of 13.5 ns while the H-F stretch excited state has a lifetime of only 0.058 ns[17]. Once again this strong mode dependence can be rationalized on the basis of the proximity of the particular intramolecular vibration to the intermolecular bond. In view of the fact that the dissociation of a binary complex can be thought of in terms of a half collision, one might expect that this difference would also be observed in the collisional relaxation data available in the literature. Indeed, Smith and co-workers[18] have carried out an extensive study on the vibrational relaxation of HCN by HF, as well as

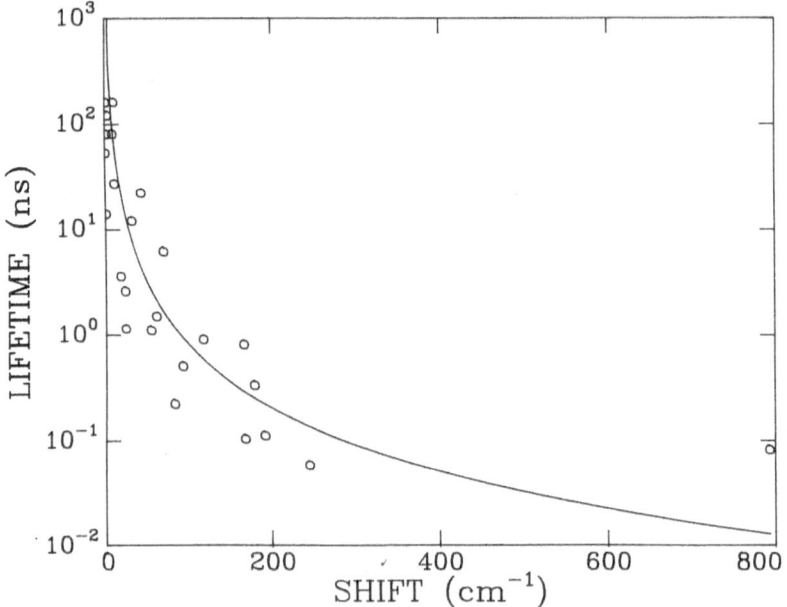

Figure 1

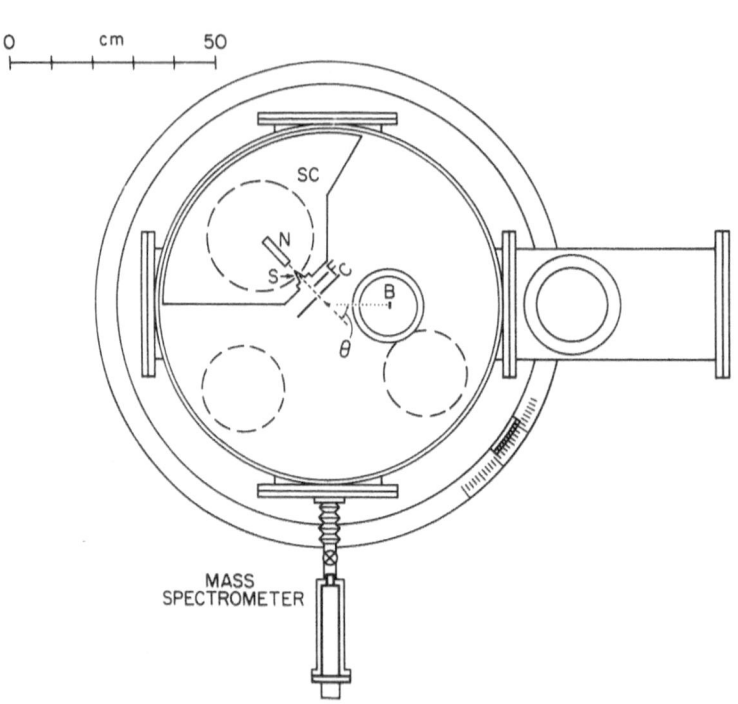

Figure 2

HF by HCN. They find that the HF stretch is relaxed by HCN 3.5 times faster than is the C-H stretch of HCN by HF. In view of the fact that the collisional data averages over all orientations, it is not surprising that the difference is smaller than that observed for the complex. What this does suggest, however, is that through the study of the vibrational predissociation of binary complexes it should be possible to determine the orientational dependence of the vibrational relaxation rates. This anisotropy can be very large, as we shall see, and is a quantity which is not available from the collision based experiments.

In order to determine the anisotropy of the vibrational relaxation rate one must be able to change the relative orientation of the two molecular partners. In other words, data must be obtained for more than one isomer of the complex. Until very recently, there have only been a few binary and ternary complexes for which more than one isomeric form has been observed, despite the fact that ab-initio calculations often indicate the existence of more than one local minimum on the potential energy surface. Recently we have begun a series of studies on various mixed HCN containing complexes. Hydrogen cyanide is an excellent candidate for this type of study since it can act both as an acid and a base. In addition, its large dipole and quadrupole moments tend to make the topology of the associated multidimensional potential surfaces even more complex, thus making the existence of more than one local minimum likely. Indeed, recently we have obtained infrared spectra for at least two isomers of each of the following systems; C_2H_2-HCN[19], N_2O-HCN[20] and CO_2-HCN[20]. For the purposes of the present discussion we will consider C_2H_2-HCN as an example.

The acetylene-hydrogen cyanide complex is of particular interest in view of the fact that the C-H stretches of both monomers can be probed using the F-center laser. Earlier microwave studies of this system showed the existence of a T-shaped isomer[21], shown in Figure 3, in which the acetylene acts as the proton acceptor. During our infrared investigation of this species, a linear isomer was also found in which the acetylene now acts as the acid. By studying the two C-H stretch vibrational modes of these two isomers it is possible to learn a great deal about the anisotropy of the vibrational

ISOMERS OF HCN-ACETYLENE

Figure 3

relaxation rate. For the linear isomer the lifetimes obtained from the homogeneous linewidths are 1.14 ns for the acetylenic stretch and >160 ns for the C-H stretch of the hydrogen cyanide. In the T-shaped isomer the lifetimes are essentially reversed for these two modes, namely 1.06 ns for the hydrogen cyanide and >160 ns for the acetylenic stretch. The limits on two of the lifetimes (>160 ns), were established by noting that the homogenous linewidths are not discerned at the available instrumental resolution, so that the best that could be done is to determine a lower limit. The first thing to note from this data is that the mode specificity is consistent with the picture given above, which suggests that the rate is dependent upon the coupling of the intramolecular vibration to the intermolecular coordinate. It is also clear that the anisotropy of the vibrational relaxation rate is very large. In scattering terms, the vibrational relaxation rates for the C-H stretches of this system are more than 100 times greater when the collision directly involves the corresponding hydrogen than when it does not. Although this collisional analogy cannot be pushed too far, since there are important energy and angular momentum differences between the collisional and predissociation experiments, these results do suggest that information on the anisotropy of the vibrational relaxation rates is now available. Relatively little has been done in this area theoretically[22], presumably because of the lack of experimental data. In view of the results now becoming available through the study of binary complexes, clearly it would be very interesting to make comparisons with theory. For example, in classical trajectory calculations this would be possible if the rates were reported prior to averaging over the orientations.

4. STATE-TO-STATE VIBRATIONAL PREDISSOCIATION

As pointed out in our previous paper on the state-to-state photodissociation of HF dimer[6], photofragment angular distributions can be used in favorable cases to determine the final internal state distribution of the fragments. The kinematics also can be used to determine the correlations between the rotational states of the two fragments. That is to say the relative rates can be determined for

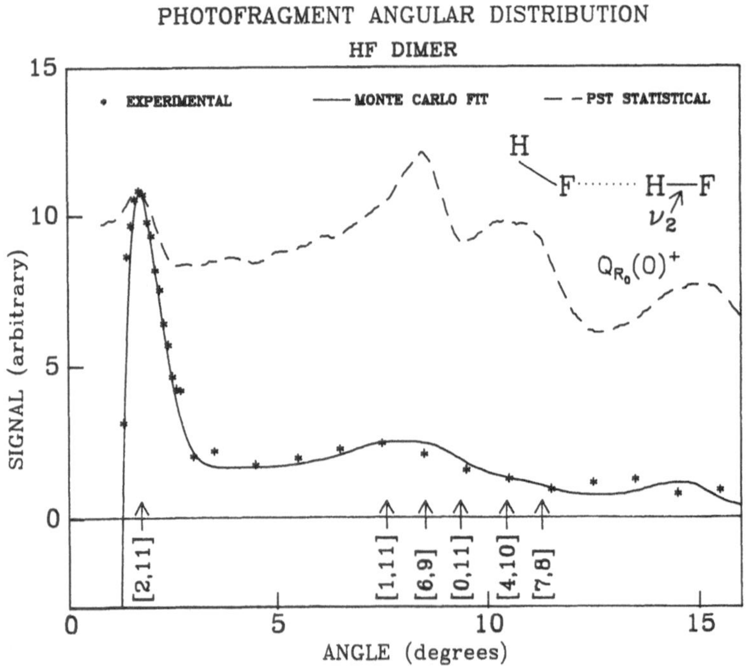

PHOTOFRAGMENT ANGULAR DISTRIBUTION
HF DIMER

Figure 4

dissociation into states defined by the rotational quantum numbers of both fragments. HF dimer is an attractive system for study using this method since the number of open fragment channels is very small. Indeed, the only open vibrational channel corresponds to both HF fragments being in the ground state. Angular distributions have now been obtained for the HF dimer corresponding to the excitation of the ν_1 $K_a=1\leftarrow0$ and $K_a=0\leftarrow0$ subbands and the ν_2 $K_a=0\leftarrow0$ subband. In all three cases the rotational and tunneling quantum numbers associated with the excited state of the parent molecule are determined.

Figure 4 shows the angular distribution corresponding to excitation of the ν_2 vibrational mode ($K_a=0\leftarrow0$). The solid line through the experimental points is a fit to the data using a Monte Carlo calculation to average over the experimental conditions. The adjustable parameters in the fit are simply the relative state-to-state probabilities and the D_0 value for HF dimer. The dissociation energy was determined from position of the (2,11) channel in the ν_2 angular distribution and was then used for all of the other data. The fits to the experiment are excellent in all cases. Also shown in the figure, as a dashed line, is the angular distribution resulting from a phase space theoretical treatment of the dissociation[23]. The dramatic failure of this theory is yet another indication of the strongly non-statistical nature of the dissociation of these systems. The statistical theory is nevertheless significant in that it can be compared with the experimental results to determine which channels are preferred and which are not.

Surprisal plots are shown in Figure 5 for all of the HF dimer ro-vibrational bands studied to date. In all cases the striking feature in these plots is the fact that the most important channels are those which correspond to one HF fragment being highly rotationally excited while the other is not. On the other hand, the channels for which J_1 $\approx J_2$ are consistently smaller. This preference can be explained if we assume that the dimer dissociates impulsively from a geometry which is essentially that of the equilibrium structure, as shown in Figure 6. In this case it is clear that there is a large torque on the proton donor while the impulsive force on the proton acceptor is essentially through its center of mass. The implication is obviously that the highly rotationally excited fragment correlates with the proton donor. Clearly it is the rotational correlations established in these experiments which enable us to obtain such a clear picture of the dissociation dynamics of this complex. Indeed, rotational distributions alone would not be sufficient to elucidate the impulsive nature of the HF dimer dissociation. It is interesting to note that Halberstadt et al.[24] have carried out calculations on the predissociation of HF dimer in which they assumed that the proton donor carried away most of the rotational energy. In this way they argued that the problem might be treated effectively as an atom-diatom system. Although this is clearly not the case, these calculations do point the way for future theoretical work on this system.

HF dimer is clearly an ideal system for the technique described here owing to the low density of fragment rotational states. For larger systems the number of channels clearly will become too great to allow resolution of the individual fragment channels. Nevertheless, in a recent study of HCN dimer we have shown that even when such structure is absent from the angular distribution, useful information on the dissociation dynamics can be obtained. For example, eventhough the angular distributions resulting from ν_1 and ν_2 excitation of HCN dimer are relatively structureless, the two angular distributions differ from

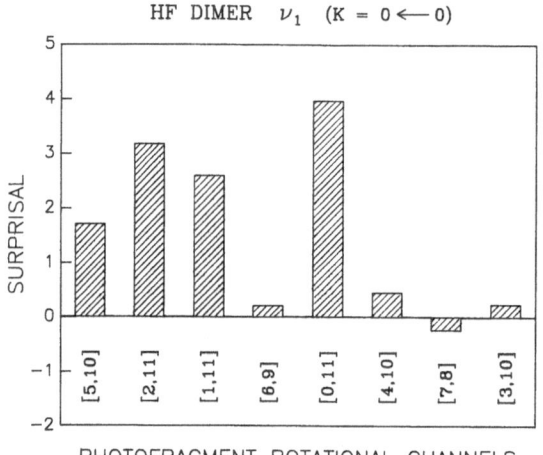

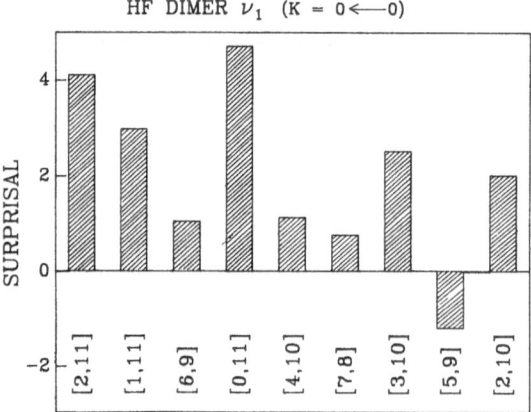

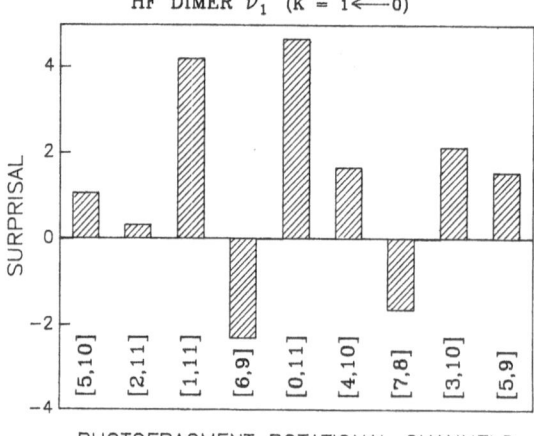

Figure 5

one another and both differ substantially from the statistical distributions. For cases such as this, it is clear that a quantitative explanation for these differences will only be possible with the aid of theory. As indicated below, however, there are a rather large number of systems which lie between these two extremes for which detailed information on the final state distribution of the photofragments can be obtained.

HF Dimer

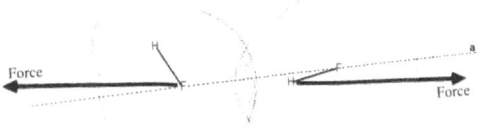

Figure 6

5. FUTURE OUTLOOKS

In keeping with the spirit of this workshop, it is appropriate to consider the systems for which the measurement of photofragment angular distributions is likely to yield detailed information on the dissociation dynamics, similar to that obtained for HF dimer. An obvious diatom-diatom system that needs to be studied in future is the mixed HF-DF dimer. Since the DF occupies the hydrogen bonding position (due to zero point energy considerations) the impulsive dissociation mechanism discussed above should give rise to highly rotationally excited DF and HF in low J states. Results of this type would clearly provide a direct confirmation of the impulsive model proposed here for the dissociation of HF dimer. Several other diatom-diatom systems are being studied in our laboratory, including H_2-HF and D_2-HF. In the case of the hydrogen containing complex, the energetics is such that both fragments must be produced in their ground vibrational states, as with HF dimer. In D_2-HF, on the other hand, the dynamics becomes somewhat more complex due to the fact that the v=1 D_2 channel is also open. From the lifetime data available for these two systems[25,26], it appears as if this intermolecular V-V channel may play an important role in the predissociation dynamics of D_2-HF. To date, however, there is no direct experimental data which confirms this. The assignment of the fragment rotational and/or vibrational structure in the angular distributions would clearly provide the information needed to determine the importance of the v=1 channel.

Understanding the role of low frequency intramolecular vibrations in the dissociation dynamics of these systems is clearly of fundamental importance. In addition to the intermolecular (across the intermolecular bond) V-V process discussed above, the possibility also exists for intramolecular (within a single monomer unit) V-V energy transfer for cases where the complex contains at least one polyatomic monomer subunit. The calculations that have been reported for Ar-C_2H_4[11] suggest that this process is also facile and hence is deserving of experimental investigation. Systems such as Ar-HCN and Ar-C_2H_2 would be ideal candidates for study using the methods discussed here.

For the case of atom-diatom systems with very high frequency intramolecular vibrations, the available results indicate that vibrational predissociation is slow, such that its rate cannot be measured using the available methods. Nevertheless, many of these systems are also of interest in the present context since rotational predissociation can also be important. Systems such as He-HF and Ne-HF, which already have been studied spectroscopically[27], are also well suited to these types of measurements.

As indicated above, the study of the vibrational predissociation dynamics of weakly bound complexes is in a very fruitful period owing to the overlap which now exists between experiment and theory. Future developments in theory most certainly will be in the area of polyatomic systems where the role of open vibrational channels are likely to be important.

6. REFERENCES

1. R.E. Miller, Science, 240:447 (1988).
2. G.D. Hayman, J. Hodge, B.J. Howard, J.S. Muenter and T.R. Dyke, Chem. Phys. Lett. 118:12 (1985).
3. D.J. Nesbitt, Chem. Rev. 88:843 (1988).
4. Z.S. Huang and R.E. Miller, J. Chem. Phys. 90:1478 (1989).
5. M.P. Cassassa, J.C. Stephenson and D.S. King, J. Chem. Phys. 85:2233 (1986).
6. D.C. Dayton, K.W. Jucks and R.E. Miller, J. Chem. Phys. 90:2631 (1989).
7. Z.S. Huang and R.E. Miller, J. Chem. Phys. 89:5408 (1988).
8. K.W. Jucks and R.E. Miller, J. Chem. Phys. 86:6637 (1987).
9. C.M. Lovejoy, M.D. Schuder and D.J. Nesbitt, J. Chem. Phys. 85:4890 (1986).
10. R.J. Le Roy and J.S. Carley, Adv. Chem. Phys. 42:353 (1980).
11. J.M. Hutson, D.C. Clary and J.A. Beswick, J. Chem. Phys. 81:4474 (1984).
12. A.D. Buckingham, P.W. Fowler and J.M. Hutson, Chem. Rev. 88:963 (1988).
13. R.J. Le Roy, M.E. Lam and M.R. Davies, private communication.

14. K.W. Jucks and R.E. Miller, J. Chem. Phys. 88:6059 (1988).
15. T.E. Gough, R.E. Miller and G. Scoles, Appl. Phys. Lett. 30:338 (1977).
16. Z.S. Huang, K.W. Jucks and R.E. Miller, J. Chem. Phys. 85:3338 (1986).
17. D.C. Dayton and R.E. Miller, Chem. Phys. Lett. 143:181 (1988).
18. G.S. Arnold, R.P. Fernando and I.W.M. Smith, J. Chem. Phys. 73:2773 (1980).
19. P.A. Block, K.W. Jucks, L.G. Pedersen and R.E. Miller, Chem. Phys., in press.
20. manuscripts in preparation
21. P.D. Aldrich, S.G. Kukolich and E.J. Campbell, J. Chem. Phys. 78:3521 (1983).
22. H.K. Shin, J. Chem. Phys. 49:3964 (1968).
23. J.C. Light, Faraday Discuss. Chem. Soc. 44:14 (1967).
24. N. Halberstadt, Ph. Brechignac, J.A. Beswick and M. Shapiro, J. Chem. Phys. 84:170 (1986).
25. K.W. Jucks and R.E. Miller, J. Chem. Phys. 87:5629 (1987).
26. C.M. Lovejoy, D.D. Nelson and D.J. Nesbitt, J. Chem. Phys. 87:5621 (1987); J. Chem. Phys. 89:7180 (1988).
27. D.J. Nesbitt, C.M. Lovejoy, T.G. Lindeman, S.V. ONeil and D.C. Clary, J. Chem. Phys. 91:722 (1989).

STRUCTURE AND DYNAMICS OF SIZE SELECTED CLUSTERS

Udo Buck

Max–Planck–Institut für Strömungsforschung

D3400 Göttingen, Fed.Rep. of Germany

1. INTRODUCTION

The infrared photodissociation of weakly bound complexes has attracted much interest in recent years.[1,2] In these experiments a vibrational mode of one molecular component is excited by an infrared photon. If the photon energy is larger than the binding energy of the complex, the complex predissociates. Typically, the clusters are prepared in a supersonic expansion and the dissociation is measured by monitoring the depletion of the molecular beam as a function of the laser frequency. The measured fraction dissociated P_{diss} is given by[3,4]

$$P_{diss} = 1 - \exp[-\sigma(\nu)\, F/(h\nu)] \quad, \tag{1}$$

where $\sigma(\nu)$ is the dissociation cross section, F the laser fluence and $h\nu$ the photon energy. In principle, these dissociation spectra contain three observables:

1) the line shift $\Delta\nu$ which is caused by the interaction of the excited oscillator with the surrounding molecules and thus gives information on the structure of the cluster;

2) the linewidth Γ which, if interpreted as homogeneously broadened, gives information on the lifetime and thus on the dynamical coupling of the molecular vibrational mode to the internal cluster modes;

Dynamics of Polyatomic Van der Waals Complexes
Edited by N. Halberstadt and K. C. Janda
Plenum Press, New York, 1990

3) the dissociation cross section σ which is directly related to the absorption and decay process.

In a quantum mechanical treatment this cross section from the initial state i to the final continuum state αE is given by[5]

$$\sigma(i \to \alpha E) = \frac{4\pi^2 \omega}{c} \mid <f|\mu \, e|i> \mid^2 \frac{\Gamma_f}{(E-E_f)^2+\Gamma_f^2} \, , \tag{2}$$

where μ is the transition dipole moment from i to the discrete state f and the coupling to the continuum is described by the golden–rule expression

$$\Gamma_f = \pi \mid <f|V|\alpha E> \mid^2 \, , \tag{3}$$

where V is the intermolecular potential. For well separated resonances one obtains the familiar Lorentzian lineshape.

The problem in these experiments is to get cluster specific information. The cluster beam is usually generated as a distribution of different cluster sizes and a liquid–He–cooled bolometer is not able to discriminate between the different masses. But even a mass spectrometer is only of limiting help because of the extensive fragmentation which occurs when these weakly bound clusters are ionized.[6] A way out of this problem is either to use very dilute mixtures in the expansion and to take the high resolution spectrum itself for identification which works mainly for dimers and a few trimers[1] or to carry out the experiments with size selected clusters. We have recently developed a method which selects smaller clusters by a scattering process with a He beam.[6,7] This technique has been successfully applied to infrared photodissociation experiments. After the first experimental results on C_2H_4 dimers were reported,[8] a series of partly different experiments were carried out for van der Waals clusters of C_2H_4 [9–11] and SF_6 [12] and the hydrogen bonded systems NH_3,[13] CH_3OH,[14,15] N_2H_4,[16] CH_3NH_2 [17] and CH_3CN.[17] In all cases vibrational modes of the molecules in the tuning range of the CO_2 laser around 10 μm were excited. Clusters up to n = 8 could be separated. It is the purpose of this contribution to review the prospects and limits of this novel experimental method and to outline the theoretical impact which is necessary to interprete the data.

We start with a description of the experimental method including the size selection, the different experimental set–ups and the variation of the internal energy of the clusters. Then theoretical concepts are presented to calculate the structure of these clusters and the corresponding lineshifts. Results will be given of recent experiments and, if possible, compared with preliminary calculations as case studies for ethylene, methanol, ammonia, hydrazine and acetonitrile. Finally, the future trends both of experimental and theoretical work will be discussed.

2. EXPERIMENT

2.1 *Size Selection*

The method of cluster separation in a scattering experiment with a secondary beam under single collision conditions is based on the fact that the heavier clusters are scattered into smaller angles with smaller final velocities compared with the higher clusters. This behaviour is best documented in a Newton diagram which correlates the different velocities and angles on the basis of conservation of momentum and energy.[18] Such a diagram is shown in Fig. 1 for CH_3OH–He scattering. The methanol beam is generated by expanding a 2.7 % mixture in Ne through a 80 μm diameter nozzle at a pressure of 1.0 bar, while the He beam is expanded through a 30 μm diameter nozzle at 30 bar.[19] The circles denote the final velocities in the center–of–mass system for elastically scattered particles. The limiting laboratory angles for the different clusters n (the tangents to the circles in Fig. 1) are 16.6o(2), 11.1o(3), 8.4o(4), 6.7o(5), 5.6o(6), 4.8o(7), and 4.2o(8). There are two modes of operation for size selection.

1) Measurements at fixed angle and mass: The choice of scattering angle excludes the larger clusters by kinematical constraints. At a scattering angle of $\theta = 6.5^o$ all $(CH_3OH)_n$ clusters n > 5 are not detected. To discriminate against the smaller ones, a mass spectrometer can be used. For this method to work at least a small fraction of the cluster must appear as an ion at a mass larger than those of the smaller clusters. For $(CH_3OH)_n$ this is indeed the case at the protonated ions $(CH_3OH)_{n-1}H^+$ that are favoured in the fragmentation process. Thus the ideal mass for the separation of n = 5

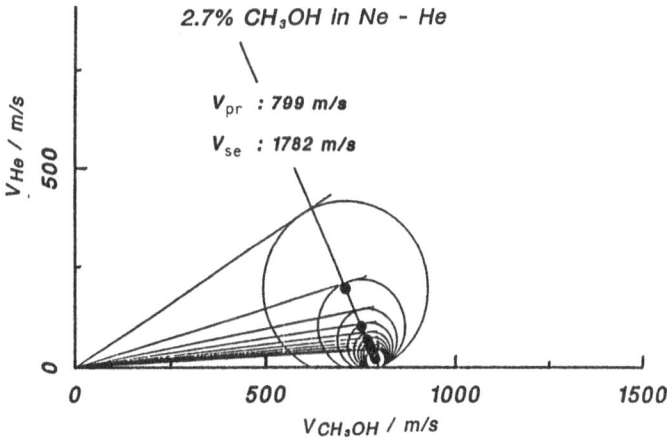

Fig. 1.   Newton diagram for the scattering of $(CH_3OH)_n$ clusters seeded in Ne by He. The positions of the elastically scattered clusters are given for n= 1–8 together with the limiting angles.

is m = 129 amu. This procedure does not always work in its idealized fashion and leads to pure size selection. However, by an appropriate choice of angle and mass and different combinations one can usually find conditions for which, at least, one cluster size dominates the spectrum.

2) Measurements at fixed angle and velocity. If, however, the cluster fragments completely to smaller masses, then different velocities help to distinguish the different cluster sizes. This method is independent of the special detection scheme.

Both methods require a crossed molecular beam apparatus with high angular and velocity resolution in order to separate the different contributions. It is noted that the scattering process with He transfers a small amount of energy to the cluster which can be measured by analyzing the scattered clusters by time–of–flight methods.[20]

2.2 *Infrared Photodissociation*

The combination of the size selection process of the last section with the IR–photodissociation is carried out in two different arrangements which are displayed in Fig. 2. In the first version the cluster beam is scattered from He for size selection. By measuring the intensity at different angles and masses different cluster sizes are selected. Then the cluster is dissociated by the infrared radiation of a pulsed line–tunable CO_2 laser (a). The dissociation is measured by monitoring the decrease of intensity as a function of laser wavelength and power. Because of the scattering process with He a certain amount of energy is transferred into internal energy of the cluster so that photodissociation takes place for internally excited, "warm" clusters. A further variant is used by replacing the pulsed CO_2 laser by a c.w. laser which reduces the power by five orders of magnitude. In order to keep the necessary laser fluence, the interaction time is increased by a collinear arrangement of laser and scattered beam. This arrangement has the advantage that the duty cycle increased by a factor of 1000 and that high–resolution lasers can also be used for excitation. In the complementary arrangement, the laser–molecular beam interaction takes place before the scattering center where the clusters are still cold (b). Then the cluster beam is dispersed by the He beam and the cluster specific detection is obtained as in the first case.

Thus the scattering method allows one to measure the photodissociation not only as a function of cluster size but also for one size as a function of the internal excitation. It is noted that in both methods exist further refinement to vary the internal energy. In the first method the amount of transferred energy can be easily changed by varying the collision energy, the scattering angle or by measuring the complete time–of–flight spectra with and without laser radiation. In this way the internal energy was varied from about 3 meV to more than 35 meV. In the second method the degree of internal energy, for the "cold" clusters can only be varied by using different carrier gases in the expansion. Thus Ne leads usually to a better degree of cooling than He.

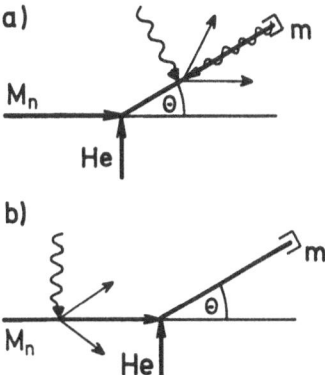

Fig. 2.   Schematic experimental arrangement: a) interaction with collisional excited clusters, b) interaction with cold clusters.

3. CALCULATIONS

3.1 *Structure*

For the interpretation of the measurements information on the structure of the clusters is very important. An ab initio calculation at one point of the total energy surface of a more than three–atom cluster including effects of correlation is a state–of–the–art task. A complete mapping of the surface is impossible. Therefore, we have used a model which is based on a reliable multi–center pair potential. Then the equilibrium geometries and total energy minima are determined by minimizing the intermolecular energy with respect to the molecular positions. To initiate a minimization run, the coordinates and angles were chosen at random. For the pair potentials we have used results calculated by the test particles method of Ahlrichs and Böhm (C_2H_4) [21] or the empirical potential data sets for N_2H_4 [22] and CH_3OH.[23] In the latter case also optimized intermolecular potential functions are available which are fitted to properties of the liquid.[24] A similar approach but based on an ab initio configuration interaction calculation for the dimer potential has been published for NH_3.[25]

3.2 Lineshift

For the calculation of the lineshift we use a method[26] originally developed by Buckingham[27] which is based on perturbation theory. The frequency shift of the ith vibrational mode is given by

$$h \, \Delta \omega_i = \frac{1}{2} \, [V''_{ii} - \sum_j (\phi^{(3)}_{iij}/\omega_j)V'_j] \ ,$$

where V' and V'' denote first and second derivatives of the intermolecular potential with respect to a set of dimensionless normal coordinates of the molecule and $\phi^{(3)}_{ijk}$ are cubic unharmonicities. The first term is the shift caused by the change of the force constant, whereas the second term arises from a shift in the equilibrium position and only plays a role if coupled to the unharmonicity of the potential. For the calculation of the derivatives the optimized potential parameters obtained in the procedures of the last section are applied.

4. RESULTS AND DISCUSSION

The dissociation spectra of the systems measured so far show a large variety of behaviour depending on the type of bonding, the excited vibrational mode and, most importantly, on the cluster size. An overview of the systems, their excited modes and the frequency shifts of their smaller clusters are presented in Table 1. The van der Waals system C_2H_4 exhibits only a small blue shift of 3 cm⁻¹ which is the same for all cluster sizes. A possible explanation for this behaviour is the chain–like structure of the larger clusters which is observed in the structure calculation described in Sec. 3.1.[21] In contrast, the two linear hydrogen bonded systems CH_3OH and CH_3NH_2 show for every cluster size a different spectrum. A red and blue shifted double peak for the dimer, blue shifted peaks for the larger clusters and for CH_3OH an interesting change again between n = 5 and n = 6. The nonlinear hydrogen bonded systems N_2H_4 and NH_3 display very large shifts between 50 and 100 cm⁻¹ with structured spectra

Table 1. Systems investigated and frequency shifts in cm^{-1}

molecule	mode	ν_0	dimer	trimer
C_2H_4	CH_2 out	949	+3.0	+3.0
CH_3OH	CO stretch	1034	−8.0,+18	+7.0
CH_3NH_2	CN stretch	1044	−1.3,−10	+1.8
CH_3CN	CC stretch	920	....	−2.0
	CH_3 bend	1041	....	+4.9
N_2H_4	NH_2 wag	937	+42,+48	55,+88
	NN stretch	1098	−16	−10
NH_3	umbrella	950	−26,+53	+66(56)

after the excitation of the correlated motions of the H–atoms. For other modes and also for CH_3CN only small shifts are observed. In the latter case, an interesting transition is observed between n = 4 and n = 5 leading to a line narrowing in one mode and a new peak in the other mode. Also the range of linewidths measured under comparable conditions varies from 39 cm^{-1} for $(N_2H_4)_5$ to 1 cm^{-1} for $(CH_3CN)_5$. Some recent characteristic results and their interpretation will be presented as case studies: the influence of internal excitation on the width Γ and the cross section for $(C_2H_4)_2$, the number of photon which are necessary for the dissociation for NH_3 clusters, the strong correlation between frequency shifts and structure in the case of CH_3OH and N_2H_4 clusters, and, finally, the transition for CH_3CN clusters.

4.1 Internal Excitation Effects: Ethylene

The photodissociation of the ethylene dimer after the excitation of the ν_7 out of plane mode has attracted much interested in the past. Broad (12 cm^{-1})[3,4] and narrow lines (10^{-4} cm^{-1})[28,29] were observed and later confirmed by measurements with size selected clusters.[10,11] The coexistence of both sharp and broad features which differs by 5 orders of magnitude in their width has prompted new theoretical efforts for the explanation. The mixing of the ν_7 vibration with a set of dark vibrational levels consisting of combinations of the van der Waals and the ethylene ν_{10} mode which is energetically closed is apparently a promising way to solve this problem.[30,31] Depending on the different coupling of the angular momentum to the continuum the broad and narrow features may arise.[31]

An interesting issue of the dimer spectrum is the strong dependence of the linewidth on the internal excitation of the cluster. We have carefully investigated this behaviour using the experimental techniques described in detail in Sec. 2. The result is presented in Fig. 3. The width jumps from a value of 12 cm^{-1} obtained for "cold" dimers[10] to more than 25 cm^{-1} and increases only slightly when the transferred energy is increased from $\Delta E = 2.5 \pm 2.0$ meV to 32 ± 2.0 meV. The explanation is given by the excitation of hot bands which have been calculated to lie between 30 cm^{-1} and 40 cm^{-1} corresponding to about 4 meV above the ground state and are separated by about $\pm$ 10 cm^{-1} from the central absorption frequency.[32] Once the hot bands are excited, the additional energy ΔE does not lead to further extensive broadening. The large linewidth is simply explained by the inhomogeneous addition of several contributions of the order of 12 cm^{-1}. For larger clusters the difference between spectra of cold and hot clusters becomes smaller and diminishes for the hexamer. Obviously the additional energy is more easily distributed among the various degrees of freedom and the excited states are closer to the ground state than in the case of the dimer. It is interesting to note that the dissociation cross section after integration over the Lorentzian profile increases with increasing internal energy ΔE. According to Eq. (2) it should be constant, since the transition dipole moment is constant for dimers. A possible explanation is the additional coupling to the ν_{10}–mode. In such a case the formula (2) is not valid and we have to multiple with the additional coupling rate from ν_{10} to the continuum[33] which can, of course, depend on the internal energy. Thus this result is another support for the two mode coupling model.

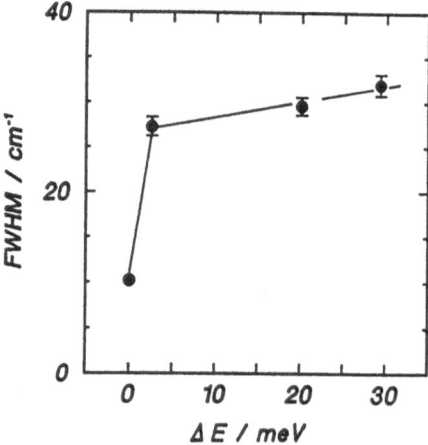

Fig. 3.   Measured width Γ of photodissociated ethylen dimers as a function of the internal energy ΔE.

The ammonia dimer dissociation spectrum was found to consist of two bands which were attributed[34] to the two non equivalent NH_3 molecules in the complex.[35] The experiments were carried out with a direct cold beam and a small laser fluence of about 0.5 mJ/cm². Therefore, one CO_2 laser photon of 125 meV is sufficient to dissociate the dimer. The measured spectrum was confirmed in recent experiment with size selected clusters.[13] In addition, it was observed that the trimer and tetramer dissociation bands were shifted to the blue. They consist each of one larger peak with a red shifted shoulder. The experiments, however, were carried out with a laser fluence of 90 mJ/cm² so that two photon excitation could not be ruled out. With the knowledge of the trimer and tetramer absorption frequencies from the experiment with size selected clusters, a new double resonance experiment with two lasers at low laser fluence and a bolometer detector revealed that both clusters can only be excited with one photon (positive signal) and a second photon shifted to the red is necessary for the dissociation (negative signal).[36] This result is in excellent agreement with the calculation of the bonding energies of these clusters which gives[25] 134 meV, 244 meV, 202 meV and 166 meV for n = 2, 3, 4, and 5, respectively.

4.2 *Lineshifts and Structure:*

Methanol: Methanol clusters have been thoroughly investigated by the scattering method using cold clusters and pulsed lasers[14] up to n = 4 and using internally excited clusters and continuous wave lasers[15] up to n = 8. The comparison of the dimer spectrum is shown in Fig. 4. Both experiments reveal a two peak structure with one peak shifted to the red and one to the blue as is listed in Table 2. As is already observed for ethylene dimers, the peak positions are nearly the same, but the linewidth is much broader in the case of the hot clusters. The clusters from n= 3 to n = 5 exhibit single peaks which are each shifted further to the blue compared with the monomer absorption frequency. The explanation is found in the structure calculation shown in Fig. 5 for the dimer and the tetramer. In the well known dimer structure, there are two non–equivalent positions of the monomer. A calculation based on the potential of Ref. 24 and using the theory presented in Sec. 3 gives for the C–O stretching mode a blue shift for the hydrogen donor (where the H participates in the bond) and a red shift for the acceptor (with the O in the bond) in surprisingly good agreement with experiment.[37] The results are given in Tab. 2. Note the much larger shift to the red of the O–H stretching mode. For the tetramer as well as for the trimer and pentamer planar cyclic structures are observed in which the C–O stretching mode is in equivalent positions and thus only one peak occurs.

Table 2. Calculated and measured lineshift for the methanol dimer in cm⁻¹

	O–H cal.	C–O cal.	C–O experiment cold	hot
Donor	− 225.2	+ 22	+ 17.6	+ 18.8
Acceptor	− 36.6	− 4.1	− 7.5	− 4.4

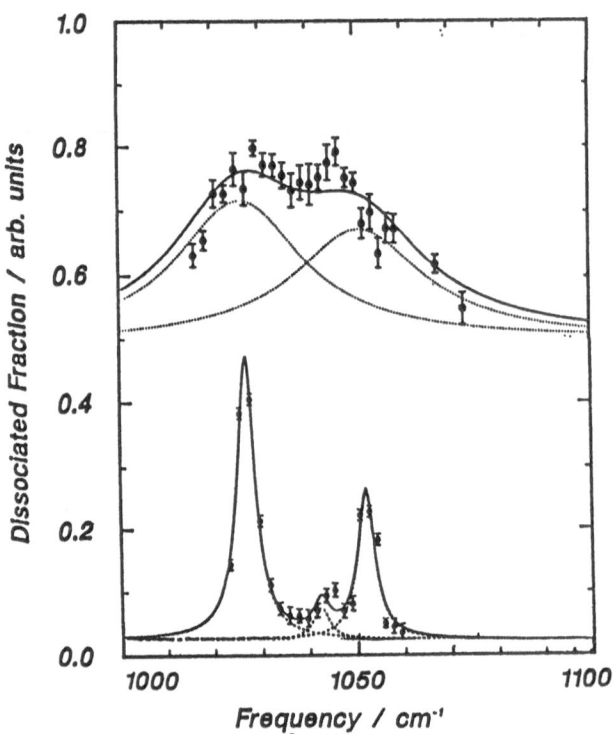

Fig. 4.   Comparison of measured photodissociation spectra of hot (upper panel) and cold (lower panel) methanol dimers.

A very interesting feature arises for the hexamer. Again a two peak structure is observed with one peak shifted to the blue and one to the red compared with the pentamer as shown in Fig. 6. The structure continues up to n = 8. This result infers a structural change for the hexamer which is observed in the calculations. Aside

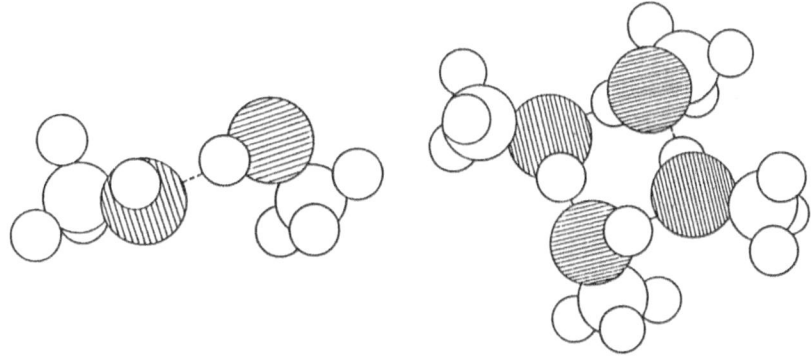

Fig. 5.   Calculated minimum energy configuration for methanol dimers and tetramers.

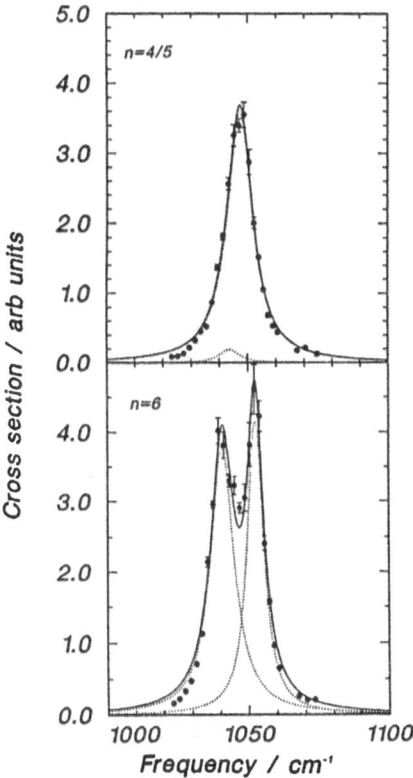

Fig. 6.   Measured photodissociation spectra of $(CH_3OH)_n$ clusters for n= 5 (upper panel) and n= 6 (lower panel).

from the planar ring structure, several isomers are found with strongly deformed rings. Therefore, we asign the two peaks either to two nonequivalent positions of the C–O or to two different isomers one of which is close to a planar ring and the other corresponds to the more distorted structure. Only further experiments and calculations can solve this problem.

Hydrazine: For this molecule two modes were excited, the antisymmetric NH_2 wagging and the N–N stretching mode.[16] The latter one shows nearly no shifts, whereas for the former one large shifts and widths are observed. The results for the dimer, internally excited (upper panel) and cold (lower panel), are displayed in Fig. 7. Both are shifted by about 45 cm⁻¹ to the blue, but only for the cold spectrum, obtained in the direct beam, a two peak structure is measured with a dramatic change in the width from 45 cm⁻¹ to 4.5 cm⁻¹. Structure calculations reveal two nonlinear hydrogen bonds which produce non–equivalent positions for the NH_2 motion but not for the N–N stretching in agreement with the experimental results. The data for the larger clusters n = 3 and n = 4 are given in Fig. 8. The trimer spectrum shows two peaks, a smaller one which is only slightly shifted to the blue with respect to the dimer spectrum. The other very large one is shifted by 88 cm⁻¹ compared to the gas phase value. The simplest explanation for this behaviour is to attribute the two peaks to two different isomers which are indeed found in preliminary structure calculations. The smaller peak close to the dimer peak is caused by a chain–like structure consisting of two dimer configurations with two hydrogen bonds in each case. The larger peak is attributed to a ring configuration with only one hydrogen bond per monomer. The equivalent positions for the excitation of the NH_2 wagging mode explain the one peak structure in the spectrum. The larger clusters exhibit also a large peak at the same position as is found for the trimer. This is an indication of the ring structure with one hydrogen bond per molecule. In addition, a second peak shifted to the blue develops.

In contrast to the finding for methanol, equivalent isomers already appear for these clusters at n = 3. In both cases there exists a close connection between the measured lineshifts and the calculated structure which can be used in refined treatments to determine the interaction potential.

4.4 A Phase Transition: Acetonitrile

This nonlinear hydrogen bonded species does not show any dimer absorption both for the C–C stretching (ν_4) and the CH_3 rocking (ν_7) mode. For the next larger clusters n = 3 and n = 4 single peaks are observed which are slightly shifted to the

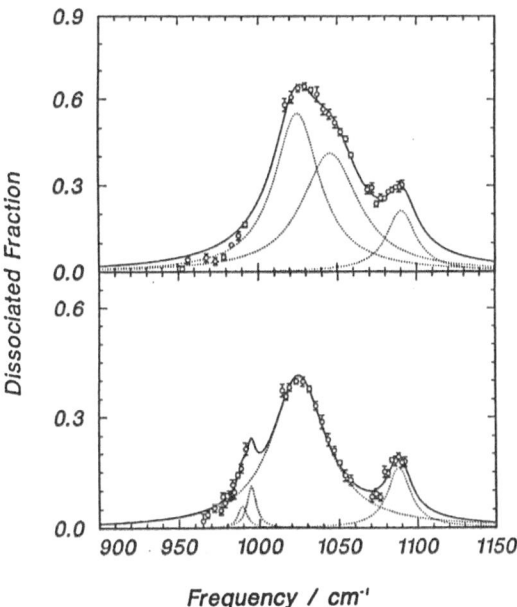

Fig. 8.   Measured photodissociation spectra of $(N_2H_4)_n$ clusters for n=3 (lower panel) and n=4 (upper panel).

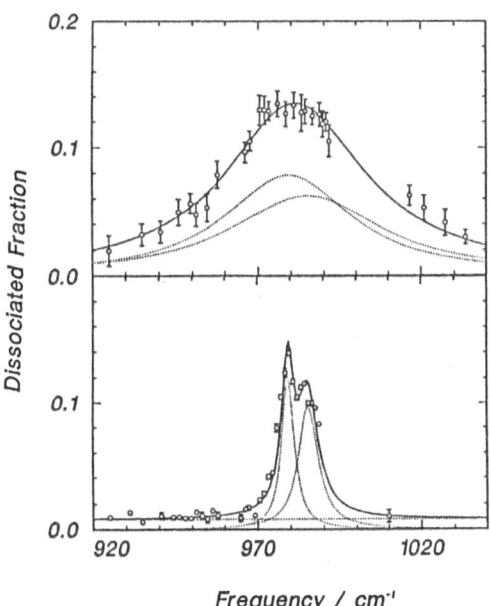

Fig. 7.   Measured photodissociation spectra of hot (upper panel) and cold (lower panel) hydrazine dimers.

red for the ν_4- and to the blue for the ν_7-mode. An interesting change is observed going from n = 4 to n = 5. The position of the ν_4-mode does not move, however, the linewidth is narrowed by about a factor of 3 to 1.0 cm^{-1}. In the ν_7-mode a new peak shows up at 1036.6 cm^{-1} which together with the main peak at 1046.6 cm^{-1} is close to the frequencies observed for the supercooled solid phase I as was already found in an experiment with non–size selected clusters.[38] The line narrowing which was not observed previously[39] and the double–peak structure could be an indication of a transition to a solid–like behaviour of the cluster.

4.5 *Concluding Remarks*

The combination of the production of size selected molecular clusters in a scattering process with the powerful infrared photodissociation of these clusters has produced a wealth of new data. From the experimental point of view, extension in several directions are possible. The size selection of clusters can be pushed to clusters larger than n = 10 by increasing the angular resolution and the intensity of larger clusters using better collimation and conical nozzles for their production. Up to now only pure clusters have been investigated. First experiments with mixed clusters were carried out already for ethylene–aceton mixtures with very interesting results. This direction opens up the study of a new class of interactions. Finally, the combination of two laser double resonance experiments with size selected clusters will help us to identify experimentally the different contribution to the observed spectrum.

From the theoretical point of view, the complete interpretation of the measured spectra are still in its infancy. Procedures for structure calculations are available. The input data, however, reliable pair potentials, are rare. Procedures for calculating lineshifts are also available. For a successful application also good calculations of the complete force field with all unharmonic coupling elements are necessary. For a realistic comparison all these calculations have to be inbedded in Molecular Dynamics or Monte Carlo simulations. The calculations of the dynamical processes which determine the linewidths are probably at best carried out by classical trajectories for these large systems.

ACKNOWLEDGEMENT

It is a pleasure to thank my coworkers Dr. Ch. Lauenstein, Dr. A. Rudolph, Dr. X.J. Gu, M. Hobein and B. Schmidt who did the work presented here. Financial support from the Deutsche Forschungsgemeinschaft (SFB 93) is gratefully acknowledged.

REFERENCES

1. R.E. Miller, J.Phys.Chem. 90: 330 (1986); Science 240: 447 (1988)

2. K.C. Janda and C.R. Bieler, preprint

3. M.P. Casassa, D.S. Bonse and K.C. Janda, J.Chem.Phys. 74: 5044 (1981)

4. M.A. Hoffbauer, K. Liu, C.F. Giese amd W.R. Gentry, J.Chem.Phys. 78: 5567 (1983)

5. J.A. Beswick, in: "Structure and Dynamics of weakly bound molecular complexes", A. Weber, ed., Reidel, Dordrecht (1987), p. 563

6. U. Buck, J.Phys.Chem. 92: 447 (1988)

7. U. Buck and H. Meyer, Phys.Rev.Lett. 52: 109 (1984); J.Chem.Phys. 84: 4854 (1986)

8. F. Huisken, H. Meyer, Ch. Lauenstein, R. Sroka and U. Buck, J.Chem.Phys. 84: 1042 (1986)

9. U. Buck, F. Huisken, Ch. Lauenstein, H. Meyer and R. Sroka, J.Chem.Phys. 87: 6276 (1987)

10. F. Huisken and T. Pertsch, J.Chem.Phys. 86: 106 (1987)

11. U. Buck, Ch. Lauenstein, A. Rudolph, B. Heijmen, S. Stolte and J. Reuss, Chem.Phys.Lett. 144: 396 (1988)

12. F. Huisken and M. Stemmler, Chem.Phys. 132: 351 (1989)

13. F. Huisken and T. Pertsch, Chem.Phys. 126: 213 (1988)

14. F. Huisken and M. Stemmler, Chem.Phys.Lett. 144: 391 (1988)

15. U. Buck, X.J. Gu, Ch. Lauenstein and A. Rudolph, J.Phys.Chem. 92: 5561 (1988)

16. U. Buck, X.J. Gu, M. Hobein and Ch. Lauenstein, Chem.Phys.Lett. (1989)

17. Ch. Lauenstein, Dissertation, University of Göttingen (1989)

18. U. Buck, in: "Atomic and Molecular Beam Methods", G. Scoles, ed., Oxford, New York (1988), Ch. 18

19. U. Buck, X.J. Gu, Ch. Lauenstein and A. Rudolph, J.Chem.Phys. (1989)

20. U. Buck, Ch. Lauenstein, R. Sroka and M. Tolle, Z.Phys. D 10: 303 (1988)

21. R. Ahlrichs, S. Brode, U. Buck, M. DeKieviet and B. Schmidt, Z. Phys. D (1989)

22. R.A. Nemenoff, J. Snir and H.A. Scheraga, J.Phys.Chem. 82: 2504 (1978)

23. F.T. Marchese, P.K. Mehrotra and D.L. Beveridge, J.Phys.Chem. 86: 2592 (1982)

24. W.L. Jorgensen, J.Phys.Chem. 90: 1276 (1986)

25. J.C. Greer, R. Ahlrichs and I.V. Hertel, Z.Phys.D. (1989), in press

26. O.O. Westlund and R.M. Lynden–Bell, Mol.Phys. 60: 1189 (1987)

27. A.D. Buckingham, J.Chem.Soc.Faraday Trans. 56: 753 (1960)

28. K.G.H. Baldwin and R.O.Watts, Chem.Phys.Lett. 129: 237 (1986); J.Chem.Phys. 87: 873 (1987)

29. B. Heijmen, C. Liedenbaum, S. Stolte and J. Reuss, Z.Phys.D 6: 199 (1987)

30. A.C. Peet, PhD thesis, University of Cambridge (1987)

31. S. Hair, A. Beswick and K. Janda, J.Chem.Phys. 89: 3 970 (1988)

32. A.C. Peet, Chem.Phys.Lett. 132: 32 (1986)

33. S.A. Rice, I. McLaughlin and J. Jortner, J.Chem.Phys. 49: 2756 (1968)

34. M. Snels, R. Fantoni, R. Sanders and W.L. Meerts, Chem.Phys. 115: 79 (1987)

35. D.D. Nelson, Jr., G.T. Fraser and W. Klemperer, J.Chem.Phys. 83: 6201 (1985)

36. B. Heijmen, A. Bizzari, S. Stolte and J. Reuss, Chem.Phys. 126: 201 (1988)

37. B. Schmidt and U. Buck, unpublished results

38. D.J. Levandier, M. Mengel and G. Scoles, in: "The Chemical Physics of Atomic and Molecular Clusters", Enrico Fermi School, Varenna, 1989

39. A.S. Al-Mubarak, G. Del Mistro, P.G. Lerthbridge, N.Y. Adul-Sattar and A.J. Stace, submitted for publication

THEORY OF PHOTODISSOCIATION AND PREDISSOCIATION PROCESSES IN VAN DER WAALS MOLECULES

G.G. Balint-Kurti

School of Chemistry
University of Bristol
Bristol, BS8 1TS, U.K.

INTRODUCTION

A van der Waals molecule consists, by definition, of two or more stable molecules held together by weak interactions. The absorption spectra of these molecules in the infrared involves the excitation of the rotational and/or vibrational state of one of the stable molecules. The energy absorbed in such an excitation process normally far exceeds the binding energy of the weak van der Waals bond which holds the complex together. The excited state created in this way therefore possesses more than enough energy to break up into its constituent stable atoms.

The infrared and UV spectra of van der Waals molecules do, however display many sharp lines.[1-4] This indicates that the excited states often have sufficiently long lifetimes to display sharp spectral features, despite the fact that they have more than enough energy to dissociate. In principle every observed spectral line corresponds to a photodissociation process. If the line is sharp the dissociation proceeds through a long lived intermediate resonance state and, in spectroscopic parlance, is termed a predissociation process. In the present brief overview I will discuss the spectra of van der Waals molecules from this view point. The main objective of the chapter will be to outline the different possible treatments of the process and their relationship to each other as well as to collect together a few key references on the theory of these processes.

PHOTODISSOCIATION THEORY

It will be sufficient for the discussion here to consider a hypothetical model van der Waals molecule for the form

$$A \xrightarrow{R} BC(r) \tag{1}$$

where R is the van der Waals stretching coordinate and joins molecule A to BC, while r denotes all the other "internal" coordinates of the BC fragment.

The initial wavefunction of the system may be expanded in the form of radial functions $\phi_j(R)$ and internal functions $Y_j(r)$ which describe the states of the separated fragments.

Dynamics of Polyatomic Van der Waals Complexes
Edited by N. Halberstadt and K. C. Janda
Plenum Press, New York, 1990

Initial Bound
state wavefunction
$$\Psi_i(R,r) = \sum_j \phi_j^i(R) \, Y_j(r) \tag{2}$$

where the label i identifies the initial bound state of the van der Waals molecule.

The final state of the system always corresponds to a dissociative or continuum state. Let us suppose that the complex breaks up to yield fragments in final internal states denoted by "f". We may then write the continuum state wavefunction in the form

$$\Psi_f^-(R,r) = \sum_{j',j} \psi_{j'}^{-f}(R) \, Y_{j'}(r) \tag{3}$$

Detailed discussions of photodissociation theory[5-7] and its application to van der Waals molecules[8] have been given elsewhere. Here we will concentrate on a brief presentation of the essential elements of the theory and on a discussion of the different approximations which are commonly invoked.

The partial integral cross section, which is a measure of the probability of absorbing light and breaking up to give a particular final state of the fragments, may be written in the form

$$\sigma_f(\nu) = \frac{8\pi^3 \nu}{3c} \sum_{\hat{\epsilon}} |\langle \Psi_f^- | \hat{\epsilon} \cdot \mu | \Psi_i \rangle|^2 \tag{4a}$$

$$= \frac{8\pi^3 \nu}{3c} \sum_{j',j} |\langle \psi_j^{-f} | \mu_{j',j} | \phi_j^i \rangle|^2 \tag{4b}$$

where $\sum_{\hat{\epsilon}}$ denotes a summation over all (i.e. 3) directions of polarization of the incident radiation. ψ_j^{-f}, $\mu_{j',j}$ and ϕ_j^i are all functions of the van der Waals distance R and the angular bracket in eq. 4b implies an integral over R. $\mu_{j',j}$ is the matrix element of the dipole moment (or transition dipole moment) function over the internal fragment functions.

$$\mu_{j',j}(R) = \int Y_{j'}(r) \, \mu(R,r) \, Y_j(r) dr. \tag{5}$$

Although it may not be apparent at first glance, the expression for the photodissociation cross section (eq. 4) provides an exact description of the photon absorption process and will readily yield the various predissociation line shapes which are observed in the spectra of van der Waals molecules.

The mathematical tools needed to calculate photodissociation cross sections were first presented by Shapiro,[9] and an exact treatment of rotational predissociation in the Ar-N_2 system was subsequently performed by Beswick and Shapiro.[10] These calculations clearly show how the theory can account for both symmetric and asymmetric resonance line shapes as well as overlapping nonisolated resonance lines. The theory has also been applied to the Ar-H_2 and Ar-HD system[8,11,12] for which benchmark calculations were performed. Several other methods of computing the exact bound-continuum integrals needed in the evaluation of the expression for the photodissociation cross section have also been presented (see refs 7 and 13).

THE ISOLATED LONG-LIVED RESONANCE APPROXIMATION AND GOLDEN RULE CALCULATIONS

The resonance line shapes arising in the photodissociation or predissociation of van der Waals molecules are often very sharp.[1] Calculations[8,11,12] indicate that for Ar-H_2 the resonances are nonoverlapping and that widths of the order of 10^{-8} - 10^{-9} cm^{-1} may be expected for vibrationally predissociation and 10^{-2} - 10^{-4} cm^{-1} for rotational predissociation processes. In such circumstances it is valid to assume that the dissociation proceeds through an intermediate resonance state. Denoting the "projection operator" which tells us how much of a particular wavefunction lies within or overlaps this resonance state by

$$\left.\begin{array}{l}\text{Projection operator onto}\\\text{intermediate resonance state}\end{array}\right\} \quad \hat{P}_r = |\Psi_r\rangle \langle \Psi_r| \tag{6}$$

We may use this to rewrite eq. 4 in a slightly approximate form which takes account of the fact that the photodissociation process is known to proceed through the resonance state Ψ_r.

$$\sigma_f(v) \approx \frac{8\pi^3 v}{3c} \sum_{\hat{\epsilon}} |\langle \Psi_f^- |\Psi_r\rangle \langle \Psi_r |\hat{\epsilon}\cdot\mu|\Psi_i\rangle|^2$$

$$= \frac{8\pi^3 v}{3c} |\langle \Psi_f^- |\Psi_r\rangle|^2 \sum_{\hat{\epsilon}} |\langle \Psi_r |\hat{\epsilon}\cdot\mu|\Psi_i\rangle|^2$$

$$\alpha |\langle \Psi_f^- |\Psi_r\rangle|^2 \tag{7}$$

The intermediate resonance state, Ψ_r, can dissociate into many possible final continuum states Ψ_f^-. The modulus squared of the bound-continuum integral on the right hand side of eq. 7 is proportional to the probability of the resonance state dissociating to yield a particular final state.

If the predissociation line is sharp, indicating only a small probability of the resonance state breaking up, then a perturbation type approach may be used. This approach is very clearly described in Shapiro's paper[9] of 1972 where he introduces and tests out numerical procedures for evaluating the bound-continuum integrals needed in both this approximate perturbation (or so called golden rule) approach and also in the exact theory of photodissociation processes.[7] The basic theory of the golden rule expressions has been presented by Levine.[14,15] It has also been carefully derived by Beswick and Jortner who have used it in a pioneering study of vibrational predissociation.[16]

The key expression in the method is that for the "half width" Γ_r of the decay of the resonance state:[9,16]

$$\Gamma_r = \pi \sum_f |\langle {}^0\Psi_f^- |V_{fr}|\Psi_v\rangle|^2 \tag{8}$$

where ${}^0\Psi_f^-$ is the continuum or scattering wavefunction which is the solution of the Schrödinger equation in a function space which excludes the space of all the localised (bound) resonance sates Ψ_r. V_{fr} is the potential which couples the bound resonance states to the dissociative scattering states. The lifetime, τ_r, of the resonance is given by the relationship

$$\tau_r = \frac{\hbar}{2\Gamma_r} \tag{9}$$

From eq. 8 we can define so called partial widths

$$\Gamma_r \quad f = \pi |\langle {}^0\Psi_f^- |V_{fr}|\Psi_r\rangle|^2. \tag{10}$$

These partial widths give the "final quantum state distributions" in that they are proportional to the probability of the resonant state decomposing to yield a particular final state "f". The total width (eq. 8) is the sum of all the partial widths (eq. 10). Recently, in an extremely thorough and elegant study, Halberstadt, Beswick and Janda[13] have applied both the complete photodisocciation theory (eq. 4b) and the golden rule approach to the study of vibrational predissociation in the Ne-Cℓ$_2$ system.

SCATTERING RESONANCE FORMALISM

The intermediate resonance states, through which we have assumed the predissociation occurs, show up in the scattering of the two partners from each other as scattering resonances. In particular the type of resonance involved is called a Feschbach resonance[17] and corresponds to the situation where some energy has been absorbed from the initial relative motion of the two partners and converted into internal (normally vibrational and rotational) energy in one or both of them. A Feschbach resonance is seen at energies when such states occur with large amplitude and when they correspond to a situation which leaves insufficient energy in the relative motion of the collision partners for the complex to break up.

It should therefore be clear that, through careful analysis of the details of the corresponding scattering resonance, we can derive knowledge about the resonant metastable predissociative state and its break-up characteristics. Ashton, Child and Hutson[18] have pioneered this approach in applications to van der Waals predissociation processes. Their paper outlines the full theory needed to extract the line widths Γ_r and the partial line widths Γ_r $_f$ from the energy dependence of the complex S scattering matrices.

The calculations used in this work are exact close coupling calculations and it is the dynamics of the system that defines the resonance state. Such calculations are more accurate than golden rule ones as no perturbation approximation has been used. They cannot yield photo-dissociation cross sections as this requires a knowledge also of the initial state of the system (eq. 4). For the few cases where comparison has been made between such close coupling scattering calculations and other ones,[12] good agreement has been obtained as regards total line widths but not on final state distributions. Le Roy, Carey and Hutson [19,20] have made extensive use of this close coupling method for calculations on H$_2$ and HD van der Waals complexes with rare gas atoms.

Another closely related method which should be mentioned is that developed by Clary[21] and applied by Hutson, Clary and Beswick[22] to the problem of vibrational predissociation in Ne-C$_2$H$_4$ and Ar-C$_2$H$_4$. In this approach the vibrational degrees of freedom plus the rotational degree of freedom about the C-C bond (i.e. the principle axis of inertia corresponding to the smallest moment of inertia) are treated using the close coupling scattering technique while the infinite order sudden approximation is used for the other rotational degrees of freedom.

NEW METHODS

Multichannel Quantum Defect Theory

A problem arises in any of the more exact types of calculation which involve the solution of large sets of coupled differential equations in that the equations must be solved many times, often at very closely spaced energies, so as to map out either the line shape of a predissociation process or so as to analyse the energy dependence of a scattering S matrix. Quantum Defect Theory (QDT) was invented to describe Rydberg states and atomic ionization processes[23] in atoms. It was subsequently extended to encompass general spectroscopic and scattering processes.[24-26] The theory is ideal for describing phenomena which involve resonances as only quantities which vary smoothly with energy are computed in the method and the resonance features emerge at the very end of the calculation through their combination.

Special features of the method are:

1) The scattering problem is solved in terms of the exact regular and irregular solutions of the uncoupled channel equations.

2) Open and Closed channels are treated on an equal footing.

3) The scattering K matrix analysis is carried out at a relatively small value of the scattering coordinate R. This is possible because the exact solutions of the uncoupled channel equations are used as a basis for solving the problem.

4) The resonance features emerge when the K matrix is transformed so as to asymptotically annihiliate the closed channels.

Figure 1 shows some calculations on the rotational predissociation of $Ar-H_2$ (see ref. 27 for further details). Note that all the basic calculated quantities in the top two panels of the figure vary smoothly with energy and they could be accurately interpolated from calculations performed at an energy separation of 1 cm^{-1}. Nevertheless combination of these smoothly varying quantities yield resonance lineshapes with half widths of the order of 10^{-2} cm^{-1}.

Time Dependent Quantum Theory

The past six years have witnessed a revolution in our ability to solve the time dependent Schödinger equation.[28,29] Such calculations have been successfully applied to photodissociation processes.[30] In general numerical difficulties arise when long lived resonance states are involved. An ingeneous numerical analytic method for overcoming these difficulties has been reported by Gray[31] at the present conference. The method, called MUSIC is well documented[32,33] and permits one to extract resonant frequency information from quite short time evolutions. Final state distributions may also be obtained in this manner without waiting for the complex to break up completely.[31] Time dependent computations may for some scattering systems be competitive with time independent techniques.[29] They have the advantage that a single calculation yields information over a wide range of energies.

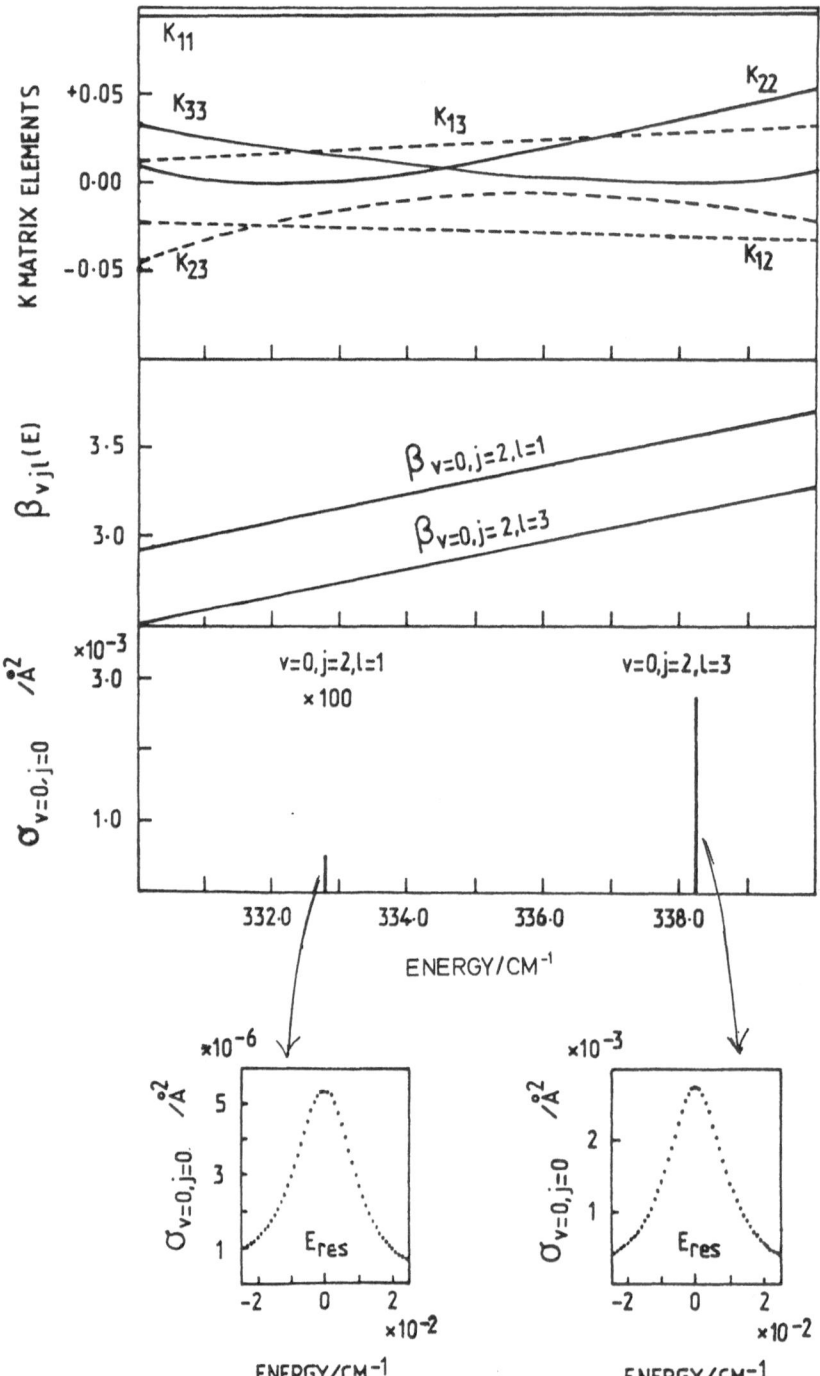

Figure 1. Photodissociation cross sections for the two lowest energy rotational predissociation processes in Ar-H₂. The top two panels show the (slow) variation of the K matrix elements and of β_{vjl}, a quantity obtained from the form of the exact uncoupled channel wavefunctions. The lower panel and the two windows show the form of the predissociation lineshape as a function of energy.

REFERENCES

1. A. R. W. McKellar, Faraday Discuss. Chem. Soc. 73:89 (1982).
2. K. C. Janda, Adv. Chem. Phys. 60:201 (1985).
3. M. T. Berry, M. R. Brustein, and M. I. Lester, J. Chem. Phys. 90:5878 (1989).
4. D. C. Clary, C. M. Lovejoy, S. V. O'Neil, and D. J. Nesbitt, Phys. Rev. Letts. 61:1576 (1988).
5. G. G. Balint-Kurti, and M. Shapiro, Chem. Phys. 61:137 (1981), erratum 72:456 (1982).
6. M. Shapiro, and R. Bersohn, Ann. Rev. Phys. Chem. 33:409 (1982).
7. G. G. Balint-Kurti, and M. Shapiro, Adv. Chem. Phys. 60:403 (1985).
8. I. F. Kidd, and G. G. Balint-Kurti, J. Chem. Phys. 82:93 (1985).
9. M. Shapiro, J. Chem. Phys. 56:2582 (1972).
10. J. A. Beswick, and M. Shapiro, Chem. Phys. 64:333 (1982).
11. I. F. Kidd, and G. G. Balint-Kurti, Chem. Phys. Letts. 105:91 (1984).
12. I. F. Kidd, and G. G. Balint-Kurti, Faraday Disc. Chem. Soc. 82:241 (1986).
13. N. Halberstradt, J. A. Beswick, and K. Janda, J. Chem. Phys. 87:3966 (1987).
14. R. D. Levine, J. Chem. Phys. 44:2029 (1966).
15. R. D. Levine, "Quantum Mechanics of Molecular Rate Processes," Clarendon, Oxford, England (1969).
16. J. A. Beswick, and J. Jortner, J. Chem. Phys. 68:2277 (1978).
17. H. Feschbach, Ann. Phys. 19:287 (1962).
18. C. J. Ashton, M. S. Child, and J. M. Hutson, J. Chem. Phys. 78:4025 (1983).
19. R. J. Le Roy, G. C. Corey, and J. M. Hutson, Faraday Disc. Chem. Soc. 73:339 (1982).
20. J. M. Hutson, and R. J. Le Roy, J. Chem. Phys. 78:4040 (1983).
21. D. C. Clary, J. Chem. Phys. 81:4446 (1984).
22. J. M. Hutson, D. C. Clary, and J. A. Beswick, J. Chem. Phys. 81:4474 (1984).
23. C. H. Greene, and Ch. Jungen, Adv. At. Mol. Phys. 21:51 (1985).
24. C. H. Greene, A. R. P. Rau, and U. Fano, Phys. Rev. A26:2441 (1982).
25. F. H. Mies, J. Chem. Phys. 80:2514 (1984), F. H. Mies, and P. S. Julienne, J. Chem. Phys. 80:2526 (1984).
26. M. Raoult, J. Chem. Phys. 87:4736 (1987).
27. M. Raoult, and G. G. Balint-Kurti, Phys. Rev. Letts. 61:2538 (1988).
28. R. Kosloff, J. Phys. Chem. 92:2087 (1988).
29. Y. Sun, R. S. Judson, and D. J. Kouri, J. Chem. Phys. 90:241 (1989).
30. R. N. Dixon, Chem. Phys. Letts. 147:377 (1988).
31. S. K. Gray, and C.E. Wozny, "Wavepacket Dynamics of van der Waals Molecules: fragmentation of $NeC\ell_2$ with three degrees of freedom", submitted for publication.
32. S. L. Marple, "Digital Spectral Analysis with Applications", Printice Hall, Englewood Cliffs, NJ (1987).
33. D. W. Noid, B. T. Brooks, S. K. Gray, and S. L. Marple, J. Phys. Chem. 92:3386 (1988).

DYNAMICS OF VAN DER WAALS COMPLEXES: BEYOND ATOM–DIATOM SYSTEMS

Jeremy M. Hutson

Department of Chemistry, University of Durham,
South Road, Durham, DH1 3LE, England

A major goal of spectroscopic studies of Van der Waals molecules is to obtain information on intermolecular forces. In order to do this, methods must be available to calculate spectroscopic properties from a proposed potential energy surface. For small Van der Waals molecules, such as the rare gas – hydrogen halide systems, such calculations can be performed using a variety of theoretical methods, based either on solution of coupled differential equations or on expanding the wavefunction in a product basis of angular and radial functions. In the atom–diatom case, there is seldom any need for dynamical approximations: the full close-coupling equations can be solved without approximation even on relatively small computers, and standard packages for doing this are available.[1,2]

For larger systems, involving either two molecules or an atom and a nonlinear molecule, the situation is not so satisfactory. The standard programs[1] have the capability to solve the resulting coupled equations, but the angular basis sets needed are very much larger, and full close-coupling calculations quickly become prohibitively expensive. Many possible approximation schemes and decoupling approximations may be envisaged, but relatively little work has been done to understand their ranges of validity and the patterns of energy levels produced.

The purpose of this paper is to explore the possibilities for accurate quantum calculations on these larger systems. The paper will begin by summarizing the methods used for atom–diatom systems, and the approximations which have been found to be accurate. This framework will then be used to discuss larger systems such as Ar_2–HCl and Ar–H_2O. The coupled equations for the more complex systems will be cast in a form that emphasises the similarities between them and the atom–diatom systems.

1. Atom–diatom systems

The coordinate system used for an atom–diatom Van der Waals complex is conventional. The vector from the centre of mass of the diatom BC to the atom A is denoted R, and has length R. The vector between the atoms B and C is r, and is usually taken to originate on the heavier of atoms B and C. The length of r is r, and the angle between R and r is θ. The unit vectors corresponding to R and r are denoted $\hat{R}$ and $\hat{r}$. The orientation of $\hat{R}$ relative to a space-fixed axis system is described by the Euler angles $(\alpha, \beta, 0)$, and these in turn define a *body-fixed* axis system x, y, z; the orientation of $\hat{r}$ relative to the body-fixed axis system is described by spherical polar coordinates (θ, ϕ). This embedding of the body-fixed angles is not the only one possible, but is the most convenient when considering generalisations to more complex systems as in the following sections.

The Hamiltonian for an atom–diatom complex is

$$H = -\frac{\hbar^2}{2\mu} R^{-1} \left(\frac{\partial^2}{\partial R^2} \right) R + \frac{\hbar^2 \hat{l}^2}{2\mu R^2} + V(R, r, \theta) + H_{\mathrm{mon}},$$

Dynamics of Polyatomic Van der Waals Complexes
Edited by N. Halberstadt and K. C. Janda
Plenum Press, New York, 1990

67

where the reduced mass μ is $m_A m_{BC}/(m_A + m_{BC})$, $\hat{l}^2$ is the angular momentum operator for end-over-end rotation of the complex as a whole, and H_{mon} is the Hamiltonian for the isolated diatomic molecule BC. The intermolecular potential $V(R, r, \theta)$ is not usually strong enough to cause significant mixing of the monomer vibrational states, so that the explicit dependence of the potential and the wavefunctions on r may usually be neglected.

The vibrationally averaged intermolecular potential $V(R, \theta)$ for an atom–diatom system is conventionally expanded in Legendre polynomials,

$$V(R, \theta) = \sum_\lambda V_\lambda(R) P_\lambda(\cos \theta). \tag{2}$$

For complexes with very small anisotropies and small reduced masses, such as those containing He or H_2, a space-fixed quantisation scheme is appropriate.[3] However, for most other Van der Waals complexes, the anisotropy is strong enough for the diatom angular momentum to be quantised along the $\boldsymbol{R}$ vector, and the total wavefunction is most conveniently expanded in body-fixed functions,

$$\psi_\alpha(\boldsymbol{R}, \boldsymbol{r}) = R^{-1} \sum_{jK} \Phi_{jK}^{JM}(\alpha, \beta, \theta, \phi) \chi_{jKJ}^\alpha(R), \tag{3}$$

where the functions $\Phi_{jK}^{JM}(\alpha, \beta, \theta, \phi)$ are eigenfunctions of the body-fixed angular momentum operator $\hat{J}_z$ with eigenvalue K,

$$\Phi_{jK}^{JM}(\alpha, \beta, \theta, \phi) = \left(\frac{2J+1}{4\pi}\right)^{\frac{1}{2}} \mathcal{D}_{MK}^{J*}(\alpha, \beta, 0) Y_{jK}(\theta, \phi). \tag{4}$$

They form a complete orthonormal set spanning the space of the angular coordinates. The function $\mathcal{D}_{MK}^J(\alpha, \beta, \gamma)$ is a rotation matrix element with the phase convention of Brink and Satchler[4] and $Y_{jK}(\theta, \phi)$ is a spherical harmonic involving the angular coordinates of the diatomic molecule in the body-fixed axis system. Physically, the end-over-end angular momentum of the complex cannot have any body-fixed projection along $\boldsymbol{R}$, so that the K quantum numbers appearing in the rotation matrix element and in the spherical harmonic must be the same.

When this representation of the wavefunction is substituted into the total Schrödinger equation of the complex, using the Hamiltonian of equation (1), the equation obtained is

$$\left[-\frac{\hbar^2}{2\mu}\frac{d^2}{dR^2} + (jKJ|V|jKJ) + \frac{\hbar^2}{2\mu R^2}(JKj|(\hat{J}-\hat{j})^2|JKj) + E_j^{\text{mon}} - E\right]\chi_{jKJ}^\alpha(R)$$

$$= -\sum_{j'}{}'(jKJ|V|j'KJ)\chi_{j'KJ}^\alpha(R)$$

$$\qquad - \sum_{K'=K\pm1}{}'\frac{\hbar^2}{2\mu R^2}(jKJ|(\hat{J}-\hat{j})^2|jK'J)\chi_{jK'J}^\alpha(R), \tag{5}$$

where the operator $\hat{l}^2$ has been replaced by its body-fixed equivalent $(\hat{J}-\hat{j})^2$, and the round bracket notation $(\ |\ |\ )$ has been adopted to indicate integration over all dynamical variables for which the associated quantum numbers are given; thus $(jKJ|V|j'K'J')$ implies integration over the angular variables only. The symbol $\sum'$ indicates summation over all $j'K' \neq jK$. This is a set of differential equations, one for each *channel* jK included in the basis set. Terms off-diagonal in jK, which couple the different equations, have been taken to the right hand side.

The potential matrix elements in the body-fixed representation are

$$(jKJ|V|j'KJ) = \sum_\lambda g_\lambda(jj'K) V_\lambda(R), \tag{6}$$

where

$$g_\lambda(jj'K) = (-)^K[(2j+1)(2j'+1)]^{\frac{1}{2}} \begin{pmatrix} j & \lambda & j' \\ 0 & 0 & 0 \end{pmatrix} \begin{pmatrix} j & \lambda & j' \\ -K & 0 & K \end{pmatrix}. \tag{7}$$

The potential matrix elements are independent of J (unlike those of the space-fixed representation) and diagonal in K. The presence of the first $3j$-symbol in this equation ensures that the matrix elements vanish unless $(j + \lambda + j')$ is even and (j, λ, j') satisfy a triangle relationship.

The matrix elements of the operator $\hat{l}^2 = (\hat{J} - \hat{j})^2$ are

$$\langle jKJ|(\hat{J} - \hat{j})^2|jKJ\rangle = J(J+1) + j(j+1) - 2K^2, \tag{8}$$

$$\langle jKJ|(\hat{J} - \hat{j})^2|jK \pm 1J\rangle = -[J(J+1) - K(K \pm 1)]^{\frac{1}{2}}[j(j+1) - K(K \pm 1)]^{\frac{1}{2}}, \tag{9}$$

with all other matrix elements zero.

The coupled equations are the most fundamental form of the Schrödinger equation for Van der Waals molecules, and most of the approximate methods used for Van der Waals complexes can be derived by making simplifying approximations to them. The most important such approximation is the *helicity decoupling* approximation, in which the off-diagonal Coriolis matrix elements of equation (9) are neglected, so that the coupled equations become diagonal in K. For high J and a given value of $j_{\max}$, the helicity decoupling approximation reduces the number of basis functions N from $(j_{\max} + 1)^2$ to $j_{\max} + 1 - K$. Since the time taken to solve the coupled equations is proportional to N^3, this results in a very considerable saving in computer time. Helicity decoupling is a good approximation provided

$$B[j(j+1) - K(K \pm 1)]^{\frac{1}{2}}[J(J+1) - K(K \pm 1)]^{\frac{1}{2}} \ll V_2[g_2(jjK \pm 1) - g_2(jjK)], \tag{10}$$

and thus is useful for low angular momentum states of most Van der Waals complexes (with the exception of the He and H_2 systems mentioned above).

The primitive body-fixed basis functions as described above do not have definite parity, except for $K = 0$. However, since parity is known to be a rigorously good quantum number, it is usually advantageous to choose basis functions which do have definite parity, and it is straightforward to define linear combinations of the primitive functions for which this is the case. Adopting the notation $\Omega \equiv |K|$, these are

$$\Phi_{j\Omega}^{JM\pm}(\alpha, \beta, \theta, \phi) = N\left[\mathcal{D}_{M\Omega}^{J*}(\alpha, \beta, 0)Y_{j\Omega}(\theta, \phi) \pm (-)^J \mathcal{D}_{M-\Omega}^{J*}(\alpha, \beta, 0)Y_{j-\Omega}(\theta, \phi)\right], \tag{11}$$

where the normalising factor N is $[(2J+1)/16\pi]^{\frac{1}{2}}$ for $\Omega = 0$ and $[(2J+1)/8\pi]^{\frac{1}{2}}$ for $\Omega > 0$. This is known as the *parity-adapted* body-fixed basis set; the matrix elements of the Van der Waals Hamiltonian between these functions are readily constructed from the matrix elements between the primitive body-fixed functions.

The coupled equations (5) take the general form

$$\frac{d^2\chi}{dR^2} = [W(R) - \epsilon]\chi(R), \tag{12}$$

where $\chi(R)$ is a column vector and W and ϵ are matrices. There are in principle an infinite number of channels (basis functions); it is usual to truncate the set to include only those channels which lie reasonably close in energy to the state(s) of interest. This is known as the *close-coupling approximation*, and calculations which make no other dynamical approximation are known as *close-coupling calculations* to distinguish them from decoupling approximations such as helicity decoupling. If N channels are included in the expansion, $W(R)$ is an $N \times N$ matrix, $\chi(R)$ is a column vector with N components, and $\epsilon = (2\mu E/\hbar^2)I$ is a constant times the unit matrix. If the boundary conditions are neglected, there are N independent solutions at each energy, so that it is actually necessary to propagate an $N \times N$ wavefunction matrix.

These equations are exactly the same as the coupled equations of molecular scattering theory, except that the boundary conditions are different for the bound state case. There are solutions of the coupled equations satisfying scattering boundary conditions for any energy greater than the dissociation energy of the complex, so that the scattering problem reduces to propagating solutions of the coupled equations from one value of R to another for a specified energy E. Many methods of doing this have been developed.[5-12] Such scattering calculations have been used to characterise predissociating states of Van der Waals complexes.[13]

The additional problem present in the bound state case, at energies below the dissociation energy of the complex, is that of locating energies which are eigenvalues of the coupled equations, where a solution may be found that satisfies bound state boundary conditions. There are several procedures available for doing this,[14-16] but the stablest are the log-derivative methods.[7,15,11] In the many-channel case, the log-derivative matrix $Y(R)$ is defined by[7]

$$Y(R) = \chi'(R)[\chi(R)]^{-1}, \tag{13}$$

where $\chi(R)$ is an $N \times N$ matrix and the prime indicates radial differentiation. The diagonal elements of the log-derivative matrix become constant when $\chi(R)$ is exponentially increasing or decreasing, so that loss of linear independence of the different solutions does not occur during propagation, and the log-derivative methods are inherently stable. In addition, the log-derivative matrix contains exactly the information needed to locate eigenvalues.[15] Incoming and outgoing solutions are propagated from the two classically forbidden regions to a matching point in the classically allowed region: if ϵ is an eigenvalue, the determinant of the difference between the two solutions is zero,

$$|Y^{\text{in}}(R_{\text{mid}}) - Y^{\text{out}}(R_{\text{mid}})| = 0. \tag{14}$$

The strategy to be adopted in searching for the zeroes of the matching determinant has been discussed in detail by Johnson[15] and Manolopoulos,[17] and has been implemented in the general-purpose program BOUND.[1]

One apparent disadvantage of the log-derivative methods is that they do not directly give explicit wavefunctions, which are needed to calculate molecular properties (via expectation values) and spectroscopic intensities (via off-diagonal matrix elements). However, the restriction is not as serious as it might appear: a finite-difference approach for extracting expectation values from coupled channel calculations has been described by Hutson,[18] and is available as an option in the BOUND program.

An alternative approach to solving the coupled equations is to use a basis set expansion for the R coordinate as well as for the angular variables. The angular basis sets used in such calculations are generally the same as in coupled channel calculations. This approach was pioneered by Le Roy and Van Kranendonk,[19] who used numerical basis sets for the radial (R) functions. Such basis sets are adequate for the rare gas–H_2 systems, but converge very poorly for more strongly anisotropic systems. An alternative basis set, based on Morse-oscillator-like functions, has been used extensively by Tennyson and coworkers.[20,2]

A recent development in this area has been the use of non-orthogonal basis sets of Gaussian functions ("distributed Gaussians").[21] These circumvent the problem of representing non-oscillatory regions of the wavefunction in terms of oscillatory functions, which is the major source of poor convergence in other types of basis-set calculation. A particularly promising approach is the combination of distributed Gaussian basis sets (DGB) for the R motion with a discrete variable representation (DVR) for the angular motion.[22]

The Van der Waals complex whose excited states have been studied in most detail is Ar–HCl, and it is interesting to consider it in some detail. The potential energy surface for this system has been through several cycles of refinement as better and better experimental information became available.[23-26] At each stage, the predictions obtained from the best-fit potential have been useful in guiding further experiments. The most recent potential is the H6(3) potential of Hutson,[26] determined from high-resolution microwave and far-infrared spectroscopy, which is reliable for all angles of approach.

The bending levels of Ar-HCl, calculated from the H6(3) potential using close-coupling calculations,[26] are shown in Figure 1a. The observed pattern of energy levels and allowed transitions may be compared with that expected for free rotation of the HCl (Figure 1b) and for a near-rigid linear molecule (Figure 1c). It may be seen immediately that the free-rotor picture is much closer to reality: complexes such as Ar-HCl are best viewed as undergoing hindered internal rotations, with quantum numbers j, Ω, J and parity.

A striking feature of Figure 1a is that the first excited Σ ($\Omega = 0$) state actually lies *below* the lowest Π ($\Omega = 1$) state. This is not at all the behaviour expected for a near-rigid molecule, shown in Figure 1c: in the language usually applied to linear triatomics, the Π state is the fundamental bending vibration, labelled 01^10, and the Σ state is its overtone, labelled 02^00. However, in the free-rotor picture, both these states correlate with $j = 1$, and are expected to be degenerate except for the potential term $V_2 g(jjK)$. Since V_2 is negative in Ar-HCl, the Σ state lies below the Π state. In physical terms, the $\Omega = 1$ states sample mainly geometries around $\theta = 90°$, where the potential well is shallowest.

2. Atom–polyatom systems

Complexes formed from atoms and linear polyatomic molecules are very similar to atom–diatom systems: the coupled equations are identical, and the same angular momentum coupling schemes apply. The only added degree of complexity is that perpendicular transitions of the polyatomic monomer are possible, and these introduce an extra quantum number (l or k) for the monomer vibrational angular momentum. Such states are analogous to those arising from $k > 0$ states of a symmetric top monomer, as discussed below.

Atom–nonlinear molecule complexes are of two basic types: atom–symmetric top and atom–asymmetric top. Several such complexes have been studied through their pure rotational spectra, but high-resolution infrared spectra, involving excitation of Van der Waals bending and stretching modes, are only just starting to become available.[27] There has also been some theoretical work on the photodissociation spectra of such systems,[28] but this has concentrated on the rates of photodissociation processes rather than on the energy level patterns.

The coordinate system needed for an atom–nonlinear molecule complex is a straightforward generalisation of that for an atom–diatom complex. The body-fixed axis system is defined as before, with Euler angles $(\alpha, \beta, 0)$ specifying the orientation of $\hat{\boldsymbol{R}}$. However, it is now necessary to define an axis system (x, y, z) fixed in the monomer; the relationship of these axes to the body-fixed axes is specified by Euler angles (ϕ, θ, χ): θ and ϕ describe the orientation of the z axis, and χ describes rotations about the z axis.

The Hamiltonian for an atom–nonlinear molecule complex is

$$H = -\frac{\hbar^2}{2\mu} R^{-1} \left(\frac{\partial^2}{\partial R^2} \right) R + \frac{\hbar^2 \hat{l}^2}{2\mu R^2} + V(R, \theta, \chi) + H_{\mathrm{mon}}, \tag{15}$$

where H_{mon} is now the Hamiltonian for the isolated nonlinear molecule, and the intermolecular potential is a function of θ and χ (but not ϕ). The intermolecular potential may be expanded in renormalised spherical harmonics,

$$V(R, \theta, \chi) = \sum_\lambda V_{\lambda\mu}(R) C_{\lambda\mu}(\theta, \chi). \tag{16}$$

Some authors[29] prefer to expand the potential in spherical harmonics $Y_{\lambda\mu}(\theta, \chi)$ rather than renormalised spherical harmonics $C_{\lambda\mu}(\theta, \chi)$, but the present choice emphasises the analogy with atom–diatom systems because $C_{\lambda 0}(\theta, \chi) \equiv P_\lambda(\cos\theta)$.

The total wavefunction is again most conveniently expanded in body-fixed functions,

$$\psi_\alpha(\boldsymbol{R}, \boldsymbol{r}) = R^{-1} \sum_{jkK} \Phi_{jkK}^{JM}(\alpha, \beta, \phi, \theta, \chi) \chi_{jkKJ}^\alpha(R), \tag{17}$$

71

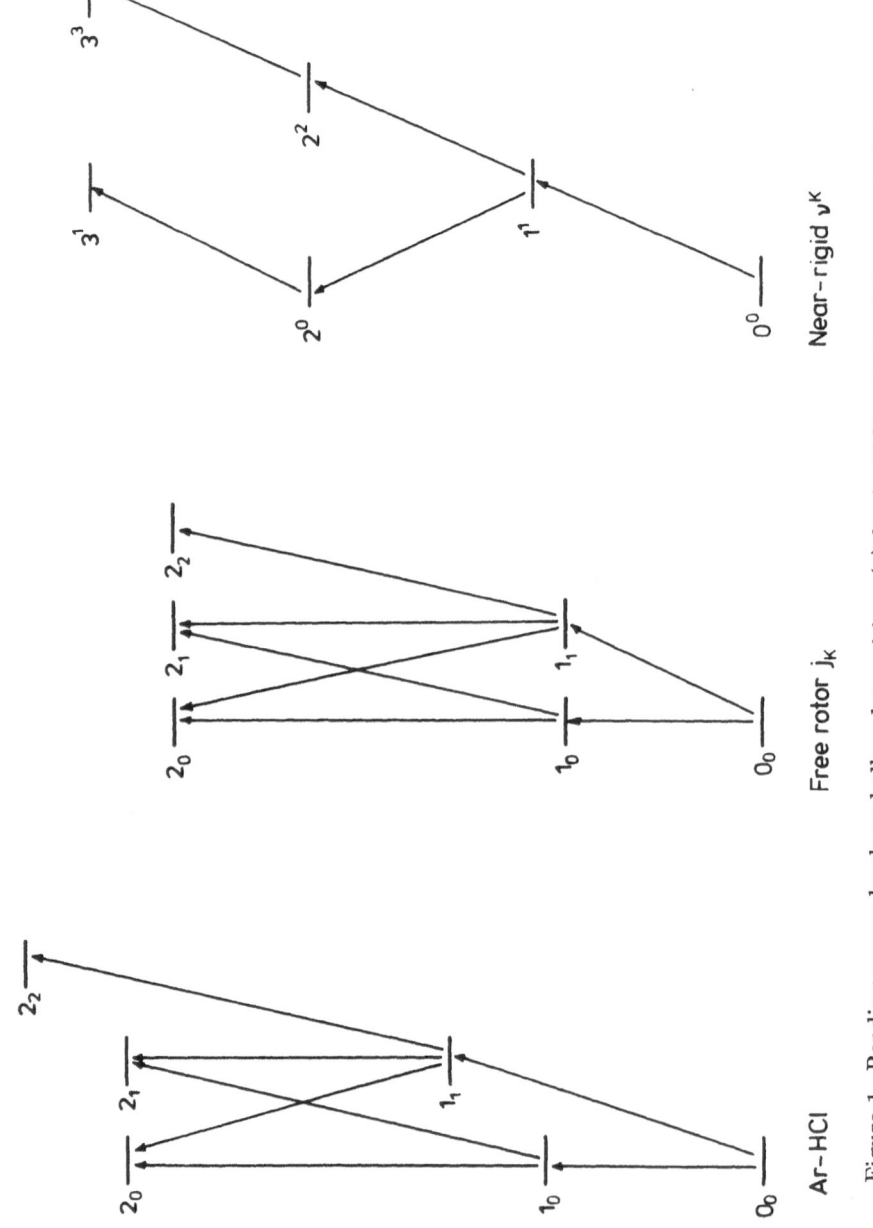

Figure 1. Bending energy levels and allowed transitions (a) for Ar–HCl, calculated from the H6(3) potential; (b) in the internal free rotor picture; (c) for a near-rigid linear molecule.

where the functions $\Phi_{jkK}^{JM}(\alpha,\beta,\phi,\theta,\chi)$ are

$$\Phi_{jkK}^{JM}(\alpha,\beta,\phi,\theta,\chi) = \left(\frac{(2J+1)(2j+1)}{32\pi^3}\right)^{\frac{1}{2}} \mathcal{D}_{MK}^{J*}(\alpha,\beta,0)\mathcal{D}_{Kk}^{j*}(\phi,\theta,\chi). \tag{18}$$

The body-fixed coupled equations are now

$$\left[-\frac{\hbar^2}{2\mu}\frac{d^2}{dR^2} + (jkKJ|V|jkKJ) + \frac{\hbar^2}{2\mu R^2}\left[J(J+1)+j(j+1)-2K^2\right] + H_{\text{mon}} - E\right]\chi_{jkKJ}^{\alpha}(R)$$

$$= -\sum_{j'k'}{}'(jkKJ|V|j'k'KJ)\chi_{j'k'KJ}^{\alpha}(R)$$

$$- \sum_{K'=K\pm1}{}' \frac{\hbar^2}{2\mu R^2}(jkKJ|(\hat{J}-\hat{j})^2|jkK'J)\chi_{jkK'J}^{\alpha}(R), \tag{19}$$

The potential matrix elements in the body-fixed representation are[29]

$$(jkKJ|V|j'k'KJ) = \sum_{\lambda} g_{\lambda\mu}(jkj'k'K)V_{\lambda\mu}(R), \tag{20}$$

where

$$g_{\lambda\mu}(jkj'k'K) = (-)^{K-k}[(2j+1)(2j'+1)]^{\frac{1}{2}} \begin{pmatrix} j & \lambda & j' \\ -k & \mu & k' \end{pmatrix} \begin{pmatrix} j & \lambda & j' \\ -K & 0 & K \end{pmatrix}. \tag{21}$$

This reduces to equation (7) for the special case $k = k' = 0$; again, the potential matrix elements are independent of J and diagonal in K. The off-diagonal matrix elements of the operator $(\hat{J}-\hat{j})^2$ in equation (19) are the same as in the atom–diatom case, and are given by equation (9) with an additional factor of $\delta_{kk'}$.

The body-fixed coupled equations are thus very similar in form to those for atom–diatom systems: the only differences are that the potential matrix elements here are a generalisation of the atom–diatom case and that H_{mon} takes a more complicated form (and may have matrix elements off-diagonal in k). As for atom–diatom systems, considerable savings in computer time may be achieved by performing helicity decoupling calculations, in which the Coriolis matrix elements off-diagonal in K are neglected.

The energy levels of the free monomer are of course more complicated than in the atom–diatom case. It is interesting to consider the specific cases of Ar–NH$_3$ and Ar–H$_2$O. NH$_3$ is a symmetric top, and it is convenient to choose the monomer-fixed z axis to lie along the C_3 axis. The interaction potential with therefore have threefold symmetry, and the only non-zero terms in equation (16) with be those for which μ is a multiple of 3. The potential can in principle couple states with $\Delta k = \pm3$, but any mixing due to such coupling will be very small because of the large a rotational constant of NH$_3$; k will remain a nearly good quantum number in the Van der Waals complex.

The inversion doubling present in free NH$_3$ requires careful treatment. The eigenstates of H_{mon} are either even or odd with respect to reflection in the plane; the states can be envisaged as arising from a tunnelling matrix element of magnitude 0.79 cm^{-1} coupling two degenerate pyramidal states. However, it is most unlikely that the two pyramidal states will remain degenerate in the presence of an Ar atom: provided the resulting splitting is large compared to the tunnelling splitting, the effect of the tunnelling will simply be to push the (already non-degenerate) states slightly further apart.

The only atom–nonlinear molecule system whose excited states have been studied in any detail is Ar–H$_2$O.[30] This system is actually quite weakly anisotropic: the anisotropy of the potential splits and shifts the H$_2$O free-rotor levels, but the free-rotor quantum numbers are

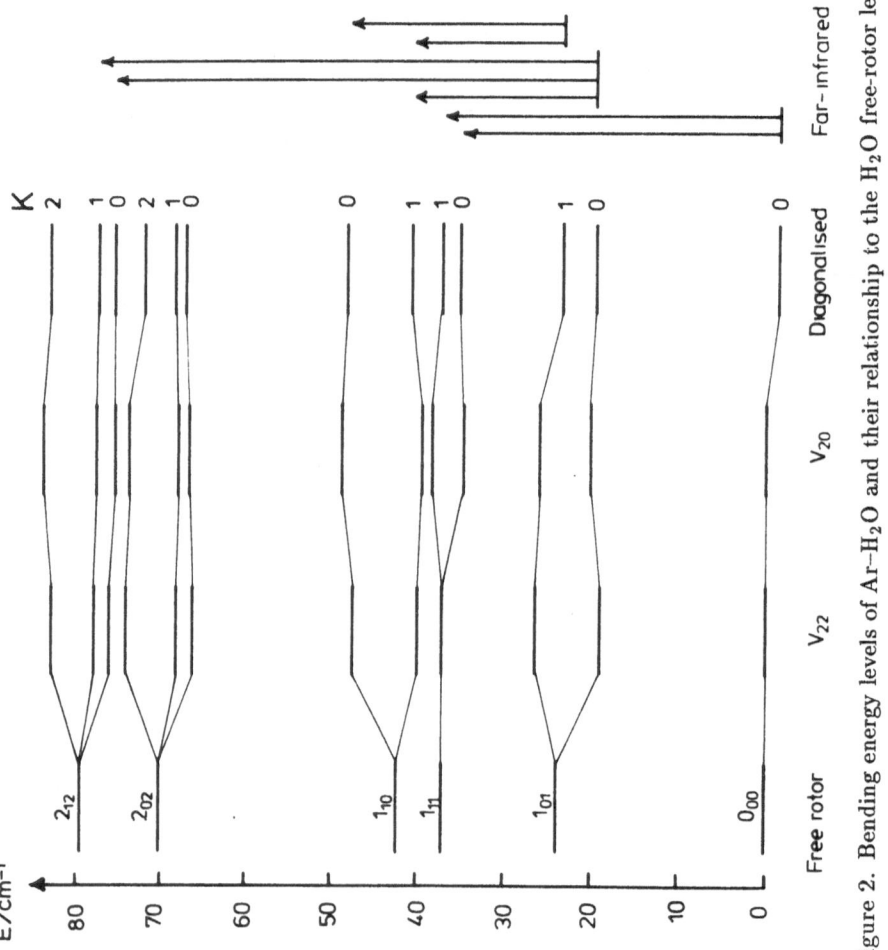

Figure 2. Bending energy levels of Ar–H_2O and their relationship to the H_2O free-rotor levels.

74

Table 1. Energy levels and wavefunctions for the lowest rotational levels of H_2O.

j_τ	$j_{k_p k_o}$	$(-)^{k_b}$	E	rotational function
1_1	1_{10}	-1	$a+b$	$\left(\frac{3}{16\pi^2}\right)^{\frac{1}{2}} \left[\mathcal{D}_{m1}^{1*}(\phi,\theta,\chi) + \mathcal{D}_{m-1}^{1*}(\phi,\theta,\chi)\right]$
1_0	1_{11}	$+1$	$a+c$	$\left(\frac{3}{8\pi^2}\right)^{\frac{1}{2}} \mathcal{D}_{m0}^{1*}(\phi,\theta,\chi)$
1_{-1}	1_{01}	-1	$b+c$	$\left(\frac{3}{16\pi^2}\right)^{\frac{1}{2}} \left[\mathcal{D}_{m1}^{1*}(\phi,\theta,\chi) - \mathcal{D}_{m-1}^{1*}(\phi,\theta,\chi)\right]$
0	0_{00}	$+1$	0	$\left(\frac{3}{8\pi^2}\right)^{\frac{1}{2}} \mathcal{D}_{00}^{0*}(\phi,\theta,\chi)$

still approximately conserved. As for NH_3, it is convenient to define the monomer-fixed z axis as lying along the symmetry axis of H_2O. Since H_2O is an asymmetric top, the monomer energy levels do not have a definite value of k, and their wavefunctions must be expanded as linear combinations of rotation matrices. In a symmetric top basis set, there are $\Delta k = \pm 2$ matrix elements of H_{mon} coupling states with $k = \pm 1$; the $j = 1$ monomer functions are listed in Table 1. In the Ar–H_2O complex, however, there are also $\mu = \pm 2$ matrix elements of the interaction potential, which also couple the $k = \pm 1$ states. Since the coupling coefficients (21) depend on K, the H_2O 1_{01} and 1_{10} states are each split by the V_{22} anisotropy into components with $K = 0$ and 1, and transitions among these states have been observed by Cohen et al.[27] There are further splittings caused by other anisotropy components; the resulting hindered rotor levels have been investigated by Hutson,[30] and are shown in Figure 2. It may be seen that, as for Ar–HCl, the energy levels are close enough to the free-rotor limit to be sensibly labelled with free-rotor quantum numbers, although there are significant shifts due to the anisotropy.

Polyatomic hydrides are a special case in that they have quite large rotational constants. It is to be expected that most Van der Waals complexes involving non-hydride monomers will be much closer to the near-rigid limit, and in some cases their spectra may be best interpreted in terms of conventional near-rigid quantum numbers, although considerable care will be needed in handling tunnelling motions.

3. Molecule–molecule systems

The general molecule–molecule case, with two polyatomic fragments, has been very little studied. Brocks et al.[31] have derived the Hamiltonian for such a system in body-fixed coordinates, but have not performed actual calculations. This is a very important area, and high-resolution spectra are available for systems such as the H_2O dimer.[32] There has been a good deal of work on understanding these spectra, especially with regard to the symmetries of the states involved.[33] However, we are still some way from understanding the relationship of the spectra to the interaction potential. Understanding the spectroscopy and angular momentum coupling in such systems remains a research topic for the future.

There has been rather more work on the specific case of diatom–diatom systems, mostly aimed at understanding the spectrum of the HF dimer. The coordinate system used is again a generalisation of the atom–diatom coordinates, with angles (α, β) defining the body-fixed axis system and angles (θ_1, ϕ_1) and (θ_2, ϕ_2) specifying the orientation of the two diatomic molecules. The interaction potential is a function of R, θ_1, θ_2 and $\phi = \phi_1 - \phi_2$.

For a diatom–diatom complex, there are three sources of angular momentum: the rotation of each of the monomers, characterised by quantum numbers j_1 and j_2, and the end-over-end rotation of the complex as a whole. There are at least two ways of formulating the diatom–diatom problem in body-fixed coordinates.[16,34] In the one which displays most clearly the similarity to atom–diatom systems,[16] j_1 and j_2 are first coupled together to form a resultant j,

$$\mathcal{Y}_{j_1 j_2}^{jK}(\theta_1, \phi_1, \theta_2, \phi_2) = \sum_{k_1 k_2} \langle j_1 j_2 k_1 k_2 | j K \rangle Y_{j_1 k_1}(\theta_1, \phi_1) Y_{j_2 k_2}(\theta_2, \phi_2), \qquad (22)$$

where $K = k_1 + k_2$ is the projection of j along R. The total wavefunction is then expanded

$$\psi_\alpha(\boldsymbol{R}, \boldsymbol{r}) = R^{-1} \sum_{j_1 j_2 K} \Phi_{j_1 j_2 j K}^{JM}(\alpha, \beta, \theta_1, \phi_1, \theta_2, \phi_2) \chi_{j_1 j_2 j K J}^\alpha(R), \tag{23}$$

where the functions $\Phi_{j_1 j_2 j K}^{JM}(\alpha, \beta, \theta_1, \phi_1, \theta_2, \phi_2)$ are

$$\Phi_{j_1 j_2 j K}^{JM}(\alpha, \beta, \theta_1, \phi_1, \theta_2, \phi_2) = \left(\frac{2J+1}{4\pi}\right)^{\frac{1}{2}} \mathcal{D}_{MK}^{J*}(\alpha, \beta, 0) \mathcal{Y}_{j_1 j_2}^{jK}(\theta_1, \phi_1, \theta_2, \phi_2). \tag{24}$$

It may be seen that this is exactly isomorphic with equation (4), except that the channels are labelled by (j_1, j_2, j) rather than by j alone. The body-fixed coupled equations are now

$$\begin{aligned}
\left[-\frac{\hbar^2}{2\mu} \frac{d^2}{dR^2} \right. & + (j_1 j_2 j K J|V|j_1 j_2 j K J) + \frac{\hbar^2}{2\mu R^2} \left[J(J+1) + j(j+1) - 2K^2 \right] \\
& \left. + E_{j_1}^{\text{mon}} + E_{j_2}^{\text{mon}} - E \right] \chi_{j_1 j_2 j K J}^\alpha(R) \\
& = - \sum_{j_1' j_2' j' K}' (j_1 j_2 j K J|V|j_1' j_2' j' K J) \chi_{j_1' j_2' j' K J}^\alpha(R) \\
& \quad - \sum_{K'=K\pm 1}' \frac{\hbar^2}{2\mu R^2} (j_1 j_2 j K J|(\hat{J}-\hat{j})^2|j_1 j_2 j K' J) \chi_{j_1 j_2 j K' J}^\alpha(R),
\end{aligned} \tag{25}$$

The potential matrix elements in the body-fixed representation have been given by Danby,[16] but are too complicated to reproduce here. The off-diagonal matrix elements of the operator $(\hat{J}-\hat{j})^2$ in equation (25) are the same as in the atom–diatom case, and are given by equation (9) with an additional factor of $\delta_{j_1 j_1'} \delta_{j_2 j_2'}$.

Once again, the body-fixed coupled equations are very similar in form to those for atom–diatom systems: the only differences are that the potential matrix elements and the monomer energies takes a more complicated form. Helicity decoupling calculations may be performed in exactly the same way as for atom–diatom and atom–polyatom systems.

4. Trimeric systems

Systems such as Ar_2HCl and Ar_2HF are of great interest, because they offer the hope of a spectroscopic determination of non-additive contributions to intermolecular potentials. Microwave spectra of these systems have been observed[35,36] and infrared spectra are likely to be feasible. A complete solution of the dynamical problem for such systems is beyond our capabilities at present, but Hutson et al.[37] have carried out a preliminary study, investigating the hindered rotation of an HCl molecule under the influence of a (fixed) pair of Ar atoms.

The coordinate system needed for a trimeric system in this approximation is a straightforward generalisation of that for an atom–diatom complex. The $\boldsymbol{R}$ vector is now defined as running from the HX centre of mass to the midpoint of the Ar_2 pair, and forms the z axis of a body-fixed axis system. An additional vector, $\boldsymbol{\rho}$, runs between the two Ar atoms. The body-fixed x axis is perpendicular to $\boldsymbol{R}$ and coplanar with $\boldsymbol{\rho}$. The relationship between the body-fixed and the space-fixed axes now requires three Euler angles (α, β, γ). Once again, the orientation of the $\hat{\boldsymbol{r}}$ vector relative to these axes is specified by two angles (θ, ϕ).

Neglecting bending and stretching motions involving the Ar–Ar pair, the Hamiltonian is

$$H = -\frac{\hbar^2}{2\mu} R^{-1} \left(\frac{\partial^2}{\partial R^2}\right) R + \frac{\hbar^2 \hat{l}^2}{2\mu R^2} + V(R, \theta, \phi) + H_{\text{mon}}, \tag{26}$$

where the intermolecular potential is now a function of R, θ and ϕ. This may be compared with the atom–nonlinear molecule case, where the potential was a function of R, θ and χ. The intermolecular potential may again be expanded in renormalised spherical harmonics,

$$V(R, \theta, \phi) = \sum_\lambda V_{\lambda\mu}(R) C_{\lambda\mu}(\theta, \phi). \tag{27}$$

The total wavefunction is again most conveniently expanded in body-fixed functions,

$$\psi_\alpha(\boldsymbol{R},\boldsymbol{r}) = R^{-1} \sum_{jkK} \Phi_{jkK}^{JM}(\alpha,\beta,\gamma,\theta,\phi)\chi_{jkKJ}^\alpha(R). \tag{28}$$

The functions $\Phi_{jkK}^{JM}(\alpha,\beta,\gamma,\theta,\phi)$ are

$$\Phi_{jkK}^{JM}(\alpha,\beta,\gamma,\theta,\phi) = \left(\frac{2J+1}{8\pi^2}\right)^{\frac{1}{2}} D_{MK}^{J*}(\alpha,\beta,\gamma)Y_{jk}(\theta,\phi), \tag{29}$$

which is very similar to equation (4) except for the additional angle γ. Since the HX hindered rotational motion is not the only source of angular momentum about the $\boldsymbol{R}$ axis, the K quantum number is not necessarily equal to k. The body-fixed coupled equations are now

$$\left[-\frac{\hbar^2}{2\mu}\frac{d^2}{dR^2} + (jkKJ|V|jkKJ) + \frac{\hbar^2}{2\mu R^2}(jkKJ|(\hat{J}-\hat{j}-\hat{j}_A)^2|jkKJ) + E_j^{\text{mon}} - E\right]\chi_{jkKJ}^\alpha(R)$$

$$= -\sum_{j'k'}{}' (jkKJ|V|j'k'KJ)\chi_{j'k'KJ}^\alpha(R)$$

$$-\sum_{K'=K\pm1}{}' \frac{\hbar^2}{2\mu R^2}(jkKJ|(\hat{J}-\hat{j}-\hat{j}_A)^2|jkK'J)\chi_{jkK'J}^\alpha(R), \tag{30}$$

where $\hat{j}_A$ is the operator for the angular momentum of the Ar$_2$ pair.

The potential matrix elements in the body-fixed representation are

$$(jkKJ|V|j'k'KJ) = \sum_\lambda g_{\lambda\mu}(jkj'k')V_{\lambda\mu}(R), \tag{31}$$

where

$$g_{\lambda\mu}(jkj'k') = (-)^k[(2j+1)(2j'+1)]^{\frac{1}{2}}\begin{pmatrix} j & \lambda & j' \\ -k & \mu & k' \end{pmatrix}\begin{pmatrix} j & \lambda & j' \\ 0 & 0 & 0 \end{pmatrix}. \tag{32}$$

This again reduces to equation (7) in the atom–diatom case, where $K = k$, $K' = k'$ and $\mu = 0$. It may also be regarded as a special case of equation (21) with $K = 0$, so that calculations on systems such as this may be performed using a program designed for atom–symmetric top calculations.

The matrix elements of the operator $\hat{l}^2 = (\hat{J}-\hat{j}-\hat{j}_A)^2$ in equation (30) are somewhat more complicated than in the atom–diatom case; however, if the $\hat{j}_A$ term is neglected, they are again given by equation (9) with an additional factor of $\delta_{kk'}$.

The dynamical differences between Ar$_2$–HX and Ar–HX complexes arise mostly from the fact that the potential expansion contains terms with $\mu \neq 0$, which cause splittings not present in the atom–diatom systems. In Ar–HX, the parity-adapted e and f states for $\Omega = 1$ are degenerate except for second-order Coriolis splittings: the Coriolis terms give rise to l-type doubling, which manifests itself in the spectrum as different rotational constants for the e and f states, but the two states share the same band origin. For Ar$_2$–HX, by contrast, potential terms with $\mu = \pm 2$ connect the $k = \pm 1$ states, and thus provide a splitting between the even and odd symmetry states which exists even for $J = 0$. The even and odd states can be considered to be hindered rotations of the HX molecule in and out of the plane of the Ar atoms, which have different frequencies because of the different bending potentials in the two directions. The different bands also have different rotational selection rules.[37]

To a first approximation, the interaction potential for Ar$_2$–HCl is just a pairwise sum of Ar–Ar and Ar–HCl potentials, both of which are accurately known. Hutson et al.[37] have used such a potential to investigate the dynamics of Ar$_2$–HCl and the dependence of its energy levels on non-additive contributions to the potential. They compared the calculated spectroscopic

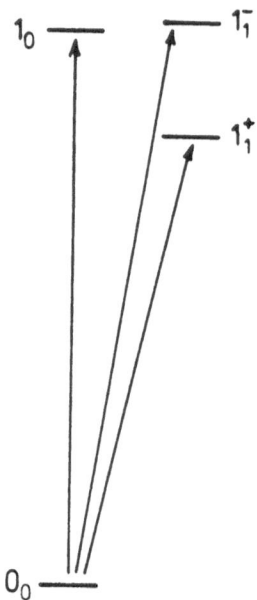

Figure 3. Bending energy levels of Ar_2–HCl.

parameters with those obtained from molecular beam microwave spectroscopy, and found that the spectra are indeed sensitive to the nonadditive part of the potential. However, they concluded that infrared spectra would be needed before unambiguous information on the nonadditive forces could be extracted. The energy level pattern calculated for Ar_2–HCl is shown in Figure 3. It may be seen that it retains the free-rotor-like structure found in Ar–HCl, despite the additional splittings arising from the ϕ-dependent anisotropy.

Conclusions

It is possible to formulate the bound state problem for a wide range of Van der Waals complexes in a form reminiscent of that for atom–diatom systems. The similarities between the different sets of coupled equations will be helpful in understanding the dynamical approximations that may be applied to the larger complexes. In particular, the helicity decoupling approximation, which has proved to be accurate for most atom–diatom systems, is equally applicable to the larger systems.

Van der Waals complexes containing monomers with large rotational constants, such as HF, HCl, H_2O and NH_3, undergo very wide-amplitude bending motions even in low vibrational states. These states are best understood using quantum numbers derived from a free internal rotor picture, rather than those derived from the conventional near-rigid picture. The anisotropy of the potential splits and shifts the monomer free-rotor states, but the qualitative pattern of energy levels and allowed transitions remains.

Acknowledgment

This work is supported by the Science and Engineering Research Council.

References

[1] J. M. Hutson, BOUND computer code, distributed via Collaborative Computational Project No. 6 of the UK Science and Engineering Research Council, on Heavy Particle Dynamics.
[2] J. Tennyson, Comp. Phys. Commun. **42**, 257 (1986).
[3] S. Bratoz and M. L. Martin, J. Chem. Phys. **42**, 1051 (1965).
[4] D. M. Brink and G. R. Satchler, *Angular Momentum*, 2nd ed., Clarendon Press, Oxford (1968).

5 R. G. Gordon, J. Chem. Phys. **51**, 14 (1969).

6 W. N. Sams and D. J. Kouri, J. Chem. Phys. **51**, 4809 (1969); J. Chem. Phys. **51**, 4815 (1969).

7 B. R. Johnson, J. Comp. Phys. **13**, 445 (1973).

8 J. C. Light and R. B. Walker, J. Chem. Phys. **65**, 4272 (1976); E. B. Stechel, R. B. Walker and J. C. Light, J. Chem. Phys. **69**, 3518 (1978).

9 G. A. Parker, T. G. Schmalz and J. C. Light, J. Chem. Phys. **73**, 1757 (1980).

10 M. H. Alexander, J. Chem. Phys. **81**, 4510 (1984).

11 D. E. Manolopoulos, J. Chem. Phys. **85**, 6425 (1986).

12 M. H. Alexander and D. E. Manolopoulos, J. Chem. Phys. **86**, 2044 (1987).

13 C. J. Ashton, M. S. Child and J. M. Hutson, J. Chem. Phys. **78**, 4025 (1982).

14 A. M. Dunker and R. G. Gordon, J. Chem. Phys. **64**, 4984 (1976).

15 B. R. Johnson, J. Chem. Phys. **69**, 4678 (1978).

16 G. Danby, J. Phys. B **16**, 3393 (1983).

17 D. E. Manolopoulos, Ph. D. thesis, Cambridge University (1988).

18 J. M. Hutson, Chem. Phys. Lett. **151**, 565 (1988).

19 R. J. Le Roy and J. van Kranendonk, J. Chem. Phys. **61**, 4750 (1974).

20 J. Tennyson and B. T. Sutcliffe, J. Chem. Phys. **77**, 4061 (1982).

21 I. P. Hamilton and J. C. Light, J. Chem. Phys. **84**, 306 (1986).

22 Z. Bačić and J. C. Light, J. Chem. Phys. **85**, 4594 (1986); J. Chem. Phys. **86**, 3065 (1987).

23 S.L. Holmgren, M. Waldman and W. Klemperer, J. Chem. Phys. **69**, 1661 (1978).

24 J. M. Hutson and B. J. Howard, Mol. Phys. **43**, 493 (1981).

25 J. M. Hutson and B. J. Howard, Mol. Phys. **45**, 769 (1982).

26 J. M. Hutson, J. Chem. Phys. **89**, 4550 (1988).

27 R. C. Cohen, K. L. Busarow, K. B. Laughlin, G. A. Blake, M. Havenith, Y. T. Lee and R. J. Saykally, J. Chem. Phys. **89**, 4494 (1988).

28 J. M. Hutson, D. C. Clary and J. A. Beswick, J. Chem. Phys. **81**, 4474 (1984); A. C. Peet, D. C. Clary and J. M. Hutson, J. Chem. Soc., Faraday Trans. II **83**, 1719 (1987).

29 S. Green, J. Chem. Phys. **64**, 3463 (1976).

30 J. M. Hutson, J. Chem. Phys. **92**, to be published (1990).

31 G. Brocks, A. van der Avoird, B. T. Sutcliffe and J. Tennyson, Mol. Phys. **50**, 1025 (1983).

32 G. T. Fraser, R. D. Suenram and L. H. Coudert, J. Chem. Phys. **90**, 6077 (1989) and references therein.

33 L. H. Coudert and J. T. Hougen, J. Mol. Spec. **130**, 86 (1988).

34 A. E. Barton and B. J. Howard, Faraday Discuss. Chem. Soc. **73**, 45 (1982).

35 H. S. Gutowsky, T. D. Klots, C. Chuang, C. A. Schmuttenmaer and T. Emilsson, J. Chem. Phys. **86**, 569 (1987).

36 T. D. Klots, C. Chuang, R. S. Ruoff, T. Emilsson and H. S. Gutowsky, J. Chem. Phys. **86**, 5315 (1987).

37 J. M. Hutson, J. A. Beswick and N. Halberstadt, J. Chem. Phys. **90**, 1337 (1989).

CLASSICAL DYNAMICS OF VAN DER WAALS MOLECULES

Stephen K. Gray

Department of Chemistry
Northern Illinois University
DeKalb, IL 60115

Abstract

We discuss the application of classical mechanics to van der Waals molecules and their reactions, with particular emphasis on the vibrational predissociation of XBC systems, where X is weakly bound to a molecular fragment BC. The role of nonlinear phenomena in influencing the classical reaction dynamics is stressed, as is the need to discover quantum and experimental consequences of such behaviour.

I. Introduction

At first glance, classical mechanics seems to be an inappropriate theory for the description of many van der Waals (vdW) systems. In this paper, for example, we will focus mostly on XBC type vdW systems where X (usually a noble gas atom) is weakly bound to a chemically bound fragment BC (usually a diatomic molecule). The total angular momentum $J = 0$ vibrational predissociation process for such a system can be denoted by:

$$XBC(s,b,v) \rightarrow X + BC(v' < v, j') , \qquad (1)$$

where the vdW metastable state, which is often probed by laser excitation, is described by effective vdW stretching and bending quantum numbers s and b, and the effective vibrational state of BC in the complex is denoted by v. First, note that the density of states, particularly for small vdW bond energies and low vibrational excitations, can be small, perhaps on the order of 1 state per 10 cm^{-1}. Therefore quantum effects such as tunneling can be important. A second, related point is that the actual energy transfer process that leads to reaction consists of the transfer of one or more large quanta of vibrational energy from BC to a weak vdW bond, with a lot of energy being available for product excitation. Classical mechanics often leads to much smaller energy exchanges, e.g. just enough energy being transferred to break the weak bond. Consequently classical mechanics can underestimate the extent of product excitation (see, e.g. Ref. 1) .

The remarks above notwithstanding, classical mechanics is still an important tool for understanding a variety of aspects pertaining to vdW reactions such as Eq. (1), as well as other processes (e.g. rotational predissociation). We may divide classical approaches to vdW dynamics up into three, overlapping categories:

Dynamics of Polyatomic Van der Waals Complexes
Edited by N. Halberstadt and K. C. Janda
Plenum Press, New York, 1990

(i) Quasiclassical simulations. By this we mean a more or less standard classical trajectory simulation of the system of interest. One would describe the XBC complex by some initial ensemble of phase points (positions and momenta) weighted in a manner somehow consistent with aspects of the actual metastable state, and then integrate classical equations of motion to generate trajectories corresponding to each intial phase point. Average lifetimes and product distrbutions can then be estimated. Woodruff and Thompson [2] were the first to show the utility of this approach in the context of vdW systems in a study of a simple collinear model of HeI_2 fragmentation. See also related work on three dimensional HeI_2, [3] $NeCl_2$, [1] T-shaped HeI_2, [4] clusters XI_2Y, [5] and model rotational predissociation. [6] As noted by Gibson and Schatz [7] in a quasiclassical study of the Ar-OCS system, experience shows classical mechanics to be reasonably quantitative (e.g. within a factor of 2 or so) if the corresponding processes are suffuciently fast (e.g. lifetimes less than 50 ps). Qualitative trends can also be predicted even when this is not the case.

(ii) Impulsive energy transfer models. Even when some aspects of the dynamics (e.g. the initial energy transfer process) might be dominated by quantum effects, other aspects might not be. Classical impulsive energy transfer models [8] begin by assuming a quantum of energy has already been transferred and then use classical trajectories to model the subsequent time evolution (or "final states interaction"). These models generally do not predict lifetimes, but can lead to interesting interpretations of product distributions, as for example, in recent work on $HeICl$ [9], $HeCl_2$ [10] and $NeCl_2$. [11] Actually, semiclassical theories, as discussed by Child in this volume, [12] represent the most rigorous melding of quantum and classical ideas.

(iii) Fundamental theoretical studies. The detailed classical phase space dynamics of vdW systems has turned out to exhibit some remarkable dynamical phenomena : not just classical chaos but structure within chaos such as separatrices and cantori. [13] The simplicity of vdW systems - for example, there usually not being any complicated reaction path curvature effects - often makes these dynamical features easier to spot in calculations. In contrast, see the more complicated phase space features in $H + H_2$ [14] and $I + HI$. [15] Such observations have suggested corrections to standard statistical theories of unimolecular processes to accomodate vdW systems, which are normally thought of as nonstatistical in nature. [16] More importantly, there is growing interest in understanding the quantum mechanical (and eventually experimental) implications of such features, which appear in many other classical mechanical systems aside from vdW systems. [17]

The view adopted in this paper is that there is intrinsic merit in understanding the detailed classical phase space structure of vdW systems. Such an understanding can explain, at the deepest level, the results of quasiclassical simulations or other classical models such as the impulsive energy transfer models. The mechanisms or ideas garnered can also be applied to the interpretation of more complicated systems. Furthermore, we advocate, whenever possible, parallel quantum dynamics (e.g. wavepacket) calculations to ascertain quantum analogues of the classical dynamics. Considerable progress has been made in the understanding of classical phase space structure as it pertains to simple, two degree of freedom models of vdW systems. [13,16,18] More limited progress has been made in in the application of such ideas to systems with three or more degrees of freedom [1,19] and in understanding the classical/quantum correspondence. [20] In our view, therefore, the two major research directions are:

1. To map out the classical phase space structure of vdW systems with three or more degrees of freedom.

2. To understand the classical/quantum correspondence for systems with two or more degrees of freedom.

Section II below outlines possible approaches for understanding the classical dynamics of vdW systems with three degrees of freedom. Section III will address aspects of the classical/quantum correspondence. Section IV consists of some brief concluding remarks.

II. Classical Phase Space Dynamics for vdW Systems with Three Degrees of Freedom

We will restrict attention to the $J = 0$ vibrational predissociation process given by Eq. (1). Results presented will pertain to a model of the $NeCl_2$ system, although the classical dynamics observed is really generic to many XBC systems. Details of the potential surface surface may be found in Ref. 1. (A more accurate surface is also available; [11] the dynamics on this surface is qualitatively similar to that discussed here.) The classical mechanics can be described by the three scattering coordinates R = distance of Ne to the center of Cl_2, r = the Cl_2 internuclear distance and γ = the angle between the **R** and **r** bond vectors. The corresponding canonically conjugate momenta will be denoted as P,p and j respectively. We may think of R and γ as being the vdW stretching and bending modes. For $J = 0$ dynamics, the angular momentum conjugate to γ, j, correlates with BC rotational action in the limit of separated X and BC fragments. A classical trajectory consists of the time evolution $\{R(t), r(t), \gamma(t), P(t), p(t), j(t)\}$ given specification of the coordinates and momenta at time t = 0. The classical analogue of Feshbach resonance decay (e.g. typical vibrational predissociation) then consists of R(t) undergoing at least one - and often many - complex oscillations followed by fragmentation: $R \rightarrow \infty$ with P and j, as well as BC vibrational energy, becoming constant in time. [4] Different initial conditions can generate quite different results due to the fact that such motion is chaotic. It may be the case, however, that certain structures in the phase space influence the flow of many trajectories and therefore determine the final, ensemble averaged properties.

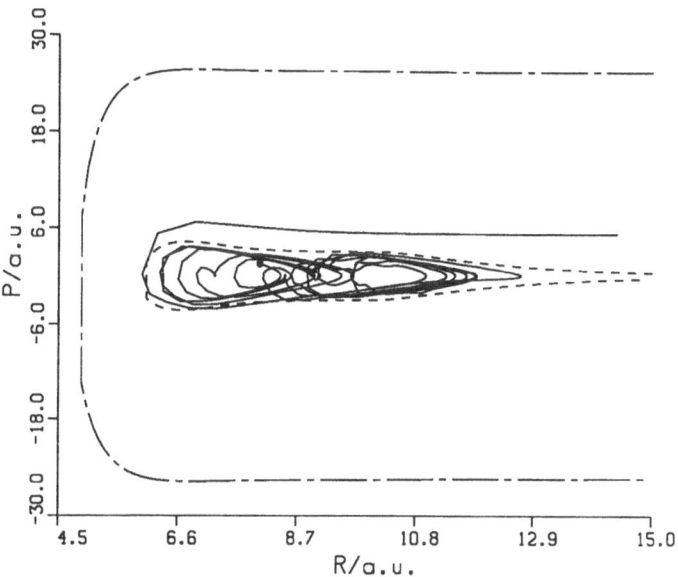

Figure 1. The variation of R and P along a typical trajectory for $NeCl_2$ with three degrees of freedom. The Cl_2 vibrational action was set initially to v =13 and zero-point energy was placed in the vdW modes. The solid curve represents the trajectory, which undergoes complex oscillations in the interaction region prior to finally breaking up. The dashed curve that closely hugs the region of complex oscillations is an analytical approximation to the intermolecular bottleneck. The outer, long-short dashed curve represents the maximum excursions in P permitted by energy conservation.

The simplest phase space structure is the intermolecular bottleneck, which was defined exactly by Davis and Gray [13] for two degree of freedom systems. In essence, this structure serves to define the region of phase space associated with complex oscillations. This structure is particularly significant for vdW systems because it then represents a much more severe constraint than energy conservation. The exact nature of the intermolecular bottleneck in three degrees of freedom is still an open question. However, we have nonetheless seen evidence of intermolecular bottlenecks in the three degree of freedom $NeCl_2$ system [1] and illustrate this feature in Figure 1. Here we display the simultaneous variations in R and P along a typical trajectory. The outer curve represents the maximum values of P permitted by energy conservation. More chemically bound molecules that display RRKM fragmentation rates [16,21] would show excursions in P out to these outer limits. Clearly the trajectory is more severely confined. The inner dashed curve represents the envelope of an approximate analytical function given elsewhere [1] to describe the intermolecular bottleneck. We may also think of the intermolecular bottleneck as an energy for reaction criterion. Trajectories with effective energy in the vdW modes sufficient to break the vdW bond are outside the intermolecular bottleneck, whereas those with effective energy less than this amount are inside the bottleneck and are therefore confined to undergo one or more complex oscillations before eventually getting enough energy to be transported outside. The intermolecular bottleneck can be used as a transition state in statistical theory descriptions of vdW dynamics, [16] and, barring intramolecular effects, determines the classical lifetime of the complex. A general goal is the development of chemically and physically appealing approximations to such structures. Actually, the intermolecular bottleneck is a special case of a more general classical structure called a separatrix. [13] Other types of separatrices, perhaps those defining the approximate bounds of resonance zones (see below), can also play a signficant role. [13]

The behavior of molecules within the intermolecular bottleneck, that is to say intramolecular dynamics, is also of interest. For example, trajectories can be trapped for signficant periods of time within resonance zones inside the intermolecular bottleneck. A classical resonance [22,23] is a region of phase space where, locally, the condition

$$s_R \omega_R + s_r \omega_r + s_\gamma \omega_\gamma = 0 , \qquad (2)$$

is met. In the above equation, s_i are any (positive,negative or zero) integers and ω_i are local fundamental frequencies associated with the respective variables. Martens, Davis and Ezra [22] have discussed how one might determine local frequencies from classical trajectories and have provided some interesting analysis of a three degree of freedom model for OCS. We have carried out a similar kind of analysis (but less extensive) with the $NeCl_2$ system. We divide up a trajectory up into several time segments and then Fourier analyse P(t),p(t) and j(t) on each interval to obtain estimates of the frequencies ω_R, ω_r, and ω_γ respectively. Because of the weakly coupled nature of vdW systems, the spectrum of, say, P(t), on a given time interval is often dominated by the local fundamental frequency associated with motion in R, ω_R. Similar remarks hold for p(t) and j(t). On a more technical note, it should be pointed out that we find it convenient to use certain alternative transform methods such as MUSIC. [24] MUSIC can yield good estimates of frequencies present in a signal even when the duration of the signal is so short that fast Fourier transform methods would be inaccurate. Noid and co-workers have applied MUSIC to a number of problems in chemical physics. [25] The result of such local frequency analysis shows that resonances play a very signficant role in the classical dynamics. In particular, "pairwise" resonances associated with just the vdW stretching and bending degrees of freedom are quite important. It turns out that pairwise resonances corresponding to either $\omega_R/\omega_\gamma = 1$ or $\omega_R/\omega_\gamma = 2$ seem to dominate the dynamics for energies corresponding to zero point excitation of the van der Waals modes and Cl_2 vibrational excitations of 11 and 13. One can denote such resonances as $(s_R,s_r,s_\gamma) = (1,0,-1)$ or $(1,0,-2)$. We also find instances of resonances between the R and r degrees of freedom, such as $\omega_r/\omega_R = 4$ or 5, which would be denoted as $(4,-1,0)$ or $(5,-1,0)$. Truly combination resonances, with none of the s_i being zero are also seen.

A given trajectory can drift into and out of various resonance zones. For example, we see instances of trajectories being in a (1,0,-1) zone and then jumping into a (1,0,-2) zone. Physically, this particular situation corresponds to the vdW stretch and bending motions vibrating with comparable frequency along one segment of the trajectory and then vibrating with frequencies that are in 1:2 Fermi resonance on a later segment. Such observations can be used to map out possible intramolecular and reactive pathways. For example, it may be possible after detailed analysis to identify modes that promote reaction into particular product states. Unlike systems with two degrees of freedom, it turns out that the resonance zones of three (or higher) degree of freedom systems are all interconnected (for arbitrarally small coupling). A single trajectory has the potential to wander into and out of infinitely many resonance zones. This phenomenon is called Arnold diffusion. [22,23] Generally, the time scale for Arnold diffusion is longer than the time scale for chaotic motion within a given resonance zone or within overlapping resonance zones. Nonetheless we believe it should be possible to find Arnold diffusion effects in vibrationally predissociating XBC systems and are currently investigating this. This effect is truly one that requires at least three coupled degrees of freedom. Thus, one does not expect any such effects in rotational predissociation of XBC type complexes where the vibrational state of BC remains essentially constant (adiabatic) throughout. See, for example, a recent classical study of ArCO with three degrees of freedom. [26] Another interesting aspect of vibrational predissociation dynamics, related to both the Arnold diffusion phenomenon and to understanding the general mechansim of fragmentation is the fact that the vibrational energy (or action) in the BC part of the molecule can fluctuate with small amplitude about some mean value for signficant periods of time and then, abruptly, change. The abrupt changes take place when the vdW stretching coordinate R gets small, reflecting a "hard wall collision" between the Ne and Cl_2. The hard wall collisions can lead to significant energy transfer to the vdW modes and consequent fragmentation. If a hard wall collision does not lead to immediate fragmentation, it may lead to signficant local frequency changes and therefore shifts into or out of various resonance zones. Recent work by Skodje [27] may assist in our understanding of such nonadiabatic jumps. See, also, some interesting work by Morales [28] on the magnitude of vibrational action jumps in vibrational predissociation.

Other, interesting phase space structures that may play a role in three degree of freedom systems include three dimensional analogues of cantori, which have been shown to be important in T-shaped HeI_2 [13] and other simple, two degree of freedom problems. Cantori may be thought of as the (fractal) remnants of certain quasiperiodic motions in a system that are particularly robust to perturbations. In a sense cantori are the very opposite of resonance zones. In two degrees of freedom, for example, they are characterized by frequency ratios that are very irrational, i.e. poorly approximated by a ratio of integers. There is, in fact numerical evidence pointing to the importance of cantori in three degree of freedom vdW problems, [19] as well as in more chemically bound systems. [22] However, there are at present no precise mathematical definitions. Progress in understanding three dimensional vdW systems will also benefit from recent work in nonlinear dynamics that has focused on 4D area preserving mappings, [29,30] since such maps are actually equivalent to the phase space dynamics of continuous, three degree of freedom systems.

III. Quantum Analogues of Classical Phase Space Structure

What are the quantum mechanical - and thus observable - consequences of the classical phase space structure discussed above? This is still a very open question but we will offer here some clues and comments. Quantum consequences can arise in two ways. First, one might see evidence of phase space structures in the metastable or resonance vdW states. That is to say, such features might be evident in the spectroscopy, as opposed to dynamics, of vdW systems. The metastable states can be thought of as almost stationary states embedded in the continuum and are generally the vdW states that are probed in energy-resolved experiments. A lot of work has been done in the related area of inferring classical phase space structure in bound state eigenfunctions for coupled oscillator problems. [31] Second, one might see classical influences in the time evolution or dynamics of such metastable states, or, more generally, in the time evolution of superpositions of such states.

Let us focus our attention first on a simple model of XBC vibrational predissociation that consists of assuming a T-shaped geometry for the complex throughout all time and treats the dynamics as a vdW stretch R being periodically forced due to the potential changes that arise as the BC molecule vibrates classically. We have shown previously that this model describes many qualitative features of the dynamics. [20] We will employ parameters consistent with the $NeCl_2$ system described in Section II and Ref. 1. Figure 2a displays the classical surface of section for the $NeCl_2$ periodically forced oscillator system with v = 13. The behaviour observed is similar to that seen in our earlier work on HeI_2. [4,13,20] We started with several points (R,P) in the interaction region, integrated the corresponding classical equations of motion, and plotted the subsequent time development each time the Cl_2 passed through a particular vibrational phase value. Thus we plot (R,P) for times that are multiples of τ, where τ is the period of vibration of a Cl_2 molecule in vibrational state v=13. The dashed curve in the figure represents the intermolecular bottleneck [13] for this system. Trajectories that cross this structure from the inside going out will continue directly on to fragment states X+BC. Furthermore, all trajectories that exit this structure exit it into the first "finger" on the left, then get propagated to the next finger, and so on. The boundary of the fingers can be generated from the intermolecular bottleneck. [13] In addition to dissociating trajectories, Fig. 2a also shows a region of trapped, quasiperiodic motion centred near R = 6.75 a.u., P = 0. a.u. Also evident in Fig. 2a are four islands around the central quasiperiodic region that can be associated with a (4,-1,0) resonance (i.e. $4\omega_R = \omega_r$).

Let us now consider the quantum behaviour of the lowest lying metastable state for the v = 13 system, $\psi_o(R)$. This state corresponds to zero-point energy in the vdW stretching coordinate and appears to be a Gaussian centred near R = 6.75 a.u. We imagine this metastable state to be prepared at t=0 and follow the subsequent time evolution, $\psi(R,t)$. To see connections with the classical phase space structure we find it useful to introduce the Wigner function [20,32] (in atomic units such that $\hbar = 1$)

$$W(R,P,t) = (1/\pi) \int_{-\infty}^{+\infty} ds \, exp(2iPs) \, \psi^*(R+s,t) \, \psi(R-s,t) . \qquad (3)$$

W is one quantum analogue of a classical phase space probability distribution function. One can, in fact, formulate quantum mechanics in terms of W instead of ψ. [32,33] Now W(R,P,t=0) shown in Fig. 2b appears as a two dimensional Gaussian centred on R=6.75 a.u. and P = 0 a.u. It contains little information about the underlying classical phase space structure (Fig. 2a) except for the fact it is peaked where, classically, one has a significant area of quasibound motion. One sees no effects of the resonance islands, for example, because the phase space area occupied by these islands is less than Planck's constant h = 2π a.u..

Consider now the time evolution $\psi(R,t)$ that results by numerically solving the Schrodinger equation, as described in Ref. 20. It turns out that the general shape of ψ around R = 6.75 a.u. remains similar to the t = 0 Gaussian shape, but eventually very small wavepackets, appear to emerge from the 6.75 a.u. peak and progress into the product R $\rightarrow \infty$ region. Fig. 2c displays the resulting Wigner transform of ψ after a time 3 τ. Remarkably, one sees W has developed fingers that closely parallel the classical fingers in Fig. 2a. We have superimposed on W in Fig. 2c the trajectory points that escape through the fingers to further demonstrate the parallels. Thus, particularly at short times, quantum "phase space" probability density appears to mimic the underlying classical dynamics associated with escape out of an intermolecular bottleneck. One can therefore estimate the observed quantum rate by considering the rate of flow of Wigner density out of the classical intermolecular bottleneck. [20] We are currently working on additional analysis of this simple model, as well as exact quantum wavepacket dynamics for more realistic, three degree of freedom vdW systems. [34] Now, the analysis

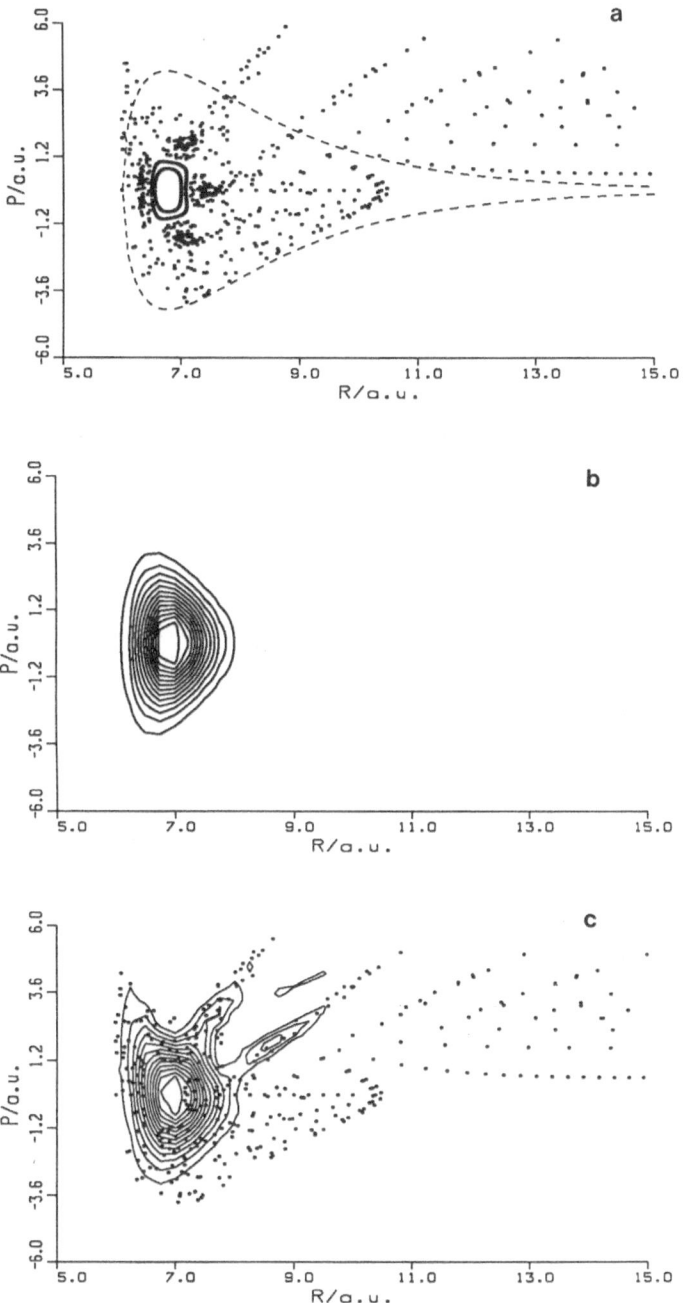

Figure 2. (a) The classical surface of section for the T-shaped NeCl$_2$ forced oscillator model, with v = 13. The dashed curve corresponds to the inter-molecular bottleneck. (b) Contours of the initial Wigner function. The contours range from 0.02 to 0.26 a.u., the 0.02 a.u. being the outermost and 0.26 a.u. the innermost contour. (c) Contours of the Wigner function after 3 τ of time evolution. For comparison, the classical trajectories that dissociate through the fingers in (a) are also superimposed.

above pertained to intermolecular bottleneck, or separatrix, effects in the dynamics of an intially prepared vdW metastable state. One might also expect instances where the metastable state itself shows evidence of separatrix structure. See, e.g., the related work by Davis.[31] In XBC systems, for example, such "separatrix" states might appear when one considers more highly excited vdW stretches.

How can other classical features, e.g. resonances, be seen in quantum systems? Perhaps the most direct and familiar consequence of a classical resonance, if it occupies a sufficient amount of phase space, is to lead to resonantly mixed quantum states (or metastable states). Typically, vdW stretching freqencies are 30 cm^{-1} and vdW bending frequencies are on the order of 15 cm^{-1}, as seen, for example, in a recent study of Ar-glyoxal complexes by Dai and co-workers.[35] This was the case for our NeCl$_2$ study and led, for example, to the importance of the (1,0,-2) resonance. Such 1:2 stretch-bend resonances, in quantum mechanics, are simply Fermi resonances. Indeed, examples of such Fermi resonances have been noted in quantum studies of the metastable states of NeCl$_2$ and ArCl$_2$.[36] vdW stretch-bend interactions have recently been invoked to explain spectral patterns observed in substituted benzene-Ar complexes.[37]

Section II also noted that classical trajectories can be in one region of phase space, associated with one particular resonance (or perhaps no simple resonance) and then jump into another region, associated with a different resonance. This represents a type of IVR (intramolecular vibrational relaxation) that occurs prior to vibrational predissociation. The quantum analogue of this behaviour might involve different metastable states being coupled to the originally prepared state. Recently, some experimental evidence of this type of behaviour has been seen in ArCl$_2$.[38] Larger vdW complexes, such as XM, where M is some pi-bonded organic molecule, have been the focus of much discussion about the competing roles of IVR and vibrational predissociation.[37,39-41] For example, Zewail and co-workers,[41] by carrying out real time measurements, have begun to unravel some interesting IVR effects in He-stilbene complexes. There is a clear need for classical studies on such more complicated vdW systems. See also the remarks by Rice in this volume.[42] A difficulty, of course, is the absence of reliable information on potential energy surfaces for such systems.

Finally, let us note that more "exotic" classical phase space features, such as the cantori briefly alluded to in Sec. II, may also have quantum analogues both in terms metastable states that could be prepared, or in terms of intermediate states (or structures) that influence the dynamics. For example, numerical work on simpler models [17,43,44] has pointed to the possibility that cantori, under certain circumstances, have even more strongly trapping effects in quantum mechanics than in classical mechanics. Conceivably, this phenomenon could be invoked in some instances to explain unusually long vibrational predissociation lifetimes.

IV. Concluding Remarks

The principle role of classical mechanics for understanding the behaviour of vdW molecules is to provide possible mechanisms and interpretations of the rich spectroscopy and dynamics observable in these systems. Classical interpretations are often useful, in a qualitative sense, even when significant quantum features are present.

Classical mechanics, however, cannot lead to very quantitative estimates of observable properties in many vdW systems studied experimentally because rather low excitation energies are being probed where the density of states is such that one is far from a classical limit. There is a clear need for experimental studies that approach this limit better. This means more experiments that probe the dynamics of vdW stretch and bend excitations, not just the zero-point vdW state which is the most commonly studied situation. Furthermore, one should also try to access simultaneously large excitation energies in the molecular part of the vdW complex.

Another important role of classical studies on vdW systems is that they can serve as useful dynamical models for complex formation and unimolecular decay in general, and also can lead to fundamental connections between classical and quantum mechanics. The issue might not be how accurately a given classical study mimics a particular system, but, rather, what dynamical phenomena can be found in the study and how such features relate to quantum mechanics. The answers to these questions relate to many problems in chemistry.

Acknowledgments

I wish to acknowledge support from the National Science Foundation (CHE-8808654) and the donors of the Petrolium Research Fund, administered by the American Chemical Society.

References

1. C.E. Wozny and S.K. Gray, Ber. Bunsenges. Phys. Chem. 92:236 (1988).

2. S.B. Woodruff and D.L. Thompson, J. Chem. Phys. 71, 377 (1979).

3. G. Delgado-Barrio, P. Villarreal, P. Mareca and G. Albelda, J. Chem. Phys. 78:280 (1983).

4. S.K. Gray, S.A. Rice and D.W. Noid, J. Chem. Phys. 84:5389 (1986).

5. G.C. Schatz, V. Buch, M.A. Ratner, and R.B. Gerber, J. Chem. Phys. 79:1808 (1983).

6. R.F. Frey, J.O. Jensen, and J. Simons, J. Phys. Chem. 89:788 (1985).

7. L.L. Gibson and G.C. Schatz, J. Chem. Phys. 83:3433 (1985).

8. R. Schinke , Ann. Rev. Phys. Chem. 39:39 (1988).

9. R.L. Waterland, J.M. Skene, and M.I. Lester, J. Chem. Phys. 89:7277 (1988).

10. J.I. Cline, B.P. Reid, D.D. Evard,N. Sivakumar, N. Halberstadt, and K.C. Janda, J. Chem. Phys. 89:3535 (1988).

11. J.I. Cline, N. Sivakumar, D.D. Evard, C.R. Bieler, B.P. Reid, N. Halberstadt, S.R. Hair, and K.C. Janda, J. Chem. Phys. 90:2605 (1989).

12. M.S. Child, this volume

13. M.J. Davis and S.K. Gray, J. Chem. Phys. 84:5389 (1986).

14. M.J. Davis, J. Chem. Phys. 86:3978 (1987).

15. R.T. Skodje and M.J. Davis, J. Chem. Phys. 88:2429 (1988).

16. S.K. Gray, S.A. Rice, and M.J. Davis, J. Phys. Chem. 90:3470 (1986).

17. G. Radons, T. Geisel, and J. Rubner, Adv. Chem. Phys. 73:891 (1989).

18. S.K. Gray and S.A. Rice, Farad. Disc. Chem. Soc. 82:307 (1986).

19. S. H. Tersigni and S.A. Rice, Ber. Bunsenges. Phys. Chem. 92:227 (1988).

20. S.K. Gray, J. Chem. Phys. 87:2051 (1987).

21. W. Forst, "Theory of Unimolecular Reactions," (Academic, New York, 1973).

22. C.C. Martens, M.J. Davis, and G.S. Ezra, Chem. Phys. Lett. 142:519 (1987).

23. A.J. Lichtenberg and M.A. Lieberman, "Regular and Stochastic Motion," Applied Mathematical Sciences, Vol 38, (Springer-Verlag, New York, 1983).

24. S.L. Marple, "Digital Spectral Analysis with Applications," (Prentice-Hall, Englewood Cliffs, N.J., 1987).

25. D.W. Noid, B.T. Broocks, S.K. Gray and S.L. Marple, J. Phys. Chem. 92:3386 (1988).

26. S.C. Farantos, J. Chem. Phys. 87:6449(1987).

27. R.T. Skodje, J. Chem. Phys. 90:6193 (1989).

28. D.A. Morales, Chem. Phys. 132:165 (1989).

29. H. Kook and J.D. Meiss, Physica D 35:65 (1989).

30. P. Gaspard and S.A. Rice, preprint.

31. M.J. Davis, J. Phys. Chem.92:3124(1988).

32. H. Hillery, R.F. O'Connell, M.O. Scully, and E.P. Wigner, Phys. Reports 106:121(1984).

33. E.J. Heller, J.Chem.Phys. 65:1289(1976).

34. S.K. Gray and C.E. Wozny, submitted to J.Chem.Phys.

35. D. Frye, P. Arias, and H.-L. Dai, J. Chem. Phys. 88:7240 (1988).

36. B.P. Reid, K.C. Janda, and N. Halberstadt, J. Phys. Chem. 92:587(1988).

37. E.J. Bieske, M.W. Rainbird, I.M. Atkinson, and A.E.W. Knight, J. Chem. Phys. 91:752(1989).

38. D.D. Evard, C.R. Bieler, J.I. Cline, N. Sivakumar, and K.C. Janda, J.Chem.Phys. 89:2829(1988).

39. B.A. Jacobson, S. Humphrey, and S.A. Rice, J. Chem. Phys. 89:5624 (1988)

40. H.-K. O, C.S. Parmenter, and M.C. Su, Ber. Bunsenges. Phys. Chem. 92:253 (1988).

41. D.H. Semmes, J. Spencer Baskin, and A.H. Zewail, Am. J. Chem. Soc. 109:4104(1987).

42. S.A. Rice, this volume

43. R.C. Brown and R.E. Wyatt, J.Phys.Chem. 90:3590(1986).

44. L.L. Gibson,G.C. Schatz, M.A. Ratner, and M.J. Davis, J. Chem. Phys. 86:3263(1987).

SEMICLASSICAL INVERSION OF VAN DER WAALS SPECTRA

M.S. Child

Theoeretical Chemistry Dept.
5 South Parks Road
Oxford, OX1 3UB
U.K.

INTRODUCTION

Semiclassical techinques have an established place in the analaysis of electronic spectra, to the extent that one often speaks of the "exact RKR" potential curves derived from experimental vibrational and rotational term values[1-2]. Similarly, when applicable, the Le Roy – Bernstein[3-4] scheme for extrapolation to dissociation limits is far superior to the traditional Birge–Sponer method.

Unfortunately these methods are seldom directly applicable to van der Waals spectra,, but the present paper shows how the underlying ideas may be adapted. One major difficulty is that Franck–Cordon restrictions almost invariably preclude the observation of a progression of van der Waals stretching bands. Hence the sequence of vibrational term values required by the conventional methods[1-4] is unavailable. Secondly, although the HX motions in A—HX complexes have useful analogies with electronic motions, their adiabatic separation from the van der Waals modes is often far from complete. Consequently inversion for the full potential function may require a preliminary deperturbation of the spectrum. Nevertheless remarkable results may be obtained in favourable cases, as exemplified below for the Ar...HF system.

Section 2 below gives an account of the recently published "rotational–RKR" procedure whereby potential curves may be deduced from single rotational progessions. It is then shown in section 3 how a sequence of such curves for different bending states of Ar...HF(v=1) may be analysed to reveal the form of the van der Waals stretching plus bending potential function. Finally Section 4 outlines a variant of the LeRoy–Bernstein extrapolation procedure applicable to weakly bound species, even when the data is restricted to a small number of v = 0 rotational term values.

THE ROTATIONAL RKR PROCEDURE

The starting point for semiclassical inversions is the Bohr quantisation condition

$$\frac{1}{\beta} \int_{a_{vJ}}^{b_{vJ}} [E(v,J) - V(R;J)]^{\frac{1}{2}} \, dR = (v+\tfrac{1}{2})\pi, \qquad (1)$$

Dynamics of Polyatomic Van der Waals Complexes
Edited by N. Halberstadt and K. C. Janda
Plenum Press, New York, 1990

for the vth vibrational level of the Jth effective potential function

$$V(R;J) = V(R) + \frac{J(J+1)\beta^2}{R^2},\qquad(2)$$

where a_{vb} and b_{vJ} are the turning–points and, with $E(v,J)$ and $V(R:J)$ expressed in wavenumber units,

$$\beta = (h/8\pi^2\mu c)\qquad(3)$$

The standard RKR analysis[1-2] then yields $V(R;J)$ via the energy dependence of its turning points, which are determined by the equations

$$f_{vJ} = b_{vJ} - a_{vJ} = 2\beta \int_{-\frac{1}{2}}^{v} \frac{dv'}{[E(v,J) - E(v',J)]^{\frac{1}{2}}}\qquad(4)$$

and

$$g_{vJ} = a_{vJ}^{-1} - b_{vJ}^{-1} = 2\beta^{-1} \int_{-\frac{1}{2}}^{v} \frac{B(v',J)dv'}{[E(v,J)-E(v',J)]^{\frac{1}{2}}},\qquad(5)$$

where

$$B(v,J) = \frac{1}{2J+1}\frac{2E(v,J)}{\partial J}.\qquad(6)$$

The problem is that the integrals in (4) and (5) require knowledge of the vibrational dependence of $E(v',J)$ and $B(v',J)$ which is unavailable in the van der Waal's context. However an approximation to the harmonic vibrational frequency, $\omega_e(J)$, in the Jth effective potential is available from the perturbation formula.

$$\omega_e(J) = [4\,B^3(v,J)/D(v,J)]^{\frac{1}{2}},\qquad(7)$$

where $D(v,J)$ is the local centrifugal stretching constant

$$D(v,J) = -\frac{1}{2}\frac{\partial^2 E(v,J)}{\partial[J(J+1)]^2}.\qquad(8)$$

This means that estimates of the effective potential minimum and the corresponding equilibrium rotational constant are available in the form

$$V_e(J) = E(v,J) - (2v+1)\,[B^3(v,J)/D(v,J)]^{\frac{1}{2}}\qquad(9)$$

$$B_e(J) = \partial\,V_e(J)/\partial J(J+1) = B(v,J) + (v+\tfrac{1}{2})\alpha_e(J)\qquad(10)$$

$$\alpha_e(J) = 6[B(v,J)\,D(v,J)]^{1/2} - H(v,J)\,[B(v,J)/D(v,J)]^{3/2}\qquad(11)$$

This knowledge of $V_e(J)$, $\omega_e(J)$, $B_e(J)$ and $\alpha_e(J)$ is sufficient to allow analytical evaluation of the integrals (4) and (5) as[5]

$$f_{vJ} = b_{vJ} - a_{vJ} = 4\beta\,[(v+\tfrac{1}{2})/\omega_e(J)]^{\frac{1}{2}}\qquad(12)$$

$$g_{vJ} = a_{vJ}^{-1} - b_{vJ}^{-1} = (4/\beta)[(v+\tfrac{1}{2})/\omega_e(J)]^{\frac{1}{2}}[B(v,J) + \tfrac{1}{6}(2v+1)\alpha_e(J)].\qquad(13)$$

Solution of (12) and (13) for the turning points a_{vJ} and b_{vJ} on the effective curve

V(R;J) then in turn yields points on the true potential V(R) by means of the identity

$$
\begin{aligned}
V(a_{vJ}) &= V(a_{vJ};J) - J(J+1)\beta^2/a_{vJ}^2 \\
&= E(v.J) - J(J+1)\beta^2/a_{vJ}^2, \quad\quad (14)
\end{aligned}
$$

with a similar equation involving b_{vJ}; see Fig.1 for an illustration of equation (14).

The final steps in the inversion are to fit a suitable potential function through the points given by equation (14), to use the effective potentials derived from it to obtain improved estimates of $V_e(J)$, $\omega_e(J)$, $B_e(J)$ and $\alpha_e(J)$, to estimate anharmonicity terms $\omega_e x_e(J)$ and hence to obtain improved estimates of the integrals in (4) and (5). Details are discussed by Child and Nesbitt[5]. The quality of the resulting inverted potential naturally depends on the form of the chosen fitting function, but the dependence is found to be remarkably weak over the range of the inverted turning points. For example the points in Fig.2 were obtained by inversion of vibrational–rotational eigenvalues for $v = 0$ and $J = 0$–30 derived from a Lennard–Jones potential and the lower panel shows the residuals after refinement with (a) a Lennard–Jones and (b) a Morse fitting function, while (c) shows these before refinement. The predictive powers of the two fits are however quite different outside the turning point range, with the Lennard–Jones dissociation energy, $D_e(LJ) = 114\ 24\text{cm}^{-1}$, being much closer to the input value, $D_e = 117.21\text{cm}^{-1}$ than the Morse prediction $D_e(M) = 81.77\text{cm}^{-1}$.

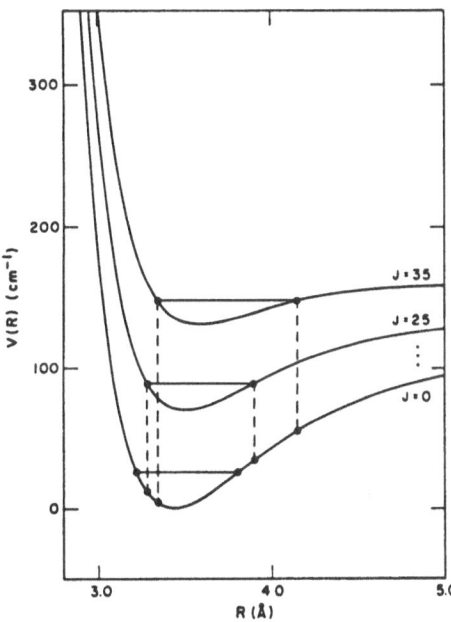

Fig.1 Schematic J dependent potentials showing how points are transferred from V(R;J) to V(R;0) (Taken from Child and Nesbitt[5] with permission).

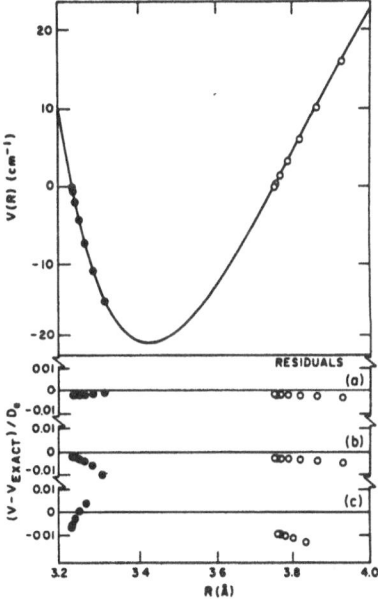

Fig.2 Test of the rotational RKR method. Solid and open dots refer to inner and outer turning points for J=0–30 in steps of 5. Residuals are given for (a) Lennard–Jones (b) Morse (c) no iterative refinement. (Taken from ref.[5] with permission).

INVERSION FOR THE Ar + HF(v=1) POTENTIAL SURFACE[6]

The above rotational RKR procedure requires a minor modification in the triatomic A..H–X van de Waals context, because part of the rotational constant comes from the parent HX molecule. Hence the proper effective potential for van der Waals stretching is

$$V(R,J) = V(R) + \beta \frac{[J(J+1)-\ell^2]}{R^2 + \rho^2} \qquad (15)$$

where

$$\rho^2 = <mr^2 \cos^2\theta>/\mu, \qquad (16)$$

r being the HX bond length and θ the angle between r and the vector R between A and the HX centre of mass. When followed through this yields[6]

$$g_{vJ} = \rho^{-1} \arctan\left\{ \frac{\rho[b_{vJ} - a_{vJ}]}{a_{vJ} b_{vJ} + \rho^2} \right\} = 2\beta^{-1} \int_{-\frac{1}{2}}^{V} \frac{B(v',J)\,dv'}{[E(v,J) - E(v',J)]^{\frac{1}{2}}} \qquad (17)$$

in place of equation (17).

Nesbitt, Child and Clary[6] applied the theory, with this modification, to the (10⁰0), (12⁰0) and (110) bands of Ar HF, using the ℓ type doubling of the (110) bands to depertub the (12⁰0) band, and thereby obtained the diatomic like potential curves in Fig.3. An adiabatic inversion procedure was then adopted, in the sense that the three potential curves were interpreted as bending vibrational eigenvalues (parametrically dependent on R) of a three term potential surface,

$$V(R,\theta) = V_0(R) + V_1(R)\,P_1(\cos\theta) + V_2(R)\,P_2(\cos\theta) \qquad (18)$$

Although the RKR turning points span only the range $3.1 < R < 4.0$ Å; the potential may be reliably extended outside this region because the term $V_0(R)$ must extrapolate to the known dissociation energy[7], $D_0 = 112$ cm⁻¹, while $V_1(R)$ and $V_2(R)$ must extrapolate smoothly to zero. Nesbitt, Child and Clary[6] give tabulated values of the $V_1(R)$ over the RKR region, together with appropriate extrapolation functions outside this range, and the remarkable success of the inversion is evidenced by the entries in table 1 in which calculated values were obtained by a converged variational calculation[8]. Not only are the data for the three input bands recovered within fractions of a percent, the spectroscopic parameters of four other bands are also predicted with remarkable accuracy.

While this success is extremely encouraging it must be noted that the near adiabatic behaviour of the Ar–HF system makes it a particularly favourable one. Detailed analysis[6] shows that the composition of the bending state wavefunctions is very insensitive to the lengths of the van der Waals bond in the observed region around 3.0–4.0Å; the two lowest states in Fig.3 are roughly equal mixtures of J=0 and J=1 asymptotic σ states, while the π bend is almost pure J=1. The breakdown of this adiabatic separation in other systems would be signalled by strong perturbations, which would have to be depertubed before or during any application of the present inversion procedure.

LONG RANGE INVERSION FOR WEAKLY BOUND COMPLEXES

Situations can exist where the binding is so weak that only a few rotational levels of the v = 0 vibrational state are bound, many He–X complexes might be cases in point. The question addressed below is to what extent knowledge of the long range functional form can be exploited to recover the potential function. It is assumed for test purposes (as in section 2) that one is concerned with a diatomic system.

Table 1. Experimental and calculated rovibrational data for Ar + HF(v=1)

State	γ_0/cm^{-1}			B/cm^{-1}		
	Exp.	Calc.	Error(%)	Exp.	Calc.	Error(%)
$10^0 0^a$	–	–	–	0.102610	0.102970	+0.35
$12^0 0^a$	57.3347	57.2925	–0.074	0.100121	0.100522	0.40
$11^{1f} 0^a$	70.3366	70.3130	–0.034	0.100325	0.100395	0.07
$11^{1e} 0$	70.3403	70.3143	–0.037	0.102609	0.102553	–0.06
$10^0 1$	41.3349	41.3455	0.03	0.093504	0.094179	1.3
$10^0 2$	71.6109	72.7959	1.7	0.082509	0.083353	1.0
$11^{1f} 1$	101.8226	102.1316	0.30	0.089666	0.089338	–0.37
$11^{1e} 1$	101.8266	102.1335	0.30	0.09218	0.09137	–0.88

ᵃ Vibrational bands used in determining the potential.

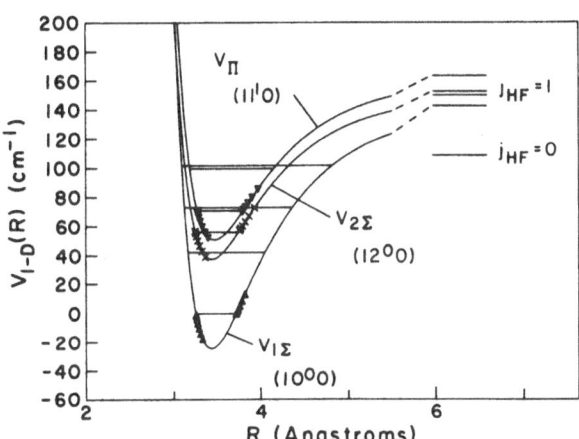

Fig.3   Effective 1D potentials obtained by rotational RKR analysis of the ($10^0 0$), ($12^0 0$) and ($11^{1f} 0$) data of ARHF. The points show the RKR turning points (taken from Nesbitt, Child and Clary[6], with permission).

The starting equation in this case is the following semiclassical estimate for the local rotational constant,

$$B(v,J) = [\partial E(v,J)/\partial J(J+1)]$$

$$= \beta^2 \int_{a_{vJ}}^{b_{vJ}} \frac{dR}{R^2[E_{vJ}-V(R;J)]^{\frac{1}{2}}} \bigg/ \int_{a_{vJ}}^{b_{vJ}} \frac{dR}{[E_{vJ}-V(R;J)]^{\frac{1}{2}}}, \qquad (19)$$

which may be obtained by differentiating equation (1) with respect to $J(J+1)$. The same equation is used by Le Roy[4], who approximates these integrals in a way that relates the vibrational dependence of $B(v,o)$ to the dissociation limit and the inverse power of the long range potential. In the present case however the vibrational dependence of $B(v,J)$ is irrelevant because only $B(o,J)$ is assumedly available; moreover if the binding is very weak one cannot neglect inner turning point contributions to the integrals in (19).

Nevertheless a suitable functional form may be assumed, say

$$V(R) = D + C_m/R^m - C_n/R^n, \tag{20}$$

and the parameters optimised by least squares comparison with the experimental $B(v,J)$ values, subject to the constraint that $v = 0$ when calculated by equation (1).

A convenient test system is provided by the recently observed $2p\sigma_u$ state of the D_2^+ ion[9], which has a binding energy of about 13cm^{-1}, sufficient to support five rotational levels of $v = 0$ and two level of $v = 1$. The test is carried out within the Born–Oppenheimer approximation, although small non Born–Oppenheimer corrections of the order of 0.002 cm^{-1} are required for comparison with experiment[9]. The potential curve is given by Peek[10], and the calculated energy levels, taken from table 6(a) of Carrington et al[9], are listed in column 2 of table 2. The energy zero is taken at the dissociation limit and the local rotational constants $B(v,J)$ were determined by a spline fit to the energies.

The proper long–range potential is of the form $D - C_4/R^4$ due to the charge induced dipole interaction; hence $n = 4$ in equation (20). Of the other parameters, m was held fixed and D and C_n were taken as variable with C_m constrained such that $v = 0$ at $E(0,0)$. As one might expect, both on the basis of the long–range theory[2-4] and in view of the paucity of the data the five $B(0,J)$ values were insufficient to determine the power of the short range repulsion because fits well within the precision of the data were available for any value of m between 7.5 and 12.0. However, as seen in Fig. 4, the shapes of the resulting curves were remarkably similar. Moreover the level of agreement with the Born–Oppenheimer potential points [10] is very encouraging bearing in mind the nature of the input data, coupled with the observation, from table 2, that the the outer turning points on the effective potentials $V(R{:}J)$ span a range of only about 1.2Å. In particular that the two curves in Fig.4 predict dissociation limits of 0.19 cm^{-1} for $m = 8.17$ and -0.15 cm^{-1} for $m = 10.0$ compared with the exact value of zero.

Given the above ambiguity in m, it is clearly desirable to find external data to resolve the uncertainty. In the present test case, discrimination was achieved by requiring that equation (1) should yield $v = 1$ at the energy $E(1,0) = -0.308$ cm^{-1}, given in table 2. This yields the potential parameters $m = 8.17$, $n = 4$, $C_m = 6.03 \times 10^7$cm^{-1}Å$^{8.17}$, $C_n = 4.69 \times 10^4$cm^{-1}Å^4, $D = 0.19$cm^{-1}, which define the solid curve in Fig.4. An attractive alternative disciminating test in other systems might based on the lifetimes of any $v = 0$ rotational states that predissociate by tunneling.

In conclusion it should be noted that, although the dissociation limit is determined to within ± 0.2cm^{-1} by this procedure, the long range attraction coefficient is much less certain because the true Born–Oppenheimer potential contains a significant C_6 term which could not be separately determined within the precision of the available data. Thus the two curves in Fig.4 give $C_4 = 4.69 \times 10^4$ cm^{-1}Å^4 for $n = 8.17$ and $C_4 = 3.95 \times 10^4$ cm^{-1}Å^4 for $n = 10.0$, compared with an input value of 3.889 cm^{-1}Å^4.

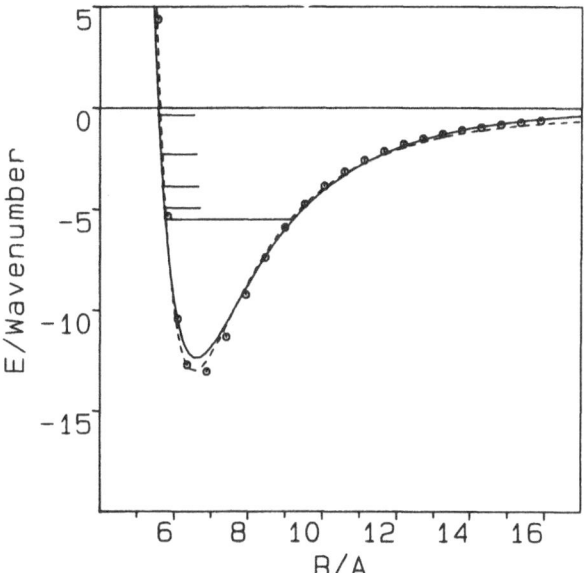

Fig.4   Inverted potentials derived from the Born–Oppenheimer $E(0,J)$ levels of D_2^+, which are marked by horizontals. Points lie on the true potential[10]; solid and dashed curves were obtained by inversion with m=8.17 and m=10 respectively.

Table 2. Input and inverted data for D_2^+ $2p\sigma_u$ Born–Oppenheimer states

v	J	E(v;J)[a] /cm⁻¹	B(v,J)[b] input /cm⁻¹	B(v,J) calc /cm⁻¹	a(v,J) /Å	b(v,J) /Å
0	0	−5.510	0.283	0.283	5.761	9.218
	1	−4.949	0.278	0.279	5.781	9.302
	2	−3.853	0.270	0.269	5.823	9.488
	3	−2.280	0.254	0.253	5.894	9.832
	4	−0.356	0.225	0.225	6.008	10.499
1	0	−0.308				
	1	−0.147				

[a] From Carrington et al[9], measured from the dissociation limit
[b] Derived from a spline fit to E(v,J).

REFERENCES

1.   J.Tellinghuisen, Computer Phys.Commun. 6, 221 (1974)

2.   R.J. Le Roy in "Semiclassical Methods" ed. M.S.Child (Reidel, NATO ASI, 1980).

3.   R.J. Le Roy and R.B. Bernstein, J.Chem.Phys., 52, 3869 (1970).

4.   R.J. Le Roy, Can.J.Phys., 50, 953 (1972)

5.   M.S. Child and D.J. Nesbitt, Chem.Phys.Lett., 149, 404 (1988)

6.   D.J. Nesbitt, M.S.Child and D.C. Clary,J.Chem.Phys., 90, 4855 (1989)

7.   G.T. Fraser and A.S. Pine, J.Chem.Phys., 85, 2502 (1986)

8.   D.C. Clary and D.J. Nesbitt, J.Chem.Phys., 90, 7000 (1989)

9.   A. Carrington, I.R. McNab, C.A. Montgomerie and R.A. Kennedy, Mol.Phys., 67, 711 (1989)

10. J.M. Peek, J.Chem.Phys., 50, 4595 (1989)

EXPERIMENTS, SMALL MOLECULES, UNPAIRED ELECTRONS

Discussion Leader: David J. Nesbitt

Joint Institute for Laboratory Astrophysics, National Institute
of Standards and Technology and Department of Chemistry,
University of Colorado, Boulder, CO 80309-0440

In this session there was representation from three different groups
making contributions to our understanding of van der Waals interactions in
open shell systems. First, Benoit Soep discussed metal atom bimolecular
cluster spectroscopy in Hg and Ca atom studies. The clusters are probed
via LIF excitation of the complex using the bare metal atom electronic
transition as the chromophore. Intermolecular energy transfer or chemical reaction in the complexes is then probed by total fluorescence and/or
state resolved detection of products. The intermolecular potentials for
these open shell systems, particularly between open shell atom+atom and
atom+diatom systems, can be directly probed by crossed beam scattering
methods. Results from this experimental approach were presented by
Piergorgio Cassevechia for a number of systems such as open shell halogen
atoms, $O(^3P)$, and NO. A third perspective on recent developments in this
field was provided by Marsha Lester with work in her group on complexes
of OH radical with Ar atoms. These studies elucidate the dynamics of vibrational predissociation in the weakly bound, Ar + $^2\Pi$ OH, ground and the
relatively strongly bound, Ar + $^2\Sigma$ OH, electronically excited state observed via laser induced florescence. A full description of these talks
is provided in the following chapters. However, in the interest of providing as complete as possible an overview of the field, a short summary
of the questions and answers that arose during the discussion periods
between the talks is reproduced below.

Benoit Soep

The CaHCℓ studies make a very nice complementary investigation to
the Rettner et al. work on Ca + HCℓ crossed beams. With regard to the
spectroscopic studies of the CaHCℓ complex, it was also noted that interesting comparisons should be possible with orientation/alignment
studies of intermultiplet electronic energy transfer by Leone and coworkers. Conference co-chairman Janda noted that certain combinations
of rare gas halogen complexes had been unobserved in his laboratories
despite several attempts. He raised the possibility that efficient electronic quenching in these complexes represents an explanation of this
lack of observation. Also noted were the studies by Lester and coworkers
on ion pair electronic states in rare gas interhalogen complexes that
yielded interesting dynamical information on high lying electronic
levels. The importance in probing final state channels in the photo-

Dynamics of Polyatomic Van der Waals Complexes
Edited by N. Halberstadt and K. C. Janda
Plenum Press, New York, 1990

dissociation of HgN_2 was raised in light of the efficient E-V quenching of Hg by CO molecules in both gas phase and matrix phase studies.

Piergorgio Cassevechia

The question was raised as to whether the atom-atom bound region of the potentials between open shell species could be more easily obtained by spectroscopic methods. The speaker agreed that if such open shell complexes could be formed in supersonic jets, that there was no reason, in principle, why such an LIF experiment could not be performed. Cassevechia cited the LIF studies on XeF by Smalley and coworkers as one such example. It was asked why the XeF ground state potentials were much deeper than other F-rare gas or F-O atom potentials. The response was that in the XeF system the greater influence of charge transfer effects were undoubtedly present, and contributed to a deeper well. It was asked whether the scattering experiments were primarily or especially sensitive to the repulsive wall. Cassavechia answered that although the diffraction oscillations are most sensitive to the repulsive wall, the rainbow and glory oscillations are a measure of well depth and R_m, respectively. Absolute cross section data, furthermore, also probe the magnitude of the potential at asymptotic separations. Finally it was noted that intermolecular potential data on $Si(^3P)$ + Ar was now obtained by Jouvet and coworkers and presented as one of the posters.

Marsha Lester

It was asked whether a more precise value of vibrational predissociation could be provided in the A electronic state of ArOH complexes. Lester responded that there is a clear trend in the spectral line broadening with <u>decrease</u> in van der Waals excitation. However, saturation effects could be playing some role, and therefore careful power broadening studies were necessary and presently under way to determine these lifetimes more quantitatively. The issue of isotopic dependence of the ArOH and ArOD vibrational relaxation in the A state was raised, based on studies performed in a Ne matrix which showed a two-fold difference (OH being faster) in the rates. Lester responded that the ArOD vibrational relaxation rates had not been measured, but commented that OH and OD are nearly free rotors in a Ne matrix, and that the longer relaxation time for OD was a consequence of a smaller rotational B constant and a V-R mechanism. The corresponding Ar matrix studies indicated relaxation rates too fast to measure for both isotopic species. Others commented that the intermolecular potential for electronically excited ArOH is similar in binding energy and degree of anisotropy to HF dimer, for which vibrational predissociation rates are known to decrease with isotopic substitution, contrary to "energy gap" predictions. It was mentioned as surprising that ArOH could be efficiently "synthesized" in the supersonic expansion since the Ar-OH binding energy was somewhat smaller than the Ar-Ar binding energy. Lester noted that there may be some important differences between a UV photolysis heated expansion and a normal 300 K pulsed expansion.

Finally, it was noted that the upper state of ArOH might have a similar intermolecular potential to the isoelectronic species ArF. A detailed response to that question was provided some weeks later to the discussion leader by Lester for inclusion in this manuscript as follows:

Our previous speaker, P. Casavecchia, just discussed the interaction potential for the open-shell $F(^2P)$ with $Ar(^1S)$ in the ground state, which was determined from crossed molecular beam scattering experiments. The ground state of ArF has an equilibrium internuclear distance of 3.5 Å and a well depth of 6.8 meV (55 cm^{-1}) [V. Aquilanti, E. Luzzati, F. Priani, and G. G. Volpi, J. Chem. Phys. _89_, 6165 (1988). The excited electronic state of ArF has a substantially reduced bond length (2.4 Å) and increased well depth of 5.5 eV (44500 cm^{-1}) [T. H. Dunning and P. J. Hay, J. Chem. Phys. _69_, 134 (1978)], which gives rise to the well known ArF excimer laser transition at 193 nm. The ArF ground state potential is qualitatively similar to the one-dimensional potential between OH ($X^2\Pi$) (center-of-mass) and Ar. The excited state potentials, however, are very different: ArF is ionic in nature and very strongly bound while the weaker bonding in OH-Ar is mainly "covalent" due to a correlation effect [A. D. Esposti and H.-J. Werner, to be published].

VAN DER WAALS MOLECULES AS PROBES FOR COLLISION PROCESSES

A. Keller and J.P. Visticot
Service des Atomes et Surfaces, CEN Saclay
91191 Gif sur Yvette, Cedex, France

S. Tsuchiya
University of Tokyo, Komba Meguro Ku,
Tokyo 153, Japan

T.S. Zwier
University of Perdue, West Lafayette,
Indiana 47907, U.S.A.

M.C. Duval, C. Jouvet, B. Soep, and C. Whitham
Laboratoire de Photophysique Moleculaire
du C.N.R.S., Bat. 213, Universite Paris-Sud,
91405, Orsay, Cedex, France

Van der Waals molecules offer a unique means to explore the intermolecular potential between interacting atoms or molecules, as it has been amply shown through this conference.

The detailed knowledge of these potentials is of great interest to collision induced processes : vibrational relaxation, electronic relaxation and chemical reactions. As an example the study of the structure and dynamics of complexes undergoing vibrational predissociation has brought into light some aspects of vibrational relaxation hidden in full collision processes[1,2]. In the same way the dissociation of a complex inducing an electronic relaxation or a chemical reaction will be the half collision analog of the related process. This relies

Dynamics of Polyatomic Van der Waals Complexes
Edited by N. Halberstadt and K. C. Janda
Plenum Press, New York, 1990

upon the fact that the <u>excited state complex (AB)* corresponds to the local excitation of A</u> within AB : (A* - B). We therefore prepare A* within the reach of B in the intermolecular field of the complex.

Thus, exploring the (A*B) potential by means of an optical excitation we can deduce the specific states, configurations and intermolecular modes that will induce the studied collision process. It should then be mentioned that there is a correspondance between the specific excitation of A*B levels and the relative movements of A* approaching B.

A* - B overall rotation is equivalent to A,B orbital motion, the excitation of this rotation will give insight into the collision impact parameter as in full collisions (Ba + HI [3]).

The A* - B stretching motion imparts translational kinetic energy along the AB coordinate in the collision.

Librational motion of B with respect to A is related to the relative movement (rotation) of B during the collisional approach.

Also specific electronic states of the complex will be excited when A* is an excited atom, as the molecular field lifts the orbital degeneracy of the atomic levels, P, for instance. In this case, within the A*B field specific Σ or Π states will be reached by the optical excitation.

The specificity which can be attained in the electronic excitation of complexes is related to orbital oriented collisions as observed by Leone and coworkers[4] in the electronic relaxation of Calcium (5p) and the reaction of Calcium (4p) with HCl by Rettner and Zare [5].

We shall report here two examples :
- The effect of the intermolecular modes and of the electronic symmetry upon the Hg-molecule intramultiplet relaxation through the process:

(Hg - Molecule) $^3P_1 \rightarrow$ Hg $^3P_0$ + Molecule

As this relaxation is rather inefficient (slow) it will be utterly dependent upon the preparation conditions. Another example of such electronic relaxation in the ICl* complex is provided by Stephenson et al[6].

- The observation of a chemical reaction induced within the (Ca*HCl) complex.

In these experiments we can distinguish the reactive surfaces corresponding to the different Ca (4p) orbital orientations and their different reactivities. We have also characterised some of the molecular movements prior to the reaction and during the reaction yielding CaCl*.

MERCURY $^3P_1$ INTRAMULTIPLET RELAXATION

The relaxation of ^3P mercury has amply been reviewed in collisions [7]. The efficiency of a collision in relaxing the initial state to the lower $^3P_0$ multiplet varies drastically with the collision partner. In the case of a monoatomic collider the efficiency is essentially 0 from $^3P_1$ to $^3P_0$ and small ($\sigma <$ 1Å^2) for $^3P_2$ to $^3P_0$, while the cross section jumps by two orders of magnitude going from atomic to molecular colliders.

The origin of this increase in efficiency has to be sought in the additional degrees of freedom provided by the molecular collider to mercury, in <u>the collision complex</u>. This is precisely the point, where exploring the dissociation of the (Hg-Molecule) $^3P_1$ complex can provide detailed information upon the effective intermolecular movements. We have studied various molecular encounters but the most significant results have been obtained for the Hg-N$_2$ and Hg-NH$_3$ complexes.

A) Hg-N$_2$

Let us summarize the results. The optical spectrum of Hg-N$_2$ close to the $^3P_1$ mercury line displays intense structure assigned to the complex inter molecular modes e.g. stretching and torsion (bend). The stretching potential (radial) has been determined by the mercury isotope effect within the rotational contours of the bands [8], while the bending potential has been determined by comparison of calculations in the hindered rotor model [9] and the observed transitions. The same model will be later discussed in this paper. We have to note that the intense progression in stretching is due to a reduction of C.a. 1 Å in the excited state equilibrium distance. On the other hand, the bending progressions arise from a geometry change from a ground state T shaped to a linear excited complex. This latter situation is of unique value to explore the bending potential.

In order to observe the relative efficiency of the electronic predissociation, we compare fluorescence excitation spectra which probe the population of the non dissociated complex and action spectra where instead, the resulting $^3P_0$ state is monitored while the laser is scanned. The latter spectra probe the efficiency of the predissociation.

Inspection of the spectra in fig.1 shows that stretching mode excitation is not very efficient in inducing the $^3P_1$ - $^3P_0$ transition but bending mode with l = 1 about the Hg-N-N axis have an efficiency close to unity as shown in the shaded areas of the figure 1a.

The measurement of the absolute value of the rate k ($^3P_1 \rightarrow \, ^3P_0$) is interesting and experimentally accessible through time decay measurements of selected vibronic levels. We can assume that, as in Hg Ar complexes the radiative lifetime of the complex is close to that of Hg $^3P_1$.
Thus I/т observed = I/т ($^3P_1$) + k ($^3P_1$ - $^3P_0$).

However, this decay is J dependent, the total angular momentum increasing the predissociation rate, thus we would have to compare rotationally averaged values of the rate. The J dependence is not only seen in decay rates but also in the inspection of the action spectrum in figure 1 which shows much broader bands that the fluorescence excitation spectra, owing to the high J lines in the action spectrum.

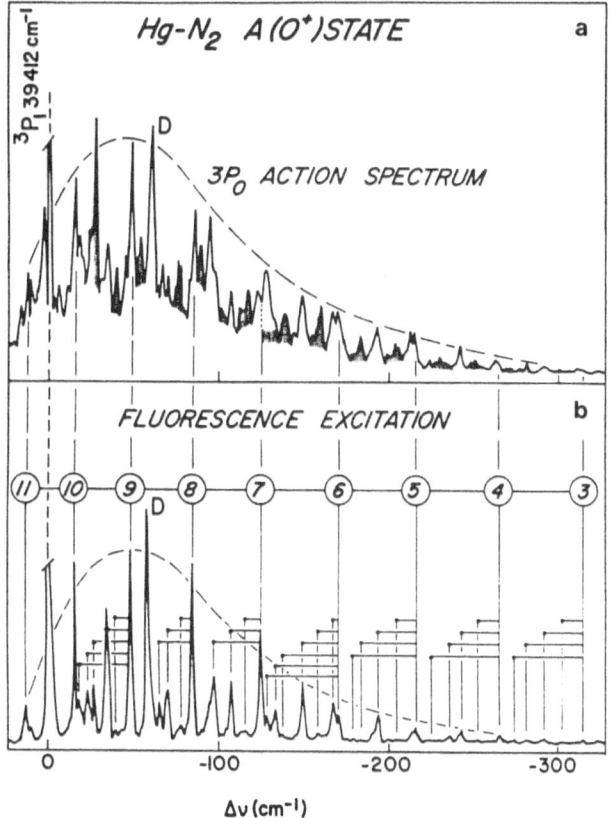

Figure 1. Hg-N_2 complex : a) Action spectrum probing the
Hg 6^3P_0 7^3S_1 transition at 4047 Å
b) Fluorescence excitation

These results i.e. the great influence on the dissociation of the pseudo rotation angular momentum l about the Hg -
N-N axis has also been demonstrated in theoretical calculations
by Jouvet and Beswick [10]. The states of the excited Hg N_2
complex are represented for the $^3P_1$ perturbed mercury by $| J$
$(= L + S)$, $\Omega >$ as $|1, 0^+ >$ and $|1, 1 >$ and as $|0,0^- >$ for the $^3P_0$
state. There the coupling between the 0^+ state ($^3P_1$) and 0^-
($^3P_0$) is essentially 0 even through higher mercury states owing
to the +, - parity difference. On the other hand, this
interdiction is lifted when l = 1 vibrations become excited as
we have observed. Nesbitt et al[11] observed the influence of
the +, - parity upon the coupling of $\Pi^{+,-}$ and Σ^+ vibrations in
the NeHF predissociation, with similar conclusions.

Jouvet and Beswick also predicted in their model that overall rotation should increase the dissociation rate as we observed and they have shown that |1,1 > and |0, 0⁻> are indirectly coupled while |1,0⁺ > and |0,0⁻ > are not.

We have been able to observe this influence of the excited electronic symmetry in the Hg NH₃ case[12].

B) Hg NH₃

In fig. 2, we have represented in the top view the action spectrum to $^3P_0$ and in the lower the fluorescence induction spectrum. The lower spectrum (LIF) reveals mainly a progression of stretching bands weakly appearing in the upper spectrum and assigned to an electronic state A. On the other hand the upper spectrum reveals another electronic state B and also transitions due to combination bands constructed on the A spectrum.

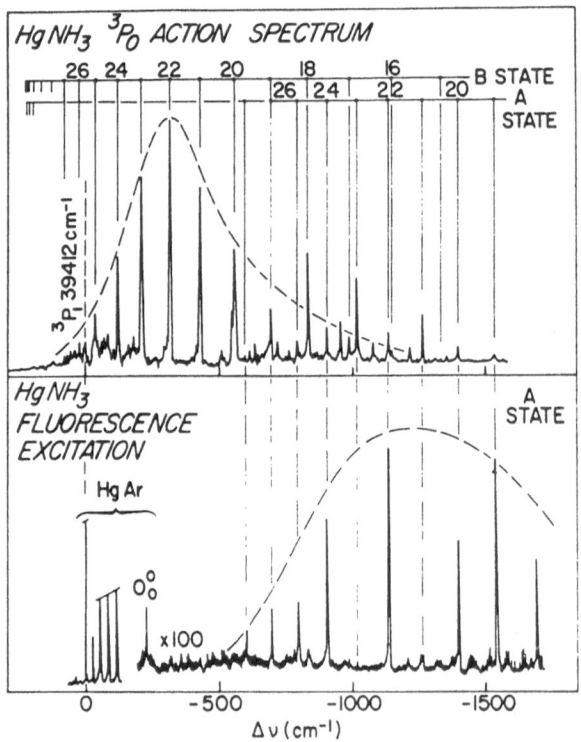

Figure 2.   Hg NH₃ complex formed in argon expansion
a) action spectrum,   b) Fluorescence excitation

Rotational contours spectra in fig.3 are of great help in assigning the A bands to a parallel transition [3a] and the B bands to a perpendicular transition [3c]. In the quasi diatomic limit the A-X transition correlates in Hg NH$_3$ to a parallel transition $0^+ - 0^+$ while the B-X transition to a perpendicular one, $1 - 0^+$.

Together the bands constructed on the A spectrum have been assigned to the excitation of the degenerate bending vibration Hg-NH$_3$ with $l = 1$ as for Hg N$_2$. This is clearly seen through the perpendicular structure of the related rotational transitions in fig. 3b. We observe by comparison of the LIF and action spectral that the excitation of this vibrational mode is highly efficient in inducing the $^3P_1 - ^3P_0$ transition. Together we observe that the A state, $|1, 0^+ >$ is prominent in fluorescence excitation and the B state $|1, 1 >$ in the action spectrum. This influence of the electronic symmetry either 1 or 0^+ upon the relaxation to $^3P_0$ is in good agreement with the aforementioned predictions[10].

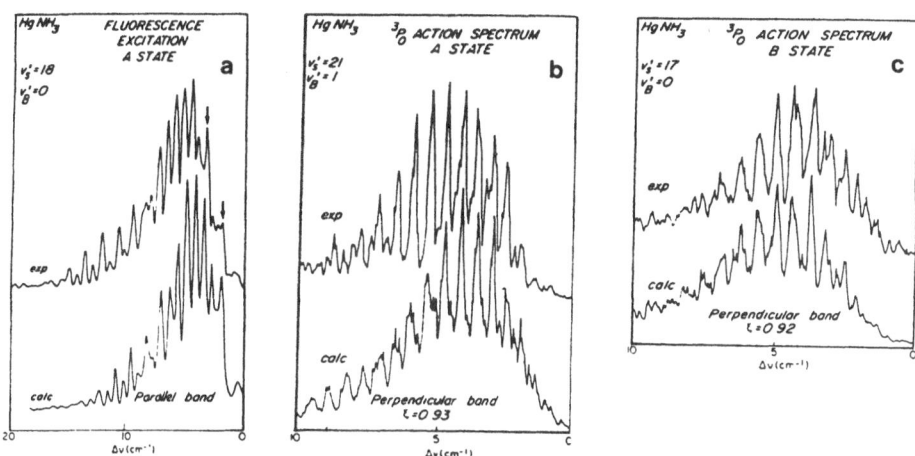

Figure 3. Hg NH$_3$ - Rotational contours observed with 0.1 cm^{-1} resolution, in each insert the top view is the experimental spectrum, the bottom represents the simulated spectrum.
- a) Fluorescence escitation, A state, parallel band
- b) $^3P_0$ action, A state, bending transition perpendicular band
- c) $^3P_0$ action, B state, perpendicular band.

Thus far we have shown that the effect of inter-molecular movements, radial or angular is important in fine structure transitions induced in collisions of atoms with molecules and can be studied with great details through the dissociation of the relevant complex. We have also observed that the electronic symmetry of the complex with the same asymptotic J can influence the transition. This effect is also observed in reactive collisions of excited atoms with molecules.

REACTION INDUCED WITHIN THE Ca-HCl EXCITED COMPLEX

The reaction of excited calcium atoms ^1P, ^1D, ^3P, ^3D with HCl has been widely studied in beam gas experiments (13-16), and Rettner et al(16) have shown that the Ca ^1P + HCl reaction is sensitive to the orientation of the calcium orbital with respect to the relative velocity in the collision. This reaction produces Ca Cl (A,B) with a high cross section, 60 $\overset{\circ}{A}{}^2$ (16) and the resulting chemiluminescence of Ca Cl is easily monitored.

We have studied the Ca + HCl reaction as produced within a complex in the same way as for electronic relaxation. We form the Ca-HCl complex in the ground state and excite it close to a transition of the calcium atom. We therefore reach an excited surface which will lead to the reaction into CaCl (A,B) + H. This reaction is believed to proceed through a crossing of the attained covalent surface and the ionic Ca^+ + HCl^- one. Tuning with a laser across the absorption of the Ca HCl system allows us to explore the entrance valley of the reaction and part of the crossing seam, through the spectroscopic analysis of the spectra.

The complex is prepared from a supersonic expansion of laser evaporated calcium and a dilute (10^{-3}) mixture of HCl in Argon at a pressure of 2.5 bar. It is verified that the complex absorption depends upon the square of the backing pressure and linearly on the concentration. Thus the results we shall report concern a 1/1 Ca-HCl complex.

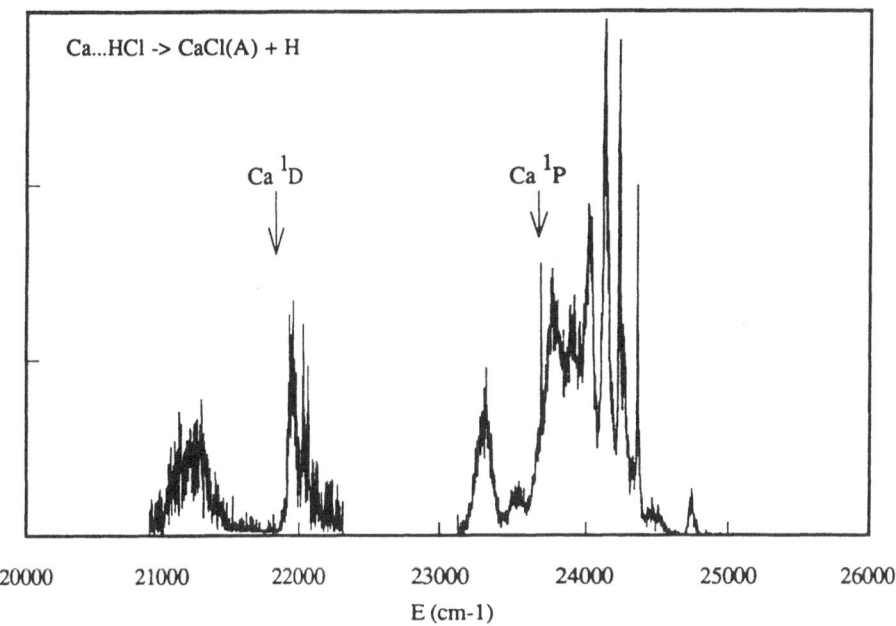

Figure 4 - Overview of the action spectrum monitoring the
A state emission from CaCl.

An overview of the spectra obtained for Ca-HCl is shown
in figure 4 where the CaCl chemiluminescence (from A or B) is
monitored as a function of the exciting laser frequency (action
spectra). We observe that the bands cluster around the ^1P and ^1D
Calcium lines. Wider regions of the optical spectrum including
the ^3D region were explored down to 5500 Å, where we expect from
the reaction energetics the chemiluminescence channel to be
closed as seen on fig.5. There resulted no signal for
frequencies below the last band observed on figure 4.

We therefore think that the absorption inducing the
reaction arises from the allowed Calcium transitions either
directly through $^1P_1$ or by state mixing in the Calcium for $^1D_2$
(It is known that $^1D_2$ Calcium atoms absorb in rare gas
matrixes[20]). However the ^1S - ^3D transition is spin forbidden
and not observed. Conversely the ionic state Ca^+ (^2S) + HCl^-
although responsible for the reaction should exhibit only a
small absorption intensity throughout the observed frequency
domain, as - In the one electron Charge Transfer, there should
be small overlap in the electron distribution between the

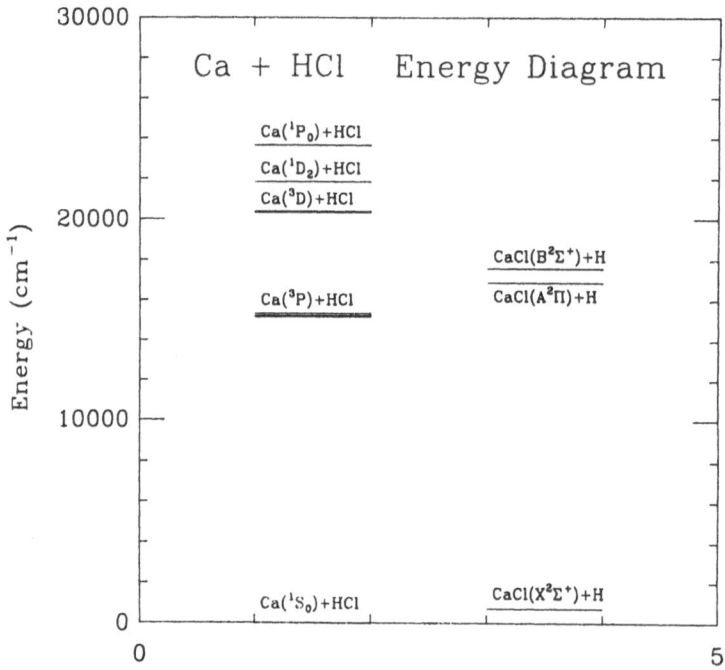

Figure 5 - Energetics of Ca and CaCl.

ground Ca HCl state, weakly ionic and the strongly ionic Ca^+-HCl^- state. The Coulomb attraction potential, known[16] to cross the covalent surface, should be almost vertical in the scanned frequency domain and therefore exhibit an extremely broad Franck Condon distribution yielding a continuous absorption. On the contrary, we see well structured transitions built on the Calcium transitions. We think that the state which is accessed by the optical absorption is a compound state α |covalent > + β |ionic> ($\alpha > \beta$). The coefficients α and β depend upon 1) the covalent transition accessed and 2) the Ca HCl distance in the same way as we proposed for the Hg $^3P_1$ + Cl_2 reaction[21].

BENDING MOTION OF HCl IN THE Ca - HCl COMPLEX

In the following we shall examine the action spectra along these views. In the spectrum represented in fig.6a two domains can be observed, one red shifted consisting in well spaced broad structures and another blue shifted consisting in

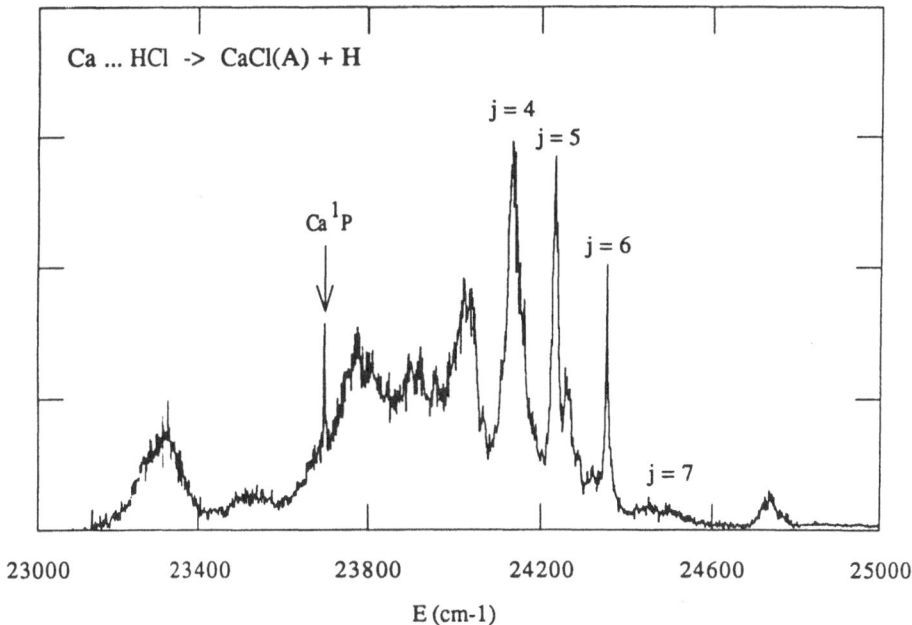

Figure 6a - Action spectrum of the Ca-HCl complex in the vicinity of the Ca ^{1}P-^{1}S transition, monitoring CaCl(A).

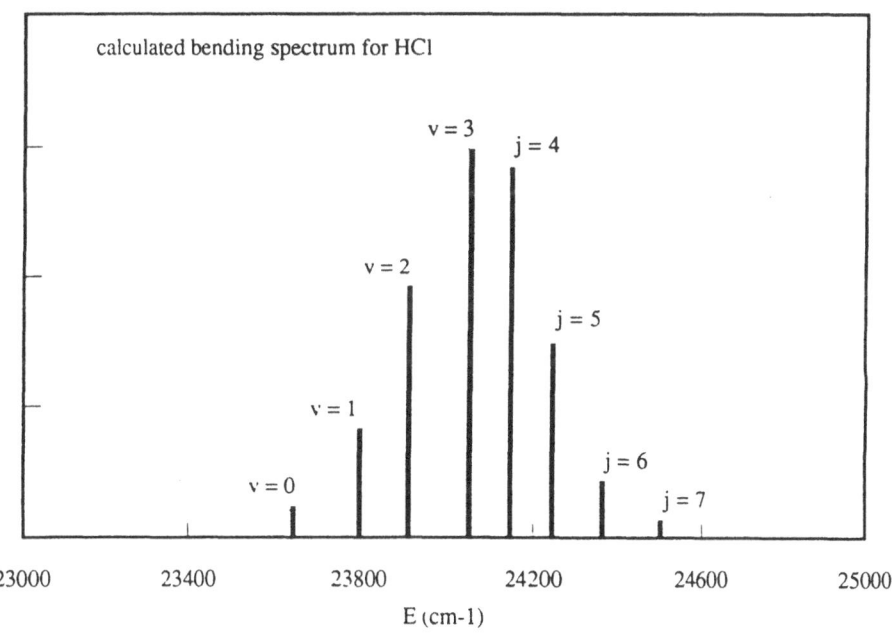

Figure 6b - Calculated bending spectrum for the Ca HCl complex with a linear ground state ($\omega_b = 70$ cm^{-1}) and a 90° T shaped excited state. The height of the bars will correspond to the area of the experimental bands.

113

narrow peaks. It is this last domain that we analyse in this section. The fact that narrow peaks are observed indicates that for these transitions the reaction is slow and permits us to model the excited state of the complex as a quasi bound state that vibrates a long time before reacting.

Various qualitative arguments tend to suggest that this part of the spectrum is characteristic of the bending motion of HCl in the complex:

- The order of magnitude of the frequency observed (200 cm^{-1}) corresponds to a van der Waals mode.

- The isotopic effect observed when replacing HCl by DCl proves that the hydrogen atom is predominantly involved in this vibration. This motion is not the van der Waals streching motion.

- More quantitatively, the relative position of the last four lines in the blue part of the spectrum correspond to the energy of the free rotation of HCl ($E_j = b_{HCl} j (j+1)$, for j = 4, 5, 6, 7 where b_{HCl} is the rotational constant of HCl molecule in its ground state). This is confirmed by the presence in the spectrum corresponding to Ca - DCl (fig.7a), of four lines corresponding to the rotational energy of the free DCl molecule ($E_j = b_{DCl} j(j + 1)$, for j = 5,6,7,8 where b_{DCl} is the rotational constant of the DCl molecule in its ground state). This suggests that these latter lines correspond to upper states with an energy greater than the potential barrier to rotation, while the other lines, being states embedded in the potential well, thereby correspond to bending motion of the HCl molecule in the Ca - HCl complex.

To make this hypothesis more quantitative, we describe the bending motion or quasi free rotation of the HCl molecule, in the same way as described in ref.8,9. We choose for the intermolecular coordinate the Jacobi coordinates (R, r, θ), where :

- r : internuclear distance between the hydrogen atom and the chlorine atom.

- R : distance between the calcium atom and the center of mass of HCl molecule.

- θ : angle between R and r.

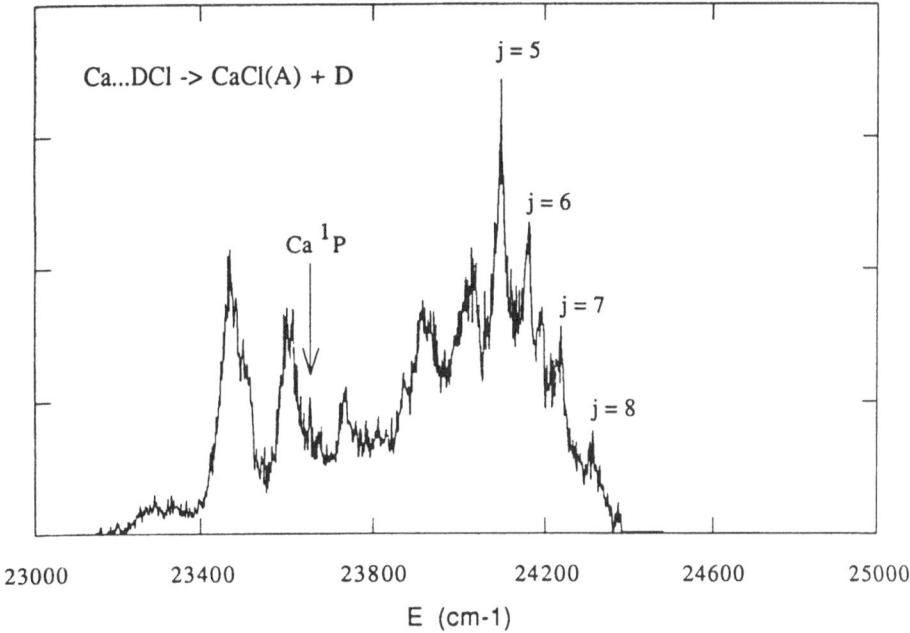

Figure 7a - Action spectrum of the Ca-DCl complex in the same
conditions as 6a.

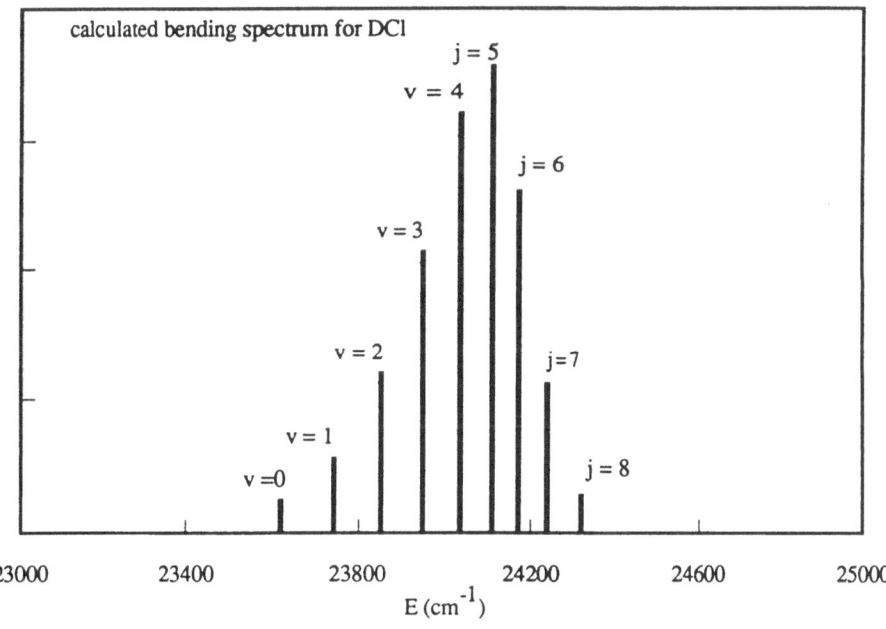

Figure 7b - Calculated bending spectrum for the Ca-DCl complex.

In this model we fix r at the value of the internuclear distance of HCl molecule in its ground state. R is fixed, but its value is irrelevant as we will se later (providing that R > 3 Å). θ is the variable coordinate describing the bending motion.

The hamiltonian for the bending motion of the HCl molecule is:

$$H = (b/h^2)\ j^2 + (B/h^2)\ 1^2 + V(\theta)$$

Where :
- j   : angular momentum of HCl molecule considered as a rigid rotator.
- 1   : angular momentum associated with R.
- V(θ) : bending potential
- b   : rotational constant of HCl in its ground state b = 10.59 cm^{-1}
- B   : rotational constant for the end-over-end of the complex B ~ 0.15 cm^{-1}.

we see that as B « b the term in 1^2 can be neglected (we neglect the centrifugal and coriolis force due to the overall rotation of the complex in space). Thus the hamiltonian reduces to :

$$H = (b/h2)\ j^2 + V(\theta)$$

To solve the eigenvalue problem, the wavefunction is expanded in a series of spherical harmonics Y_{j,m_j} (θ,φ). But now the hamiltonian commutes with J_z. Thus mj is a good quantum number. If we assume that in the ground state the complex has a linear equilibrium geometry (like Hg-HCl and Mg-HF [17,18]). Then only the state corresponding to mj = 0 will be populated, leading to the excitation of states with mj = 0 only. We thus choose Y_{jo} (θ,0) as basis functions, and write V(θ) as a linear combination of Legendre polynomials :

$$V(\theta) = \Sigma\ C_k\ P_k\ (\cos(\theta))$$

All the matrix elements of H between two spherical harmonics may be calculated analytically. Diagonalizing this matrix we obtain the energy levels and the wavefunction describing the bending motion of HCl in the potential V(θ).

For all these bending states to be excited from the ground electronic state, it is necessary that the equilibrium geometry of the complex be very different in its ground and excited electronic states. Assuming that the complex is linear in its ground state, the complex in its excited state must have a bent equilibrium geometry.

The potential is written as the following two poly-nominals :

$$V(\theta) = h_1 + c2 \; \theta^2 + c3 \; \theta^3 + c4 \; \theta^4 \qquad \text{for } \theta \text{ in } [0,\theta_0]$$
$$V(\theta) = -a2(\theta - \pi)^2 - a_3 \; (\theta - \pi)^3 - a4 \; (\theta - \pi)^4 \; \text{for } \theta \text{ in } [\theta_0,\pi]$$

with :

$$c_2 = k/2 - 6h_1/\theta_0{}^2 \qquad\qquad a_2 = -k/2$$
$$c_3 = 8h_1/\theta_0{}^3 - k/\theta_0 \qquad a_3 = k/(\pi - \theta_0) - 4h_2/(\pi - \theta_0)^3$$
$$c_4 = 3h_1/\theta_0{}^4 + k/\theta_0{}^2 \qquad a_4 = -k/2(\pi - \theta_0)^2 + 3h_2/(\pi - \theta_0)^4 .$$

In that way, we have a more or less flexible form for the potential depending on 4 physical parameters : h_1, h_2 determining the potential barriers at 0 and π, θ_0 the equilibrium angle and k the force constant which can be approximatively related to the bending frequency :

$$\omega \approx (2kb_{HC1})^{1/2}$$

We have adjusted these parameters to obtain a satis-factory agreement between the position of calculated energy levels and experimental lines.

We found $\theta_0 = \pi/2$, ω = 190 cm-1 (k = 1750 cm^{-1}), h1 = h2 = 610 cm^{-1}. The potential is shown in fig.8 and the calculated spectrum is compared with the experimental one in fig.6b. Moreover we calculated with the same bending potential, the energies corresponding to the bending motion of DC1 (fig.7b) and we found also a good agreement with the experimental spectrum. The fact that in this potential the barrier heights h_1 and h_2 are equal may seem surprising. We nevertheless obtained the best fit of the line positions and intensities with it.

To explain this behaviour we have to imagine that the HCl bender moves in an effective "adiabatic", potential which does not correspond to a cut at fixed R of V (R, θ). Instead, during the motion across the barrier the Ca-HCl bond will stretch resulting in a lowering of the barrier h_1.

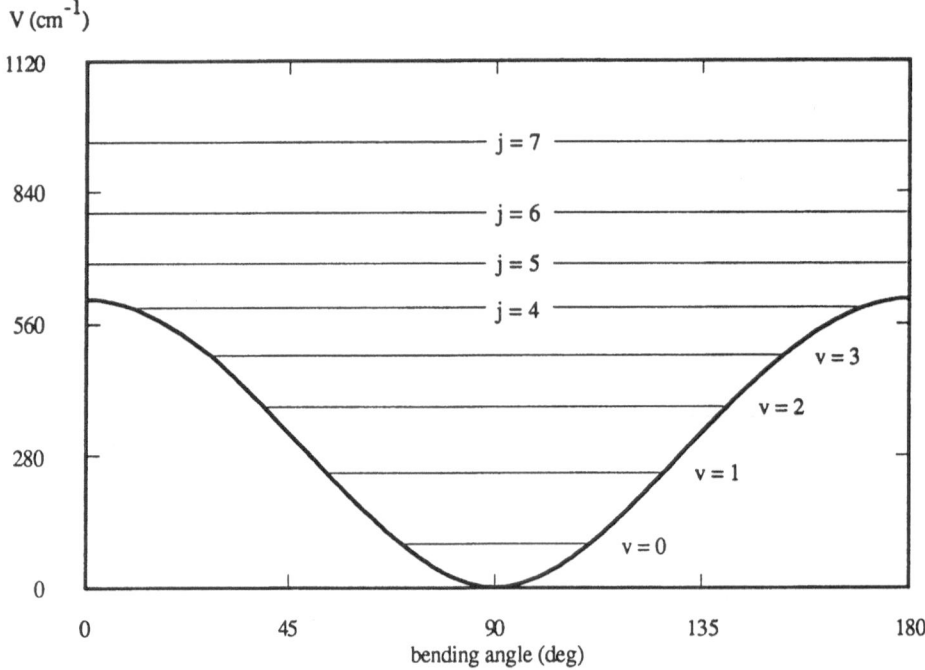

Figure 8 - Potential curve for the bending motion of the HCl molecule in the Ca-HCl complex. The energy levels labeled J = 4,5,6,7 correspond to the quasi free rotation of HCl

Thus most of the transitions in fig.6a should correspond for the blue domain of the spectrum to bending motion of HCl in the excited complex. Now if we assume that the wave function describing the bending motion of HCl molecule in the electronic ground state is approximatively a gaussian centered on θ = 0, we can adjust the width of the gaussian to reproduce the relative intensities of experimental transitions. Calculating the overlap between the gaussian function and the wave function corresponding to the four states above the barrier (which have experimental intensity that may be easy to measure) and adjust the width of the gaussian by a least square fitting, we obtain a width corresponding to a bending frequency in the ground state : ω = 70 cm^{-1}.

Moreover, we see in the calculated spectrum that the first bending state (v = 0) of DCl is 30 cm^{-1} red-shifted from the first bending state of Ca-HCl. Assuming that the three coordinates (R,r,θ) are separable and that isotopic exchange does not affect the frequency of the van der Waals stretch (R motion), we may deduce that the frequency of the HCl vibration (r motion) in the excited electronic state of the complex is only $\approx$ 200 cm^{-1} less than the corresponding frequency in the ground electronic state of the complex.

Thus, if we assume that the streching vibration of HCl ($\omega_{HCl} \approx 3000$ cm^{-1}) in the ground electronic state of the complex is only weakly perturbed by the presence of calcium atom then, this must also be true for the excited electronic state. This is consistent with the model, considering HCl as a rigid rotator hindered by the potential $V(\theta)$.

ORBITAL ORIENTATION AND REACTIVITY IN Ca-HCL

In the Ca DCl spectrum in figure 7a there appears conspicuously two domains, one of which having an origin at 23800 cm^{-1}. The intermolecular field should split up the Calcium ^1P state into 2 states Σ and Π in the linear configuration and 3 states 2A' + A" in Cs geometry. We therefore tentatively assign the bending vibrational progression to a linear to bent transition $\Sigma \rightarrow$ (A' + A"). On the same spectrum in fig.7a the other domain consisting of 3 well spaced broader bands should then correspond to a Σ - Σ transition. This transition should exhibit diffuse bands owing to a strong interaction between the Ca 4pσ orbital and the σ^* bond in HCl. This interaction enhances the chemical reaction and leading to a rapid decay of the excited complex into the products H + CaCl. The change in the width of the Ca HCl transitions is even more drastic and indicative of an important H/D isotope effect on the reactivity.

The observation of the local bending mode of the HCl moiety in the Σ - (A' + A") transition is indicative of the local excitation of the calcium atom in a 4pπ configuration, together with the observation of an imperceptible weakening of the HCl bond in the excited state. Conversely the important isotope shift in the $\Sigma - \Sigma$ origin, combined with broad transitions is indicative of a weakening of the HCl bond for this transition. The progressions we have observed (fig.6a,7a) remain unassigned although we feel that they should involve a resonance of the departing H from the Ca-Cl, in the terms described by Neumark et al.[19].

WELL DEPTHS OF THE Ca-HCl COMPLEXES

The potential well of all the states should be quite deep : the ground state of CaHCl must have greater depth than 1400 cm^{-1} (De (Ca$_2$)) as Ca$_2$ formation disappears where minute quantities of HCl are added. The extension of the observed spectra and their displacement from the Calcium line confirm that the radial coordinate Ca-HCl must provide a well of depth > 1400 cm^{-1} also in the excited states, correlated to $^1P_1$ or to $^1D_2$. Therefore state interactions should be important within the vibrational manifolds of each electronic state correlating to either ^1P or to $^1D_2$ Calcium.

CONCLUSION

We have shown several examples where the excitation of the complex yields direct and precise information on a collision process :
- influence of the intermolecular motion in Hg-N$_2$, Hg-NH$_3$ complexes on the electronic relaxation
- influence of the orbital symmetry on the reactivity of the CaHCl complex. This example is striking as intense vibrational structure could be observed describing the relative motion of the complex prior to reaction. In this instance the harpoon character of the reaction would have led to a structureless spectrum indicative of a direct dissociation. The excitation of the reactive complex is to yield direct information on the intimate couplings between the neutral and ionic surfaces.

REFERENCES

1)   - D.H. Levy, Adv. Chem. Phys. **47**, (1981) 323
2)   - K.C. Janda, Adv. Chem. Phys. **LX** (1985) 201
3)   - C. Noda, J.S. Mc Killop, M.A. Johnson, J. Waldeck
     and R.N. Zare, J. Chem. Phys. **85** (1986) 856
4)   - W. Bussert, D. Neushäfer and S.R. Leone
     J. Chem. Phys. **87** (1987) 3833
5)   - C.T. Rettner and R.N. Zare
     J. Chem. Phys. **77** (1982) 2416
6)   - T.A. Stephenson and M. Lester, in this volume.
7)   - A.B. Callear, Chem. Rev. **87** (1987) 335
8)   - K. Yamanouchi, S. Isogai, S. Tschuchiya, M.C. Duval,
     C. Jouvet, O. Benoist d'Azy and B. Soep,
     J. Chem. Phys. **89** (1988) 2975
9a) - G.E. Ewing, Acc. Chem. Res. **8** (1975) 185
  b) - S.L. Holmgren, M. Waldman and W. Klemperer
     J. Chem. Phys. **67** (1977) 4414
10) - C. Jouvet and A. Beswick
     J. Chem. Phys. **86** (1987) 5500
11) - D.C. Clary, C.M. Lovejoy, S.V. O'Neil and D.J. Nesbitt
     Phys. Rev. lett. **61** (1988) 1576
12) - M.C. Duval, B. Soep, R.D. Von Zee, W.B. Bosma
     and T.S. Zwier, J. Chem. Phys. **88** (1988) 2148.
13) - H. Telle and U. Brinkmann
     Mol. Phys. **39** (1980) 361
14) - U. Brinkmann, V.H. Schmidt and H. Telle
     Chem. Phys. Lett. **73** (1980) 530
15) - N. Furio, M.L. Campbell and P.J. Dagdigian
     J. Chem. phys. **84** (1986) 4332
16) - C.T. Rettner and R.N. Zare
     J. Chem. phys. **77** (1982) 2416
17) - J.A. Shea and E.J. Campbell
     J. Chem. Phys. **81** (1984) 5326
18) - J.D. Augspurger and C.E. Dyskra
     Chem. Phys. lett. **158** (1989) 399
19) - A. Weaver, R.B. Metz, S.E. Bradforth and P.M. Neumark
     J. Phys. Chem. **92** (1988) 5558
20) - J.E. Francis and S.E. Weber
     J. Chem. Phys. **56** (1972) 5879
21) - C. Jouvet, M. Boivineau, M.C. Duval and B. Soep
     J. Phys. Chem. **91** (1987) 5416

POTENTIAL ENERGY SURFACES FOR OPEN SHELL SPECIES

Piergiorgio Casavecchia

Dipartimento di Chimica
Università di Perugia
06100 Perugia, Italy

ABSTRACT. In this paper we examine the present status and future prospects in the area of determination of potential energy surfaces for weakly interacting open shell systems, i.e., van der Waals complexes where one of the partners is an open shell atom or molecule, possessing both orbital and spin degeneracy in its ground state. Extension of molecular beam scattering measurements beyond the closed shell atom-atom case to include open shell atoms and molecules with closed shell species is analyzed. The theoretical framework and computational procedures needed for deriving potential energy surfaces for open shell systems from scattering data are discussed. The complementary information obtainable from spectroscopic investigations is also considered.

INTRODUCTION

The interaction between two closed shell atoms, such as two rare gas atoms ($^1S_0$), is isotropic and described by a single potential energy curve depending only on the internuclear distance, $V(R)$. The interaction between two atoms becomes anisotropic when at least one of the two partners is an open shell atom not in S-state, such as an atom of the groups III, IV, VI or VII of the periodic system. Figure 1 depicts typical anisotropic electronic cloud distributions of ground state open shell (P-state) atoms leading to interatomic anisotropy. Group I, II, V and VIII atoms are also open shell species in their electronically excited states. In all these cases the interaction is described by a manifold of electronic potential energy curves correlating asymptotically with the different states of the open shell atom.

The atom-molecule (diatomic) interaction is always anisotropic, even when both partners are closed shells, as for instance, rare gas($^1S_0$)-N_2($^1\Sigma$). The potential for the interaction between a closed shell atom and a P-state atom or a diatomic molecule, assumed as a rigid rotor, can be represented in both cases by formally using the same potential expansion in Legendre polynomials. In Fig. 2 an intuitive picture of this fact is given. This leads to a unified treatment of the collision dynamics of such systems. However, if the atom is not

Dynamics of Polyatomic Van der Waals Complexes
Edited by N. Halberstadt and K. C. Janda
Plenum Press, New York, 1990

an S-state atom or the diatom is not a Σ-state molecule (as NO(²Π) for example), additional complications in the anisotropy of the interaction can occur. Specifically, in addition to anisotropy effects arising from molecular rotations, at each internuclear distance several electronic states are also possible due to the presence of an internal angular momentum;

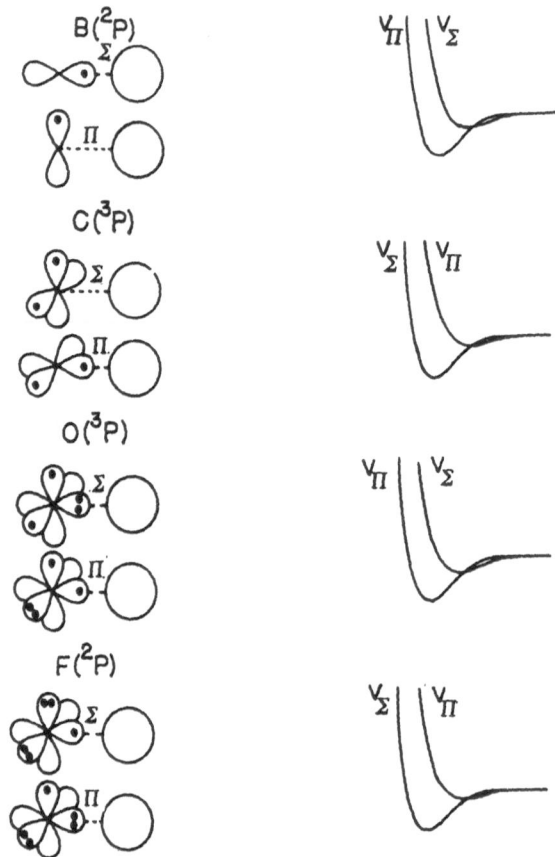

Fig. 1. Adapted from Ref. 30: For typical electronic cloud distributions of open shell atoms illustrated in the first column, the qualitative behavior of the electrostatic V_Σ and V_Π interactions with closed shell systems can be anticipated (second column). By taking into account also the spin-orbit interaction, three and six potential curves in the case of a ²P atom and a ³P atom, respectively, will describe the interaction (see text).

the most complex case being represented by the open shell atom - open shell molecule interaction. Generally, the interaction

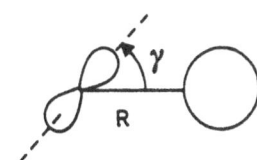 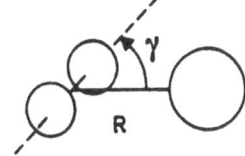

$$V(R,\gamma)=V_0(R)+V_2(R)P_2(\cos\gamma) \qquad V(R,\gamma)=V_0(R)+\sum_{n=2}^{\infty}V_n(R)P_n(\cos\gamma)$$

Fig. 2. Showing (see Ref. 30 and references therein) that the electrostatic interaction of a P-state atom and a diatomic molecule with a closed shell atom can both be represented by the same potential expansion, describing the atom by the symmetry of its electronic cloud and assuming the molecule as a rigid rotor. The relationship of V_0 and V_2 with the perhaps more familiar electrostatic interaction curves V_Σ (= $V_{\gamma=0}^{\circ}$) and V_Π (= $V_{\gamma=90}^{\circ}$) are $V_0=1/3(V_\Sigma+2V_\Pi)$ and $V_2=5/3(V_\Sigma-V_\Pi)$.

between two open shell species leads to the formation of a chemical bond.

The aim of this paper is to examine the present status and future prospects in the area of determination of potential energy surfaces (PES's) for van der Waals complexes where one of the partners is an open shell atom or molecule possessing both orbital and spin degeneracy in its ground state. For this purpose it is convenient to consider the two cases separately.

OPEN SHELL ATOM - CLOSED SHELL ATOM (MOLECULE) INTERACTION

From a theoretical point of view, the calculation of reliable PES's for van der Waals systems is still a big problem: even for isotropic systems the most advanced quantum chemical methods[1] have not yet reached the stage of quantitative reliability, with the exception of a few light cases. Our most detailed knowledge about intermolecular forces comes from experiment. The crossed molecular beam scattering method has proven to be a powerful tool to give quantitative information on many adiabatic isotropic atom-atom potentials not readily accessible to spectroscopic investigation.[2-4] The method, which is direct and universal, covers the entire energy range. In particular, since about 1970, major advances in elastic scattering techniques, coupled to theoretical and spectroscopic and bulk methods, have led to determinations of potentials energy curves for the benchmark homonuclear and heteronuclear rare gas pairs which are very accurate (better than 1% in the minimum position and about 1%-2% in the well depth) and conclusive.[5-7] Here the theory connecting the potential with experimental observables is essentially exact, this being true not only for scattering properties, but also for spectroscopic results and gaseous properties.

The first attempt of measuring the anisotropy of the van der Waals force was carried out about 30 years ago in pioneering beam work by Schlier and coworkers[8] on just open shell systems: Ga(^{2}P)-He,Ar,Xe. However, not until the late 70's did a systematic study of ground state open shell atom interactions begin, notably with the halogen (^{2}P$_{3/2,1/2}$) and oxygen (^{3}P$_{2,1,0}$)-rare gas atom systems.[9-15] The development of UV rare gas halide excimer lasers and the interest in testing current theories of molecular bonding, angular momentum coupling, and potential energy surface modeling, prompted a great deal of experimental[9-14, 16-19] and theoretical[20] work on these open shell species. Again, most information has been provided by scattering experiments.[9-15,21,22] In some favorable cases, complementary information has been obtained from spectroscopy.[16-19] Traditional bulk data, as second virial and transport coefficients, are instead lacking for such systems. Very little reliable information has been obtained to date from ab initio calculations[20] on the ground state interaction of rare gas halides and oxides, while useful information has been provided for the excited states. In the past, information on P-state atoms was mainly provided from bulk experiments on collisional quenching of resonance radiation, pressure shift and broadening of linewidths, depolarization, orientation, and so on, on excited ^{2}P alkali atoms with rare gases.

Scattering experiments without state selection

The full matrix of halogen atom-rare gas atom interactions was studied in Berkeley by performing differential cross section measurements at different collision energies.[9-11,13,14] In these experiments the relative population of the fine structure components of the open shell atoms, produced by thermal dissociation in supersonic expansion, was estimated by Boltzmann and degeneracy weights by assuming equilibrium of translational and electronic temperatures in the beam (electronic to translational relaxation being very small under the experimental conditions). The effect of the atom-atom anisotropy is that of quenching the rainbow and diffraction structure in the differential cross section and the glory ondulations in the velocity dependence of the integral cross section, with respect to what is expected for isotropic interactions. The extraction of interaction potentials from scattering data for systems containing non S-state open shell atoms is complicated by the fact that there is more than one PES involved. Nevertheless, it was shown[9-14] that, if nonadiabatic coupling between the relevant states is weak, it is possible to obtain meaningful results by a suitable and fast decoupling scheme of analysis, to be discussed below.

Accurate potentials in the well region for F-Xe and Cl-Xe X$^2\Sigma_{1/2}$ (in Hund's case (a) notation) were obtained from differential scattering data,[9,11] which were in very good agreement with spectroscopically determined potentials. The same molecular beam experiments provided, for the first time, information on the F-Xe and Cl-Xe $^2\Pi_{3/2}$ and $^2\Pi_{1/2}$ potentials in the attractive and low repulsive regions.[9,11] These potentials are inaccessible to spectroscopic study. The X$^2\Sigma_{1/2}$, $^2\Pi_{3/2}$ and $^2\Pi_{1/2}$ potentials were also obtained for F(^{2}P)-Ne,Ar, and Kr,[10] Br(^{2}P)-Ar, Kr, and Xe,[13] and I(^{2}P)-Kr and Xe.[14]

Scattering experiments with state selection

In order to obtain more detailed information on the anisotropy, scattering experiments with state selection are required. When the open shell atom under study is an excited state atom, specific sub-level preparation can be achieved by excitation with proper laser light or by selective quenching of non-selectively excited species. This approach is well documented, for instance, for alkali[2,1] and alkaline earth atoms in the excited ^2P_j and $^{3,1}P_j$ states,[23,24] respectively, colliding with rare gas atoms. Selective quenching has been instead used for excited metastable rare gas atoms.[25] When the open shell under study is a ground state atom, a selective preparation by photons cannot be employed. In these cases, a molecular beam technique which employs a controlled population of atomic sub-levels using inhomogeneous magnetic fields can be applied. Significant results from absolute integral cross section measurements involving $F(^2P_j)$, $O(^3P_j)$, and $N(^2D, ^2P)$ atoms scattered from rare gases and several molecules have been obtained in the Perugia laboratory during the last few years.[21,22,26-30] Orientation effects in the scattering of Ga atoms by rare gas atoms have also been reported.[31] In the Perugia experiments, effusive beams of $F(^2P)$ and $O(^3P)$ are produced by microwave discharge in CF_4 and O_2, respectively. The relative population of the j fine structure components of the ground state P-atoms, determined by magnetic analysis, is found to be in substantial agreement with that calculated for Maxwell-Boltzman distribution at the measured beam temperature.[21,22] These scattering results have been analyzed using an adiabatic decoupling scheme to derive the interaction as a spherical part, $V_0(R)$, and an anisotropic component, $V_2(R)$ (see Fig. 2), from which information has been obtained on the six lowest states of the rare gas oxides[21] and the three lowest states of the rare gas fluorides,[22] and also on nonadiabatic coupling terms. An improvement of the F-rare gas potentials has been possible.

The extension of magnetic selection to differential scattering experiments would be desiderable, but is presently limited by the low beam intensities achievable when using a Stern-Gerlach type of magnet.

Analysis of scattering data

Although close-coupling calculations for exactly solving the open shell atom scattering problem are readily feasible today,[12] they become impractical for trial-and-error best-fit analysis of experimental data. Hence the need of decoupling approximations.

The analysis of open shell atom scattering data follows recent advances in the quantum mechanical treatment of P-state atom collisions, which exploits alternative angular momentum coupling schemes and useful decoupling approximations.[12,32-34] Basically, one has to solve a multichannel Schroedinger equation where the potential energy is the sum of three contributions: (i) the electrostatic interaction with its anisotropy, (ii) the term which contains the centrifugal interaction due to collisional angular momentum, and (iii) the spin-orbit (S-O) interaction (the diatom rotational interaction in the atom-rigid rotor case). According to the relative importance of these three terms, five alternative representations are possible, corresponding to different coupling schemes for the angular momenta involved (Hund's cases). The choice of representations and recipes for simplifications based on decoupling schemes have been given in the literature.[26,32,33]

When the S-O splitting is large with respect to centrifugal effects, we can restrict our attention to only two cases: the molecular case (a) and the diatomic case (c), valid at short and long-range, respectively, when the electrostatic interaction is stronger or, respectively, weaker than the S-O splitting. A proper label for scattering states is then $|j\Omega\rangle$, where j is the atomic angular momentum and Ω is the absolute value of its projection along the internuclear axis R. In this centrifugal sudden or coupled states (CS) decoupling approximation, the Ω quantum number is conserved and the scattering is considered to take place elastically separately on each of the adiabatic potential energy curves $V_{|j\Omega\rangle}$ correlating with the magnetic sublevels of the open shell atom at infinite distance.

Following Ref. 12, here we report explicitly the expressions of the three effective adiabatic potential curves $V_{|j\Omega\rangle}$ correlating with the different $|jm_j\rangle$ states of an halogen atom ($\Omega = |m_j|$):

$$V_{|3/2,1/2\rangle} = [V_\Sigma + V_\Pi - \Delta - D]/2$$

$$V_{|3/2,3/2\rangle} = V_\Pi - \Delta \qquad\qquad (1)$$

$$V_{|1/2,1/2\rangle} = [V_\Sigma + V_\Pi - \Delta + D]/2$$

where

$$D = [(V_\Sigma - V_\Pi)^2 + \Delta^2 - 2/3\ \Delta\ (V_\Sigma - V_\Pi)]^{1/2}$$

and Δ is the S-O constant. The electrostatic potentials V_Σ and V_Π (i.e., the eigenvalues of the electrostatic Hamiltonian, without S-O interaction) are related to the spherical component V_0 and the anisotropic component V_2 by:

$$V_\Sigma = V_0 + 2/5\ V_2 \quad ; \qquad V_\Pi = V_0 - 1/5\ V_2\ .$$

Σ and Π stand for $\Lambda = 0$ and $\Lambda = 1$, respectively, where Λ is the projection of the electronic angular momentum L on the internuclear axis.

Within the Ω-conserving approximation described above, the differential cross section will be given by a weighted sum of the cross sections[9-14] $\sigma_{j\Omega}(\theta)$ for scattering by the potentials $V_{|j\Omega\rangle}$:

$$\sigma(\theta) = \sum_{j\Omega} W_{j\Omega}\ \sigma_{j\Omega}(\theta)$$

where the weights $W_{j\Omega}$ represent the relative population of the $|j\Omega\rangle$ sublevels. A similar formula has been properly extended also to integral cross sections.[21,22,26]

The Ω-conserving approximation is identical to the

coupled–states approximation in the theory of rotational excitation. Its accuracy has been demonstrated[12] by comparison with rigorous coupled–channel calculations for integral and differential cross sections for $F(^2P)$–Ar, Xe and $Cl(^2P)$–Xe at the thermal energies of the experiments. The inelastic cross sections were shown to be much smaller than the elastic ones and all the features of the experimental differential cross sections were well reproduced within the adiabatic approximation, which neglects nonadiabatic coupling between the $V_{1,0}$ potentials. The approximation is expected to hold even better for systems where the open shell atom has larger S–O splitting than in F and Cl. Aquilanti et al.[21] have shown, by computing nonadiabatic coupling terms between the different adiabatic curves, that this is also valid for $O(^3P_{2,1,0})$ scattering with rare gases, where $O(^3P)$ has a smaller S–O splitting than $F(^2P)$.

For systems with very small S–O splitting, such as excited $Li(^2P)$, both centrifugal and fine structure effects can be neglected with respect to the electrostatic anisotropy, giving rise to the so-called molecular elastic decoupling scheme,[32,33] which is the analog of the infinite-order-sudden (IOS) approximation in the theory of rotational excitation. This approximation is based on the assumption that, during the collision, the electron orbital angular momentum and spin remain uncoupled, and that coupling is weak between the adiabatic states of the electrostatic electronic Hamiltonian.

In conclusion, the experimental scattering techniques and the theory necessary to extract the potential energy surfaces for open shell atomic systems is well established and can be extended, at least in principle, to many other open shell atoms, the difficulties being more experimental than theoretical.

Spectroscopic experiments

Standard spectroscopic techniques could only be applied to the $X^2\Sigma_{1/2}$ state of XeF[16] and XeCl,[19] which are the only rare gas halides where bound-bound transitions have been observed.

Laser induced fluorescence (LIF) techniques in supersonic free jets can also yield useful information on the potential energy curves of open shell atomic systems. This type of studies has provided high quality data on the ground and excited states of NaAr type of molecules.[35] The LIF technique was also successfully applied for probing the potential surfaces of XeF.[36] The B←X fluorescence excitation spectrum of XeF in a supersonic free jet is sufficiently simplified that rotational analysis and accurate vibrational spacing are readily obtained, overcoming the complexity of gas phase emission spectroscopy, mainly due in this case to isotopic richness of natural Xe.

Recently, a similar approach has been used to investigate van der Waals complexes of $Al(^2P)$ with Ar[37] and $In(^2P)$ with Ar, Kr, and Xe.[38] Laser vaporization of the refractory atom followed by supersonic expansion in rare gas carrier was used to sinthesize van der Waals complexes. Gardner and Lester[37] then used mass resolved resonance-enhanced multiphoton ionization (REMPI) for obtaining rovibration excitation spectra for specific masses, getting rid of the cluster contamination problem. Lower limits for the ground state $X^2\Pi_{1/2}$ and the $B^2\Sigma$ state[38] binding energies of AlAr were obtained. Callender et al.[38] recorded LIF excitation and emission spectra, after having identified the spectrum by also using mass resolved REMPI. From the measured vibrational spacings, the bond energies of the $X^2\Pi_{1/2,3/2}$ states and $B^2\Sigma_{1/2}$ state were determined, and, by exploiting the similarity of In-rare gas molecules with $Na(^2P)$-rare gas molecules[38] and using the correlation rules of Liuti and Pirani,[39] the equilibrium bond distances were also estimated. Very recently, a spectroscopic investigation of the complex $Si(^3P)$-Ar formed in a free jet expansion has also been reported.[40]

Clearly, these mass resolved spectroscopic techniques can be extended in a natural way to the investigation of a variety of van der Waals complexes.

Concluding remarks

Among the ground state open shell species, so far only halogen, oxygen and gallium atoms have been used in scattering studies. Improvement in atomic beam source technologies will hopefully allow in the near future to perform scattering measurements with also other ground state open shell atoms of the periodic system.

An interesting and unifying aspect of van der Waals interactions emerges from an examination of potential parameters for the spherical averages V_0 of a large variety of open shell systems:[30] they appear to follow the systematics already established[39,41] for interaction between closed shell systems. The anisotropy of interaction, V_2, most generally changes sign as a function of R because is the sum of two contributions: the anisotropy of the long range potential determined by the polarizability anisotropy of the open shell atom, and the short range repulsive anisotropy interaction which decays exponentially with distance[20] and is generally related to charge transfer effects with upper ionic states. These two contributions have different signs. It would be very interesting and useful to establish correlations in terms of simple properties of separated atoms also for the anisotropic component of the interaction: this would allow us to predict reasonably potential parameters also for open shell systems which remain difficult to investigate experimentally.

The additional complexity arising in the open shell atom-closed shell molecule interaction has been scarcely investigated experimentally to date, although these interactions are chemically very interesting. The scattering theory for fine structure and rotational transitions in atom (^2P)-Σ state rigid rotor has been developed[42] and coupled-channel calculations reported for $F(^2P)$ + $H_2(^1\Sigma_g)$.[42] Recently, integral cross section measurements with analysis of magnetic sublevels has provided information on the S-O dependence of the long-range PES for $O(^3P)$ with H_2 and CH_4,[29] and also $F(^2P)$ with H_2 and CH_4.[43] Under the experimental conditions of these studies (high rotational temperature of the molecule), empirical evidence was found for excluding contributions from molecular rotations to anisotropy effects which were observed and attributed to the open shell structure of the atom. The interaction of open shell atoms (as halogens and oxygen) with simple diatomic molecules, in which both types of anisotropy effects will need to be accounted for, can be investigated in differential scattering experiments by crossing supersonic beams of the two colliding species, in which the molecule is rotationally cold. This type of experiments is being planned in our laboratory using the same crossed molecular beam apparatus employed for the study of rare gas-rare gas,[3,4] rare gas-diatom,[4,44-46] and rare gas-NO($^2\Pi$) interactions.[44,47,48]

A joint experimental and theoretical effort should extend considerably our knowledge of open shell atom interactions in the near future.

OPEN SHELL MOLECULE - CLOSED SHELL ATOM (MOLECULE) INTERACTION

The interaction of Σ-state molecules with closed shell atoms and related energy transfer processes have been extensively studied and are well understood, on almost a quantitative level (see Ref. 49, and also the article by Beneventi et al.[50] in this Volume). For open shell molecules not in Σ-state the situation is more involved: in addition to rotational excitation, transitions between spin-orbit states and/or Λ-doublet levels may occur. Ab initio calculations of the relevant PES's are obviously even more challenging than for closed shell species.

Excited Π molecules are not discussed here. However, energy transfer in electronically excited $^2\Pi$ states has been recently studied.[51] In this case a complete preparation of different initial quantum states is possible by laser excitation. The subsequent collisional redistribution can be studied by dispersing the fluorescence or probing the neighboring levels with a second laser. In this way state-to-state data can be obtained. Systems investigated include ZnH, CdH and CaF(A$^2\Pi_{1/2,3/2}$) + rare gases.[51]

Ground state Π-molecules instead, as was occurring with ground state P atoms, cannot be prepared in specific initial states by lasers. Initial state preparation is usually achieved by jet cooling in supersonic beam expansions.[44,47,52-56]

Among Π-state molecules, the kinetics of molecular interactions involving NO($^2\Pi$) has received much experimental and theoretical attention.[47,54-59] NO is the only stable diatomic which has in the ground state non-zero spin and electronic angular momentum. The transient species OH($^2\Pi$) and CH($^2\Pi$) are other ground state Π-molecules which also have attracted large experimental and theoretical interest in relation to their importance in upper atmosphere and interstellar space physics. In particular, Dixon and Field,[60] following the formulation of Arthurs and Dalgarno[61] for the structureless particle-rigid rotor problem, have presented the general theory for the rotational energy transfer induced in an open shell molecule (both orbitally and non-orbitally degenerate) by collision with an open shell atom. Applications were reported for H($^2S_{1/2}$) + CN(X$^2\Sigma$), and H($^2S_{1/2}$) + OH(X$^2\Pi$) and CH(X$^2\Pi$) in relation to collisional pumping as an excitation mechanism for interstellar maser action of OH and CH through the inversion of Λ-doublet populations.[60] The theory of inelastic collisions of Π-molecules with structureless particles has also been formulated by other authors.[62-64] In particular, Alexander[63] reported the full close-coupling formulation of the collision between a $^2\Pi$ diatomic and a closed shell atom; decoupling CS and IOS approximations were also developed and applications given for inelastic collisions of NO(X$^2\Pi$) with Ar and He.[59,65] These studies were partially motivated by molecular beam[66] and laser double resonance experiments[67] on integral cross section and rate measurements, respectively, for S-O and rotational transitions in NO($^2\Pi$)-rare gas collisions. While the CS approximation is always found to be accurate also for Ar-NO down to low energies, the IOS decoupling scheme is accurate for intramultiplet Λ-averaged rotational transitions for He-NO, but less accurate for Ar-NO.[59] The neglect of the 123 cm^{-1} S-O splitting within IOS leads to an overestimation of the $\Omega=1/2\rightarrow\Omega=3/2$ inelasticity.[59]

The NO($^2\Pi$)-Ar system has been the most investigated. Nielson et al.,[58] using the electron gas model, have calculated an angle-dependent PES. Comparison with experiment has shown that this PES is only qualitatively correct.[54,56] The scattering studies of Andresen and coworkers[54] have provided a critical test of the proposed electron-gas PES, but no potential was fitted to their state-resolved integral cross section data. A series of sophisticated spectroscopic investigations has yielded a substantial understanding of the ground state (and also of several excited states) of NO-Ar.[68-72] In particular, limits were placed on the dissociation energy of both ground and excited states (see Ref. 71 for an up to date comparison of ground state dissociation energies obtained from different experimental and theoretical sources): substantial agreement with scattering determinations was found. Using microwave and radiofrequency spectroscopy, Howard and coworkers[72] have analyzed the rotation spectrum of NO-Ar to yield the ground state geometry: the results show a near T-shaped molecule with a vibrationally averaged minimum position distance of 3.71 Å.

The anisotropy in NO($^2\Pi$)-rare gas systems was firstly investigated by Reuss and coworkers[55,56] analyzing, with the sudden approximation, the orientational dependence of the glory structure of the integral cross section (ICS) in experiments with state selection and without state selection of NO. An estimate of the anisotropy for NO-He was also reported from low resolution total differential cross section (DCS) measurements.[73]

For atom-diatom systems, measurements of the DCS for single rotational transitions provide the most detailed information on the anisotropy of the repulsive part of the potential.[49,74] When such measurements are not possible because of very small rotational spacing and presently available experimental resolution, as for NO-rare gas systems, a reliable PES can still be obtained from total (elastic + inelastic) DCS measured under high resolution conditions, within a multiproperty analysis of microscopic and macroscopic data sensitive to different portion of the full PES.[44-48,50]

In our laboratory we have recently determined a full PES for NO($^2\Pi$)-He, Ne, Ar, and Kr from high resolution total DCS measured at different collision energies.[44,47,48] The analogy of the NO($^2\Pi$)-rare gas system with the atom(^{2}P)-rare gas problem discussed in the previous section has been exploited in the data analysis. This work is part of a more systematic molecular beam study of rare gas-rare gas and rare gas-molecule interactions carried out in Perugia in these last few years.[3,4,44-48,50] Since analysis of scattering data for $^2\Pi$ molecules is not well established as for ^{2}P atoms, it is useful to outline the procedure of analysis, which can also be extended to other $^2\Pi$-S van der Waals colliding systems.

The interaction potential for NO($^2\Pi$)-R(^{1}S)

While the interaction between a rare gas atom R(^{1}S$_0$) and a $^3\Sigma$ or $^1\Sigma$ molecule (as O$_2$ and N$_2$, respectively) is described by a single PES, the interaction between a rare gas atom and a NO($^2\Pi$) molecule gives rise to two PES's. In fact, when a rare gas (^{1}S) approaches NO($^2\Pi$), the degeneracy of the $^2\Pi$ state is removed: two surfaces, $E_{A'(+)}$ and $E_{A''(-)}$, describe the interaction, where the designations A'(+) and A''(-) correspond to whether the wave function is symmetric or antisymmetric under reflection with respect to the plane of the three atoms. This description is adequate for $^1\Pi$ molecule with ^{1}S atoms; but NO in the ground $^2\Pi$ state has a S-O splitting Δ of 123 cm^{-1} and its effects should be taken into account for collisions taking place at thermal energies. Then, including S-O interaction (assumed to be independent of the internuclear distance) and solving the secular equation in a procedure similar to the ^{2}P atom problem (see previous Section), one obtains the two adiabatic PES's $V_{\mp}$ which are expressed by[58]

$$V_{\mp} = 1/2(E_{A'(+)} + E_{A''(-)})$$
$$\pm \left\{ [1/2(E_{A'(+)} - E_{A''(-)})]^2 + (1/2\ \Delta)^2 \right\}^{1/2} \qquad (2)$$

where V_- and V_+ correlate with NO($^2\Pi_{3/2}$)+R(^{1}S$_0$) and NO($^2\Pi_{1/2}$)+R(^{1}S$_0$), respectively. The situation is analogous to atom(^{2}P$_{1/2,3/2}$)+atom(^{1}S$_0$): The only difference, in the present case, is that the three level system which is encountered in atomic collisions reduces to a two level system.

Alexander and coworkers, in a fully quantum calculation of inelastic cross sections for collisions of NO($^2\Pi$) with Ar and He have shown that coupling is weak between the adiabatic states of the total electronic Hamiltonian including S-O interaction.[59,65] Therefore, since inelastic cross sections are much smaller than elastic cross sections, retaining only S-O effects and neglecting centrifugal coupling effects, we can describe the collision in terms of elastic scattering occurring separately on each of the adiabatic PES's $V_{\mp}$. This model corresponds to the Ω conserving approximation of the atomic case, where Ω (the projection of the total electronic angular momentum along the internuclear axis) is a good quantum number and Hund's case (c) affords a good description of the situation. Our scattering problem can also be described in terms of Hund's (c) and (e) coupling case, in which the nature of the two molecular states of NO, characterized by Ω=1/2 and 3/2, is preserved, since they correlate adiabatically with the V_+ and V_- surfaces, respectively. As shown in the theoretical treatment[63] of collisions of $^2\Pi$ molecules in Hund's case (a) (as is NO at low angular momenta), the sum potential (E$_{A'}$+E$_{A''}$)/2 primarily induces the S-O multiplet conserving

rotational transitions, while the difference potential
$(E_{A'}-E_{A''})/2$ determines the intermultiplet rotational
transitions. Theoretical work[58] for NO-Ar has shown that the
difference potential is very small and in most of the van der
Waals region is dominated by the S-O coupling. The experimental
finding[54,66,75] for NO-He, Ne, and Ar that intermultiplet
transitions are much smaller than intramultiplet transitions
appears to support the theoretical results. Therefore,
neglecting in equation (2) the first term under square root,
the two relevant potentials reduce to two identical potentials
given by the sum potential (the average of $E_{A'}$ and $E_{A''}$) and
separated by the S-O splitting. Within this simplified
adiabatic picture, it is possible to parametrize $V_{\mp}$ smoothly,
without need for the full equation (2). We remark that the
entrance channel in our experimental conditions[44] is
represented by $NO(^2\Pi_{1/2}, j=1/2-9/2) + R(^1S_0)$. Hence, scattering
can be considered to take place on a single PES.
Clearly, if the difference potential cannot be neglected with
respect to the S-O interaction, the full equation (2) for $V_{\mp}$
has to be used, which is similar to equation (1) expressing
$V_{|j,\Omega\rangle}$ in terms of V_Σ, V_Π and the S-O splitting in the atom(^2P)
+ rare gas case. This would be required for treating the
scattering of $OH(^2\Pi) + H_2$ for instance, where a probability for
fine structure transitions comparable to that for
intramultiplet transitions indicates a large difference
(comparable to the sum) between the $E_{A'}$ and $E_{A''}$ potential
surfaces, which differ essentially in the orientation of the
unpaired pΠ lobe relative to the collision partner.[53] In this
case one needs to parametrize $E_{A'(+)}$ and $E_{A''(-)}$, as in the ^2P
atom case one parametrizes V_Σ and V_Π (or V_0 and V_2).

The other two relevant interactions to be considered in
our scattering problem are the electrostatic anisotropy, due to
the relative orientation of NO with respect to the atom, and
the rotational coupling. The rotational spacing in NO is quite
small, being the rotational constant 1.7 cm^{-1}. Since the
collision energies of our experiments are substancially greater
than the rotational spacing and the molecules are prepared in
their lowest rotational levels by supersonic expansion, the IOS
approximation[76] can be used for calculating the total DCS. The
reliability of the IOS decoupling scheme for total and
rotationally inelastic cross sections has been widely
discussed.[45,74,77] For systems as O_2 and N_2 interacting with He
and Ne, total DCS computed in the IOS approximation have been
found to be in agreement with rigorous close-coupling
calculations, the IOS method failing only for large rotational
transitions, which contribute negligibly to the total
DCS.[45,77,78] For interactions with Ar, although less
satisfactory, IOS can still be used as a first
approximation.[59,79]

In our analysis a potential model in which the angle
dependence is obtained by making the size parameters (well
depth ε and minimum position R_m) angle dependent was
used,[44,45,47,50] since the usual expansion into Legendre
polynomials converges very slowly[45,80] for systems different
from rare gas-H_2. The same reduced form has been assumed for
all orientations:

$$V(R,\gamma) = \varepsilon(\gamma)\, f(x) , \qquad x = R/R_m(\gamma) ,$$

where the parameters $\varepsilon(\gamma)$ and $R_m(\gamma)$ are given by

$$\varepsilon(\gamma) = \bar{\varepsilon}[1 + A_1 \, P_1(\cos \gamma) + A_2 \, P_2(\cos \gamma)] \ ,$$

$$R_m(\gamma) = \bar{R}_m[1 + B_1 \, P_1(\cos \gamma) + B_2 \, P_2(\cos \gamma)] \ . \tag{3}$$

We have neglected the slight eccentricity of NO and retained only an effective P_2 term, since the total DCS data do not warrant the use of a more elaborate description of the anisotropy (i.e., a parametrization which includes also a P_1 term).[44,47] Recent experimental results[54,66,75] show a marked propensity for even rotational transitions in NO($^2\Pi_{1/2}$)+Ar collisions, which indicates that the NO-Ar potential is dominated by the even Legendre terms. The constraints were imposed that the long-range anisotropy had to be consistent with the anisotropy of the NO polarizability and that the most stable configuration occurs for the T-shaped NO-R complexes.

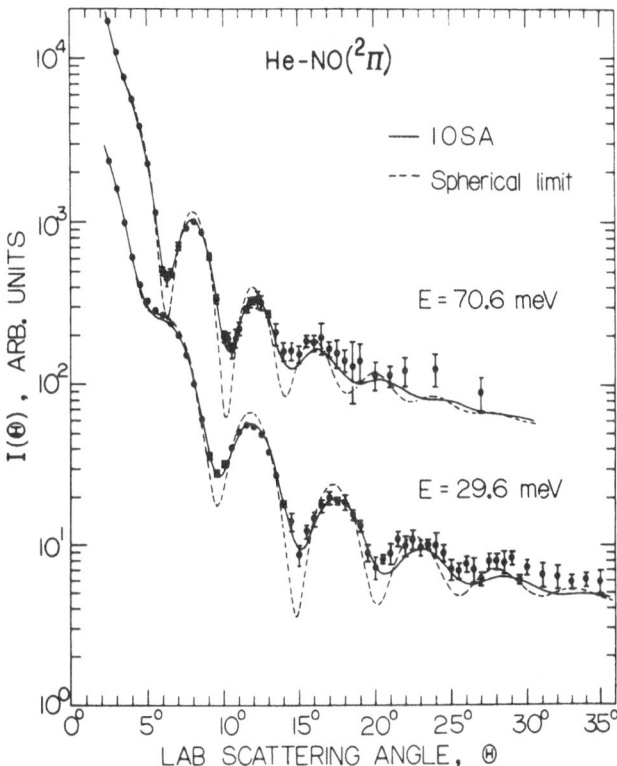

Fig. 3. Total differential cross sections for He-NO at two collision energies (Ref. 44). The experimental results are compared with IOS calculations performed using a fully anisotropic PES, and with calculations performed with an effective spherical potential (obtained from the full PES by setting equal to zero the anisotropy parameters in Equation (3). Due to the effect of the anisotropy, a clear quenching of the diffraction structure is present.

Scattering results for $NO(^2\Pi)-R(^1S)$

In Fig. 3 we report typical total differential scattering data measured for $He-NO(^2\Pi)$ at two collision energies. The quenching of the quantum diffraction oscillations in the angular dependence of the total DCS with respect to what is expected from a spherical interaction contains information on the anisotropy of the repulsive wall of the potential. The analysis of these data has permitted us to derive a full anisotropic PES,[44] which represents a significant improvement with respect to a previous determination. Second virial coefficients were also included in the analysis.

In Fig. 4 we report total DCS data for $Ne-NO(^2\Pi)$ at E=77.2 meV. These data have been analyzed along the lines followed for He-NO, O_2 and N_2,[44] $He-CO_2$,[46] $Ne-N_2$[45] and $Ne-O_2$.[81] The spherical average $V_0(R)$ (the first term in the Legendre expansion of the full PES) and two cuts of the full PES for the parallel and perpendicular geometries are depicted in Fig. 5, where they are compared with the results of a previous determination[55] based on the measurement of integral cross sections with state selected NO. As can be seen, the PES of Thuis et al.[55] has a spherical component which is too deep and too inward located on the R scale, the product εR_m being almost the same. Moreover, the anisotropy measured by the $\Delta\sigma$ ($=\sigma_\parallel - \sigma_\perp$) parameter (where σ is the distance at which the potential crosses the zero), is much smaller than the present results.[55] A comparison of the prediction of the surface of Thuis et al.[55] with the total DCS data (see Fig. 4) reflecting these deficiences: The calculated diffraction oscillations superimposed on the main rainbow structure are more widely spaced and shifted with respect to the experiment because of the incorrect absolute location of the minimum of the spherical potential; moreover, they are not sufficiently quenched because of the too weak anisotropy.

Potential surfaces derived from similar experiments[47] for $Ar-NO(^2\Pi)$ and $Kr-NO(^2\Pi)$ were instead found to be in agreement with the results of Thuis et al..[55,56] For Ar-NO the well depth (D_e value) for the T-geometry of 111.6 ± 5 cm^{-1}, corresponding to a D_0 of 96 ± 5 cm^{-1}, compares well with recent spectroscopic determinations.[71] The minimum position $R_m=3.61\pm0.08$ Å for the same geometry compares satisfactorily with the molecular beam electric resonance results of Mills et al.,[72] who derived a vibrationally averaged value of 3.71 Å, which is known to be slightly larger than the R_m value.

Outlook

Improvement of some of the $NO(^2\Pi)$-rare gas PES's may already be possible with the existing experimental data by including all the information in a multiproperty type of analysis, which uses, if necessary, more rigorous schemes of analysis than the IOS prescription, with possibly also the inclusion of P_1-terms in the potential expansion for scattering calculations. Differential state-to-state cross sections for rotational and S-O transitions would be desirable. At the present, this could only be obtained by employing laser techniques.

Very recent and promising work on the interaction and

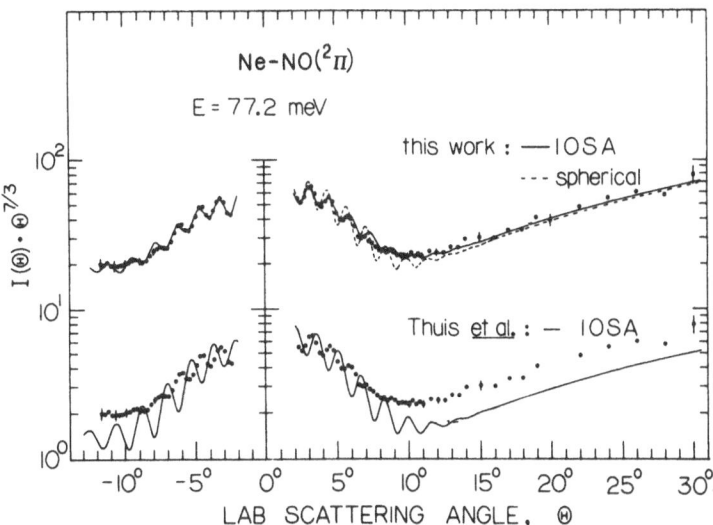

Fig. 4. Total differential cross section data for Ne-NO plotted as $I(\Theta)\Theta^{7/3}$. Upper plot: comparison with the prediction of a full anisotropic PES in the IOS approximation and of a spherical effective potential (Ref. 48). Lower plot: the same data compared with the prediction of the PES of Thuis et al. (Ref. 55).

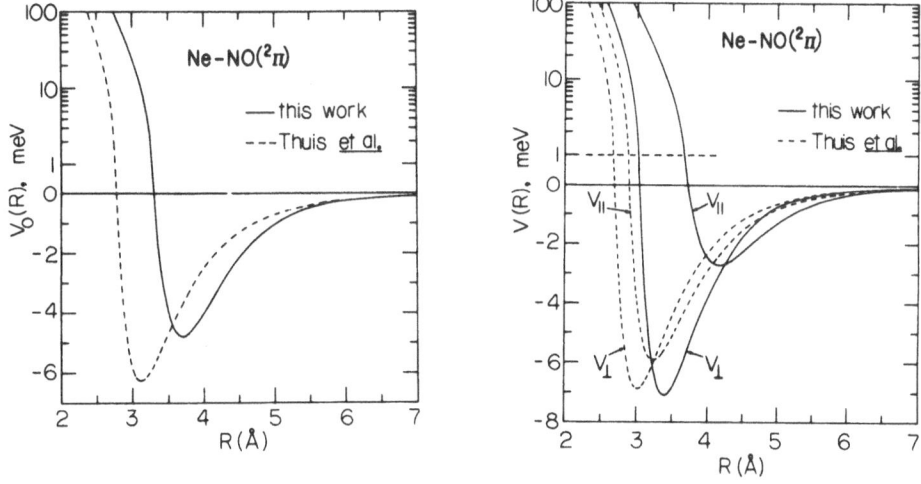

Fig. 5. Best-fit spherical average potential (V_0) and anisotropic potential for parallel ($V_\parallel$) and perpendicular ($V_\perp$) configurations (Ref. 48) compared with the corresponding results of Thuis et al. (Ref. 55).

dynamics of open shell species is represented by the studies of Lester and coworkers,[82] who use spectroscopic methods to derive detailed information on the attractive well region of ground and excited states of $OH(^2\Pi)$-Ar, a kind of prototype system amenable also to theoretical investigation. The vibrational overtone spectrum of OH-Ar is obtained using a novel infrared pump – ultraviolet probe scheme. The ground electronic state correlating with $OH(X^2\Pi_{3/2})$+Ar appear to be linear, in contrast to NO-Ar. Interestingly, but non surprisingly, OH-Ar appear to exhibit properties similar to those of the XH-Ar systems, where X is a halogen atom.

Andresen and coworkers[53] have reported a crossed beam study on $OH(^2\Pi_{3/2}, j=3/2) + H_2$ which uses LIF detection to measure integral cross sections for single rotational transitions, intra- and inter-multiplet. State selected beams of OH produced in supersonic expansions can permit in the future to measure total and also state-to-state integral (and possibly also differential) cross sections, which will provide detailed information on the relevant PES's for OH interacting with rare gases and other species.

Meanwhile, energy transfer is being investigated in $CH(X^2\Pi)$-D_2 and He collisions by state-to-state integral cross section measurements in crossed beam experiments with LIF detection.[83] Finally, theoretical studies on the interaction of open shell atoms with open shell molecules, $O(^3P_{2,1,0})$ + $OH(^2\Pi_{3/2,1/2})$, in relation to fine structure effects in chemical reactions dominated by long-range forces, are also being reported.[84]

Again, a fruitful coupling of scattering, spectroscopic and theoretical studies should allow in the coming years the determination of potential surfaces for an ever growing number of open shell molecular systems.

ACKNOWLEDGMENTS

The author wishes to thank all his colleagues of the Molecular Beam Group of Perugia for useful discussion and continuous encouragement along the years. Special recognition goes to Prof. V.Aquilanti for illuminating discussions about open shell interactions. It is also a pleasure to thank in particular Dr. L.Beneventi and Prof. G.G.Volpi who participated in the NO-rare gas experiments. Financial support from the EEC Science program, the CNR bilateral agreements, ENEA and the Italian Ministry of Public Education is gratefully acknowledged.

REFERENCES

1. W.Meyer, P.C.Hariharan, and W.Kutzelnig, J.Chem.Phys. 73, 1880 (1980); J.Schaefer and W.Meyer, J.Chem.Phys. 70, 344 (1979); H.J.Bohm and R.Ahlrichs, Mol.Phys. 55, 1159 (1985).
2. U.Buck, in "Atomic and Molecular Beam Methods", G.Scoles, ed., Oxford, New York (1987), Vol. 1., and references therein.
3. L.Beneventi, P.Casavecchia, and G.G.Volpi, J.Chem.Phys. 84, 4828 (1986).

4. L.Beneventi, P.Casavecchia, and G.G.Volpi, in "Structure and Dynamics of Weakly Bound Molecular Complexes", A.Weber, ed., NATO ASI Ser. C, Reidel, Dordrecht (1987), Vol. 212, p. 441.
5. R.A.Aziz, in "Inert Gases", M.L.Klein, ed., Springer, Berlin (1984), Chap. 2.
6. G.C.Maitland, M.Rigby, E.B.Smith, and W.A.Wakeham, "Intermolecular Forces", Clarendon, Oxford (1981).
7. G.Scoles, Annu.Rev.Phys.Chem. $\underline{31}$, 81 (1980).
8. K.Berkling, Ch.Schlier, and P.Toschek, Z.Physik $\underline{168}$, 81 (1962).
9. C.H.Becker, P.Casavecchia, and Y.T.Lee, J.Chem.Phys. $\underline{69}$, 2377 (1978).
10. C.H.Becker, P.Casavecchia, and Y.T.Lee, J.Chem.Phys. $\underline{70}$, 2986 (1979).
11. C.H.Becker, J.J.Valentini, P.Casavecchia, S.J.Sibener, and Y.T.Lee, Chem.Phys.Lett. $\underline{61}$, 1 (1979).
12. C.H.Becker, P.Casavecchia, Y.T.Lee, R.E.Olson, and W.A.Lester, Jr., J.Chem.Phys. $\underline{70}$, 5477 (1979).
13. P.Casavecchia, G.He, R.K.Sparks, and Y.T.Lee, J.Chem.Phys. $\underline{75}$, 710 (1981).
14. P.Casavecchia, G.He, R.K.Sparks, and Y.T.Lee, J.Chem.Phys. $\underline{77}$, 1878 (1982).
15. V.Aquilanti, G.Liuti, F.Pirani, F.Vecchiocattivi, and G.G.Volpi, J.Chem.Phys. $\underline{65}$, 4751 (1976); V.Aquilanti, E.Luzzatti, F.Pirani, and G.G.Volpi, J.Chem.Phys. $\underline{73}$, 1181 (1980).
16. J.Tellinghuisen, P.C.Tellinghuisen, J.A.Coxon, J.E.Velazco, and D.W.Setser, J.Chem.Phys. $\underline{68}$, 5187 (1978).
17. A.L.Smith and P.C.Kobrinsky, J.Mol.Spectrosc. $\underline{69}$, 1 (1978).
18. D.L.Monts, L.M.Ziurys, S.M.Beck, M.G.Liverman, and R.E.Smalley, J.Chem.Phys. $\underline{71}$, 4057 (1979).
19. A.Sur, A.K.Hui, and J.Tellinghuisen, J.Mol.Spectrosc. $\underline{74}$, 465 (1979); J.Tellinghuisen, J.M.Hoffman, G.C.Tisone, and A.K.Hays, J.Chem.Phys. $\underline{64}$, 2484 (1976).
20. T.H.Dunning and P.J.Hay, J.Chem.Phys. $\underline{69}$, 134 (1978); P.J.Hay and T.H.Dunning, J.Chem.Phys. $\underline{69}$, 2209 (1978); M.Krauss, J.Chem.Phys. $\underline{29}$, 350 (1976).
21. V.Aquilanti, R.Candori, and F.Pirani, J.Chem.Phys. $\underline{89}$, 6157 (1988).
22. V.Aquilanti, E.Luzzatti, F.Pirani, and G.G.Volpi, J.Chem.Phys. $\underline{89}$, 6165 (1988).
23. K.Bergmann, in "Atomic and Molecular Beam Methods", G.Scoles, ed., Oxford, New York (1987), Vol. 1; R.Duren and E.Hasselbrink, J.Chem.Phys. $\underline{85}$, 1880 (1986).
24. P.J.Dagdigian, in "Atomic and Molecular Beam Methods", G.Scoles, ed., Oxford, New York (1987), Vol. 1.
25. M.J.Verheijen and H.C.W. Beijerinck, Chem.Phys. $\underline{102}$, 255 (1986).
26. V.Aquilanti, G.Grossi, and F.Pirani, "Electronic and Atomic Collisions", Invited Papers XIII ICPEAC, J.Eichler, I.V.Hertel, and N.Stolterfoht, eds., Berlin (1983), p. 441.
27. F.Vecchiocattivi, Comments At.Mol.Phys. $\underline{17}$, 163 (1986).
28. V.Aquilanti, F.Pirani, and F.Vecchiocattivi, in "Structure and Dynamics of Weakly Bound Molecular Complexes", A.Weber, ed., NATO ASI Ser. C, Reidel, Dordrecht (1987), Vol. 212, p. 423.
29. V.Aquilanti, R.Candori, L.Mariani, and F.Pirani, J.Phys.Chem. $\underline{93}$, 130 (1989).
30. V.Aquilanti, G.Liuti, F.Pirani, and F.Vecchiocattivi, J.Chem.Soc.Faraday Trans. 2, $\underline{85}$, 955 (1989).
31. N.Hishinuma and O.Sueoka, Chem.Phys.Lett. $\underline{121}$, 293 (1985).

32. V.Aquilanti and G.Grossi, J.Chem.Phys. 73, 1165 (1980).
33. V.Aquilanti, P.Casavecchia, G.Grossi, and A.Lagana',
    J.Chem.Phys. 73, 1173 (1980).
34. V.Aquilanti, G.Grossi, and A.Lagana', Nuovo Cimento 63b, 7
    (1981).
35. R.E.Smalley, D.A.Auerbach, P.S.H.Fitch, D.H.Levy, and
    L.Wharton, J.Chem.Phys. 66, 3778 (1977); R.Ahmad-Bitar,
    W.P.Lapatovich, D.E.Pritchard, and I.Renhorn,
    Phys.Rev.Lett. 39, 1657 (1977).
36. D.L.Monts, L.M.Ziurys, S.M.Beck, M.G.Liverman, and
    R.E.Smalley, J.Chem.Phys. 71, 4057 (1979).
37. J.M.Gardner and M.I.Lester, Chem.Phys.Lett. 137, 301
    (1987).
38. C.L.Callender, S.A.Mitchell, and P.A.Hackett, J.Chem.Phys.
    90, 2535 (1989).
39. G.Liuti and F.Pirani, Chem.Phys.Lett. 122, 245 (1985).
40. C.Lardeux-Dedonder, C.Jouvet, M.Richard-Viard, and
    D.Solgadi, J.Chem.Phys., to be published.
41. K.T.Tang and J.P.Toennies, Z.Phys.D 1, 91 (1986); G.Ihm,
    M.W.Cole, F.Toigo, and G.Scoles, J.Chem.Phys. 87, 3995
    (1987).
42. F.Rebentrost and W.A.Lester, Jr., J.Chem.Phys. 67, 3367
    (1977); L.D.Thomas, W.A.Lester, Jr., and F.Rebentrost,
    J.Chem.Phys. 69, 5489 (1978).
43. V.Aquilanti, R.Candori, D.Cappelletti, and F.Pirani, to be
    published.
44. L.Beneventi, P.Casavecchia, and G.G.Volpi, J.Chem.Phys. 85,
    7011 (1986).
45. L.Beneventi, P.Casavecchia, F.Vecchiocattivi, G.G.Volpi,
    D.Lemoine, and M.H.Alexander, J.Chem.Phys. 89, 3505 (1988).
46. L.Beneventi, P.Casavecchia, F.Vecchiocattivi, G.G.Volpi,
    U.Buck, Ch.Lauenstein, and R.Schinke, J.Chem.Phys. 89, 4671
    (1988).
47. P.Casavecchia, A.Lagana', and G.G.Volpi, Chem.Phys.Lett.
    112, 445 (1984).
48. L.Beneventi, P.Casavecchia, and G.G.Volpi, to be published.
49. U.Buck, Comments At.Mol.Phys. 17, 143 (1986); and
    references therein.
50. L.Beneventi,P.Casavecchia, and G.G.Volpi, this Volume.
51. J.Dufayard and O.Nedelec, Chem.Phys. 71, 279 (1982); 84,
    167 (1984); Chem.Phys.Lett. 119, 234 (1985); C.Dufour,
    B.Pinchemel, M.Douay, J.Schamps, and M.H.Alexander,
    Chem.Phys. 98, 315 (1985).
52. H.W.Lulf and P.Andresen, in "Rarefied Gas Dynamics",
    Academic, New York (1985), Vol. 2, p.911.
53. P.Andresen, D.Hausler, and H.W.Lulf, J.Chem.Phys. 81, 571
    (1984).
54. H.Joswig, P.Andresen, and R.Schinke, J.Chem.Phys. 85, 1904
    (1986).
55. H.H.W.Thuis, S.Stolte, and J.Reuss, Chem.Phys. 43, 351
    (1979).
56. H.H.W.Thuis, S.Stolte, J.Reuss, J.J.H. van den Biesen, and
    C.J.N. van den Meijdenberg, Chem.Phys. 52, 211 (1980).
57. J.Kosanetaky, U.List, W.Urban, H.Vormann, and E.H.Fink,
    Chem.Phys. 50, 361 (1980), and references therein.
58. G.C.Nielson, G.A.Parker, and R.T Pack, J.Chem.Phys. 64,
    2055 (1976); 66, 1396 (1977).
59. G.C.Corey and M.H.Alexander, J.Chem.Phys. 85, 5652 (1986),
    and references therein.
60. R.N.Dixon and D.Field, Proc.R.Soc.Lond. A.366, 225 (1979);
    A.366, 247 (1979); A.368, 99 (1979).

61. A.Arthurs and A.Dalgarno, Proc.R.Soc.Lond. A.256, 54 (1960).
62. S.Green and R.N.Zare, Chem.Phys. 7, 62 (1975).
63. M.H.Alexander, J.Chem.Phys. 76, 5974 (1982).
64. K.Klar, J.Phys.B 6, 2139 (1973); M.Bertojo, A.C.Cheung, and C.H.Townes, Astrophys. J. 208, 914 (1976); M.Shapiro and H.Kaplan, J.Chem.Phys. 71, 2182 (1979); D.P.Dewangan and D.R.Flower, J.Phys.B 14, 2179 (1981); 16, 2157 (1983).
65. T.Orlikowski and M.H.Alexander, J.Chem.Phys. 79, 6006 (1983).
66. P.Andresen, H.Joswig, H.Pauly, and R.Schinke, J.Chem.Phys. 77, 2204 (1982).
67. Aa.S.Sudbo and M.M.T.Loy, J.Chem.Phys. 76, 3646 (1982).
68. P.R.R.Langridge-Smith, E.Carrasquillo M. and D.H.Levy, J.Chem.Phys. 74, 6513 (1981).
69. K.Sato, Y.Achiba, and K.Kimura, J.Chem.Phys. 81, 57 (1984); K.Sato, Y.Achiba, H.Nakamura, and K.Kimura, J.Chem.Phys. 85, 1418 (1986).
70. J.C.Miller and W.C.Cheng, J.Phys.Chem. 89, 1647 (1985); J.C.Miller, J.Chem.Phys. 86, 3166 (1987).
71. J.C.Miller, J.Chem.Phys. 90, 4031 (1989).
72. P.D.A.Mills, C.M.Western, B.J.Howard, J.Phys.Chem. 90, 4961 (1986).
73. M.Keil, J.T.Slankas, and A.Kuppermann, J.Chem.Phys. 70, 541 (1979).
74. M.Faubel, Adv.At.Mol.Phys. 19, 345 (1983).
75. A.W.Smith and A.W.Johnson, Chem.Phys.Lett. 93, 608 (1982).
76. R.T Pack, J.Chem.Phys. 60, 633 (1974); D.Secrest, J.Chem.Phys. 62, 710 (1975); R.Goldflam, S.Green, and D.J.Kouri, J.Chem.Phys. 67, 4149 (1977); G.A.Parker and R.T Pack, J.Chem.Phys. 68, 1585 (1978); D.J.Kouri, in "Atom-Molecule Collision Theory", R.B.Bernstein, ed., Plenum, New York (1979), p. 301.
77. F.A.Gianturco and A.Palma, J.Phys.B 18, L519 (1985).
78. L.Beneventi, P.Casavecchia, G.G.Volpi, D.Lemoine, and G.C.Corey, to be published.
79. M.S.Bowers, M.Faubel, and K.T.Tang, J.Chem.Phys. 87, 5687 (1987).
80. R.T Pack, J.J.Valentini, and J.B.Cross, J.Chem.Phys. 77, 5486 (1982); R.T Pack, E.Piper, G.A.Pfeffer, and J.P.Toennies, J.Chem.Phys. 80, 4940 (1984); R.T Pack, Chem.Phys.Lett. 55, 197 (1978).
81. L.Beneventi, P.Casavecchia, F.Pirani, F.Vecchiocattivi, G.G.Volpi, A.van der Avoird, and J.Reuss, to be published.
82. M.T.Berry, M.R.Brustein, J.R.Adamo, and M.I.Lester, J.Phys.Chem. 92, 5551 (1988); K.M.Beck, M.T.Berry, M.R.Brustein, and M.I.Lester, J.Chem.Phys., to be published; M.I.Lester, this Volume.
83. R.G.Macdonald and K.Liu, in "1989 Conference on the Dynamics of Molecular Collisions", Asilomar, Ca, USA, July 16-21, 1989, Book of Abstracts, B32.
84. M.M.Graff and A.F.Wagner, in "1989 Conference on the Dynamics of Molecular Collisions", Asilomar, Ca, USA, July 16-21, 1989, Book of Abstracts, A19.

1988-89: THE YEAR OF OH-Ar

Marsha I. Lester[*]

Department of Chemistry
University of Pennsylvania
Philadelphia, Pennsylvania 19104-6323
USA

ABSTRACT. Fluorescence excitation spectra of OH-Ar complexes are observed in the vicinity of several OH A $^2\Sigma^+$ - X $^2\Pi_{3/2}$ transitions. Intermolecular potentials derived from these spectra illustrate the dramatic change in the OH-Ar potential upon electronic excitation of the OH moiety. The OH (center-of-mass) to Ar distance is substantially reduced and Ar becomes much more tightly bound to OH. The differences in the OH-Ar potentials are also reflected in the rate of vibrational predissociation, which is at least a thousand times faster in the excited electronic state. Much smaller changes in the potentials are found upon vibrational excitation of OH, as shown in the ground electronic state by vibrational overtone spectroscopy.

INTRODUCTION

The past year has seen fervent activity in a variety of studies on a new type of van der Waals (vdW) molecule, the open shell OH-Ar complex. Prior to this recent activity, little had been known about the intermolecular potential between the hydroxyl radical and argon, though a wealth of information on interaction potentials has been derived from spectroscopic studies of weakly bound vdW complexes in closed shell systems. Complexes of the hydroxyl radical with argon represent a model case for determining the interaction potential in an open-shell system since OH-Ar is both experimentally observable and theoretically tractable.

Goodman and Brus[1] first identified a low frequency vibrational mode associated with the OH-Ar complex upon electronic excitation of hydroxyl radicals imbedded in a solid argon matrix. An anharmonic vibrational progression consisting of six members was observed in the region about each of the OH A $^2\Sigma^+$(v'=0-2) - X $^2\Pi$(v"=0) transitions. The vibrational progression in the matrix data was attributed to an OH-Ar stretching mode in the excited electronic state correlating with OH A $^2\Sigma^+$+ Ar. The vibrational motion was associated with OH interacting with a single Ar nearest neighbor, and was largely unaffected by the addition of other Ar neighbors.

---

[*]Alfred P. Sloan Research Fellow and Camille and Henry Dreyfus Foundation Teacher-Scholar.

Dynamics of Polyatomic Van der Waals Complexes
Edited by N. Halberstadt and K. C. Janda
Plenum Press, New York, 1990

143

Collisional studies of hydroxyl radicals in the A $^2\Sigma^+$ state with Ar and a variety of other partners have shown that attractive, long-range forces dominate the relaxation dynamics.[2] Vibrational energy transfer cross sections were found to be on the order of gas kinetic, and decreased sharply with increasing rotational level. The high efficiency for inelastic energy transfer processes has been attributed to the formation of a transitory "collision complex" in the entrance channel of the potential energy surface, while the rotational level dependence pointed to the anisotropic nature of the attractive forces.[3]

In this laboratory, we have taken a more detailed look at the intermolecular potential between OH and Ar in the gas phase by aggregating the collision partners together in a vdW complex, and studying the spectroscopy and dynamics of the prepared OH-Ar complex. We have performed spectroscopic experiments on the OH-Ar complex using laser-induced fluorescence[4-5] and infrared overtone spectroscopy[6] to examine the attractive well regions of the interaction potential between argon and the hydroxyl radical in the ground X $^2\Pi_{3/2}$ and excited A $^2\Sigma^+$ electronic states. Vibrational excitation of the OH moiety induces the unimolecular dissociation of the complex, providing us with a means to probe the dynamics taking place on these potential energy surfaces.[7] The results show surprising differences, in both the potentials and the dynamics, between the ground and excited electronic states. An overview of the latest developments in this rapidly evolving topic, both experimental and theoretical, are presented in this paper.

EXPERIMENTAL

The experimental method used to produce and detect complexes of the hydroxyl radical with argon has been described in detail elsewhere.[5] Nitric acid in argon carrier gas exits from a pulsed valve and flows through a quartz capillary tube affixed to the valve. An ArF excimer laser (193 nm) irradiates the gas pulse either within or just beneath the capillary, thereby producing hydroxyl radicals. Subsequent collisions in the expanding gas serve to both cool the OH radicals to the lowest few rotational levels and facilitate the formation of OH-Ar complexes. The photolysis products are probed by laser-induced fluorescence about various OH A $^2\Sigma^+$- X $^2\Pi_{3/2}$ transitions in the 280-360 nm region using the frequency-doubled or fundamental output of a XeCl excimer-pumped dye laser. The fluorescence is imaged through a monochromator and filters to reduce background signals. The monochromator is scanned to record dispersed fluorescence spectra.

Vibrational overtone spectra are obtained using an infrared pump-ultraviolet probe technique.[6] The OH-Ar complexes are prepared with two quanta of OH stretch ($v_{OH}=2$) by direct overtone pumping using tunable infrared radiation at 1.4 μm (~7000 cm^{-1}). The infrared is generated by Raman shifting (second Stokes) the output of a Nd:YAG pumped dye laser. The OH-Ar ($v_{OH}=2$) complexes are then probed by ultraviolet laser-induced fluorescence on OH-Ar transitions located near the OH 1-2 transition, as outlined above.

SPECTROSCOPIC RESULTS AND ANALYSIS

In this laboratory, we have recorded the fluorescence excitation spectra of OH-Ar complexes in the vicinity of the OH A $^2\Sigma^+$- X $^2\Pi_{3/2}$ 0-0, 1-0, 1-1, 1-2, 2-1, and 2-2 transitions.[4-6] As an example, a scan about the

OH 1-0 region is shown in Fig. 1. In addition to the well-characterized OH $P_1(1)$, $Q_1(1)$, and $R_1(1)$ lines, we observe a series of new spectroscopic features attributed to complexes of OH with Ar. Each of the transitions in this region involve the promotion of the OH moiety from the ground electronic state with no vibrational excitation to the excited electronic state with one quantum of OH vibration, $v_{OH}=1$.

The main progression features (labeled 1 to 6) have been assigned to a vibrational progression in the OH-Ar stretching mode in the excited electronic state which correlates to OH A $^2\Sigma^+(v'=1)$ + Ar $(^1S_0)$. Each of these bands exhibit similar rotational structure. The weak origin band of the 1-0 progression (not shown) has been identified at 34749.4 cm^{-1}. The features labeled A-D are also attributed to OH-Ar but have different rotational structure and likely involve bending vibrations. The OH-Ar bands in Fig. 1 are analogous in relative position and rotational structure to features previously identified in the OH 0-0 region.[4,5] Shifted to lower energy from peaks 1-5 are broad satellite features (labeled 1b through 5b) whose intensities are maximized by more severe cooling conditions (higher backing pressures and photolyzing higher in the capillary), suggesting that they arise from higher order clusters, OH-Ar$_n$ (n≥2).

A spectroscopic analysis of the rotation-vibration structure of the main progression features has been performed to evaluate the OH-Ar stretching potential for OH radicals in the A $^2\Sigma^+$ and X $^2\Pi_{3/2}$ states. Birge-Sponer plots of the main progression features are linear, indicating that a simple anharmonic oscillator characterizes the vibrational motion over most of the potential well. The Birge-Sponer analysis is used to provide an estimate of the binding energy of Ar to OH A(v'=0), as well as the fundamental vibrational frequency ω_e' and anharmonicity $\omega_e'x_e'$ of the OH-Ar vdW stretch.[5] Some of the derived potential parameters depend on the vdW

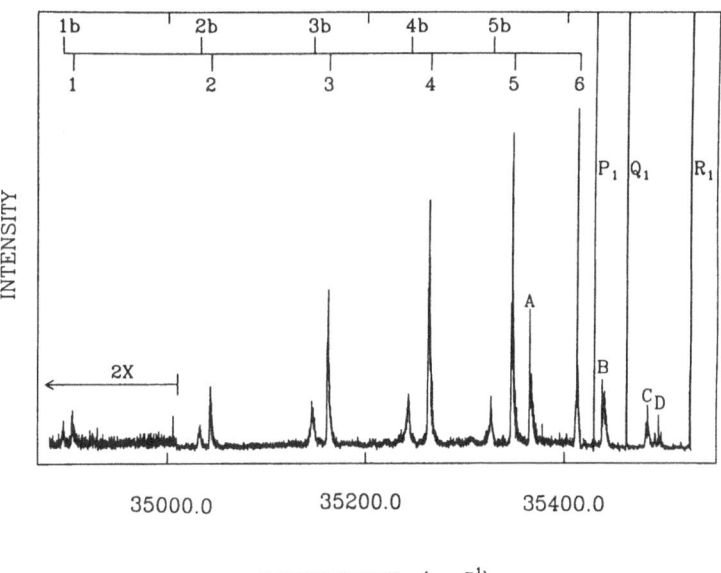

WAVENUMBER (cm^{-1})

Fig. 1. Fluorescence excitation spectrum of OH-Ar in the OH A-X 1-0 region.

Table 1. Vibrational Parameters (cm^{-1}) Evaluated from Birge-Sponer Analysis of OH-Ar Stretching Mode in the Excited Electronic State

	$v_{OH} = 0$	$v_{OH} = 1$	$v_{OH} = 2$
ω_e	170	174	171
$\omega_e x_e$	9	9	9
D_0	718	–	–
$\Delta D_0 (0 \rightarrow v_{OH})$[a]	–	+30	+21

[a]Derived from spectroscopic shifts of vibrational origin bands ($v_{vdW}=0$).

vibrational quantum number assignments. Our quantum number assignments have been made by analogy to a similar vibrational progression observed upon electronic excitation of hydroxyl radicals imbedded in an argon matrix.[1] A summary of the vibrational constants for the OH-Ar vdW stretch with $v_{OH}=0$ to 2 is given in Table 1.

The binding energy of Ar to OH A($v'=0$) is estimated from a linear Birge-Sponer extrapolation to be $D_0' = 718$ cm^{-1}. The binding energy of Ar to OH in the X($v''=0$) state is evaluated from the binding energy of OH-Ar in the excited state and the spectroscopic shift of the vdW vibrational origin ($v_{vdW}=0$) in the 0-0 progression (extrapolated) relative to the $P_1(1)$ line of free OH, yielding $D_0'' = 69$ cm^{-1}. Thus, electronic excitation of the OH radical results in an order of magnitude increase in the binding energy of Ar to OH.

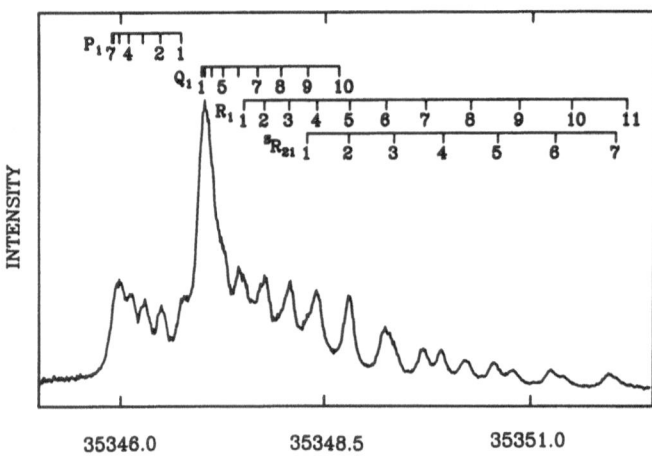

WAVENUMBER (cm^{-1})

Fig. 2. High resolution scan of the OH-Ar $v_{vdW}=5$ feature in the 1-0 progression. Superimposed are the calculated line positions from the rotational analysis.

Table 2. Rotational Constants (cm^{-1}) for OH-Ar in the
Ground and Excited Electronic States

	$v_{OH} = 0$	$v_{OH} = 1$	$v_{OH} = 2$
X $v_{vdW} = 0$	0.104	0.102	0.101[a]
A $v_{vdW} = 0$	-	0.167	-
1	0.164	0.159	-
2	0.157	0.153	-
3	0.148	0.145	-
4	0.139	0.135	(0.135)[b]
5	0.128	0.128	(0.125)[b]
6	0.114	0.114	-

[a] Obtained from vibrational overtone spectrum.
[b] Obtained from contour fits of low resolution spectra.

The vibronic bands in the OH-Ar stretching progressions exhibit analyzable rotational structure.[5,8,9] The rotational analysis is performed by assuming a linear geometry for OH-Ar in the ground and excited states. The linear model readily accounts for the experimentally observed rotational structure. A simultaneous least squares fit of the calculated P_1, Q_1, R_1, and $^SR_{21}$ rotational line positions to experimental positions in the 0-0, 1-0, and 1-1 regions yields the rotational constants given in Table 2 (columns labeled $v_{OH}=0$ and 1). The quality of the fit is illustrated in Fig. 2, where the calculated line positions are displayed above an experimental spectrum for the transition to the $v_{vdW}=5$ level in the 1-0 region. Rotational constants for vdW levels correlating to OH A(v'=2) have been obtained from contour fits of low resolution spectra in the 2-1 region. The compilation of rotational constants given in Table 2 shows a systematic trend of decreasing B_v with OH vibrational excitation in the ground and excited electronic states. This corresponds to a trend of increasing OH (center-of-mass) to Ar distance upon vibrational excitation of the OH moiety.

OH-Ar INTERACTION POTENTIALS

A Morse potential correlating with OH A(v'=0) + Ar ($^1S_0$) can be defined from the vdW vibrational constants and the vdW bond length at the equilibrium position of the potential well, $r_e' = 2.9$ Å, determined by an extrapolation of the vibrationally averaged rotational constants. A Morse potential, which approximates the actual excited state OH-Ar potential, is plotted in Fig. 3 as a function of the OH-Ar separation distance. The vibrational levels predicted for the OH-Ar vdW stretching mode are displayed; only the vdW vibrational levels 1-6 have been experimentally observed in this well.

Much less information is available about the ground state OH-Ar potential which correlates with OH X(v"=0) + Ar ($^1S_0$). The Morse potential shown in Fig. 3 is constructed using the ground state binding energy, the average vdW bond length at the zero point level (3.7 Å), and an estimate[5] for the vibrational frequency of the OH-Ar stretch. Although one excited vdW vibrational level is predicted to exist in the shallow ground state well, no vibrationally excited vdW levels have been detected to date.

The spectroscopic studies[4,5] have illustrated the dramatic change in the OH-Ar potential upon electronic excitation of the OH moiety from the $X\,^2\Pi_{3/2}$ to the $A\,^2\Sigma^+$ state. Two striking features are evident by comparison of the resultant potential curves. First, the binding energy of Ar to OH increases by an order of magnitude, from approximate values of 70 to 720 cm^{-1}, upon electronic excitation of the OH radical. Second, the OH (center-of-mass) to Ar separation distance at the equilibrium position of the OH-Ar potential decreases from 3.6 to 2.9 Å when OH is promoted from the X to the A states. The vdW bond length change yields spectroscopic access to a large portion of the excited state OH-Ar potential well.

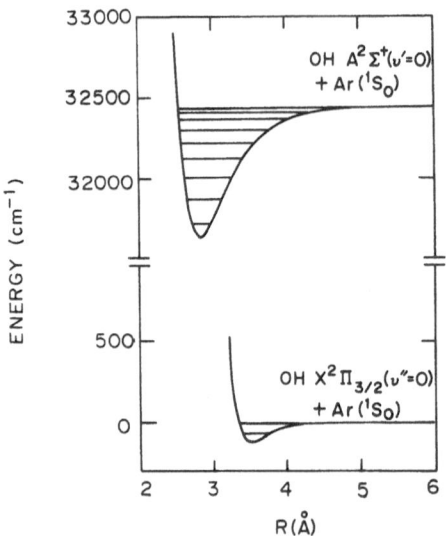

Fig. 3. Morse potential curves representing interaction potential between Ar and hydroxyl radicals in the A and X states as a function of Ar to OH (center-of-mass) distance.

Ab initio CALCULATIONS

Recently, Esposti and Werner[10] have calculated three dimensional *ab initio* interaction potentials for OH ($A\,^2\Sigma^+$) + Ar and OH ($X\,^2\Pi_i$) + Ar. In the excited electronic state correlating to OH(v=0) + Ar, the potential exhibits deep minima in the linear hydrogen-bonded OH-Ar and anti-hydrogen-bonded Ar-OH configurations. A contour diagram of the *ab initio* potential for OH ($A\,^2\Sigma^+$) + Ar is shown in Fig. 4. The potential is strongly anisotropic, with a high barrier separating the two isomers. The OH-Ar isomer has a well depth of 1100 cm^{-1} at an equilibrium distance of 2.8 Å (recall that the experimental value for r_e' is 2.9 Å), while the Ar-OH isomer has a comparable well depth (1000 cm^{-1}) but at a significantly different equilibrium distance of 2.2 Å. Charge transfer contributes to the strong bonding of Ar-OH, while the OH-Ar interaction is mainly "covalent" due to a correlation effect. The bound state levels supported by this potential have been calculated by Chakravarty and Clary.[11] Preliminary results suggest a high frequency excited state bending vibration (~400 cm^{-1}), which would make a sizable contribution to the zero point energy. The vibrational constants derived for the vdW stretch of the OH-Ar conformer are in good agreement with the experimental values presented above.

OH (A' $^2\Sigma^+$) + Ar

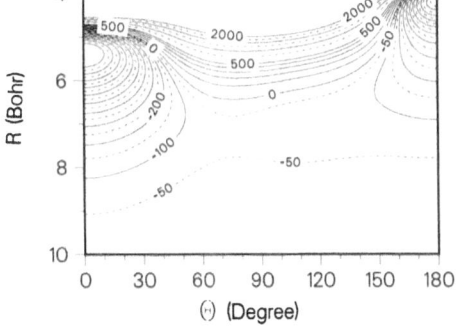

Fig. 4. Contour diagram of the *ab initio* potential energy surface for OH A $^2\Sigma^+$ + Ar (Ref. 10). Energy contours are labeled in in cm^{-1}, R is the Ar to OH (center-of-mass) distance, and θ the orientation angle between R and the OH axis. θ is 0° for the linear OH-Ar configuration.

Two different OH-Ar potentials correlate with OH (X $^2\Pi$) + Ar, depending on the orientation of the orbital containing the unpaired electron of the OH radical with respect to the OH-Ar plane. The *ab initio* calculations[10] indicate a double minimum potential in the ground electronic state (A', orbital of unpaired electron in OH-Ar plane), with a primary minimum for the linear OH-Ar configuration (~100 cm^{-1}) at a separation distance of 3.8 Å (similar to the experimental value for r_0'' of 3.7 Å). The second anti-hydrogen-bonded isomer is predicted to have a low barrier to conversion in the ground state. Thus, in contrast to the intermolecular potential in the A state, the ground state potential is shallow and slowly varying with respect to angle. At the zero point level, the complex is expected to undergo a wide amplitude bending motion. The *ab initio* results confirm the experimental observations of a strongly bound excited state potential with a minimum that is substantially displaced from the ground state minimum.

Chakravarty and Clary[11] have also simulated the positions and intensities of features in the ultraviolet excitation spectrum, based solely on the *ab initio* potential surfaces. All of the experimentally observed members of the OH-Ar stretching progression in the OH A-X 0-0 region are reproduced by the theoretical calculations. Additional bands are predicted to appear in the high energy region of the spectrum, analogous to the non-progression features identified in the experimental spectra. The latter are attributed to Fermi resonances, which mix bend or bend-stretch combinations with pure stretching vibrations.

VIBRATIONAL PREDISSOCIATION DYNAMICS

The major changes observed in the OH-Ar potential energy surface upon electronic excitation of OH prompted us to examine the dynamics occurring on these surfaces. The unimolecular dissociation dynamics of OH-Ar complexes were investigated by preparing the complexes with one quantum of OH vibrational excitation (v_{OH}), which is more than sufficient energy to break the OH-Ar bond. We have found that the resultant vibrational predissociation dynamics of OH-Ar differ enormously in its ground and excited electronic states.[7]

Preparation of OH-Ar complexes with one quantum of OH vibration in the excited electronic state results in a strong emission signal on the OH 0-0 transition, indicating rapid vibrational predissociation. Since no emission is observed from OH-Ar (v_{OH}'=1), an upper limit of 10 ns can be placed on the vibrational predissociation lifetime in the excited electronic state.[7] OH-Ar transitions in the 1-0, 1-1, and 1-2 regions result in emission from OH A $^2\Sigma^+$ v=0 photofragments. OH-Ar transitions in the OH 2-1 and 2-2 regions result exclusively in OH A $^2\Sigma^+$ v=1 products, as evidenced by emission only on the OH 1-0 and 1-1 transitions. Thus, OH-Ar complexes undergo vibrational predissociation with the loss of one quantum of OH vibrational excitation, following the Δv=-1 propensity rule observed in many other systems.

OH-Ar fluorescence excitation features in the vicinity of OH 1-1 and 2-1 transitions originate from OH-Ar complexes containing one quantum of OH vibrational excitation in the ground electronic state. Complexes of Ar with OH X(v"=1)† readily form and are apparently long-lived, persisting at least 30 μs after production.[7] From the pump-probe experiment described below, we know that OH-Ar ($v_{OH}"$=2) is similarly long-lived, with a vibrational predissociation lifetime of $\geq$ 6 μs.[6] Hence, the rate of vibrational predissociation is at least three orders of magnitude slower in the ground electronic state than in the excited electronic state. The marked increase in the rate of vibrational predissociation of OH-Ar upon electronic excitation points to an excited state interaction which effectively couples the OH vibration to the OH-Ar stretching mode. Analogous results have been obtained for vibrational relaxation of the OH radical, where vibrational energy transfer cross sections in the excited A state were found to be 10-1000 times greater than in the ground state with a variety of molecular collision partners.[13]

Rotationally resolved dispersed fluorescence spectra of the OH A$^2\Sigma^+$ (v'=0) products have been recorded following excitation of OH-Ar features (v_{vdW}=3-5) in the OH 1-0 region.[7] The emission profile shown in Fig. 5 is labeled to indicate the OH 0-0 transitions which contribute to the spectrum. Although the rotational distribution cannot be uniquely determined from the emission spectrum at the present resolution (60 cm^{-1}), the population of OH

---

†The present results differ from earlier work (Ref. 12) in which little vibrational excitation of the OH product was reported upon photolysis of HNO$_3$ at 193 nm.

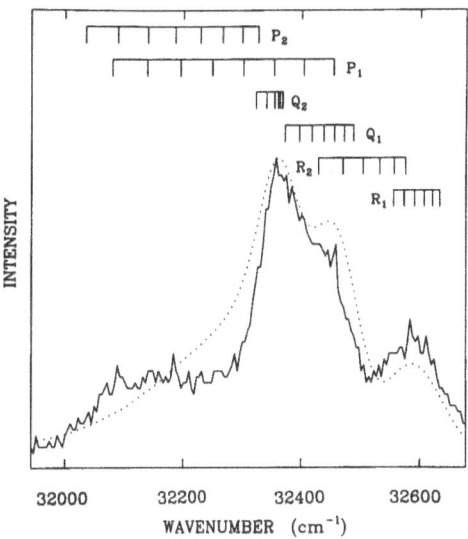

Fig. 5. Dispersed fluorescence spectrum from OH A(v'=0) products
following vibrational predissociation of OH-Ar complexes
with one quantum of OH vibrational excitation.

product rotational levels clearly deviates significantly from a Boltzmann
distribution (dotted line, 600 K). From the termination of the P_2 branch,

it is evident that the OH product rotational distribution abruptly ends
after N = 7 (944 cm^{-1}). Since rotational levels as high as N = 12
(2598 cm^{-1}) are energetically accessible, the cutoff suggests a restricted
zero-point bending motion in the complex. Using a purely impulsive model
based on the *ab initio* potential (a rotational sudden approximation), we
estimate that the N = 7 cutoff corresponds to a ± 20° range of Ar departure
angles with respect to a linear OH-Ar configuration.

OH VIBRATIONAL EXCITATION

Vibrational excitation of the OH subunit introduces the possibility of
vibrational predissociation of the OH-Ar complex as discussed above.
Vibrational excitation of OH-Ar complexes also induces small changes in the
intermolecular potentials which can be detected through detailed
spectroscopic studies. We have performed two types of experiments to
examine the influence of OH vibration on the intermolecular potentials -
laser induced fluorescence[8] and vibrational overtone spectroscopy.[6]

The effect of OH vibration on the OH A $^2\Sigma^+$+ Ar potential is determined
by comparing the OH-Ar fluorescence excitation spectra observed in the OH
A-X 0-0, 1-0, 1-1, and 2-1 regions. The vibrational frequency of the
fundamental OH stretch in OH-Ar is evaluated from the vibrational origin
bands (v_{vdW}=0) of the 0-0 (extrapolated) and 1-0 progressions, yielding
2957 cm^{-1}. This is approximately 30 cm^{-1} lower than the fundamental
vibrational frequency in uncomplexed OH, indicating a relatively strong
coupling between the OH and OH-Ar vibrational modes. It is this coupling

151

which leads to the rapid vibrational predissociation observed in the excited electronic state. The spectral shift also shows that Ar is 30 cm^{-1} more strongly bound to OH A $^2\Sigma^+$ in v'=1 than v'=0. Using a similar procedure in the 1-1 and 2-1 regions, we find that OH-Ar v_{OH}=2 is more deeply bound than OH-Ar v_{OH}=0 (by 21 cm^{-1}), although somewhat less strongly bound than OH-Ar v_{OH}=1.

The changes in the binding energy of OH-Ar upon vibrational excitation of OH within the ground electronic state is obtained through comparison of the transition energies for OH-Ar in the OH A-X 1-0, 1-1 and 1-2 regions.[6] The differences in the positions of excitation features originating from v_{OH}''=0, 1 and 2 and terminating on common vdW levels (v_{vdW}) of v_{OH}'=1 yield the vibrational frequencies of the OH fundamental and overtone transitions in the ground electronic state of OH-Ar, 3567.8 and 6970.4 cm^{-1}, respectively. These transitions are shifted 0.6 ± 0.1 and 0.9 ± 0.2 cm^{-1} towards lower energy of the corresponding transitions in free OH, indicating a weak perturbation of the OH vibration upon complexation with Ar. This is consistent with the very long vibrational predissociation lifetimes measured for v_{OH}=1 and 2 in the ground electronic state. The shifts are a measure of the relative binding energy of Ar to OH in the v=0, 1, and 2 levels of the X $^2\Pi_{3/2}$ state. The OH-Ar binding energy increases only slightly upon vibrational excitation of OH in the ground electronic state.

The vibrational overtone spectrum of OH-Ar complexes shown in Fig. 6 has been obtained by fixing the ultraviolet probe laser on an OH-Ar excitation feature associated with the OH A(v'=1) - X(v''=2) transition, while the infrared pump laser was scanned. The ultraviolet laser was fixed on the Q_1 bandhead, probing the N=1 to 4 levels of the vibrationally excited complex OH-Ar (v_{OH}=2). The infrared spectrum exhibits a distinctive P, Q, R branch structure, centered at 6970.4 ± 0.2 cm^{-1}, characteristic of the vibration-rotation spectrum of a molecule with electronic angular momentum ($\Lambda > 0$) in its ground electronic state. A linear OH-Ar complex will have the same ground state symmetry label as OH, namely $^2\Pi$.

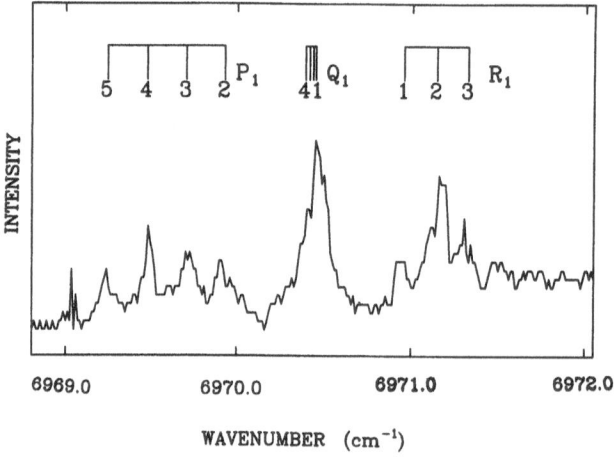

Fig. 6. Vibrational overtone spectrum of OH-Ar.

A rotational analysis of the overtone spectrum yielded vibrationally averaged rotational constants of 0.104 ± 0.003 cm^{-1} and 0.101 ± 0.003 cm^{-1} for OH-Ar in the v_{OH}=0 and v_{OH}=2 levels, respectively.[6] The rotational assignments and calculated line positions from the least squares fit are displayed above Fig. 6. The rotational constant for the vibrationless level of the ground state derived from the rotational analysis of the infrared spectrum is in agreement with the value independently obtained by rotational analysis of the features in the laser induced fluorescence spectra about OH A-X 0-0 and 1-0, $B_0'' = 0.104 ± 0.003$ cm^{-1}. The rotational constant decreases by 0.003 ± 0.003 cm^{-1} upon overtone excitation, equivalent to an increase of less than 0.1 Å in the OH-Ar bondlength.

A comparison of the experimental and theoretical results compiled to date on the open-shell OH-Ar complex with earlier work on hydrogen halide-rare gas systems indicates that OH-Ar in its ground electronic state is much like these closed-shell systems. Upon vibrational excitation of the OH bond, OH-Ar exhibits a long vibrational predissociation lifetime, an increased well depth, and a slight lengthening of the vdW bond. These are much the same effects observed for the hydrogen halide-rare gas complexes.[14] The *ab initio* intermolecular potential for OH-Ar is also similar to those determined for ArHF and ArHCl.[15] The similarities between OH-Ar and ArHCl or ArHF are not surprising in that the vibrational frequency, rotational constants, and dipole moments of the OH radical are comparable to those of the HCl and HF molecules.

OD-Ar EXPERIMENTS

Fawzy and Heaven[9,16] have examined the spectroscopic changes of the OH-Ar stretching progressions in the 0-0 and 1-0 regions upon H/D substitution. The fundamental vibrational frequency ω_e, anharmonicity $\omega_e x_e$, and rotational constant at the equilibrium position of the excited state potential well B_e, were found to decrease in OD-Ar relative to OH-Ar, as expected for the heavier isotope. The magnitude of the isotope effect on vibrational and rotational parameters, however, did not simply follow the isotope relationships of a pseudo-diatomic molecule (the OH molecule considered to be a point mass at its center-of-mass). This indicates that the OH-Ar stretching motion cannot be fully described by a pseudo-diatomic model.

Rostas and coworkers[17] have observed an excimer-type emission spectrum following laser photolysis of D_2O in high pressure argon which they ascribe to OD-Ar. The emission is attributed to transitions from the lower vibrational levels in the potential well correlating with OD A $^2\Sigma^+$ + Ar to dissociative regions of the ground state surface. The OD-Ar emission spectrum is very different from the dispersed fluorescence spectra recorded in this laboratory following excitation of the OH-Ar complex in the OH A-X 0-0 region (v_{vdW}=5).[8] In the latter case, emission occurs from the OH-Ar complex on bound-bound transitions terminating in the two spin-orbit components of the ground electronic state as well as on bound-free transitions to the ground state continuum.

CONCLUSION

The extensive experimental and theoretical work on the OH-Ar system in the past year has provided a wealth of new information on the intermolecular potential between argon and the hydroxyl radical in the ground X $^2\Pi_{3/2}$ and excited A $^2\Sigma^+$ electronic states. The intermolecular potential undergoes a dramatic change with electronic excitation of the OH moiety. The spectroscopic studies demonstrate that the equilibrium bond length decreases by 0.7 Å and the Ar becomes much more tightly bound to the OH, as evidenced by an increase in the well depth from approximately 70 to 720 cm^{-1}. The *ab initio* calculations are in excellent accord with the experimental results, indicating a much stronger interaction in the excited electronic state and a substantially reduced OH-Ar separation distance at the minimum of OH A $^2\Sigma^+$ state potential. Much smaller changes are observed in the potentials upon vibrational excitation of the OH bond. The resultant vibrational predissociation dynamics of OH-Ar complexes are strikingly different in the ground and excited electronic state. The rate of vibrational predissociation is found to be at least three orders of magnitude faster in the excited A state than in the ground state. The changes in the OH-Ar intermolecular potential upon electronic excitation of the OH radical are manifested in the full- and half-collision dynamics occuring on these surfaces. Theoretical calculations of the vibrational predissociation dynamics have not yet been performed, but are anticipated in the near future. These detailed experimental results provide insight into the physical origin of the OH-Ar interaction as well as present a rigorous test of the *ab initio* potentials and the theoretical methods used to evaluate the collision dynamics in OH-Ar and other open-shell systems.

ACKNOWLEDGEMENTS

The research performed in my laboratory has involved several colleagues, whose role in the experiments, analysis, and the continually evolving interpretation has been essential. I am pleased to acknowledge Mary T. Berry, Mitchell R. Brustein, Kenneth M. Beck, and Joseph R. Adamo for their participation. This research was supported by the Division of Chemical Sciences, Office of Basic Energy Sciences of the Department of Energy. Partial equipment support was provided by the University of Pennsylvania Research Foundation. I gratefully acknowledge the Natural Science Association at the University of Pennsylvania for a Young Faculty Award. I thank H.-J. Werner and D.C. Clary for stimulating discussions and W.M. Fawzy and F. Rostas for sharing their results with me prior to publication.

REFERENCES

1. J. Goodman and L.E. Brus, J. Chem. Phys. 67:4858 (1977).
2. R.K. Lengel and D.R. Crosley, J. Chem. Phys. 68:5309 (1978).
3. D.R. Crosley, J. Phys. Chem. 93:6273 (1989) and references cited therein.
4. M.T. Berry, M.R. Brustein, J.R. Adamo, and M.I. Lester, J. Phys. Chem. 92:5551 (1988).
5. M.T. Berry, M.R. Brustein, and M.I. Lester, Chem. Phys. Lett. 153:17 (1988).
6. K.M. Beck, M.T. Berry, M.R. Brustein, and M.I. Lester, Chem. Phys. Lett., in press.
7. M.T. Berry, M.R. Brustein, and M.I. Lester, J. Chem. Phys. 90:5878 (1989).
8. M.T. Berry, M.R. Brustein, and M.I. Lester, to be published.
9. W.M. Fawzy and M.C. Heaven, J. Chem. Phys. 89:7030 (1988).
10. A.D. Esposti and H.-J. Werner, to be published.
11. C. Chakravarty and D.C. Clary, to be published.
12. A. Jacobs, K. Kleinermanns, H. Kuge, and J. Wolfrum, J. Chem. Phys. 79:3162 (1983).

13. K.J. Rensberger, J.B. Jeffries, and D.R. Crosley, J. Chem. Phys. 90:2174 (1989).

14. Z.S. Huang, K.W. Jucks, and R.E. Miller, J. Chem. Phys. 85:6905 (1986); J.M. Hutson, J. Chem. Phys. 81:2357 (1984); G.T. Fraser and A.S. Pine, J. Chem. Phys. 85:2502 (1986) and references cited therein.

15. J.M. Hutson and B.J. Howard, Mol. Phys. 45:769 (1982); ibid., 45:791 (1982); J.M. Hutson, J. Chem. Phys. 89:4550 (1988) and references cited therein.

16. W.M. Fawzy and M.C. Heaven, to be published.

17. J.L. Lemaire, W.-Ü L. Tchang-Brillet, N. Shafizadeh, J. Rostas, and F. Rostas, to be published.

DISCUSSION FOLLOWING THE TALK BY DON TRUHLAR

P.R. Bunker

Herzberg Institute of Astrophysics
National Research Council
Ottawa, Ontario, Canada

The results of new calculations of the rotation-vibration energy levels of $(HF)_2$, HFDF and $(DF)_2$, from the ab initio potential of Kofranek et al[1,2], were presented by Bunker. It was shown that the minimum energy path connecting the two symmetrically equivalent minima also takes the molecule through the fully linear (HFHF) configuration. The barrier to straightening is almost exactly the same height as the tunnelling barrier. It was emphasized that over this part of the minimum energy path the rotational constant A changes very much with position along the path. Calculations of the rotational-tunnelling energy levels allowing for the motion through the linear configuration predict large changes in the trans-bending tunnelling energies with the K-rotational constant. These predictions closely match experimental results, where these are available, and lead to useful predictions for the tunnelling excited states that have not yet been observed. Modelling the dipole moment function for the dimer as the vector sum of two HF dipole moments leads to predictions for the intensities of the K-subbands which should be of use to experimentalists in their search for them. Bunker also reported calculations of the two HF stretching frequencies as a function of position along the minimum energy path. These calculations showed that the higher HF stretching frequency hardly changed whereas the lower frequency had its largest value at the saddle point. In his talk Truhlar presented similar results. These results do not lead to an explanation of the very small tunnelling splitting that has been observed[3] in the spectrum of both of these vibrationally excited states. Bunker reported preliminary calculations of the diagonal adiabatic correction to these effective tunnelling potentials, based on the ideas of Mills[4]. These do not show a large enough effect to account for the discrepency but further work in this direction is in progress. Bunker reported the results of a new ab initio potential for the HC1 dimer that is being calculated by Karpfen and colleagues. This potential has a minimum energy path that does not pass through the linear configuration and along which the A-rotational constant hardly changes. This leads to the prediction that the tunnelling energy levels will not change significantly with K. Nesbitt reported that this agrees with the experimental observations, and Klemperer said that he liked the idea which could be called a centrifugal distortion effect. Bunker expressed the hope that it would not be necessary to make such detailed `Super-Molecule´ calculations in order to get a potential function that was useful for assigning experimental spectra; the hope being that it might be simple to construct such a potential from the individual molecular moments without

Dynamics of Polyatomic Van der Waals Complexes
Edited by N. Halberstadt and K. C. Janda
Plenum Press, New York, 1990

doing such elaborate ab initio calculations. Stone said that this might not be so easy.

References

1. M. Kofranek, H. Lischka, and A. Karpfen, Chem. Phys. 121, 137 (1988).
2. P.R. Bunker, M. Kofranek, H. Lischka and A. Karpfen, J. Chem. Phys. 89, 3002 (1988).
3. A.S. Pine and W.J. Lafferty, J. Chem. Phys. 78, 2154 (1983).
4. I.M. Mills, J. Phys. Chem. 88, 532 (1984).

THE HF DIMER: POTENTIAL ENERGY SURFACE AND DYNAMICAL PROCESSES

Donald G. Truhlar

Department of Chemistry and Supercomputer Institute
University of Minnesota, Minneapolis, Minnesota 55455, U.S.A.

Abstract. In this paper I give an overview of the current status of knowledge of the four-body potential energy function and dynamics of the HF dimer. The discussion of potential energy functions includes both single-center expansions and multi-site functions. The discussion of dynamics includes both intramolecular processes of the van der Waals dimer and diatom-diatom energy transfer collisions.

1. INTRODUCTION

The dynamics of $(HF)_2$ has received considerable attention, both experimental and theoretical, but we are still very far from a complete understanding. This paper will review some of what has been learned.

In deciding the scope of this paper in relation to the title of the Workshop, I made a very broad interpretation of what constitutes the "HF dimer". In particular HF-HF collision processes are considered to be relevant since the wave functions describing these processes are the continuum solutions of the same Hamiltonian whose discrete solutions describe the bound van der Waals dimer. I even prefer to turn this around: in my own mind, one of the most intriguing aspects of van der Waals chemistry is that it provides information on the bound and quasibound states of potentials that describe interesting collision systems!

The present paper does *not*, however, include interactions of $(HF)_2$ with other species, such as occur in $(HF)_3$ and $HCN-(HF)_2$ complexes or in condensation to the liquid.

This paper is not a complete review of $(HF)_2$ structure or dynamics and in particular it does not include a full history or bibliography of the subject or a detailed discussion of the spectroscopy. The goal rather is to discuss a subset of the most important issues in a way that may be useful in planning further theoretical work. The emphasis is on papers

Dynamics of Polyatomic Van der Waals Complexes
Edited by N. Halberstadt and K. C. Janda
Plenum Press, New York, 1990

159

that, in the opinion of the author, are particularly relevant to current attempts to gain a complete understanding of the potential energy surface and dynamics of HF dimer. The subjects considered are the equilibrium geometry (section 2), the potential energy surface (section 3), the degenerate tunneling rearrangement (section 4), predissociation of quasibound states and photofragmentation of bound states of the dimer (section 5), and energy transfer collisions (section 6). The information about these aspects of $(HF)_2$ has been derived from thermodynamic analysis,[1,2] electronic structure calculations and potential energy surface modelling,[3-86] infrared and far infared spectroscopy,[87-110] radiofrequency and microwave spectroscopy,[111-115] vibrational relaxation and energy transfer measurements,[116-155] rotational relaxation and population transfer experiments,[144,153,156-163] crossed molecular beam studies,[164] photofragment spectroscopy,[165,166] and dynamical calculations of collisional energy transfer[14-17,28,43-46,57,58,71,72,83,117,123,124,155,167-199], isomerization,[93,200-207] and predissociation.[85,102,208-213]

2. EQUILIBRIUM GEOMETRY

The HF dimer is a nonlinear, planar hydrogen bonded complex in which both monomers exhibit large amplitude bending motions. Its equilibrium geometry, i.e., the minimum of the ground-state Born-Oppenheimer potential energy surface, is best known from theoretical calculations. The best calculations are by Frisch et al.[26] with second-order Møller-Plesset (MP2) perturbation theory and a triplet-zeta basis set (6-311G) augmented by diffuse functions on all atoms (++) and a double set of polarization functions (2d,2p) and by Kofranek et al.[35] with the coupled pair functional (CPF) technique and an extended valence basis (E) augmented by diffuse functions (+) and a double polarization set on F (2d) and a single polarization set (p) on H. The geometries calculated at these two levels are compared in Table 1, which also lists values calculated by the approximate-coupled-clusters-doubles (ACCD) calculation of Michael, Dykstra, and Lisy[70] with a triple-zeta(TZ)-plus-single-polarization (d,p) basis set. The geometrical quantities are the H-F bond lengths in the monomer (r_m), the hydrogen donor (r_D), and the hydrogen acceptor (r_A), the changes in the H-F distances upon association (Δr_D and Δr_A), the F-F distance (r_{FF}), and the angle of the F-to-H vector with the F-to-F vector in the donor (θ_D) and the acceptor (θ_A). The energetic quantities in this table will be discussed in Section 3.1.

Comparison of the geometry to the rotational constants of the microwave experiments is complicated by vibrational averaging over the large-amplitude vibrations in the latter. Howard et al.[113] found average r_{FF} values of 2.791 Å for $(HF)_2$, 2.783 Å for HFDF, and 2.778 Å for $(DF)_2$. They concluded that the van der Waals stretching motion is too small to account for this difference and suggested bending motions as its primary cause. A simple extrapolation yielded an equilibrium value of 2.72 Å with an uncertainty of at least 0.02 Å. Barton and Howard,[63] based on a more complete model, had obtained 2.68 Å. Michael et al.[70] estimated that averaging over only the van der Waals mode increases r_{FF} by 0.02 Å, which is consistent in direction, and Gutowsky et al.[114] estimate that their improved treatment of the average angular structure reduces the best estimate of r_{FF} by 0.004 Å. These analyses are roughly consistent with 2.69 ± 0.02 Å, and one concludes that the ab initio values in Table I may be too large by 0.05-0.12 Å. Dykstra's molecular-mechanics-for-clusters (MMC) was parametrized[84] to yield 2.74 Å, a value that does not take account of the effect[63,113,114] of bending motions on the F-F distance.

The theoretical calculations show a slight lengthening of the H-F bonds upon association, 0.004-0.006 Å for the hydrogen donor and 0.000-0.003 Å for the acceptor.

Howard et al.[113] estimated equilibrium values of $\theta_D = 10 \pm 6$ deg and $\theta_A = -117 \pm 6$ deg, whereas Gutowsky et al.,[114] including experimental hyperfine structure in the analysis, estimated $\theta_D = 7 \pm 3$ deg and $\theta_A = -120 \pm 2$ deg. These data are reasonably consistent with the best ab initio calculations as listed in Table I.

3. POTENTIAL ENERGY SURFACE

3.1. Near Equilibrium

Table I also lists, again for the three best ab initio studies, the equilibrium dissociation energy (D_e), zero point change upon association (ΔZPE), ground-state dissociation energy (D_0), and standard-state heat of association at 0 K (ΔH_0^0). The latter three quantities are all based on a harmonic treatment of vibrations.

The three best ab initio values for D_e range from 4.3 to 5.0 kcal/mol.[26,35,70] The best "purely" experimental value is 4.6 ± 0.2 kcal/mol (1622 ± 65 cm^{-1}), obtained by combining the value of $D_0 = 3.04 \pm 0.01$ kcal/mol (1065 ± 5 cm^{-1}), obtained Dayton et al.[165] by photofragment spectroscopy, with the empirical estimate of $\Delta ZPE = 1.6 \pm 0.2$ kcal/mol (557 ± 60 cm^{-1}) of Pine and Howard,[94] which they obtained using experimental values for the H-F stretch perturbations and the results[63] of Barton and Howard's surface modelling for the contribution of the van der Waals modes to ΔZPE. The theoretical values of D_0 in Table I are consistent with the experimental value within 0.0-0.6 kcal/mol (0.0-0.4 kcal/mol if we ignore the experimental uncertainty).

Leaving the equilibrium structure the first deviations of the potential surface are contained in the harmonic force field which is most conveniently characterized by the harmonic frequencies. Table II shows that the three best ab initio calculations do not agree nearly as well for the frequencies as they do for the geometry and dissociation energy in Table I.

Table I. *Ab initio* equilibrium geometries and variational total energies

	MP2 6-311++G(2d,2p)	CPF E+(2d,p)	ACCD TZ(d,p)
E (E$_h$)	-200.657	-200.615	-200.570
r$_m$	0.917	0.919	0.922
r$_D$ (Å)	0.923	0.924	0.926
Δr$_D$ (Å)[a]	0.006	0.004	0.004
r$_A$ (Å)	0.917	0.922	0.924
Δr$_A$ (Å)[a]	0.000	0.003	0.002
r$_{FF}$ (Å)	2.759	2.792	2.768
θ$_D$ (deg)	5.5	6.8	6.4
θ$_A$ (deg)	b	-114.45	-120.1
D$_e$ (kcal/mol)	5.0	4.3	4.6
ΔZPE (kcal/mol)[a,c]	1.9	1.7	1.5
D$_0$ (kcal/mol)[c]	3.1	2.6	3.0
ΔH$_0^0$ (kcal/mol)[a,c]	-3.7	-3.3	-3.6

[a]In this table Δ refers to the change upon association
[b]not given
[c]based on harmonic vibrational analysis

Table II. *Ab initio* harmonic frequencies

	harmonic				anharmonic	
	MP2 6-311++G(2d,2p)	CPF E+(2d,p)	ACCD TZ(d,p)	empirical	empirical	Ar matrix
ν$_m$ (monomer stretch)	4170	4135	4167	4139	3958	3919
ν$_1$ (r$_A$ stretch)	4127	4103	4103	4113	3931	3826
Δν$_1$	-53	-32	-64	-25	...	-93
ν$_2$ (r$_D$ stretch)	4054	4052	4056	4063	3868	3702
Δν$_2$	-16	-83	-111	-76	...	-217
ν$_3$ (symmetric bend)	582	510	420	520	304	561
ν$_4$ (antisymmetric bend)	231	216	127	337	160	400
ν$_5$ (r$_{FF}$ stretch)	163	150	167	178	148	263
ν$_6$ (torsion)	516	413	a	430	370	446

[a]not calculated

It is important in comparing theoretical and experimental frequencies to put them on the same basis, harmonic or anharmonic. Table II gives both kinds of frequencies as obtained empirically. The values of υ_3, υ_4, and υ_5 are from Barton and Howard,[63] based on quantum mechanical vibrational energy calculations for a model empirical potential energy surface, and the values of υ_1 and υ_2 and the anharmonic value of υ_6 are based on the spectra of Puttkamer and Quack.[105,108a] The harmonic υ_6 is based on the fit of Hancock et al.[81] to Barton and Howard's[63] out-of-plane bend potential. The empirical υ_5 is based[63] in part on the centrifugal distortion constants; a later analysis[113] gave 153 cm^{-1}. I have not included an early report[87] of a dimer absorption at 381 cm^{-1}.

The CPF harmonic vibrational frequencies are in best overall agreement with the empirical harmonic ones.

The final column of Table II shows matrix isolation spectra from Redington and Hamill[102] and Andrews and Johnson[99] as reassigned by Redington and Hamill.[102] The frequencies appear to be highly perturbed and for some reason are closer to the gas-phase harmonic values than the gas-phase anharmonic ones.

Dykstra[84a] has recently estimated gas-phase harmonic transition moments for υ_3-υ_6, but only for υ_5 has the effect of anharmonicity on these quantities been estimated.[70]

3.2. Saddle Point Properties

Moving along to a larger amplitude deviation from equilibrium, we turn our attention to the saddle point for hydrogen bond switching. The need for a very extensive basis set to treat (HF)$_2$ accurately has been discussed by many authors.[29,67,69] Nevertheless, as we summarize the state of our knowledge about the whole potential energy surface, we must continually assess the reliability of features that have been studied explicitly only with dangerously incomplete basis sets. Table III summarizes several calculations on the geometry and harmonic vibrational frequencies of the hydrogen-bond-switching saddle point. This saddle point has a C_{2h} parallelogram geometry completely specified by giving two distances and one angle. Four results are shown. The first two based on Hartree-Fock self-consistent-field (SCF) calculations, one[23] with a double-zeta-plus-polarization, DZ(d,p), basis set and one[30] with a quadruple-zeta-plus-double-polarization-on-F-and-single-polarization-on-H, QZ(2d,p), basis set. The other two calculations, the ACCD/TZ(d,p) calculations,[81] and CPF/E+(2d,p) calculations,[35] include electron electron and are in good agreement with each other, although the latter gives a larger (and presumably more accurate since the basis set is bigger) value for r_{FF} by 0.04 Å. The SCF calculation is less accurate for the monomer stretches, but gives reasonable results for the other quantities.

There is little empirical information about the C_{2h} saddle point with which to compare. The most reliable empirical estimates are probably those derived from the model potential surface of Barton and Howard.[63] They obtained $r_{FF} = 2.70$ Å, $\theta_{HFF} = 61.5$ deg, $D^{\ddagger} = 4.0$ kcal/mol, and $V^{\ddagger} = 0.9$ kcal/mol. The fit of Hancock et al.[81] to their out-of-plane bend potential gives $\upsilon_4 = 522$ cm^{-1} at the ACCD/TZ(d,p) saddle point geometry,

106-153 cm^{-1} higher than the *ab initio* values. However, Howard stated[63] that the torsional potential is poorly determined by their fit.

The (HF)$_2$ potential surface also contains a higher-energy D$_{2h}$ saddle point which is the transition state for the exchange reaction HF...H'F' → H'F'...HF. The best information on the saddle point comes from the SCF/DZ(d,p) calculations of Gaw *et al.*[23] They obtained a potential energy barrier of 52.7 kcal/mol relative to the equilibrium dimer. SCF calculations are notorious for overestimating exchange barriers, so this value should be re-examined with correlation energy and a better one-electron basis set.

3.3. Global Surface Properties

Several global potential functions have been proposed. Yarkony *et al.*,[20] Cournoyer and Jorgensen,[47] and Jorgensen[48] carried out extensive calculations at the SCF

Table III.   *Ab initio* geometries, binding energies, potential energy barriers, and harmonic vibrational frequencies at the C$_{2h}$ saddle point

	SCF DZ(d,p)	SCF QZ(2d,p)	ACCD TZ(d,p)	CPF E+(2d,2p)
r_{HF}(Å)	0.915	a	0.925	0.922
r_{FF} (Å)	2.721	2.80	2.757	2.796
θ_{HFF} (deg)	53.4	57	54.7	54.2
$D^{\neq}$ (kcal/mol)	3.9	2.7	3.5	3.3
$V^{\neq}$ (kcal/mol)	0.8	1.1	1.1	1.0
$V^{\neq}$ (cm^{-1})	276	385	385	357
υ_1 (H-F stretch)	4376	b	4099	4078
υ_2 (symmetric bend)	565	b	564	520
υ_3 (r_{FF} stretch)	152	b	155	132
υ_4 (torsion)	414	b	c	367
υ_5 (H-F stretch)	4394	b	4114	4097
$\upsilon^{\neq}$ (antisymmetric bend)	189i	b	198i	203i

^{a}not reported
^{b}not calculated
^{c}not calculated at ACCD/TZ(d,p) level

level with fixed monomer separations, and these have been fit to various functional forms.[38,47,48] The Yarkony *et al.* calculations involved a DZ(d,p) basis, but the Cournoyer-Jorgensen and Jorgensen calculations did not include polarization functions, which makes them unreliable.

There have also been some attempts to extend the Yarkony *et al.* calculations. Klein *et al.*[50] reported a fit to which they added a dispersion interaction, retaining the restriction to fixed monomer distances, and Redington and Hamill[102] generalized their result to non-fixed monomer bond lengths by adding Morse curves. Gianturco *et al.*[58] and Schwenke and the author[73] extended the Alexander-DePristo fit[38] to non-fixed monomer bond lengths by making assumptions about logarithmic derivatives but did not include dispersion. The latter of these extensions is called the modified Alexander-DePristo (MAD) surface. Poulsen, Billing, and Steinfeld[44,46] fit the Yarkony *et al.* results to a functional form designed to allow variable monomer bond lengths *and* they included dispersion. Three of the surfaces[38,44,50] based on the Yarkony *et al.* SCF calculations have been compared to the Barton-Howard[63] empirical surface, which also involves the fixed monomer restriction, by Nyeland *et al.*[45]

Brobjer and Murrell[62] also created a potential for fixed monomer bond lengths by combining SCF results with dispersion energies.

Cournoyer and Jorgensen[49] parametrized a site-site interaction model to simulations on liquid hydrogen fluoride. Because this simulation neglected the many-body effects and the potential has a restrictive functional form, the resulting potential is not particularly accurate for the dimer. Halberstadt *et al.*[85] defined an extension of this to allow variation of the monomer bond lengths, but the extension was not calibrated against independent information. E.g., the charges were not redistributed to give the correct dependence of the dipole and quadrupole moment on r_{HF}. Another deficiency of the Cournoyer-Jorgensen[49] and extended Cournoyer-Jorgensen (ECJ)[85] potentials is that the well depth is too large[83,85]

Additional globally defined surfaces in the literature include the pairwise potential of Berard and Thomarrson,[28] the semiempirical valence bond potential of Wilkins,[43] and the modified Stockmayer potential of Coltrin *et al.*[57]

The above potential surfaces all have serious deficiencies. Some of them, however, may be useful for model studies, and it is important to understand their accuracy (or lack thereof) to learn as much as possible from dynamical simulations which have already been performed using them.

The most accurate surfaces for $(HF)_2$ are the Barton-Howard (BH) empirical surface[63] and four surfaces[27,36,81,83] based on correlated electronic structure calculations with polarized one-electronic basis sets. The BH surface is defined only for fixed monomer bond lengths. Both the BH surface and the surface of Hancock, the authors, and Dykstra (HTD)[81] are calibrated only in the van der Waals region. The surfaces of Redmon and Binkley (RB),[27] Schwenke and the author (RBST),[83] and Bunker, Kofranek, Lischka, and Karpfen (BKLK)[36] all include *ab initio* data about the higher repulsive walls as well. However the BKLK surface is only claimed to be valid for $r_{FF} > 1.85$ Å, which is the range covered by the *ab initio* data.

The BH surface was fit to rotation and centrifugal distortion constants, the average dipole moment and its centrifugal distortion, and the tunneling splittings for K = 0 and 1. The van der Waals stretch was treated as a one-dimensional problem with an effective potential given by an adiabatic treatment; rotation was included by perturbation theory.

The HTD surface[81] is a fit to ACCD/TZ(d,p) calculations[70,81] for planar geometries augmented by a global representation of the BH empirical out-of-plane potential. The in-plane potential is based on a ten-site model. The in-plane potential was fit in two steps, taking advantage of the fact that correlation energy is a smoother function of geometry than is the SCF energy. The root-mean-square fitting error in the final step was only 0.03 kcal/mol at 378 points. This is very good, but it is much harder to achieve this kind of fit when more geometries with large repulsion energies are included.

The RB potential[27] is based on 1332 points at the Møller-Plesset fourth-order (MP4) level with 6-311G(d,p) basis set. The fit involves 174 parameters. One deficiency of this surface is that none of the 1332 distances has *both* monomers simultaneously displaced from equilibrium.

The RBST surface[83] is based on 1449 *ab initio* points at the MP4/6-311G(d,p) level, with most of the 117 new points having both monomers simultaneously displaced from equilibrium. The fit was accomplished in two steps, motivated by model studies[71] showing the importance of the vibrational forces at the translational turning points in determining vibrational energy transfer probabilities. First the points with fixed monomer bond lengths (i.e., $r_{HF} = r_m$) were fit. Then the forces along the HF bonds were fit, and finally the second derivatives along the HF bonds were fit, including the cross second derivative. Since the emphasis in this fit was on repulsive walls and the vibrational forces, the van der Waals well is reprsented with only moderate accuracy: $r_D = 0.915$ Å, $r_A = 0.925$Å, $r_{FF} = 2.551$ Å, $\theta_D = 50.7$ deg, $\theta_A = -108.8$ deg, $D_e = 5.1$ kcal/mol (compare Table I).

The BKLK surface[36] is based on 1061 *ab initio* points at the CPF/E+(2d,p) level. The final fitting involved 42 adjustable parameters and 7 constrained parameters. The standard deviation for the final *weighted* fitting, which emphasized energies within 6.3 kcal/mol of the minimum, was 0.08 kcal/mol.

The BH and ST potentials are expressed as truncated spherical harmonic expansions. This type of representation must be used cautiously since it is known that such expansions are slowly convergent.[83,191,192]

Recent work[42c] has produced more accurate values of the dispersion coefficients than were available previously for $r_D = r_A = r_m$ for both isotropic and anisotropic terms through R^{-10}, where R is the separation distance of the monomers. It would be desirable to constrain future analytic representations to agree with these leading terms, as well as with the leading terms in the static-moment and induction multiple expansions, and if possible the dependence of the long-range potential on monomer separation should be converged as well.

Although great progress has been made, no one potential appears to have all the attributes we would like to see in the "final" function.

4. DEGENERATE TUNNELING REARRANGEMENT

The tunneling interconversion of HF...HF to FH...FH was first observed by Dyke et al.[111] in molecular beam electric resonance studies of the radiofrequency ($\Delta M_J = \pm 1$) and microwave ($\Delta J = \pm 1$, $\Delta M_J = 0$) spectra. They also observed the splitting for the perdeutero dimer. (The mixed dimer, HF...DF, has two different isotopomers[114,115] rather than a splitting.) The splittings which have been observed so far for HF...HF are summarized in Table IV.[93,106-108,111,113,115] (Note that HF dimer is a nearly symmetric top so the figure-axis angular momentum quantum number K is a useful quantum number.)

Table IV shows several very interesting trends. Excitation of either monomer stretch (υ_1 or υ_2) lowers the tunneling probability, whereas excitation of the torsion (υ_6) increases it. The rotational dependence is very large. There have been several explanations.

Mills[200] used a model based on zero-order adiabatic stretches with a Born-Huang-type diagonal contribution $B_{v_1 v_2 \ldots}(r)$ arising from the kinetic energy along the tunneling coordinate r. He concluded that $B_{10\ldots}(r)$ and $B_{01\ldots}(r)$ are larger than $B_{00\ldots}(r)$ since the second-derivative kinetic energy operator is sensitive to the change in form of the high-frequency vibrations as the system progresses along s where the vibron must be transferred along with the hydrogen bond. A very simple estimate yielded a peak of about 100 cm^{-1} in $B_{10\ldots}(r)$. This argument was discussed further by Pine et al.[93] who concluded that a 100 cm^{-1} shift could indeed be consistent with the observed dependence of the splitting on v_1 and v_2. They[93] also pointed out that a small shift in tunneling path could also explain the effect, and they raised the question of whether oscillating transition dipoles may effect the barrier. They concluded that the latter effect is small.

Pine et al.[93] also discussed the rotational dependence of the tunneling splitting and said it could be explained as either an effect of centrifugal distortion or a difference in the effective A rotational constants for the two tunneling levels. They explained the dependence on stretch excitation as an effective increase in the tunneling barrier and/or length of the pathway.

Puttkamer, Quack, and Suhm[106,107] also explained the dependence of tunneling splittings on rotational and vibrational excitation with simple models. They explained the effect of K as a centrifugal distortion effect that tends to move the H-F bonds into positions closer to perpendicular to the F-F axis, which is the same direction as required to initiate

Table IV.   Tunneling splittings (cm^{-1}) as a function of vibrational quantum numbers (row headings) and the figure-axis angular momentum quantum number K

	000000[a]	100000	010000	000001
HF...HF				
K = 0	0.66	0.216	0.233	...
K = 1	1.06	0.35	0.34	1.62
K = 2	2.00	0.71	...	3.44
K = 3	3.8	1.4	...	...
K = 4	...	3.3	...	...
DF...DF				
K = 0	0.053	0.0163	0.0164	...
K = 1	0.069	0.0223	...	...
K = 2	0.115	...	...	...

[a]The column headings are the vibrational quantum numbers $v_1 v_2 ... v_6$.

tunneling. They suggested that the variations of υ_1 and υ_2 along the interconversion coordinate might be sufficient to explain the dependence on v_1 and v_2, if these charges are considerably larger than the small (≤ 26 cm^{-1}) changes predicted by the CPF calculations. They also pointed out that calculations neglecting vibration-rotation coupling might overestimate the tunneling splittings for $v_1 = 1$ or $v_2 = 1$ since they experimentally *did* observe some vibration-rotation mixing. Puttkamer and Quack explained the dependence of the tunneling splitting on torsional mode excitation as due to the weaker, and thus longer, hydrogen bond in excited torsional states.

Fraser[205] has recently presented additional modelling calculations using the adiabatic model of Mills without $B_{v_1 v_2}...(r)$ and also using a model based on localized vibrational modes coupled by transition dipoles, which he called the coupled diabatic potential (CDP) model. The adiabatic model was further simplified, and it yielded 0.85-0.90 cm^{-1} for $v_1 = 1$ and 0.48-0.51 cm^{-1} for $v_2 = 1$, much larger than the experimental results of 0.22-0.23 cm^{-1}. (The adiabatic barrier for $v_2 = 1$ was assumed to be 63 cm^{-1} higher than for $v_1 = 1$, whereas the ACCD and CPF harmonic values, from Tables II and III, are 26-43 cm^{-1}. However, anharmonic values and values for fit surfaces may be significantly different; see below.) The CDP model is roughly consistent with experiment if the transition dipoles in the complex are 1.4-2 times larger than in the unperturbed monomer. The CDP tunneling splittings for $v_1 = 1$ are 1.08-1.81 times larger than those for $v_2 = 1$, in poor agreement with the experimental ratio of 0.92. Further discussion of the CDP model and $B_{v_1 v_2}...(r)$ has been provided by Sibert,[206] who stressed the importance of the non-Born-Oppenheimer terms.

There have been four more complete attempts to calculate the tunneling splittings, in which multidimensional potential energy surfaces were used. Three of the calculations, by Barton and Howard,[63] by Hancock, the author, and coworkers,[203] and by Bunker *et al.*,[207] considered only $v_1 = v_2 = 0$. One set of calculations, by Hancock and the author,[204] also considered the effect of exciting υ_1 or υ_2.

Barton and Howard[63] used their calculation as part of their modelling effort to obtain an empirical potential energy surface. They treated the HF bonds as rigid and obtained the tunneling effect implicitly by diagonalizing a four-dimensional Hamiltonian.

Hancock, the author, and coworkers[203] used the resulting (BH) potential surface in a semiclassical calculation of the tunneling splitting. An advantage of the semiclassical approach is that it enables one to interpret the tunneling event in terms of an effective potential and effective reduced mass as functions of distance s along a semiclassical tunneling path. The effective potential is completely adiabatic without $B_{v_1 v_2 \ldots}(s)$, and the effective reduced mass is less than the mass to which the coordinates are scaled to account for path shortening effects (negative centrifugal effects) by which the hydrogens move along a dynamically optimum tunneling path shorter than the minimum energy path. The calculations reveal that the actual tunneling path is shifted from the minimum-energy path toward geometries with smaller r_{FF} values. The semiclassical calculation yields a tunneling splitting 41% lower than the experimental one, indicating that the semiclassical adiabatic picture is qualitatively correct.

In a second paper, Hancock and the author[204] used the HTD potential surface and performed calculations for both the ground state as well as the stretch-excited states with υ_1 and υ_2 excited. These calculations include all six vibrational degrees of freedom. The calculated tunneling splittings are 0.61, 0.36, and 0.52 cm^{-1}, in comparison with the experimental values of 0.66, 0.22, and 0.23 cm^{-1}. Thus the decrease of the tunneling probability upon stretch excitation is accounted for adiabatically, but the equivalence of the tunneling probability for the two possible excitations is not. We concluded that a dominant reason for the decrease in tunneling probability is a raising of the effective barrier for tunneling when a monomer is excited. The effective barriers obtained in these calculations, including anharmonicity, were 292 cm^{-1} for the ground state, 527 for υ_1 excited, and 371 for υ_2 excited, as compared to the classical barrier height of 385 cm^{-1} for this potential surface. (The anharmonic values of υ_1 and υ_2, calculated by subtracting the fundamental from the overtone, for this surface are 3983 and 3928 cm^{-1}, somewhat higher than the empirical anharmonic values in Table II.) The coordinates were scaled to a reduced mass of 1 amu in all cases, but—because the minimum-energy paths are curved—the effective reduced masses are 0.48-0.77 amu at various points in the tunneling regions; this accounts in the small-curvature approximation for the deviation of the tunneling path from the minimum energy path due to reaction-path centrifugal effects. (The tunneling probabilities per vibration along the tunneling coordinate are $1\text{-}3 \times 10^{-4}$.) An interesting observation from these calculations is that for v_1 or $v_2 = 1$, tunneling occurs at energies 8818-8878 cm^{-1} (25.2-25.4 kcal/mol) above the dimer classical equilibrium energy, so it is necessary to know the potential energy surface for this energy range and a bit higher as well to calculate the process reliably. Other significant conclusions include: (i) The curvature of the MEP is very important, increasing the tunneling probabilities by a factor of about $2\frac{1}{2}$. (ii) The tightening of the symmetric bend is important on the ACCD surface, and it raises the effective barrier for tunneling. (Interestingly the symmetric bend frequency does not change significantly on the CPF surface.) (iii) Anharmonicity is very important, both in determining the height and shape of the effective barrier and also by determining the energy at which the tunneling occurs.

In the most recent study Bunker et al.[207] used a quantum mechanical reaction path involving approximations to the effective moment of inertia and refitting the potential along a pre-selected path. The ground-state tunneling splitting was calculated as a function of K, yielding 0.65, 0.98, 1.98, and 4.4 cm^{-1} for K = 0-3, in good agreement with the values in Table IV. For (DF)$_2$ they obtained 0.04, 0069, and 0.115 cm^{-1} for K = 0-2. It is not clear whether the agreement with experiment is fortuitous since the monomer stretches and the torsion degree of freedom are fixed in these calculations, they did not include the energetic effects of vibrations orthogonal to the selected path, and they did not consider the deviation of the tunneling path from the selected path due to reaction-path centrifugal effects. These would appear to be serious approximations for quantitative work. But the calculations do show that a significant part of the K dependence comes from an internal centrifugal potential proportional to K^2.

5. PREDISSOCIATION AND PHOTOFRAGMENTATION

The best estimates of the predissociation lifetime come from observations of spectral line widths larger than can be accounted for by pressure or Doppler broadening. If $\Delta\bar{v}$ is the contribution of the lifetime τ to the full width at half maximum of the spectral line (after removing pressure and Doppler effects as well as saturation effects and power broadening), then $\tau = (\pi c \Delta\bar{v})^{-1}$. The best data to date for the v_1 and v_2 fundamentals are apparently those of Pine and Fraser,[96] and the lifetimes inferred from their work are summarized in the first three data rows of Table V. These lifetimes correspond to $\Delta\bar{v}$ in the range 0.002-0.01 cm^{-1}. (Earlier measurements by Pine et al.[93] and Huang et al.[109] gave $\tau = 24$ ns for v_1 and $\tau = 0.8$ or 1.0 ns for v_2 but did not resolve any dependence on K or the tunneling state.)

Puttkamer and Quack[104,105] have obtained lower bounds for the lifetimes of the first and second overtone states and these are listed in rows 4 and 5 of Table V. In a very recent paper[108a] they revised the estimate for $v_1 = 2$, K = 0 to > 0.05 ns and pointed out a "less likely" assignment by which the lifetime might be as short as 0.01 ns for $v_1 = 2$, K = 2. Most recently, Fraser and Pine[97] measured the predissociation lifetime of HF...DF

Table V. Predissociation lifetimes (ns)

mode	K	tunneling level 0	tunneling level 1
HF...HF			
υ_1	0	50	34
υ_1	1	31	27
υ_2	0	1.0	1.0
$2\upsilon_1,2\upsilon_2$	...	>0.04[a]	
$3\upsilon_1,3\upsilon_2$	...	>0.04[a]	
HF...DF			
υ_1	1-0	14[b]	

[a]No information on tunneling level dependence
[b]There are no tunneling split levels in this sytem

complexes in which the hydrogen accepting HF monomer bond is excited, and these results are also given in Table V.

Early theoretical work on predissociation employed a vibration-to-translation (V-T) mechanism, which predicted 8×10^5 ns,[208] which is clearly inconsistent with Table V. In later work, Ewing introduced the rotational channels of the HF fragments with a vibration-to-translation-and-rotation (V-T,R) mechanism. A first-order perturbation theory treatment involving decoupled channels, which is not particularly reliable, showed an increase of only one order of magnitude in the predissociation rate with inclusion of rotation;[209] in fact, however, τ was now calculated[209,210] to be 2×10^{10}-10^{11} ns due to other changes in the model. In another modification of the model,[211,212] a localized interaction between the $v_1 = 1$ and $v_3 = 1$ vibrationally adiabatic potential curves was postulated, and the calculated lifetime was reduced to 0.1 ns. Since υ_3 is a low-frequency mode, it is of course quantitatively unreliable to treat it as adiabatically decoupled, but this calculation does indicate that inclusion of rotation can have an enormous effect.

Halberstadt *et al.*[85] performed coupled-channel calculations of both the resonance width and its partial widths, which determine the decay probabilities into individual product states. Their calculations were based on the ECJ potential energy surface (abbreviations for potential energy surfaces are defined in Section 3) and treated both the orientation and the stretching coordinate of the hydrogen accepting monomer as fixed. The calculated lifetime of the complex was 6.4 ns ($\Delta \bar{v} = 0.0017$ cm^{-1}), in rough agreement with the current best experimental value of 1.0 ns. The most populated final rotational levels of the hydrogen donating HF fragment were predicted to be the highest energetically accessible ones.

A very valuable aspect of the study of v_2 predissociation by Halberstadt et al.[85] is that is provides tests of three approximate decoupled theories. It is important to understand the validity of such models since, if valid, they provide appealing and useful physical pictures of the dynamic event. Neglecting rotation raised the calculated lifetime to 3.5×10^7 ns, decoupling the diabatic rotational states gave $\tau = 0.004$ ns, and decoupling the vibrationally adiabatic states gave $\tau = 0.003$ ns. Not only do the decoupling approximations lead to too short lifetimes; they also lead to incorrect final rotational distributions. This shows that consideration of exit channel interactions is critical for understanding the final rotational distributions.

Pine[95] provided a rationalization of the v_1 vs. v_2 dependence of τ in terms of mass-weighted projections of the two H-F stretching motions on the hydrogen bond. However this model is inconsistent with the direction of the later observed K dependence of τ.[96]

Dayton et al.[165] measured photofragmentation angular distributions for both v_1 and v_2 excitations. In the case of v_2 excitation they found a large peak associated with the highest energetically allowed state (which is consistent with earlier studies[91] showing that little energy appears as product translation), and this state, the $(j_1',j_2') = (2,11)$ state, is barely energetically allowed. This may suggest that this near resonance is one reason why $\tau(v_2) \ll \tau(v_1)$, although a similar momentum gap argument fails[97] to predict $\tau(v_1,K=0) > \tau(v_1,K>0)$, which is a much less dramatic effect. Surprisal plots[166] of the final state distributions indicate that the preferred channels are those in which one monomer fragment is in high j', while the other is in low j' states. For v_1 excitation the low j_1'-high j_2' propensity is much stronger than the energetics or than found in the calculations of Halberstadt et al. The propensity observed is consistent with an impulsive dissociation that tends to excite the hydrogen donating monomer much more than the hydrogen accepting one.[166]

6. ENERGY TRANSFER COLLISIONS

Vibrational relaxation may occur by vibration-to-vibration (V-V), vibration-to-rotation (V-R), and vibration-to-translation (V-T) energy transfer, or usually by some combination. Rotational relaxation occurs by rotation-to-rotation (R-R) and rotation-to-translation (R-T) energy transfer.

6.1. Rotational Energy Transfer

Rotational energy transfer in HF is very fast, with rotational relaxation occurring at about gas kinetic rates. When the rotational distribution in HF gas is perturbed by an excess population in the $v = 2, j = 3$ state, it comes to rotational equilibrium in about one half the average time for an HF molecule to suffer a hard sphere collision.[162] It is inferred from fitting schemes that multiquantum transitions are significant, contributing about 25% to the overall rate.[163] Rotational relaxation is also reasonably rapid for higher j, in the range j = 11-13, but it is several times slower than for low j.[153,153a] Taatjes and Leone[153a] measured a relaxation rate for HF($v = 0, j = 13$) by HF of 1.8×10^{-10} cm^{3}molecule^{-1}s^{-1}.

Copeland and Crim extracted a matrix of R-R rate constants from their experiments by three different fitting schemes.[162,163] The final rate coefficients range from 8×10^{-10} cm^{3}molecule^{-1}s^{-1} for $j = 0, j' = 1$ to 2×10^{-12} cm^{3}molecule^{-1}s^{-1} for $j = 0$ or $1, j' = 6$. It is customary to convert energy transfer coefficients into collision efficiencies (or their inverses, the collision numbers). Collision efficiency is defined as the rate coefficient for relaxation divided by the collision rate coefficient, which is ambiguous, but is most commonly defined by the hard-sphere collision formula $k_{col} = \left(8\tilde{k}T / \pi\mu\right)^{1/2} \pi d_{AB}^2$. A reasonable value for d_{AB} is 2.65 Å, as obtained from the spherical average of the potential,[195] and this yields $k_{col} = 1.75 \times 10^{-10}$ cm^{3}molecule^{-1}s^{-1}. In this review we will use this value for all initial states. This yields collision efficiencies for state-to-state R-R energy transfer as large as 480% (i.e., almost five times larger than the "gas kinetic" rate of hard sphere collisions).

Vohralik and Miller[164] used a crossed molecular beam apparatus to study HF-HF collisions. Laser excitation was used to state select one beam, and the depletion of the excited state gave evidence of resonant rotational energy transfer. Using a kinetic model involving the reasonable assumption that exactly resonant R-R processes dominate the depletion process, they obtained cross sections for one- and two-quantum resonant R-R processes, HF(j_1) + HF(j_2) →HF(j_2) + HF(j_1). Their results are given in Table VI.

Table VI. Cross sections (Å^2) for resonant R → R processes at a relative translational energy of 0.176 eV (1420 cm^{-1})

j_1	j_2	σ
0	1	320
1	2	256
2	3	247
0	2	40
1	3	40

These cross sections are very large, at first seeming inconsistent with the results of Copeland and Crim discussed in the previous paragraph. However the difference may be a consequence of the fact that Copeland and Crim observed the slightly nonresonant collisions $HF(v_1=2,j_1)$ + $HF(v_2=0,j_2)$ whereas Vohralik and Miller observed $HF(v_1=0,j_1)$ + $HF(v_2=0,j_2)$. This is suggested by approximate close coupling calculations of Vohralik et al.[174a] which show a dramatic decrease in cross section with increasing vibrational mismatch.

DePristo and Alexander[171] made exploratory coupled channels calculations of rotational energy transfer cross sections for rigid diatoms. Unfortunately convergence checks showed that the cross sections were still not well converged with the largest channel sets employed. For the nearly resonant (40 cm^{-1} endoergic) cross section, $(j_1,j_2) = (11) \rightarrow (02)$ however, a converged value of 3.5 Å^2 was calculated for a relative translational energy of 918 cm^{-1} (0.114 eV). The Born approximation overestimates this cross section by a factor of 46.

Alper et al.[181] performed similar calculations with the quasiclassical trajectory (QCT) method, in which classical mechanics is combined with quantized initial conditions. Despite the uncertainties in the quantal results,[171] they concluded that the QCT method leads to good agreement with quantum mechanics.

Alexander and DePristo[172] concluded that an adiabatically corrected sudden approximation based on straight-line paths and the dipole-dipole interaction will provide similarly accurate results for the $(j_1,j_2) = 00 \rightarrow 11, 02, 22$ and $11 \rightarrow 02$ transitions at hyperthermal energies.

Alexander[173] then extended the coupled channel basis set for rigid HF-rigid HF collisions to convergence for rotationally inelastic collisions out of the ground state at collision energies 0.5-1.5 eV (3900-12500 cm^{-1}). Calculations were performed for two potentials, both expressed in a laboratory-frame coordinate system as

$$V(\hat{r}_1,\hat{r}_2,\vec{R}) = \sum_{m=1}^{M} A_m(R) \mathcal{Y}_m(\hat{r}_1,\hat{r}_2,\hat{R}) \tag{1}$$

where $\hat{r}_1$ and $\hat{r}_2$ denote diatom orientations, $\vec{R}$ is vector from one diatom to another, and $\mathcal{Y}_m(\hat{r}_1,\hat{r}_2,\hat{R})$ is a symmetrized angular function. The first potential had $M = 2$, and the second had $M = 6$. Calculations were performed with up to 76 coupled channels.

Alexander[173] found that the largest cross sections are associated with nearly resonant dipole-allowed processes, with $j_1,j_2 \rightarrow j_1 \pm 1, j_2 \mp 1$ R-R process and with the 00 $\rightarrow$ 11 T-R process. He concluded, in addition, that the total inelastic cross section is determined mainly by the dipole-dipole interaction, but that short-range anisotropic forces are probably more important for smaller state-to-state cross sections than are long-range forces involving multipole-multipole interactions higher order than the dipole-dipole one.

Bosananc et al.[198] also carried out close coupling calculations for rigid HF-rigid HF collisions. Their calculations refer to a collision energy of 300 cm^{-1} (0.372 eV). They expanded the Brobjer-Murrell potential using a molecule-frame expansion in symmetrized harmonic $Y_m(\hat{r}_1 \cdot \hat{R}, \hat{r}_2 \cdot \hat{R})$ with M terms, and they used values of M up to 11. They treated up to 58 coupled channels. They found significant differences between the partial cross sections for the $M = 11$ and 7 potentials and an order of magnitude or larger difference from results with $M = 3$. For $M = 11$, they found the largest cross section for $01 \rightarrow 11$.

Although the potentials used for the early quantum calculations described above are now known to be quite inadequate, these calculations did show that simple perturbation theory ideas based on long-range multipole moments are far from adequate for most aspects of rotationally inelastic HF-HF collisions.

Takayanagi and Wada[199] calculated the purely resonant transitions observed by Vohralik and Miller,[164] and—using straight-line trajectories and the pure dipole-dipole interaction—they obtained 400 Å^2 for $10 \rightarrow 01$, at 0.15 eV in reasonable agreement with the value in Table VI. Billing[180] combined a similar calculation for impact parameters greater than 12 Å with coupled channels calculations for close collisions and obtained a value about 10% lower, in better agreement with experiment. Vohralik et al.[174a] have also obtained good agreement with experiment for the $10 \rightarrow 01$ cross section, as well as for the $20 \rightarrow 02$, using approximate close coupling calculations and the Alexander-DePristo[38] potential. Cross sections with $|\Delta j| \geq 2$ calculated using the full potential were found to be significantly smaller than those obtained using the dipole potential, whereas cross sections for the first order, dipole-coupled resonant transitions like $10 \rightarrow 01$ are not very sensitive to other terms in the potential.

Schwenke and the author studied HF-HF collisions with a more realistic representation of the potential, and we obtained converged quantum cross sections for rigid HF-rigid HF collisions with total angular momentum J equal to 0.[190-193] In collaboration with Coltrin these were used to test QCT calculations.[191,193]

The first set[190] of converged dynamical calculations used the full Alexander-DePristo fit[38] to the SCF interaction energies of Yarkony et al.[20] The potential has the form of eq. (1) with $M = 6$, but has $M = 9$ when re-expressed in the molecule-frame expansion. For collision energies in the range 0.076-1.550 eV (613-12500 cm^{-1}), convergence was typically achieved with the number of channels N equal to about 200 (calculations were performed with up to 285 coupled channels). The first-order dipole transitions $00 \rightarrow 11$ has the largest inelastic transition probability at each energy, ranging from 0.05 to 0.22. The $00 \rightarrow 02$, $00 \rightarrow 22$, and $00 \rightarrow 32$ transitions have probabilities in the 0.02-0.08 range at the two highest energies. None of the other transitions has a probability in excess of 0.05 at any of the energies.

The second study[192] was much more realistic. The potential surface of Brobjer and Murrell was expanded in molecule-frame harmonics, and M was increased until convergence. Simultaneously the number of channels N was also increased to convergence. The final converged calculations, for J = 0 at collision energies 0.076-0.657

eV (613-5300 cm^{-1}), involved $M = 525$ and $N = 440$. The high value of M shows that high-order anisotropic terms in the spherical harmonic expansion are very important. The calculations showed extensive rotational excitation. Collisions with molecules initially in the ground rotational state preferentially populate states with large values of the sum j'_{sum} of the final rotational quantum numbers, j_1' and j_2'. As the collision energy increases, the maximum transition probability moves to higher values of j'_{sum}. At 0.567-0.657 eV, the most probable transition out of the ground state is to $j_1' = j_2' = 7$, which has a probability 0.09-0.12.

Our third study[193] of rigid HF-rigid HF collisions used the even more accurate potential surface of Redmon and Binkley. Again molecule-frame harmonics were used, and M and the number of channels N were increased to convergence, yielding in this case $M = 825$ and again $N = 440$. Converged quantal dynamics evaluations were performed for $J = 0$ and collision energies of 0.076 and 0.322 eV (613 and 2597 cm^{-1}), and they were compared to QCT calculations. At the lower energy for the ground initial state the QCT calculations predict less excitation to high j' states than the quantal calculations do; the QCT transition probabilities peak at $j'_{sum} = 2$ whereas the quantal ones peak at $j'_{sum} = 6$. Similar trends were found for the higher energy and for excited initial states.

6.2. Vibrational Energy Transfer

Vibrational relaxation of HF is very efficient. Using the value $k_{col} = 1.75 \times 10^{-10}$ cm^{3}molecule^{-1}s^{-1} (see Sect. 6.1), the relaxation collision efficiency for HF(v=1) is about 10^{-2} at room temperature, decreasing to $\sim 2 \times 10^{-3}$ at 1000 K, and then increasing again.[15,121,123,133,135,139,146,148] The inverse T dependence at 300-1000 K is characteristic of a process controlled by long-range attractive forces.

The results for the thermal relaxation process summarized briefly in the previous paragraph are dominated by relaxation of the v = 1 state. Relaxation of higher levels has been studied using exothermic reactions, sequential photon absorption, and single-photon excitation of overtones to produce HF(v ≥ 2). Dzelzkalns and Kaufman[151] have reported the most complete set of vibrational quenching rate coefficients k_q for HF-HF collisions, covering the first seven excited vibrational levels. Their results are in good agreement with several other less complete data sets[143-147,153] The quenching collision efficiencies obtained from the data of Dzelzkalns and Kaufman[151] with our nominal value of k_{col} are given in Table VII. A power law fit[151] yields $k_q(v) \sim v^{2.7\pm0.2}$.

Table VII also lists the rotationless endoergicities ΔE for the V-V processes HF(v) + HF(0) → HF(v-1) + HF(1), where the values in parentheses are vibrational quantum numbers.

Since the V-V process is becoming significantly endothermic at large v (compare the ΔE values in Table VII to $\tilde{k}T$, which equals 207 cm^{-1} at 298 K), where the quenching process is faster than gas kinetic, the V-T,R process must be very efficient for high v. In a subsequent study Dzelzkalns and Kaufman[152] were able to partition the relaxation for $v \geq 2$ into a V-V,R,T component, corresponding to $HF(v,j_1) + HF(0,j_2) \rightarrow HF(v-1,j_1') + HF(1,j_2')$, and a V-R,T component, corresponding to $HF(v,j_1) + HF(0,j_2) \rightarrow HF(v-1,j_1') + HF(0,j_2')$. Their result for $v = 2$ is in good agreement with an earlier measurement by Copeland et al.[147] Table VII gives Dzelzkalns and Kaufman's V-V,R,T fractions f_v as well as the derived collision efficiencies for both types of energy transfer. The table shows the decrease for f_v with v and also the increasing efficiency of the V-R,T process as v increases. Robinson et al.[148] found that the efficiency of HF(v=2) relaxation decreases more weakly as T is increased than does the $v = 1$ relaxation. They interpreted this in terms of the opening of the V-V,R,T pathway for relaxing this level and concluded that the slightly endoergic V-V,R,T route has a markedly weaker inverse temperature dependence than the exoergic V-R,T one.

Haugen et al.[153,153a] studied the state-to-state dynamics of V-R,T relaxation of HF(v=1) by HF by infrared pulse-probe transient absorption spectroscopy, i.e., laser double resonance, following earlier work of Hinchen.[160] and Crim and coworkers.[147-161] They found that a substantial fraction of the relaxation occurs to high-lying rotational states of v=0; in particular relaxation to $j = 10$-13 states comprises $30 \pm 10\%$ of the total relaxation, with $j = 14$ contributing $0.20 \pm 0.15\%$. The nascent population transfer distribution is a strongly decreasing function of j in the 10-14 range, suggesting a maximum for $j \approx 8$-10. In contrast, for relaxation of $v = 3$, about 95% of the nascent relaxated molecules have $j \leq 5$.[143] The most complete theoretical studies of vibrational relaxation are those of Wilkins and Kwok,[43,183] Billing, Poulsen, and Steinfeld[44,46,175,176,179,178] and Coltrin, Koszykowski, and Marcus.[57,184-186] Theory and experiment have been compared in detail in several theoretical[44,46,57,175,183,184,186] and experimental[143,145-152] papers.

Table VII. Quenching collision coefficient efficiencies for HF($1 \leq v \leq 7$) by HF(v=0)

v	$k_q + k_{col}$	ΔE (cm^{-1})	f_v	$k_{V\text{-}V,R,T} + k_{col}$	$k_{V\text{-}R,T} + k_{col}$
1	0.010	0	...	...	0.010[a]
2	0.11	173	0.55	0.06	0.05
3	0.18	340	0.30	0.05	0.13
4	0.42	503	0.15	0.06	0.36
5	0.80	633	≲0.1	≲0.08	~0.8
6	1.7	819	≲0.1	≲0.2	~1.7
7	2.6	973	≲0.1	≲0.3	~2.6

[a]The V-V,R,T route does not relax the vibrational energy for $v = 1$

An important point concerning vibrational relaxation in pure HF concerns the role of HF collisions with $(HF)_2$ and whether this dominates HF-HF collisions in vibrational relaxation.[154,155] Rensberger et al.[155] measured a rate coefficient 75 times larger for vibrational relaxation of HF by $(HF)_2$ than for relaxation by HF. However at 1 torr pressure there are only 250 ppm of dimer present so only 1.8% of the relaxation is estimated to be due to collisions with dimers at this pressure at room temperature, and we can safely interpret the experimental data summarized above as reflecting the dynamics of HF-HF collisions rather than of $(HF)_3$ systems.

Wilkins and Kwok[43,183] performed trajectory calculations on a poorly calibrated potential surface (in particular D_e = 2.7 kcal/mol vs. the experimental value of 4.6 kcal/mol discussed in section 2, and no electronic structure calculations were performed to adjust the shape of the analytic function). Nevertheless their V-V,R,T fractions f_v are in good agreement with experiment,[149,152] and their power law exponent is nearly 2, again in reasonable agreement with experiment.[150] These calculations appear however, to overestimate the role of multiquantum vibrational transitions.[143,145,146] Poulsen and Billing[179] were unable to reproduce Wilkins' or Alper et al.'s rotational distributions with trajectory calculations on a different potential surface.

Poulsen , Billing, and Steinfeld[44,175-179] performed semiclassical calculations based on a classical path for relative translation and rotation and an approximate coupled levels treatment of vibration. Their calculations also predict reasonable V-V,R,T fractions f_v.[149,152] Their results agree well with the experimental temperature dependence of the v = 1 relaxation;[44] they also agree with the experimental temperature dependence of the V-R,T rate coefficient for v = 2 better than the Wilkins-Kwok or Coltrin-Marcus studies.[148]

Coltrin and Marcus used a QCT moment method.[57,184-186] Their results[184] for the V-V,R,T fraction and the power law scaling are in poor agreement with experiment,[150,151] but the magnitude and temperature dependence of their V-V,R,T rate coefficients for v = 2 are in excellent agreement with experiment, whereas the Wilkins-Kwok and Billing-Poulsen results seriously underestimate this rate.[148]

The Poulsen-Billing and Coltrin-Marcus[185] calculations agree in predicting that orbiting states are important and multiquantum vibrational energy transfer is not important. However the well depth is 6.9 kcal/mol on the potential surface they used, which is too large and certainly taints the conclusions about orbiting states. An important difference between these two sets of calculations though is the role of V-R energy transfer. Coltrin and Marcus[185] repeated their trajectory moment calculations with rigid rotor trajectories as used for the classical path calculations of Billing and Steinfeld. This neglect of rotation-translation coupling in the trajectories significantly decreased the total deactivation and V-R,T energy transfer rate coefficients at large v, indicating that this is a serious dynamical approximation. This may explain why the Poulsen-Billing calculations underestimate the deactivation rates at high v.[145] The Coltrin-Marcus calculations, however, may also be criticized on dynamical grounds in that when the moment method was tested[187] against accurate quantum dynamics for V-V energy transfer in breathing spheres, the transition probabilities obtained by this method were found to be only semiquantitative.

In more recent calculations,[46,180] Billing included vibration-rotation coupling, although still approximately. His deactivation cross sections from HF(v=1) still populate too low a final j distribution as compared to experiment.[153]

The first converged quantum dynamics calculations[73,190,194] for V-V,R energy transfer in HF-HF collisions were carried out for the MAD potential (see Sect. 3.3). The results showed very little coupling of the V-V energy transfer process to rotation. Since the dynamics are exact for the assumed surface and since we believe this result is probably not correct, we believe this indicates a qualitative deficiency of the potential surface, in particular that it is insufficiently anisotropic. This indicates that even qualitative aspects of the vibration-to-vibration energy transfer are sensitive to the higher-order anisotropy of the potential energy surface.

Although these calculations play an important role in defining the issues and illustrating the possibilities, they lack in credibility due to the potential surfaces used and/or the methods used for the dynamics. Schwenke and the author[83,195] have, however, performed some large coupled channels calculations with more accurate surfaces. These calculations, which were carried out for total angular momentum zero, correspond to almost converged quantum dynamics for realistic potential energy surfaces, and we finish this section with a description of the results.

First, calculations with 694-948 coupled channels were performed for the RB and RBST surfaces. Both surfaces were found to predict that V-V energy transfer proceeds with highest or nearly highest probability into the highest energetically allowed values of j'_{sum} at two different relative translational energies (0.002455 and 0.076 eV, i.e., 20 and 613 cm^{-1}, respectively, corresponding to energies 1811 and 2404 cm^{-1} above the classical equilibrium energy of the dimer for this potential surface). Then a set of large-scale V-V energy transfer calculations was presented for the RBST surface at three energies in this range. These calculations show that the most important j'_{sum} for the energy transfer process $2HF(v=1,j=0) \rightarrow HF(v_1'=2,j_1') + HF(v_1'=0,j_2'= j'_{sum} - j_1')$ is within one of the maximum value allowed energetically at all 3 energies.

7. THE FUTURE

There are many areas of $(HF)_2$ structure and dynamics about which we remain completely ignorant. There has been, for example, no work on the potentials or dynamics for electronically excited states. Even for the ground electronic state there are many unresolved questions.

We need a potential surface that combines the best features of the HTD, RB, RBST, and BKLK surfaces and is equally as valid for forces on the repulsive wall for scattering and predissociation dynamics as for energies in the well region. We need further calculations to pin down the barrier for the four-center exchange reaction.

We need anharmonic energy level calculations with all six vibrational degrees of freedom. Although we have a qualitatively correct semiclassical picture of the tunneling splitting and its dependence on monomer stretch and rotational excitations, the details are far from settled. We need calculations of the effect of υ_6 excitation on the tunneling splitting.

Vibrational predissociation lifetimes, branching ratios, and infrared intensities should be calculated by coupled channels calculations with both H-F stretching degrees of freedom and one or more of the most accurate potential energy surfaces.

We must explore rotational and vibrational energy transfer in HF-HF collisions by reliable dynamical methods with accurate potential energy surfaces not only at low total angular momentum such as occurs for the dimer dynamics that have been studied spectroscopically, but also for the glancing, high-angular-momentum collisions that must dominate the large energy transfer cross sections.

This should keep us busy for a while.

8. ACKNOWLEDGMENTS

I am grateful to Frank Brown, Michael Coltrin, Cliff Dykstra, Gene Hancock, Paul Rejto, Roland Schweitzer, David Schwenke, Rozeanne Steckler, Devarajan Thirumalai, and Michael Unekis for collaboration on HF-HF projects, to Millard Alexander, Phil Bunker, Fleming Crim, Gerry Fraser, Stephen Leone, Roger Miller, Alan Pine, and Martin Quack for helpful discussions or sending recent results, and to the National Science Foundation for financial support.

REFERENCES

1.   R. L. Redington, *J. Phys. Chem.* 86:552 (1982).
2.   L. A. Curtiss and M. Blander, *Chem. Rev.* 88:827 (1988).
3.   P. A. Kollman and L. C. Allen, *J. Chem. Phys.* 52:5085 (1970).
4.   L. C. Allen and P. A. Kollman, *J. Amer. Chem. Soc.* 92:4108 (1970).
5.   P. Kollman, A. Johansson, and S. Rothenberg, *Chem. Phys. Lett.* 24:199 (1974).
6.   P. Kollman, J. McKelvey, A. Johansson, and S. Rothenberg, *J. Amer. Chem. Soc.* 97:955 (1975).
7.   L. C. Allen, *J. Amer. Chem. Soc.* 97:6721 (1975).
8.   P. Kollman, *J. Amer. Chem. Soc.* 99:4875 (1977).
9.   P. A. Kollman, *J. Amer. Chem. Soc.* 100:2974 (1978).
10.   R. C. Kerns and L. C. Allen, *J. Amer. Chem. Soc.* 100:6587 (1978).
11.   P. A. Kollman, *in:* "Chemical Applications of Atomic and Molecular Electrostatic Potentials," D. G. Truhlar, ed., Plenum, New York (1981), p. 243.

12. G. H. F. Diercksen and W. P. Kraemer, *Chem. Phys. Lett.* 6:419 (1971).
13. F. J. Zeleznik and R. V. Svehla, *J. Chem. Phys.* 53:632 (1970).
14. H. K. Shin, *Chem. Phys. Lett.* 10:81 (1971); errata: 11:628 (1971).
15. J. F. Bott and N. Cohen, *J. Chem. Phys.* 55:3698 (1971).
16. H. K. Shin, *J. Chem. Phys.* 59:879 (1973).
17. H. K. Shin, *J. Amer. Chem. Soc.* 98:5765 (1976).
18. J. E. Del Bene and J. A. Pople, *J. Chem. Phys.* 55:2296 (1971).
19. W. A. Latham, L. A. Curtiss, W. J. Hehre, and J. A. Pople, *Progr. Phys. Org. Chem.* 11:175 (1974).
20. D. R. Yarkony, S. V. O'Neil, H. F. Schaefer III, C. P. Baskin, and C. F. Bender, *J. Chem. Phys.* 60:855 (1974).
21. L. A. Curtiss and J. A. Pople, *J. Mol. Spectrosc.* 61:1 (1976).
22. J. A. Pople, *Faraday Discuss. Chem. Soc.* 73:7 (1982).
23. J. F. Gaw, Y. Yamaguchi, M. A. Vincent, and H. F. Schaefer III, *J. Amer. Chem. Soc.* 106:3133 (1984).
24. M. J. Frisch, J. A. Pople, and J. E. Del Bene, *J. Phys. Chem.* 89:3664 (1985).
25. A. E. Reed, F. Weinhold, L. A. Curtiss, and D. J. Pochatko, *J. Chem. Phys.* 84:5687 (1986).
26. M. J. Frisch, J. E. Del Bene, J. S. Binkley, and H. F. Schaefer III, *J. Chem. Phys.* 84:2279 (1986).
27. M. J. Redmon and J. S. Binkley, *J. Chem. Phys.* 87:969 (1987).
28. G. C. Berend and R. L. Thommarson, *J. Chem. Phys.* 58:3203 (1973).
29. H. Lischka, *J. Amer. Chem. Soc.* 96:4761 (1974).
30. H. Lischka, *Chem. Phys. Lett.* 66:108 (1979).
31. A. Karpfen, *Chem. Phys.* 47:401 (1980).
32. A. Beyer and A. Karpfen, *Chem. Phys.* 64:343 (1982).
33. A. Karpfen, A. Beyer, and P. Schuster, *Chem. Phys. Lett.* 102:289 (1983).
34. A. Beyer, A. Karpfen, and P. Schuster, *Top. Curr. Chem.* 120:1 (1984).
35. M. Kofranek, H. Lischka, and A. Karpfen, *Chem. Phys.* 121:137 (1988).
36. P. R. Bunker, M. Kofranek, H. Lischka, and A. Karpfen, *J. Chem. Phys.* 89:3002 (1988); errata: in ref. 207.
37. G. A. Parker, R. L. Snow, and R. T Pack, *Chem. Phys. Lett.* 33:399 (1975).
38. M. H. Alexander and A. E. DePristo, *J. Chem. Phys.* 65:5009 (1976).
39. H. Umeyama and K. Morokuma, *J. Amer. Chem. Soc.* 99:1316 (1977).
40. K.-C. Ng, W. J. Meath, and A. R. Allnat, *Mol. Phys.* 33:699 (1977).
41. K.-C. Ng, W. J. Meath, and A. R. Allnat, *Mol. Phys.* 38:449 (1979).
42. F. Mulder, G. F. Thomas, and W. J. Meath, *Mol. Phys.* 41:249 (1980).
42a. A. Kumar and W. J. Meath, *Mol. Phys.* 54:823 (1985).
42b. P. J. Knowles and W. J. Meath, *Mol. Phys.* 60:1143 (1987).
42c. W. Rijks and P. E. S. Wormer, *J. Chem. Phys.* 90:6507 (1989).
43. R. L. Wilkins, *J. Chem. Phys.* 67:5838 (1977).
44. L. L. Poulsen, G. D. Billing, and J. I. Steinfeld, *J. Chem. Phys.* 68:5121 (1978).
45. C. Nyeland, L. L. Poulsen, and G. D. Billing, *J. Phys. Chem.* 88:5858 (1984).
46. G. D. Billing, *J. Chem. Phys.* 84:2593 (1986).
47. W. L. Jorgensen and M. E. Cournoyer, *J. Amer. Chem. Soc.* 100:4942 (1978).
48. W. L. Jorgensen, *J. Chem. Phys.* 70:5888 (1979).
49. M. E. Cournoyer and W. L. Jorgensen, *Mol. Phys.* 51:119 (1984).
50. M. L. Klein, I. R. McDonald, and S. F. O'Shea, *J. Chem. Phys.* 69:63 (1978).
51. M. L. Klein and I. R. McDonald, *J. Chem. Phys.* 71:298 (1979).
52. F. H. Stillinger, *Int. J. Quantum Chem.* 14:649 (1978).
53. C. W. David, *Chem. Phys.* 53:105 (1980).
54. D. Maillard and B. Silvi, *Mol. Phys.* 40:933 (1980).
55. C. Girardet, A. Schriver, and D. Maillard, *Mol. Phys.* 41:779 (1980).
56. P. N. Swepston, S. Colby, H. L. Sellers, and L. Schäfer, *Chem. Phys. Lett.* 72:364 (1980).
57. M. E. Coltrin, M. L. Koszykowski, and R. A. Marcus, *J. Chem. Phys.* 73:3643 (1980).
58. F. A. Gianturco, U. T. Lamanna, and F. Battaglia, *Int. J. Quantum Chem.* 19:217 (1981).
59. J. T. Brobjer and J. N. Murrell, *Chem. Phys. Lett.* 77:601 (1981).

60. J. T. Brobjer and J. N. Murrell, *J. Chem. Soc. Faraday Trans.* 2 78:1853 (1982).
61. J. T. Brobjer, *Faraday Discuss. Chem. Soc.* 73:123:128 (1982).
62. J. T. Brobjer and J. N. Murrell, *Mol. Phys.* 50:885 (1983).
63. A. E. Barton and B. J. Howard, *Faraday Discuss. Chem. Soc.* 73:45:121:122 (1982).
64. A. D. Buckingham and P. W. Fowler, *J. Chem. Phys.* 79:6426 (1983).
65. A. D. Buckingham and P. W. Fowler, *Can. J. Chem.* 63:2018 (1985).
66. P. Hobza and J. Sauer, *Theor. Chim. Acta* 65:279 (1984).
67. M. M. Szcześniak and S. Scheiner, *J. Chem. Phys.* 80:1535 (1984).
68. Z. Latajka and S. Scheiner, *J. Comput. Chem.* 8:663 (1987).
69. D. W. Schwenke and D. G. Truhlar, *J. Chem. Phys.* 82:2418 (1985); errata: 84:4113 (1986), 86:3760 (1987).
70. D. W. Michael, C. E. Dykstra, and J. M. Lisy, *J. Chem. Phys.* 81:5998 (1984).
71. D. G. Truhlar, F. B. Brown, D. W. Schwenke, R. Steckler, and B. C. Garrett, *in:* "Comparison of Ab Initio Quantum Chemistry with Experiment for Small Molecules," R. J. Bartlett, ed., Reidel, Dordrecht (1985), p. 95.
72. C. E. Dykstra and J. M. Lisy, *ibid.*, p. 245.
73. D. W. Schwenke and D. G. Truhlar, *Theor. Chim. Acta* 69:175 (1986).
74. C. E. Dykstra, S.-Y. Liu, and D. J. Malik, *J. Mol. Struc. (Theochem.)* 135:357 (1986).
75. S.-Y. Liu and C. E. Dykstra, *Chem. Phys.* 107:343 (1986).
76. S. K. Loushin, S.-Y. Liu, and C. E. Dykstra, *J. Chem. Phys.* 84:2720 (1986).
77. S.-Y. Liu and C. E. Dykstra. *J. Phys. Chem.* 90:3097 (1986).
78. S.-Y. Liu, C. E. Dykstra, and D. J. Malik, *Chem. Phys. Lett.* 130:403 (1986).
79. C. E. Dykstra and S.-Y. Liu, *in:* "Structure and Dynamics of Weakly Bound Molecular Complexes" (NATO ASI Series, Vol. C212), A. Weber, ed. Reidel, Dordrecht (1987), p. 319.
80. C. E. Dykstra, *J. Phys. Chem.* 91:6216 (1987).
81. G. C. Hancock, D. G. Truhlar, and C. E. Dykstra, *J. Chem. Phys.* 88:1786 (1988).
82. C. E. Dykstra, *Acc. Chem. Res.* 21:355 (1988).
83. D. W. Schwenke and D. G. Truhlar, *J. Chem. Phys.* 88:4800 (1988).
84. C. E. Dykstra, *J. Amer. Chem. Soc.* 111:6168 (1989).
84a. C. E. Dykstra, *J. Phys. Chem.*, in press.
85. N. Halberstadt, Ph. Bréchignac, J. A. Beswick, and M. Shapiro, *J. Chem. Phys.* 84:170 (1986).
86. W. A. Sokalski, A. H. Lowrey, S. Roszak, V. Lewchenko, J. Blaisdell, P. C. Hariharan, and J. J. Kaufman, *J. Comput. Chem.* 7:693 (1986).
87. P. V. Huong and M. Cuozi, *J. Chim. Phys.* 66:420 (1969).
88. J. L. Himes and T. A. Wiggins, *J. Mol. Spectrosc.* 40:418 (1971).
89. J. M. Lisy, M. F. Vernon, A. Tramer, H.-S. Kowk, D. J. Krajnovich, T. R. Shen, and Y. T. Lee, paper presented at the Fifth International Conference on Laser Spectroscopy, Alberta, Canada, 29 June-3 July 1981 [Lawrence Berkeley Laboratory report LBL-12981, University of California, July 1981].
90. J. M. Lisy, A. Tramer, M. F. Vernon, and Y. T. Lee, *J. Chem. Phys.* 75:4733 (1981).
91. M. F. Vernon, J. M. Lisy, D. J. Krajnovich, A. Tramer, H.-S. Kowk, Y. R. Shen, and Y. T. Lee, *Faraday Discuss. Chem. Soc.* 73:387 (1982).
92. A. S. Pine and W. J. Lafferty, *J. Chem. Phys.* 78:2154 (1983).
93. A. S. Pine, W. J. Lafferty, and B. J. Howard, *J. Chem. Phys.* 81:2939 (1984).
94. A. S. Pine and B. J. Howard, *J. Chem. Phys.* 84:590 (1986).
95. A. S. Pine, *in:* "Structure and Dynamics of Weakly Bound Molecular Complexes" (NATO ASI Series, Vol. C212), A. Weber, ed., Reidel, Dordrecht (1987), p. 93.
96. A. S. Pine and G. T. Fraser, *J. Chem. Phys.* 89:6636 (1988).
97. G. T. Fraser and A. S. Pine, *J. Chem. Phys.* 91:633 (1989).
98. L. Andrews and G. L. Johnson, *Chem. Phys. Lett.* 96:133 (1983).
99. L. Andrews and J. L. Johnson, *J. Phys. Chem.* 88:425 (1984).
100. L. Andrews, V. E. Bondybey, and J. H. English, *J. Chem. Phys.* 81:3452 (1984).
101. R. D. Hunt and L. Andrews, *J. Chem. Phys.* 82:4442 (1985).

102. R. L. Redington and D. F. Hamill, *J. Chem. Phys.* 80:2446 (1984).
103. R. L. DeLeon and J. S. Muenter, *J. Chem. Phys.* 80:6092 (1984).
104. K. v. Puttkamer and M. Quack, *Chimia* 39:358 (1985).
105. K. v. Puttkamer and M. Quack, *Faraday Discuss. Chem. Soc.* 82:377 (1986).
106. K. v. Puttkamer and M. Quack, *Mol. Phys.* 62:1047 (1987).
107. K. v. Puttkamer, M. Quack, and M. A. Suhm, *Mol. Phys.* 65:1025 (1988).
108. K. v. Puttkamer, M. Quack, and M. A. Suhm, *Infrared Phys.* 29:535 (1989).
108a. K. v. Puttkamer and M. Quack, *Chem Phys.* 139:31 (1989).
109. Z. S. Huang, K. W. Jucks, and R. E. Miller, *J. Chem. Phys.* 85:3338 (1986).
110. C. M. Lovejoy and D. J. Nesbitt, *Rev. Sci. Instrum.* 58:807 (1987).
111. T. R. Dyke, B. J. Howard, and W. Klemperer, *J. Chem. Phys.* 56:2442 (1972).
112. W. Klemperer, *Ber. Bunsenges. Phys. Chem.* 78:128 (1974).
113. B. J. Howard, T. R. Dyke, and W. Klemperer, *J. Chem. Phys.* 81:5417 (1984).
114. H. S. Gutowsky, C. Chuang, J. D. Keen, T. D. Klots, and T. Emilsson, *J. Chem. Phys.* 83:2070 (1985).
115. W. J. Lafferty, R. D. Suenram, and F. J. Lovas, *J. Mol. Spectrosc.* 123:434 (1987).
116. G. Flynn and E. Weitz, *Annu. Rev. Phys. Chem.* 25:275 (1974).
117. S. Ormonde, *Rev. Mod. Phys.* 47:193 (1975).
118. S. R. Leone, *J. Phys. Chem. Ref. Data* 11:953 (1982).
119. J. R. Airey and S. F. Fried, *Chem. Phys. Lett.* 8:23 (1971).
120. J. R. Airey and I. W. M. Smith, *J. Chem. Phys.* 57:1669 (1972).
121. S. S. Fried, J. Wilson, and R. L. Taylor, *IEEE J. Quantum Elect.* 9:59 (1973).
122. P. R. Poole and I. W. M. Smith, *J. Chem. Soc. Faraday Trans. 2* 73:1434 (1977).
123. J. F. Bott, *J. Chem. Phys.* 57:96 (1972).
124. J. F. Bott and N. Cohen, *J. Chem. Phys.* 58:934 (1973).
125. J. F. Bott, *Chem. Phys. Lett.* 23:335 (1973).
126. J. F. Bott, *J. Chem. Phys.* 61:3414 (1974).
127. J. F. Bott, *J. Chem. Phys.* 70:4123 (1979).
128. J. K. Hancock and W. H. Green, *J. Chem. Phys.* 56:2474 (1972).
129. W. H. Green, and J. K. Hancock, *IEEE J. Quantum Elect.* 9:50 (1973).
130. J. K. Hancock and W. H. Green, *IEEE J. Quantum Elect.* 11:694 (1975).
131. R. R. Stephens and T. A. Cool, *J. Chem. Phys.* 56:5863 (1972).
132. J. L. Ahl and T. A. Cool, *J. Chem. Phys.* 58:5540 (1973).
133. R. A. Lucht and T. A. Cool, *J. Chem. Phys.* 60:1026 (1974).
134. R. A. Lucht and T. A. Cool, *J. Chem. Phys.* 63:3962 (1975).
135. J. A. Blauer, W. C. Solomon, and T. W. Owens, *Int. J. Chem. Kinet.* 4:293 (1972).
136. D. L. Thompson, *J. Chem. Phys.* 57:2589 (1972).
137. J. J. Hinchen, *J. Chem. Phys.* 59:233 (1973).
138. J. J. Hinchen, *J. Chem. Phys.* 59:2224 (1973).
139. L. S. Blair, W. D. Breshears, and G. L. Schott, *J. Chem. Phys.* 59:1582 (1973).
140. K. Ernst, R. M. Osgood, Jr., A. Javan, and P. B. Sackett, *Chem. Phys. Lett.* 23:553 (1973).
141. R. M. Osgood, P. B. Sackett, and A. Javan, *J. Chem. Phys.* 60:1464 (1974).
142. M. A. Kwok and R. L. Wilkins, *J. Chem. Phys.* 63:2453 (1975).
143. D. J. Douglas and C. B. Moore, *Chem. Phys. Lett.* 57:485 (1978).
144. J. K. Lampert, G. M. Jursich, and F. F. Crim, *Chem. Phys. Lett.* 71:258 (1980).
145. G. M. Jursich and F. F. Crim, *J. Chem. Phys.* 74:4455 (1981).
146. T. J. Foster and F. F. Crim, *J. Chem. Phys.* 75:3871 (1981).
147. R. A. Copeland, D. J. Pearson, J. M. Robinson and F. F. Crim, *J. Chem. Phys.* 77:3974 (1982); errata: 78:6344 (1983).
148. J. M. Robinson, D. J. Pearson, R. A. Copeland, and F. F. Crim, *J. Chem. Phys.* 82:780 (1985).
149. J. M. Robinson, K. J. Rensberger, and F. F. Crim, *J. Chem. Phys.* 84:220 (1986).
150. L. S. Dzelzkalns and F. Kaufman, *J. Chem. Phys.* 77:3508 (1982).
151. L. S. Dzelzkalns and F. Kaufman, *J. Chem. Phys.* 79:3836 (1983).
152. L. S. Dzelzkalns and F. Kaufman, *J. Chem. Phys.* 79:3363 (1983).
153. H. K. Haugen, W. H. Pence, and S. R. Leone, *J. Chem. Phys.* 80:1839 (1984).

153a. C. A. Taatjes and S. R. Leone, *J. Chem. Phys.* 89:302 (1988).
154. E. L. Knuth, H.-G. Rubahn, J. P. Toennies, and J. Wanner, *J. Chem. Phys.* 85:2653 (1986).
155. K. J. Rensberger, J. M. Robinson, and F. F. Crim, *J. Chem. Phys.* 86:1340 (1987).
156. L. M. Peterson, G. H. Lindquist, and C. B. Arnold, *J. Chem. Phys.* 61:3480 (1974).
157. J. J. Hinchen, *Appl. Phys. Lett.* 27:672 (1975).
158. J. J. Hinchen and R. H. Hobbs, *J. Chem. Phys.* 65:2732 (1976).
159. J. J. Hinchen and R H. Hobbs, *J. Appl. Phys.* 50:628 (1979).
160. J. J. Hinchen, *in*: "Gas Lasers," E. W. McDaniel and W. Nighan, eds. (Vol. 3 of *Applied Atomic Collision Physics*, H. S. W. Massey, E. W. McDaniel, and B. Bederson, eds.), Academic, New York (1982), p. 191.
161. R. A. Copeland, D. J. Pearson, and F. F. Crim, *Chem. Phys. Lett.* 81:541 (1981).
162. R. A. Copeland and F. F. Crim, *J. Chem. Phys.* 78:5551 (1983).
163. R. A. Copeland and F. F. Crim, *J. Chem. Phys.* 81:5819 (1984).
164. P. F. Vohralik and R. E. Miller, *J. Chem. Phys.* 83:1609 (1985).
165. D. C. Dayton, K. W. Jucks, and R. E. Miller, *J. Chem. Phys.* 90:2631 (1989).
166. R. E. Miller, personal communication.
167. H. K. Shin, *Chem. Phys. Lett.* 26:450 (1974).
168. H. K. Shin, *J. Chem. Phys.* 63:2901 (1975).
169. H. K. Shin, *J. Chem. Phys.* 64:3634 (1976).
170. H. K. Shin, *Chem. Phys. Lett.* 50:377 (1977).
171. A. E. DePristo and M. H. Alexander, *J. Chem. Phys.* 66:1334 (1977).
172. M. H. Alexander and A. E. DePristo, *J. Phys. Chem.* 83:1499 (1979).
173. M. A. Alexander, *J. Chem. Phys.* 73:5135 (1980).
174. A. E. DePristo, *J. Chem. Phys.* 74:5037 (1981).
174a. P. F. Vohralik, R. O. Watts, and M. H. Alexander, *J. Chem. Phys.* 91:7563 (1989).
175. G. D Billing and L. L. Poulsen, *J. Chem. Phys.* 68:5128 (1978).
176. L. L. Poulsen and G. D. Billing, *Chem. Phys.* 36:271 (1979).
177. G. D. Billing and L. L. Poulsen, *Chem. Phys. Lett.* 66:177 (1979).
178. L. L. Poulsen and G. B. Billing, *Chem. Phys.* 46: 287 (1980).
179. L. L. Poulsen and G. D. Billing, *Chem. Phys.* 53:389 (1980).
180. G. B. Billing, *Chem. Phys.* 112:95 (1987).
181. J. S. Alper, M. A. Carroll, and A. Gelb, *Chem. Phys.* 32:471 (1978).
182. R. L. Wilkins, *J. Chem. Phys.* 70:2700 (1979).
183. R. L. Wilkins and M. A. Kwok, *J. Chem. Phys.* 73:3198 (1980).
184. M. E. Coltrin and R. A. Marcus, *J. Chem. Phys.* 73:4390 (1980).
185. M. E. Coltrin and R. A. Marcus, *J. Chem. Phys.* 73:2179 (1980).
186. M. E. Coltrin and R. A. Marcus, *J. Chem. Phys.* 76: 2379 (1982).
187. D. W. Schwenke, D. Thirumalai, D. G. Truhlar, and M. E. Coltrin, *J. Chem. Phys.* 78:3078 (1983).
188. D. W. Schwenke and D. G. Truhlar, *in*: "Supercomputer Applications," R. W. Numrich, ed., Plenum, New York (1985), p. 215.
189. D. W. Schwenke and D. G. Truhlar, *in* proceedings of the 1985 Cray Science and Engineering Symposium, Bloomington, Minnesota, 14-17 April 1985 [University of Minnesota Supercomputer Institute research report UMSI85/5, Minnesota Supercomputer Institute, April 1985].
190. D. W. Schwenke and D. G. Truhlar, *in*: "Supercomputer Simulations in Chemistry" (Lectures Notes in Chemistry, Vol. 44), M. Dupuis, ed., Springer-Verlag, Berlin (1986), p. 165.
191. D. W. Schwenke, D. G. Truhlar, and M. E. Coltrin, *in*: proceedings of the First Symposium on Computational Chemistry on Cray Supercomputers, Minneapolis, 11-13 September 1986 [University of Minnesota Supercomputer Isntitute research report 86/50, Minnesota Supercomputer Institute, September 1986].
192. D. W. Schwenke and D. G. Truhlar, *J. Comput. Chem.* 8:282 (1987).
193. D. W. Schwenke, D. G. Truhlar, and M. E. Coltrin, *J. Chem. Phys.* 87:983 (1987).

194. D. W. Schwenke and D. G. Truhlar, *Theor. Chim. Acta* 72:1 (1987).
195. D. W. Schwenke and D. G. Truhlar, *in*: "Supercomputer Research in Chemistry and Chemical Engineering" (A.C.S. Symposium Series, Vol. 353), K. F. Jensen and D. G. Truhlar, eds., American Chemical Society, Washington (1987), p. 176.
196. D. W. Schwenke, K. Haug, D. G. Truhlar, R. H. Schweitzer, J. Z. H. Zhang, Y. Sun, and D. J. Kouri, *Theor. Chim. Acta* 72:237 (1987).
197. R. J. Gordon, *J. Chem. Phys.* 74:1676 (1981).
198. S. Bosanac, J. T. Brobjer, and J. N. Murrell, *Mol. Phys.* 51:313 (1984).
199. K. Takayanagi and T. Wada, *J. Phys. Soc. Japan* 54:2122 (1985).
200. I. M. Mills, *J. Phys. Chem.* 88:532 (1984).
201. J. T. Hougen and N. Ohashi, *J. Mol. Spectrosc.* 109:134 (1985).
202. J. T. Hougen, *in*: "Structure and Dynamics of Weakly Bound Complexes" (NATO ASI Series, Vol. C212), A. Weber, ed., Reidel, Dordrecht (1987), p. 191.
203. G. C. Hancock, P. Rejto, R. Steckler, F. B. Brown, D. W. Schwenke, and D. G. Truhlar, *J. Chem. Phys.* 85:4997 (1986).
204. G. C. Hancock and D. G. Truhlar, *J. Chem. Phys.* 90:3498 (1989).
205. G. T. Fraser, *J. Chem. Phys.* 90:2097 (1989).
206. E. L. Sibert III, *J. Phys. Chem.* 93:5022 (1989).
207. P. R. Bunker, T. Carrington, Jr., P. C. Gomez, M. D. Marshall, M. Kofranek, H. Lischka, and A. Karpfen, *J. Chem. Phys.* 91:5154 (1989).
208. G. Ewing, *Chem. Phys.* 29:253 (1978).
209. G. E. Ewing, *J. Chem. Phys.* 72:2096 (1980).
210. G. E. Ewing, *in*: "Potential Energy Surfaces and Dynamics Calculations," D. G. Truhlar, ed., Plenum, New York (1981), p. 75.
211. G. E. Ewing, *Chem. Phys.* 63:411 (1981).
212. G. E. Ewing, *Faraday Discuss. Chem. Soc.* 73:122, 325 (1982).
213. G. E. Ewing, *J. Phys. Chem.* 91:4662 (1987).

DISCUSSION ON THE DYNAMICS OF LARGER MOLECULES

(MOSTLY EXPERIMENTS)

André Tramer

Laboratoire de Photophysique Moléculaire
Bâtiment 213
Université de Paris-Sud
91405 Orsay Cédex
France

DISCUSSION

The discussion was focused on the problem of the vibrational predissociation (VP) and of the intramolecular vibrational redistribution (IVR) in
molecular complexes. Some of the preliminary conclusions may be summarized
as follows:

1. The interest of studies on relaxation mechanisms of van der Waals
complexes is general: they may be considered as models for a larger class of
unimolecular reactions. The specificity of van der Waals systems relies on
the exceptionally weak coupling between intra- and inter-molecular modes,
much weaker than typical intra-state coupling in molecules. The time scales
for IVR and VP processes are therefore significantly longer and more convenient for real-time experiments.

2. The energy (or momentum) gap laws are extremely useful for the estimation of VP (IVR) rates as long as the potential energy surfaces of the initial and final states are not very different. They may fail in the case of a
relatively strong coupling between intra- and inter-molecular modes, i.e.,
for the systems for which the frequency of one of the intra-molecular modes
is strongly shifted by complex formation (e.g., X-H stretching modes in
hydrogen-bonded systems).

3. The amount of data concerning the dependence of the VP (IVR) rates on
the excitation of different vibrations is still very limited, and some of the
observed effects cannot be easily explained. A more extended study of this
dependence would be of interest, especially in hydrogen-bonded complexes
where the frequency shift is very different for different modes.

4. The relation between IVR and/or VP rates in ground and electronically
excited states of molecular complexes has not been systematically studied.
Such studies would be important for a better understanding of their mechanisms.

5. It has been shown that the relaxation of vibrational levels of the
tetrazine-argon complex is sequential (IVR prior to VP). It would be important to know whether such a behavior is characteristic for a large class of

Dynamics of Polyatomic Van der Waals Complexes
Edited by N. Halberstadt and K. C. Janda
Plenum Press, New York, 1990

systems. It may be expected that in complexes of large molecules, where the statistical approach is appropriate, the IVR rate increases and that of VP decreases with increasing level density of the final state. If this is the case, the sequential (IVR → VP) relaxation will be a rule in the large-molecule limit. For smaller systems, the determining factor would be the relative strength of coupling between intra-molecular modes as compared to that of intra- to inter-molecular ones.

VAN DER WAALS MOLECULES AS A VEHICLE FOR THE STUDY OF UNIMOLECULAR REACTIONS

Stuart A. Rice

Department of Chemistry and The James Franck Institute
The University of Chicago, Chicago IL,   60637

I.   INTRODUCTION

The general aspects of the dynamics of thermal unimolecular reactions were elucidated in the 1920's by the introduction of three key ideas. First, the Lindemann hypothesis[1] provided an explanation, using the phenomenological formulation of reaction kinetics, of how collisional excitation of the molecules can be consistent with first order reaction kinetics in the high pressure regime while being, of course, second order in the pressure in the low pressure regime. Although qualitatively satisfactory, the application of the Lindemann hypothesis in its simplest form predicts that the transition in the reaction kinetics as a function of pressure from first order to second order occurs at a much higher presure than is observed.[2] Second, the Hinshelwood theory[3] introduced the notion that the rate constant for the energization process can be much greater for a complex molecule than for a simple molecule because the energy of a complex molecule can be distributed over a larger number of degrees of freedom. The Hinshelwood analysis removes the above mentioned problem with respect to the pressure dependence of the unimolecular reaction rate. Third, the Rice-Kassel-Ramsberger (RKR) theory[4] provided a framework for interpreting the variation of the rate of reaction of the energized molecule with energy and the number of degrees of freedom of the molecule. The RKR theory bypassed the analysis of the complicated intramolecular dynamics by assuming that the rate of reaction depends only on

Dynamics of Polyatomic Van der Waals Complexes
Edited by N. Halberstadt and K. C. Janda
Plenum Press, New York, 1990

the energy of the molecule, which is equivalent to assuming that the rate of intramolecular energy transfer greatly exceeds the rate of reaction. All of the more modern analyses of the rate of a unimolecular reaction incorporate the RKR assumption.[5] When this assumption is used in the description of the evolution of molecular systems excited by nonthermal means, e.g. by absorption of a photon, it implies that nontrivial control of selectivity of product formation is impossible. In fact, the so called statistical theory of unimolecular reaction rate provides a very satisfactory description of the observed behaviour of many systems, including cases when the activation of the molecules is by collision and when it is more selective with respect to the initial state generated. Yet we still know very little of the details of the competition between intramolecular energy redistribution and chemical reaction, and we do not understand the limits of validity of the rapid energy redistribution hypothesis in terms of the characteristic properties of the molecular Hamiltonian. In particular, very little attention has been focussed on possible interference effects in the evolution of the initial state of the excited molecule. It is arguable that a better understanding of the dynamics of intramolecular energy redistribution will permit definition of the limits of controllability of product formation in a reaction. Indeed, much of the recent interest in testing the limits of validity of the theory of unimolecular reaction rate has, as underlying motivation, the goal of developing a method for influencing the selectivity of product formation.

van der Waals molecules are useful vehicles for the study of intramolecular vibrational redistribution and various aspects of unimolecular fragmentation. In the typical van der Waals molecule the coupling between the partners is sufficiently weak that the electronic and vibrational states of the molecule are very little different from the states of the free partners, except for the states uniquely associated with the stretching and bending of the van der Waals bond. Given these conditions, one can uniquely associate states of the van der Waals molecule with states of the free partners and one can selectively excite well defined initial states of the van der Waals molecule and follow their evolution. It also follows that quite modest levels of excitation of the van der Waals molecule permit probing of its dynamical behaviour over a considerable range of energy in excess of its binding energy.

Measurements of intramolecular vibrational redistribution and vibrational predissociation have been reported for a variety of van der Waals molecules,[6] and for a considerable range of excitation energy (relative to the van der Waals molecule binding energy). In addition, several theoretical analyses of the initial state dependence of the rate of fragmentation of van der Waals molecules have been reported,[7] some of which are very sophisticated, others rather simple. While much has been learned about the several phenomena cited, as shown by the contributions to this Workshop, even more remains unknown.

This paper reviews aspects of the dynamics of unimolecular reactions with emphasis on what has been, and can be, learned from studies of van der Waals molecule fragmentation. The organizing theme of the presentation is the search for, and validation of, an active method for influencing the selectivity of product formation in a photoinduced reaction. As already noted above, it is commonly accepted that competition between intramolecular vibrational redistribution and the processes that lead to reaction influences the selectivity of product formation in a unimolecular reaction. Our examination of the dynamical basis of that competition leads to consideration of an alternative form for the classical mechanical theory of the rate of a unimolecular reaction, the search for conditions under which there is possible a decoupling of a part of an N degree of freedom system from the rest of the system on a time scale of interest with respect to subsystem internal energy transfer and chemical reaction, the beginnings of an understanding of the influence of classical mechanical chaos on the corresponding quantum mechanical description of an unbounded system, and the development of a paradigm for the active control of selectivity of a photoinduced reaction.

II. SOME SELECTED EXPERIMENTAL DATA

For present purposes, namely, learning about the irreducible limitations to active control of the selectivity of product formation in a reaction and formulating a scheme consistent with those limitations, it suffices to describe only a limited subset of the available data concerning intramolecular vibrational redistribution and vibrational predissociation in van der Waals molecules. Of course, in these predissociation reactions the different products correspond to different final states of the separated partners that formed the van der Waals molecule. For that reason, the dynamics of intramolecular energy transfer and fragmentation of van der Waals molecules in which at least one of the partners is a polyatomic molecule are of most interest. Nevertheless, because of its important role in stimulating other experiments and theoretical analyses, we shall briefly describe some of the properties of the van der Waals molecules I_2He_n and a few similar molecules.

The predissociation lifetime of I_2He_n is found to be dependent on the iodine vibrational level initially excited, and for n = 1 ranges from 38 ps (v = 26) to 221 ps (v = 12), with a variation that is a nonlinear function of the iodine vibrational quantum number v[8]. The predissociation lifetimes for the neon[9], hydrogen[10] and deuterium[10] van der Waals complexes with iodine are found to be somewhat shorter. As for the product iodine vibrational state distribution, for I_2He_n the $\Delta v = -n$ channel is dominant, with a branching ratio of greater than 0.95 for v = 20. Similar "selection rules" are observed to dominate the iodine product vibrational state distributions for the other van der Waals complexes mentioned[9,11], though in general the minority dissociation channels are more significant than in the case of I_2He_n. In contrast, for the reaction $I_2He \longrightarrow I_2 + He$ the product rotational state distribution does not show any pattern of preferential population[12], which is consistent with theoretical considerations.[13] The vibrational predissociation of IClHe from v = 2,3 yields ICl rotational state distributions with maxima at j = 7 and j = 16.[14] It is suggested that the maxima in the product rotational state distribution correspond to rotational rainbows in the half collision scattering of He from I and Cl, respectively. Analogously, the vibrational predissociation of IClNe from v ~ 10-25 yields an ICl rotational state distribution which is peaked at values of j that increase as the available energy increases, a result which is attributed to half collision scattering of Ne in the anisotropic potential of ICl. In general, if the moment of inertia of the diatom is large (as in

I_2) one expects little rotational excitation following vibrational predissociation since the product diatom is unable to accept a significant amount of rotational energy without suffering a large change in angular momentum.

Studies of state-to-state reaction probabilities in polyatomic-rare gas atom van der Waals molecules have been carried out for several systems, an incomplete listing of which contains: s-tetrazine-Ar[16]; glyoxal-H_2, Ar and Kr[17]; benzene-He and Ar[18]; several species of deuterated benzene-He and Ar[19]; aniline-He[20]; p-difluorobenzene-Ar[21,22]; pyrimidine-Ar and deuterated pyrimidine-Ar[23].

We consider, first, the results of studies of glyoxal-H_2 van der Waals molecules, carried out by Halberstadt and Soep[17]. For given initial state, the distribution of final state vibrational energy in the glyoxal fragment is found to be highly selective. Halberstadt and Soep use a "spectator mode" model to interpret the relaxation dynamics of combination mode levels. According to this model the dissociation channels active for a combination mode will be a superposition of the dissociation channels active for the constituent fundamental vibrations. For example, glyoxal-H_2 excited to the 8^1 level dissociates to yield vibrationless bare glyoxal with a branching ratio of unity, while complexes excited to 5^1 dissociate to yield primarily 7^1. The spectator model predicts that the dominant dissociation pathways for the $5^1 8^1$ level of glyoxal-H_2 will populate the 5^1 and $8^1 7^1$ levels. This is the behaviour observed. Taken at face value, the spectator model implies that anharmonic interaction between vibrational modes of the polyatomic molecule does not play a prominent role in the vibrational predissociation process.

The structures of all of the aromatic ring-rare gas atom van der Waals molecules are similar: the rare gas atom is situated on the axis perpendicular to the plane of, and through the center of, the aromatic ring, at a distance of about 3.4 Å[24]. Furthermore, the binding energies of these molecules are rather similar, all being of the order of a few hundred cm^{-1}, so that excitation of almost any vibration in the S_1 electronic manifold of the aromatic partner leads to vibrational predissociation.

In the best studied case, tetrazine-Ar, time resolved measurements[16b,16c] show that the vibrational predissociation mechanism involves sequential processes, one or more of which are intramolecular vibrational energy redistribution steps. The initial step in the fragmentation process is an energy scrambling that leaves the van der Waals molecule in some metastable bound state. A second step then transfers energy to the van der Waals bond, leading to bond breaking, and leaving the tetrazine ring with some reduced amount of vibrational energy. In contrast with the glyoxal-H_2 system behaviour mentioned above, this observation implies that anharmonic mixing of the aromatic ring vibrations is a very important element in the predissociation mechanism. It is not known whether the vibrational predissociation mechanisms of the other aromatic ring-rare gas atom systems mentioned above are also sequential, but given the similarities in the structures, the vibrational frequencies and the likely vibrational couplings it seems plausible that some are. It is then very striking that, despite an initial intramolecular vibrational energy redistribution step, the common feature in the results of the studies of the initial state dependence of the product state distribution in the fragmentation reaction is that vibrational predissociation is quite selective; in all of the fragmentation reactions only a few of the energetically accessible final states are populated, and which states are populated depends on the initial state prepared in a nontrivial fashion.

As an example of the behavior observed, we display in Tables I and II the branching to final states in the fragmentation of $C_6H_{6-n}D_n$-He and $C_6H_{6-n}D_n$-He$_2$ following excitation of vibration 6^1 of the van der Waals molecule.[19] None of the common vibrational coupling mechanisms appears to adequately explain the selectivity of vibrational energy redistribution, or the change in selectivity with change in isotopic composition. An independent mode coupling theory, such as that due to Beswick and Jortner,[7a] can explain some but not all of the qualitative features of the observations.

Keeping the focus of the discussion on the qualitative features of the van der Waals molecule fragmentation reaction, Weber and Rice[16e] have shown that S_0 tetrazine-Ar dissociates at least an order of magnitude more slowly than does S_1 tetrazine-Ar, implying that coupling between the electronic state of the polyatom and the vibrational motion of the adatom cannot be ignored. The direct measurements of the rates of vibrational predissociation from different vibrational levels of p-difluorobenzene-Ar reported by Jacobson, Humphrey and Rice[22] contain several surprising features, of which we mention two at

TABLE IA. RELAXATION BRANCHING PROBABILITIES FROM 6^1-He$_1$

Level	C_6H_6	C_6H_5D	$C_6H_4D_2$ 1,2	$C_6H_4D_2$ 1,3	$C_6H_4D_2$ 1,4	$C_6H_3D_3$ 1,3,5	$C_6H_2D_4$ 1,3,4,5	C_6HD_5	C_6D_6
10^1	...	...	...	0.30	...	0.60	0.33	0.33	0.36
16^2	0.71	0.31	0.31	0.24	0.14	0.15	0.22	0.13	0.07
4^1	0.0					0.11			
11^1	0.20	0.69	0.69	0.46			0.45	0.54	0.54
16^1				0.0	0.86	0.10	0.0		0.0
0^0	0.09	0.0	0.0	0.0	0.0	0.04	0.0	0.0	0.03

From Ref. 19.

this point. First, the vibrational predissociation rate does not necessarily increase with additional quanta in a given ring vibrational mode (Table III). Second, direct excitation of the presumed reaction coordinate, namely the van der Waals bond stretching motion, may decrease the vibrational predissociation rate. These results are inconsistent with the simple theories of van der Waals molecule vibrational predissociation that have been proposed; they are also inconsistent with rather sophisticated calculations of state-to-state reaction rates for a model of p-difluorobenzene-Ar which has only harmonic motions.[7j]

TABLE IB. RELAXATION BRANCHING PROBABILITIES FROM 6^1-He$_1$ RENORMALIZED TO EXCLUDE LEVEL 10^1.

Level	C_6H_6	$C_6H_5D_1$	$C_6H_4D_2$ 1,2	$C_6H_4D_2$ 1,3	$C_6H_4D_2$ 1,4	$C_6H_3D_3$ 1,3,5	$C_6H_2D_4$ 1,3,4,5	C_6HD_5	C_6D_6
16^2	0.71	0.31	0.31	0.34	0.14	0.37	0.33	0.20	0.12
4^1	0.0					0.28			
11^1	0.20	0.69	0.69	0.66	0.86	0.24	0.67	0.80	0.84
16^1				0.0			0.0		0.0
0^0	0.09	0.0	0.0	0.0	0.0	0.11	0.0	0.0	0.04

From Ref. 19.

TABLE IIA. RELAXATION BRANCHING PROBABILITIES FROM 6^1-He$_2$.

Level	C_6H_6	$C_6H_5D_1$	$C_6H_4D_2$ 1,2	$C_6H_4D_2$ 1,2	$C_6H_4D_2$ 1,4	$C_6H_3D_3$ 1,3,5	$C_6H_2D_4$ 1,3,4,5	C_6HD_5	C_6D_6
10^1	...	...	...	0.12	...	0.13	0.11	0.0	0.04
16^2	0.19	0.17	0.15	0.14	0.35	0.10	0.18	0.18	0.12
4^1	0.0					0.04			
11^1	0.51	0.55	0.51	0.05	0.47	0.51	0.03	0.42	0.0
16^1				0.49			0.37		0.46
0^0	0.30	0.28	0.34	0.20	0.18	0.22	0.31	0.40	0.38

From Ref. 19.

All of the preceding points to the need for rather detailed information about the potential energy surfaces of the ground and electronic excited states of polyatom-atom van der Waals molecules. In particular, we need to know, for each potential energy surface, the fundamental vibrational frequencies and (at least) the lowest order anharmonic coupling constants in order to interpret the observed selectivity of product state formation in the van der Waals molecule fragmentation. It must be emphasized that the information needed includes the diagonal and

TABLE IIB. RELAXATION BRANCHING PROBABILITIES FROM 6^1-He$_2$ RENORMALIZED TO EXCLUDE LEVEL 10^1.

Level	C_6H_6	$C_6H_5D_1$	$C_6H_4D_2$ 1,2	$C_6H_4D_2$ 1,3	$C_6H_4D_2$ 1,4	$C_6H_3D_3$ 1,3,5	$C_6H_2D_4$ 1,3,4,5	C_6HD_5	C_6D_6
16^2	0.19					0.12			
4^1	0.0	0.17	0.15	0.16	0.35	0.04	0.20	0.18	0.12
11^1				0.06			0.05		0.0
16^1	0.51	0.55	0.51	0.55	0.47	0.58	0.41	0.42	0.48
0^0	0.30	0.28	0.34	0.23	0.18	0.26	0.34	0.40	0.40

From Ref. 19.

off diagonal anharmonicities of what in the zeroth order uncoupled mode approximation are the polyatomic molecule modes, the intermolecular van der Waals bond modes and, of course, the coupling between these classes of zero order modes. Very little such information is available. Even for a well studied polyatomic molecule such as benzene there is very little known about the anharmonicity of the potential energy surface. As to the van der Waals bond modes in polyatom-atom systems, experimental studies of intermolecular vdW vibrational

TABLE III. P-DIFLUOROBENZENE-ARGON PREDISSOCIATION RATES.

Level	Vibrational energy (cm-1)	$1/k_{VP}$ (ns)
$\overline{30}^2$	244	too small to measure
$\overline{8}^2$	375	too small to measure
$\overline{27}^1$	401	too small to measure
$\overline{6}^1$	410	4.3 ± 0.5
$\overline{5}^1$	817	4.1 ± 0.7
$\overline{6}^2$	820	20.0 ± 2.0
$\overline{5}^1\overline{\sigma}^a$	859	9.0 ± 0.5
$\overline{6}^2\overline{\sigma}^1$	862	3.6 ± 0.5
$\overline{5}^1\overline{6}^1$	1227	5-9
$\overline{6}^3$	1232	5-8
$\overline{3}^1$	1251	<1.5
$\overline{5}^2$	1634	3.7 ± 1

[a] σ denotes the stretching mode of the van der Waals bond.

From Ref. 22.

frequencies have been reported for the S_1 potential energy surfaces of p-difluorobenzene-Ar[21,22,25], carbazole-Kr[26], benzene-Ar[27], and for both the S_0 and S_1 potential energy surfaces of s-tetrazine-Ar, Kr and Xe[16a,16b,16f]. Theoretical studies of van der Waals bond vibrations in polyatom-atom species have usually been based on the assumption that the potential energy surface is adequately approximated as a superposition of atom-atom interactions. This assumption appears to give only a zeroth order approximation to the potential energy surface.

Knight and coworkers[25] have interpreted their measurements of the bending and stretching of the van der Waals bonds in the (S_1) molecules fluorobenzene-Ar, chlorobenzene-Ar, phenol-Ar and aniline-Ar in terms of Fermi resonance between the overtones of the two bending vibrations and the stretching vibration. They assumed the lower part of the vibrational manifold is adequately described by a 3 x 3 Hamiltonian with only cubic anharmonic coupling. The off-diagonal matrix elements necessary to reproduce the observed frequencies and intensities of the transitions are found to have values in the range $0.5 - 4.5$ cm^{-1}; an application of the same model to s-tetrazine-Ar leads to off diagonal matrix elements in the range $1.2 - 3.5$ cm^{-1} (Table IV). This model, which appears to give an excellent description of the experimental date (Table V), implies that the van der Waals bond stretching and bending motions are so heavily mixed that even the lowest lying vibrational states are not accurately described in a normal mode basis, and that all semblance of distinct stretching or bending motion is lost as the vibrational energy increases.

In contrast, Weber and coworkers[16f] have interpreted their measurements of the bending and stretching of the van der Waals bonds in the (S_0) molecules tetrazine-Ar, tetrazine-Kr and tetrazine-Xe without appeal to Fermi resonance between the bending and stretching modes. Their assignments rest on the reasonable (but far from perfect) fit to the frequencies of the observed transitions obtained using a harmonic basis for the description of the vibrations (Table VI), on the observation of very consistent trends in the distributions of intensities of overtone transitions of the three van der Waals molecules, (Table VII) which implies that all three van der Waals molecules have the same amount of vibrational mixing, and on the ratio of the values of the symmetric stretching frequency in Xe-tetrazine-Xe and the stretching frequency in tetrazine-Xe. A comparison of the van der Waals bond modes for the S_0 and S_1 potential energy surfaces shows that for the three tetrazine-X molecules the frequencies are very similar, those built on the S_1 surface being larger than those built on the S_0 surface by one or two wavenumbers. There is a remarkable similarity in the values of the several frequencies for the different rare gas atoms.

Table IVA. FREQUENCIES OF VAN DER WAALS VIBRATIONS IN AROMATIC-Ar COMPLEXES BASED ON CONVENTIONAL ASSIGNMENTS.

Molecule	β_x	β_y	σ_z
benzene-Ar	15.5		40
pDFB-Ar	?	17	42
tetrazine-Ar (S_1)	17	19	44
tetrazine-Ar (S_0)	15.5	17.6	42.1
chlorobenzene-Ar	16	17	45
fluorobenzene-Ar	21	17	47
phenol-Ar	21	19	47
aniline-Ar	22	19	49

From Ref. 25.

TABLE IVB. ZERO-ORDER ENERGIES AND COUPLING MATRIX ELEMENTS USED IN THE FERMI RESONANCE MODEL FOR VAN DER WAALS BOND VIBRATIONS.

	H_{11}	H_{22}	H_{33}	H_{13}[a]	H_{23}[a]
Aniline	46.0	39.5	43.5	4.5	0.5
Phenol	41.0	37.4	43.0	4.5	0.5
Fluorobenzene	41.7	35.0	42.2	4.5	2.5
Chlorobenzene	31.0	36.0	41.5	3.0	4.5
Tetrazine	33.5	36.0	40.0	1.2	3.5

[a]Cubic coupling matrix elements proportional to $Q^2_\beta Q_\sigma$.
From Ref. 25

TABLE V. CALCULATED EIGENVALUES, EIGENVECTORS AND BAND INTENSITIES USING THE ZERO-ORDER PARAMETERS AND MATRIX ELEMENTS GIVEN IN TABLE IV COMPARED WITH OBSERVED BAND POSITIONS AND INTENSITIES.

	ψ_1	ψ_2	ψ_3
Aniline-Ar			
$a(2\nu\beta_x)$	0.25	0.55	-0.79
$a(2\nu\beta_y)$	0.89	-0.45	-0.03
$a(\nu\sigma_z)$	-0.37	-0.70	-0.61
Calc. eigenvalues	39.3	40.3	49.4
Obs. vib. energy	39.0	41.0	49.0
Intensity (calc.)[a]	0.14	0.49	0.37
Intensity (obs.)	0.15	0.49	0.35
Phenol-Ar			
$a(2\nu\beta_x)$	0.53	0.57	0.62
$a(2\nu\beta_y)$	0.71	-0.70	0.04
$a(\nu\sigma_z)$	-0.46	-0.42	0.78
Calc. eigenvalues	37.1	37.7	46.6
Obs. vib. energy	37.0	37.0	47.0
Intensity (calc.)[a]	0.21	0.18	0.61
Intensity (obs.)	sum = 0.37		0.63
Fluorobenzene-Ar			
$a(2\nu\beta_x)$	0.23	-0.72	0.66
$a(2\nu\beta_y)$	0.89	0.42	0.16
$a(\nu\sigma_z)$	-0.39	0.55	0.74
Calc. eigenvalues	33.9	38.3	46.7
Obs. vib. energy	34.0	38.0	47.0
Intensity (calc.)[a]	0.15	0..30	0.54
Intensity (obs.)	0.17	0.29	0.54
Chlorobenzene-Ar			
$a(2\nu\beta_x)$	0.91	-0.36	0.19
$a(2\nu\beta_y)$	0.24	0.85	0.46
$a(\nu\sigma_z)$	-0.33	-0.37	0.87
Calc. eigenvalues	29.9	34.1	44.5
Obs. vib. energy	30.0	34.0	45.0
Intensity (calc.)[a]	0.11	0.14	0.75
Intensity (obs.)	0.12	0.14	0.57
Tetrazine-Ar			
$a(2\nu\beta_x)$	0.84	-0.52	0.12
$a(2\nu\beta_y)$	0.41	0.77	0.49
$a(\nu\sigma_z)$	-0.35	-0.37	0.86
Calc. eigenvalues	33.0	34.3	42.2
Obs. vib. energy	33.0	33.0	42.1
Intensity (calc.)[a]	0.12	0.13	0.74
Intensity (obs.)	sum = 0.24		

[a] The calculated intensity is obtained from the square of the coefficient a(vsz) of the eigenvector that corresponds to stretch character.

From Ref. 25.

TABLE VI. OBSERVED VAN DER WAALS MODE FREQUENCIES AND HARMONIC PROGRESSIONS IN THE SPECTRA OF THE CLUSTERS.

	Tetrazine-argon		Tetrazine-krypton		Tetrazine-xenon	
	observed shift	harmonic shift	observed shift	harmonic shift	observed shift	harmonic shift
σ_1	42.1	42.1[a]	37.1	37.1[a]	36.5	36.5[a]
σ_2	81.2	84.2	(71.0)	74.2	71.8	73.0
β_{x2}	33	31.0	32	32.2	36	34.0
β_{x4}	62.1	62.1[a]	64.5	64.5[a]	68.1	68.1[a]
β_{x6}	94.9	93.2	95.8	96.6	98.8	102.2
β_{x8}	...	...	...	...	126.1	136.2
β_{y2}	33	35.2	...	...	36	36.7
β_{y4}	70.3	70.3[a]	(74.6)	74.6[a]	71.3	71.3[a]
β_{y6}	101.7	105.5	...	...	103.4	107.0
β_{y8}	...	...	...	...	132.5	142.5

[a] This line was taken to define the fundamental frequency.

From Ref. 16f.

TABLE VII. **SUMMARY OF SHIFT FREQUENCIES, INTENSITIES AND ASSIGNMENTS OF THE VAN DER WAALS VIBRATIONAL MODES IN SEVERAL MOLECULES.**

	Tetrazine-argon		Tetrazine-krypton		Tetrazine-xenon	
	Shift	Intensity[a]	Shift	Intensity[a]	Shift	Intensity[a]
σ_1	42.1	19	37.1	38	36.5	80
σ_2	81.2	0.9	(71.0)	1.9	71.8	...
β_{x2}	33	3	32	3.7	36	...
β_{x4}	62.1	1.3	64.5	1.9	68.1	...
β_{x6}	94.9	0.3	95.8	0.6	98.8	1.2
β_{x8}	...	...	...	...	126.1	0.2
β_{y2}	33	3	...	...	36	...
β_{y4}	70.3	0.8	(74.6)	1.2	71.3	...
β_{y6}	101.7	0.2	...	...	103.4	0.5
β_{y8}	...	...	...	...	132.5	0.1

[a] Intensity with 0^0 excitation. 0_0^0 emission = 1000.

From Ref. 16f.

There is some further relevant information. Weber and Rice[16d] have studied the dephasing of vibrational excitations in both tetrazine and tetrazine-X van der Waals molecules. They find that, even when there is considerable excess vibrational energy in the polyatom ring modes relative to the van der Waals molecule binding energy, the dephasing is dominated by the macroscopic interference associated with the heterogeneous structure of the Q-branch transitions. The observed macroscopic dephasing times are in the range 300 ps to 2 ns, which implies that the lower limit for the intramolecular dephasing time falls in the same range. Clearly, this set of observations implies that the tetrazine ring mode-ring mode couplings and the tetrazine ring mode-van der Waals mode couplings are weak. Although the observations do not imply anything about the van der Waals bond stretching and bending mode mixing, it does seem strange that such mixing is strongly insulated from the ring vibration excitations.

It is difficult to reconcile the somewhat discordant interpretations of the spectrum of the van der Waals bond stretching and bending motions in tetrazine-Ar. On balance, it is more plausible that there is nonnegligible interaction between these motions than that they are completely independent, but then we must assert that said interaction is surprisingly independent of the species of adatom in the tetrazine-X molecule.

Before considering how the intramolecular dynamics determines the absolute value of the unimolecular reaction rate, and how van der Waals molecules can serve as a vehicle for the study of those intramolecular processes that compete with reaction, we ask if the characteristic features of the fragmentation reactions described in this section can be interpreted using perturbation theory. This approach is at the opposite end of the spectrum from the statistical theory of unimolecular reaction rate, since it focuses attention on state-to-state transitions. We shall see that such an analysis has some successes and some failures.

The literature provides several approaches to understanding vibrational predissociation of polyatom-atom van der Waals

molecules.[7a,7i,17-28] The basic physical picture underlying all of
these approaches is a separation between the fast intrapolyatom
motion and the slow intermolecular motion. The motion in the
fast coordinates is assumed to depend negligibly on
displacements in the slow coordinates, while the fast motion is
averaged to provide a potential surface for the slow motion. In
other words, one makes an adiabatic approximation. The problem
is formally identical to the separation of electronic and
nuclear coordinates in a molecule, except that now the
vibrational states of the polyatomic moiety play the role of
electronic states and the vibrational states of the polyatom-
atom bond play the role of the vibrational states. In both
cases predissociation represents a breakdown of the adiabatic
approximation. The variant approaches which fall under the name
"adiabatic" have been characterized clearly in an excellent
article by Balhausen and Hansen, and we will use their
nomenclature.[29] These approaches differ in two related ways: (1)
the basis, or zeroth order wave functions, chosen to describe
the motion of the fast coordinates; and (2) the exact potential
which is then used to find the wave function which describes the
motion in the slow coordinates.

All the analyses of van der Waals molecule predissociation
begin with the same system Hamiltonian. We will denote the
mass-weighted van der Waals coordinates by $\tilde{r}$. Then, neglecting
the influence of rotation on the system dynamics, the
Hamiltonian is

$$H = -\frac{\hbar^2}{2}\nabla_Q^2 - \frac{\hbar^2}{2}\nabla_r^2 + V_{tot}(\tilde{Q},\tilde{r}),$$
(II.1)

In this equation $V_{tot}(\tilde{Q},\tilde{r})$ represents the entire potential energy,
where we have already integrated over the electronic
coordinates. The simplest of the various adiabatic
representations is the "crude adiabatic" representation, in
which the aromatic wave functions are found at a single value of
$\tilde{r}$, which we will call $\tilde{r}_0$. One separates V_{tot} into two portions,
$V^{(0)}(\tilde{Q}) = V_{tot}(\tilde{Q},\tilde{r}_0)$ and $\Delta V(\tilde{Q},\tilde{r}) = \tilde{V}_{tot}(\tilde{Q},\tilde{r}) - V^{(0)}(\tilde{Q})$, and then solves an $\tilde{r}$-
independent Schrödinger equation for the aromatic wave function
$\psi^0_{\{v\}}(\tilde{Q})$:

$$\{-\frac{\hbar^2}{2}\nabla_Q^2 + V^{(0)}(\tilde{Q})\}\psi^0_{\{v\}}(\tilde{Q}) = \varepsilon_v\psi^0_{\{v\}}(\tilde{Q}).$$
(II.2)

In this equation we label each aromatic wave function by the set of vibrational quantum numbers $\{v\}$. Again one chooses an appropriate wave function in $\tilde{r}$ and writes the full adiabatic wave function $\psi^{(0)}_{\{v\}}(\tilde{Q})\phi_{\{v\},E}(\tilde{r})$. In this representation the off-diagonal elements are defined by the term $\Delta V(\tilde{Q},\tilde{r})$. The van der Waals kinetic energy operator, while neglected in the aromatic Hamiltonian, cannot couple $\tilde{r}$-independent wave functions and thus cannot give transitions between aromatic vibrational levels. If one uses these crude adiabatic wave functions to represent the initial and final levels of the vibrational predissociation process, the Golden Rule expression for the rate constant is

$$k_{VP} = \frac{2\pi}{\hbar} \sum_{\{v'\}\neq\{v\}} |\langle \psi^0_{\{v'\}}(\tilde{Q})\phi_{\{v'\},E}(\tilde{r})|\Delta V(\tilde{Q},\tilde{r})|\psi^0_{\{v\}}(\tilde{Q})\phi_{\{v\},E}(\tilde{r})\rangle|^2 \rho_{\{v'\}}(E). \qquad (II.3)$$

In Eq. (II.3) the initial vibrational level of the aromatic is denoted by the set of quantum numbers $\{v\}$. The subscript E on the van der Waals coordinate wave functions is the total energy of the complex. It indicates a discrete level in the case of the initial state and a continuum level in the case of the final state. $\rho_{\{v'\}}(E)$ is the density of continuum levels associated with aromatic vibrational state $\{v'\}$ at total energy E. We can integrate Eq. (3) over $\tilde{Q}$ to obtain

$$k_{VP} = \frac{2\pi}{\hbar} \sum_{\{v'\}\neq\{v\}} |\langle \phi_{\{v'\},E}(\tilde{r})|\Delta V^0_{\{v'\},\{v\}}(\tilde{r})|\phi_{\{v\},E}(\tilde{r})\rangle|^2 \rho_{\{v'\}}(E), \qquad (II.4)$$

$$\Delta V^0_{\{v'\},\{v\}}(\tilde{r}) = \langle \psi^0_{\{v'\}}(\tilde{Q})|\Delta V(\tilde{Q},\tilde{r})\psi^0_{\{v\}}(\tilde{Q})\rangle, \qquad (II.5)$$

where the angle brackets indicate integration over $\tilde{r}$ in Eq. (II.4) and over $\tilde{Q}$ in Eq. (II.5).

This crude adiabatic representation is the starting point for all the theoretical approaches referred to earlier. These approaches differ in the construction of $\Delta V(\tilde{Q},\tilde{r})$, the inter-molecular potential which provides the coupling. Beswick and Jortner calculate the potential as a sum of atom-atom interactions, and then expand the interaction in the atom-atom coordinates.[7a] Ewing also calculates the potential as a sum of atom-atom interactions, but then expands the interaction in the normal modes of the free polyatom.[7i] Weber and Rice's measurements[13e] of vibrational predissociation rates in S_0 s-tetrazine-Ar show very different dynamics from those observed in S_1, indicating that the electronic structure in the delocalized orbitals is important and that surfaces constructed from atom-atom potentials can be misleading. Therefore, they follow the lead of Halberstadt and Soep[17] and simply expand the interaction in the polyatom normal modes, relying on symmetry principles or spectroscopic data to estimate the relative magnitudes of the expansion coefficients where possible.[13e,28] We will follow this last approach as far as it takes us.

It seems sensible to expand the intermolecular potential up to quadratic terms, since the higher terms are generally neglected in the free molecule:

$$\Delta V(\widetilde{Q},\vec{r}) = V_0(\vec{r}) + \sum_i [\frac{\partial \Delta V(\widetilde{Q},\vec{r})}{\partial Q_i}]_{\widetilde{Q}=0} Q_i + \frac{1}{2}\sum_{i,j} [\frac{\partial^2 \Delta V(\widetilde{Q},\vec{r})}{\partial Q_i \partial Q_j}]_{Q=0} Q_i Q_j. \tag{II.6}$$

In this expression $\widetilde{Q}=0$ at the equilibrium configuration of the free polyatom. The first term gives the van der Waals potential at the free molecule equilibrium positions of the polyatom nuclei. This term cannot couple different polyatom vibrational states, however, so it does not contribute directly to vibrational predissociation. The second and third terms are written as derivatives of the van der Waals potential with respect to displacements in the polyatom coordinates, but they can also be viewed as changes in the intramolecular potential in the presence of the adatom. We can make the physical meaning of Eq. (II.6) clearer by rewriting it as

$$\Delta V(\widetilde{Q},\vec{r}) = \sum_i \{-\omega_i^2 \delta_i(\vec{r}) Q_i + \frac{1}{2}\Delta\omega_i^2(\vec{r}) Q_i^2\} + \sum_{i<j} \gamma_{ij} Q_i Q_j, \tag{II.7}$$

where ω_i is the vibrational frequency of mode i in the free molecule, $\delta_i(\vec{r})$ is the displacement in mode i, and $\Delta\omega_i^2(\vec{r})$ is the change in the square of the vibrational frequency of mode i. In principle one could also rewrite the second derivative cross terms in terms of mode-mixing coefficients, but this becomes cumbersome. Therefore we will simply write

$$\gamma_{ij} \equiv \left[\frac{\partial^2 \Delta V(\widetilde{Q},\vec{r})}{\partial Q_i Q_j}\right]_{\widetilde{Q}=0}$$

In some cases symmetry restrictions can help us estimate
the magnitudes of some of the terms in Eq. (II.7). For example,
the displacements, or first derivatives, vanish for nontotally
symmetric modes. Likewise, the cross derivatives are nonzero
only when Q_i and Q_j have the same symmetry. For the purpose of
evaluating matrix elements, however, one must use the symmetry
of the complex with the adatom atom located at $\tilde{r}$. For example, for
a given nontotally symmetric mode the first term in $\Delta V^0_{\{v'\},\{v\}}(\tilde{r})$
has a node in any plane with respect to which the mode is
antisymmetric. Since the integral over $\tilde{r}$ in Eq. (II.4) heavily
weights the portion of $\Delta V^0_{\{v'\},\{v\}}(\tilde{r})$ near the equilibrium value of $\tilde{r}$,
we can expect that terms with such nodes will give smaller
vibrational predissociation rates than terms without such nodes.
These terms will not be zero, however, as long as the final
state wave function in the $\tilde{r}$ coordinates is nontotally symmetric
in the equilibrium point group. Since such final state wave
functions necessarily have nonzero orbital angular momentum of
the fragments, we might also expect the effect of angular
momentum constraints on such pathways to be different from their
effect on pathways with dissociation along, say, the C_{2v} axis of
an aromatic-rare gas atom van der Waals molecule.

If we use the interaction potential of Eq. (II.7) and
harmonic oscillator wave functions for the aromatic wave
functions, we can easily evaluate the integral[30] in Eq. (II.5) to
obtain the coupling functions for one- and two-quantum changes
in a single mode. Two-quantum changes arise from the second-
derivative or frequency-shift terms:

$$\Delta V^0_{\{v_{j\neq i}, v_i - 2\},\{v\}}(\tilde{r}) = \frac{1}{2}\Delta\omega_i^2(\tilde{r})[v_i(v_i - 1)]^{1/2}\frac{\hbar}{2\omega_i} \cong \frac{1}{2}[v_i(v_i - 1)]^{1/2}\hbar\Delta\omega_i(\tilde{r}). \tag{II.8}$$

In principle these terms may be directly evaluated by vibronic
spectroscopy, since $\hbar\Delta\omega_i(\tilde{r})$ is just the "vibrational shift". For
example, the frequency of mode 8 is 7 cm^{-1} higher in pDFB-Ar than
in pDFB. Then Eq. (II.8) gives $\Delta V^0_{\{0^0\},\{8^2\}}(\tilde{r}) = 2^{-1/2} \times 7$ cm^{-1} = 5.0cm^{-1} at
the equilibrium value of $\tilde{r}$. Remarkably few assumptions are
required to relate $\Delta\omega_i$ to the rate of vibrational predissociation
of a van der Waals molecule with mode i excited.[13e,28] While the
exact form of Eq. (II.8) relies on the harmonic approximation

and the truncated expansion of the potential in Eq. (II.6),
similar expressions will arise even without these approximations. In the harmonic approximation, one-quantum transitions
arise from the first terms in Eq. (II.7). Using harmonic
oscillator wave functions for the polyatom wave functions, we
can evaluate the coupling for a one-quantum loss in mode i. We
obtain

$$\Delta V^0_{\{v_{j\neq i'}v_{i-1}\},\{v\}}(\vec{r}) = -\omega_i^2\delta_i(\vec{r})[\frac{v_i\hbar}{2\omega_i}]^{1/2} = -[\frac{v_i}{2}]^{1/2}a_i(\vec{r})\hbar\omega_i,$$

(II.9)

where the dimensionless quantity $a_i(\vec{r})$ is the ratio of $\delta_i(\vec{r})$ to the
turning point of coordinate Q_i in the lowest vibrational level of
the free molecule. Three interesting points come out of this
analysis of one-quantum vibrational predissociation processes.
First, these one-quantum transitions are allowed only for
totally symmetric modes. Second, the rate of one-quantum
transitions is proportional to the square of the displacement.
Equation (II.9) indicates that displacements of only 1% of the
zero-point level vibrational turning point can produce matrix
elements on the order of a few cm^{-1}. Such displacements are
quite plausible. If we define $\Delta\varepsilon_i(\vec{r})$ as the polyatom energy
change associated with a displacement $\delta_i(\vec{r})$, then

$$\Delta\varepsilon_i(\vec{r}) = \frac{1}{2}\omega_i^2\delta_i^2(\vec{r}), a_i(\vec{r}) = [\frac{2\Delta\varepsilon_i(\vec{r})}{\hbar\omega_i}]^{1/2}.$$

(II.10)

For example, let $\hbar\omega_i = 500$ cm^{-1}, a typical value. It seems
plausible that the polyatom moiety's potential energy in this
coordinate should change at least on the order of 0.1 cm^{-1} upon
formation of a complex. Then $a_i(\vec{r}) = (0.0004)^{1/2} = 0.02$. Substitution
of these values into Eq. (II.9) gives a matrix element of 7 cm^{-1}
for $v_i = 1$. Third, the vibrational predissociation rate constant
for a one-quantum transition should vary linearly with the
quantum number of the initial state. Since the energy lost to
the van der Waals coordinate in a one-quantum change is
independent of the initial state quantum number, the vibrational
predissociation rate should vary with the square of the coupling
function in Eq. (II.9). This equation indicates that the
vibrational predissociation rate should then be proportional to
the initial state vibrational quantum number v_i.

We now consider some of the successes and some of the failures of the perturbation theory analysis of van der Waals molecule predissociation.

Weber and Rice have extracted from the perturbation theory analysis a set of rules intended to describe the qualitative features of the vibrational energy redistribution which accompanies van der Waals molecule fragmentation.[28] These rules are:

1. The energy that is transferred in the first step of the sequence that leads to fragmentation, denoted ΔE^{IVR}, is bounded above and below. The allowed energy transfer range is determined by the density of states on the low energy side and the energy gap law on the high energy side. For the tetrazine-Ar molecule, Weber and Rice estimate these bounds on ΔE^{IVR} to be about 100 and 300 cm^{-1}, respectively.

2. Reactions that alter the quantum number in a vibration that has a large vibrational band shift are faster than others.

3. The fewer ring quanta exchanged in a reaction, the faster the reaction.

4. Some terms in the Taylor expansion (II.6) are zero because of symmetry restrictions. Reactions that would be determined by such terms are forbidden in the lowest approximation. They may proceed, though, via a coupling of the van der Waals bend vibrations with the ring vibrations. Such reactions are expected to be slower than unrestricted reactions, provided all the other circumstances are comparable.

Weber and Rice have tested these rules by examining the rates and branching ratios for products of the initial step in the vibrational redistribution of T-Ar excited to various vibrational states in the S_1 manifold. We shall consider only two examples.

$6a^1$ *Excitation.* $6a^1$ excitation leaves the T-Ar cluster with 703 cm^{-1} of vibrational energy. Given the requirement that 100-300 cm^{-1} of energy may be redistributed in the first IVR step, there are three product states available: $16b^1$ ($\Delta E^{IVR} = 297$ cm^{-1}), $16a^2$ ($\Delta E^{IVR} = 178$ cm^{-1}), and 4^1 ($\Delta E^{IVR} = 125$ cm^{-1}).

According to rule 2, the $16a^2$ product should dominate, and comparison with Table VIII confirms this result. Note that rule 2 (maximization of the modulation factor) appears stronger than rule 3 (minimization of the total number of quanta exchanged), which would favor the $16b^1$ or 4^1 product.

The $16b^1$ product, while only second to $16a^2$, still is favored compared to 4^1 because of the better modulation generated by 16b and the symmetry restriction on the pathway toward 4^1. Indeed, as Table VIII shows, only the 16b product has been observed. Reactions leading to 4^1 are too slow to be observed.

Heppener and Rettschnick[16c] report a small amount of $16a^1$ $16b^1$ product. The reaction to form this product would deposit only 33 cm^{-1} of energy into the van der Waals bond. It is conceivable that this reaction derives its efficiency from an accidental degeneracy: two quanta of bend vibration appear to have just about 34 cm^{-1} of energy.[16b]

6a^2 Excitation. Upon excitation of level $6a^2$ the tetrazine ring contains 1406 cm^{-1} of vibrational energy. Twenty vibrational levels fall within the range of 100 cm^{-1} < ΔE^{IVR} < 300 cm^{-1} (see Table IX). In spite of this abundance of pathways, Brumbaugh et al. report only two IVR products. The assignment of product $5^1 6a^1$ is rather uncertain, and only a minor fraction of the molecules end up in this level. The path that leads to $16a^2 6a^1$ is, therefore, the only important one.

If only the minimization of Δn_{tot} were important, the preferred pathway would lead to $16b^1 6a^1$ or $4^1 6a^1$. On the other hand, if only the maximization of Δn_{16a} were important, one would expect to observe levels $16a^3 6b^1$ and $16a^3 16b^1$. None of the product levels just cited are observed. Instead, the dominant pathway leads toward $16a^2 6a^1$, which represents a compromise product: Δn_{16a} is large, even though not the largest possible value, and Δn_{tot} is small, although not the smallest possible value. We find that the observed reaction maximizes the ratio $\Delta n_{16a}/\Delta n_{tot}$. Of the 20 conceivable pathways, only that one is observed which best satisfies both rules 2 and 3.

TABLE VIII. SUMMARY OF THE EXPERIMENTAL DATA FOR THE SYSTEM S_1 TETRAZINE–ARGON[a].

excitation	product states	branching ratio[b]	rate const[c]
$16a^2$	$16a^1$	0.11	4.6×10^8
	$16b^1$	0.22	–
4^1	–	–	$\leq 2 \times 10^8$
$6a^1$	$16a^2$	0.03	3.5×10^8
	$16b^1$	0.003	–
	$16a^1 16b^1$	–	5×10^7
$6b^2$	$16a^1 16b^1$	0.003	1.6×10^9
	$16a^2$	0.003	d
$16a^2 6a^1$	$16a^1 6a^1$	0.04	–
$6a^2$	$16a^2 6a^1$	0.05	–
	$5^1 6a^1$?	0.01	–

a. The product states are the first in the sequence of reactions leading to fragmentation.

b. From Ref. 16a.

c. From Ref. 16c.

d. Reported to be a secondary product.

TABLE IX.   CONCEIVABLE PRODUCT STATES IN FRAGMENTATION
OF S_1 TETRAZINE-Ar (100 cm^{-1} $<$ ΔE^{IVR} $<$ 300
cm^{-1}) AFTER EXCITATION TO LEVEL $6a^2$ AT 1406
cm^{-1} OF VIBRATIONAL ENERGY.

vibration	ΔE^{IVR}, cm^{-1}	Δn_{16a}	(Δn_{16b})	Δn_{tot}
$16b^16a^1$	297		(1)	2
$6b^216b^1$	276		(1)	5
$16a^36b^1$	257	3		6
4^2	250	0		4
$6b^116b^2$	231		(2)	5
$16a^316b^1$	213	3	(1)	6
$16a^25^1$	205	2		5
$16a^16b^14^1$	204	1		5
$16a^217b^1$	200	2		5
$16b^3$	187		(3)	5
$16a^26a^1$	178	2		3
$16a^116b^14^1$	160	1	(1)	
$16a^26b^2$	157	2		6
4^15^1	152	0		4
4^117b^1	147	0		4
$16a^11^1$	139	1		4
4^16a^1	125	0		2
$16a^26b^116b^1$	113	2	(1)	
$16a^16b^15^1$	106	1		5
$16a^16b^117b^1$	101	1		5

From Ref. 28.

In contrast with the qualitative success of the perturbation theory analysis of tetrazine-Ar fragmentation, such an analysis fails to describe many of the features of p-difluorobenzene-Ar fragmentation. For example, we have argued that reactions that alter the quantum number in a vibration that has a large vibrational band shift are faster than others (Rule 2), but the predissociation of p-difluorobenzene-Ar initially excited to $\overline{8^2}$ is very slow although $\Delta\omega_8$ is large. As to the prediction that the vibrational predissociation rate is proportional to the initial state vibrational quantum number, fragmentation of $\overline{5^1}$ pDFB-Ar and $\overline{5^2}$ pDFB-Ar fit the prediction well if we assume that $\overline{5^2}$ predissociates predominantly to $\overline{5^1}$, but the results for $\overline{6^1}$ and $\overline{6^2}$ vary in exactly the opposite direction. The ratio of the rate of $\overline{6^2} \to 6^1$ to the rate of $\overline{6^1} \to 0^0$ is 0.22 ± 0.04. Since the perturbation theory model predicts that this ratio should be 2, we are faced with an order of magnitude disparity. We consider these disparities to reflect a fundamental flaw in the analysis.

One could argue that $\overline{6^2}$ p-DFB-Ar is anomalous because it is mixed with the $\overline{5^1}$ level 3 cm^{-1} away, but no scheme using this assumption has been found to successfully explain the observations.[22]

Can we modify the standard crude adiabatic theory to better explain the data? The simplest modification is to replace the crude adiabatic approach with what Ballhausen and Hansen[29] call a "Herzberg-Teller adiabatic" approach. As applied to p-DFB-Ar this means including the effects of mode mixing induced by the stationary argon atom. The analysis is tedious and ultimately unsuccessful.[31] In Ewing's approach to van der Waals molecule predissociation the initial level is a mixture of all bound crude adiabatic states sufficiently nearby in energy.[7h] The predissociation rate is then dominated by those small components of this mixture which possess large amounts of vibrational excitation in the van der Waals bond. For example, Ewing points out that, given a reasonable model for the van der Waals potential of pDFB-Ar, $\overline{6^2}$ is nearly isoenergetic with $\overline{17^2\sigma^8}$. In this case one might expect strong mixing between the two levels even if the interaction is weak. Clearly, if one includes bending vibrations one can find a large number of such resonances, with the details quite sensitive to the exact form of the van der Waals potential. Ewing calculates that such Fermi resonances, with mixing coefficients on the order of 0.1, should be common in aromatic-argon complexes.[71] This mixing is determined largely by unpredictable resonances of such levels with the zeroth-order level; therefore, the theory can easily rationalize peculiar variations in the vibrational predissociation rates of particular states. Ewing has pointed out that if such resonances are critical it will probably not be possible to develop a predictive theory of vibrational predissociation in such large complexes above the lowest energies. These resonances can give rise to relaxation within the van der Waals modes under particular conditions involving the coupling strength, the density of van der Waals vibrational levels, and the optical excitation pulse profile.

Ewing's theory correctly locates the basic flaw in the crude adiabatic approaches described above: while the zeroth-order states employed give a good description of the final scattering levels, they give a poor description of the initial bound level. Ewing's work has shown that small changes in the description of the initial state can dramatically affect the calculated vibrational predissociation rate. The wave functions representing one of these more complex initial states cannot be factored into two functions where one depends only on the van der Waals coordinate, and therefore these states are nonadiabatic. It may be possible to revise the traditional approach to vibrational predissociation to allow nonadiabatic corrections. The role of nonadiabatic mixing in optical transitions due to vibrational motion in the slow coordinate has been studied at some length. The ratio of the nonadiabatic mixing to the adiabatic Herzberg-Teller mixing is often found to be proportional to $\hbar\omega/\Delta E$, where ω is the slow coordinate vibrational frequency and ΔE is the spacing between fast coordinate energy levels.[32] This ratio is not at all small in polyatom-atom van der Waals molecules. Moreover, one cannot carry the criterion stated directly into vibrational predissociation theory because in the optical transition problem the zeroth-order crude adiabatic states are determined for the equilibrium value of the slow coordinate. In conventional

vibrational predissociation theory the crude adiabatic states
are taken with the slow coordinate at infinite displacement from
equilibrium. A representation which uses the complex levels as
zeroth-order levels, and treats the dissociated levels as mixed,
might be better able to treat adiabatic and nonadiabatic mixing
in a unified manner.

III.   AN ALTERNATIVE FORM FOR CLASSICAL UNIMOLECULAR
REACTION RATE THEORY

We now examine some examples of the use of van der Waals
molecule fragmentation to learn more about the generic
characteristics of unimolecular reactions. The spectrum of
possible descriptions of the rate of a unimolecular reaction
runs from calculation of state-to-state transition
probabilities, which depend on many detailed features
of the potential energy surface and the associated molecular
dynamics, to the fully statistical theory which describes the
extreme situation wherein only the energy of the excited
molecule determines the reaction rate. Although the
fragmentation of a van der Waals molecule is a case where
disparity in vibrational frequencies is expected to lead to
restricted intramolecular energy flow and then to a
nonstatistical decay process, computer simulations of the
classical fragmentation process $I_2He \rightarrow I_2 + He$ (for a T-shaped model
of the I_2He molecule) yield the result that the decay of the
population of I_2He is nearly exponential over a wide range of
initial energy.[33,34] This observation suggests that a statistical
analysis of the fragmentation process, albeit different from the
standard RRKM analysis, can accurately predict the unimolecular
reaction rate even in cases where the molecular frequencies are
very different.

Of course, a proper description of the fragmentation of a
van der Waals molecule must be based on quantum mechanics and
must account for the competition between intramolecular
vibrational energy redistribution and reaction. However,
approximate statistical theories of the reaction rate based on
classical mechanics can be very useful in the construction of a
physical picture of the relevant molecular dynamics. For that
reason we examine how the classical mechanical theory of

unimolecular reaction rate can be improved, using a model of van der Waals molecule fragmentation as the vehicle for our analysis. Although we know that in the real case coupling of van der Waals bend and stretch motions is likely of some importance, and that there is nontrivial coupling of molecular rotation and vibration leading to rotational excitation of the product polyatom, at first we shall neglect all these effects. Indeed, we shall focus attention on the relative importance of bottlenecks to intramolecular energy transfer and to formation of fragments in a nonrotating molecule. For simplicity, for most of the discussion we shall adopt a model of van der Waals molecule fragmentation that has only two degrees of freedom: the van der Waals bond stretching motion and a vibrational excitation of the diatom or polyatom. The influence of other vibrational degrees of freedom and of rotation on the van der Waals molecule fragmentation will be examined later, but only cursorily.

The key ideas necessary to the improvement of the classical theory of unimolecular reaction rate come from nonlinear dynamics. These ideas are: first, that a general dynamical system supports both quasiperiodic and chaotic motion, with the two forms of motion intermingled in a complex pattern in the system phase space; second, that in a system with two degrees of freedom the rate of intramolecular energy redistribution can be obtained from consideration of the breakup of the last torus supporting quasiperiodic motion; and, third, that the transition state should be identified with the phase space curve separating bounded and free fragment motion, not with a surface in configuration space. The tools necessary to compute the breakup of the last torus supporting quasiperiodic motion and to calculate the flux of phase points through the remnant cantorus were provided by the work of MacKay, Meiss and Percival,[35] and Bensimon and Kadanoff;[36] those necessary for the calculation of the phase space curve dividing bounded and free fragment motion (partial separatrix) were provided by the work of Channon and Lebowitz.[37] Davis and Gray[34] and Gray, Rice and Davis[38] assume that the crossing of the cantorus and of the partial separatrix are independent processes, and that the transition state for a unimolecular fragmentation should be identified with the partial separatrix. The fluxes of phase points crossing these independent bottlenecks are computed from the interpretation of motion in phase space as a Hamiltonian induced mapping of the phase space into itself. Specifically, one iteration in time of the set of phase points defining the cantorus or partial separatrix generates a new curve that differs from the original in one localized region, where it defines a turnstile consisting of a curve bounding phase points outside and inside the original curve. The number of phase points outside the original curve, divided by the period of the iteration, defines the flux leaving the original bounded region. For the case of a van der Waals molecule such as I_2He, modelled as a T-shaped system with only the I_2 stretching and I_2-He stretching motions considered, the slow degree of freedom phase space (which is also a Poincaré surface of section) is shown in Fig.1 for several energies. Fig. 2 shows, for one particular energy, the system cantorus and partial separatrix, which are bottlenecks to the flow of phase points. Fig. 3 shows a comparison of the computed rate of fragmentation and the "observed" rate of fragmentation, where "observed" refers to exact numerical solution of the classical equations of motion for the T-shaped I_2He model used for the theoretical analysis. It is found that the rates of crossing the cantorus and the separatrix are comparable, so intra-molecular redistribution of vibrational energy plays an important role in determining the reaction rate. We note that for this model system conventional RRKM theory predicts a rate of fragmentation about three orders of magnitude larger than that observed. We also note that despite the large difference in the van der Waals stretching and I_2 frequencies of vibration (28 and 128 cm^{-1}, respectively), and the consequent inhibition of intramolecular energy redistribution, the modified statistical theory of reaction rate provides an excellent description of the data.

Gray, Rice and Davis[38] and Gray and Rice[39] have further developed the theory sketched above, with emphasis on analytical approximations that reduce the complexity of calculations and that permit the interpretation of variation of fragmentation rate with parameters such as vibrational frequency difference, van der Waals well depth, etc. The theory has been applied to the study of isomerization,[140] where it successfully resolves the problem of "surface recrossing." An estimate of the effect of

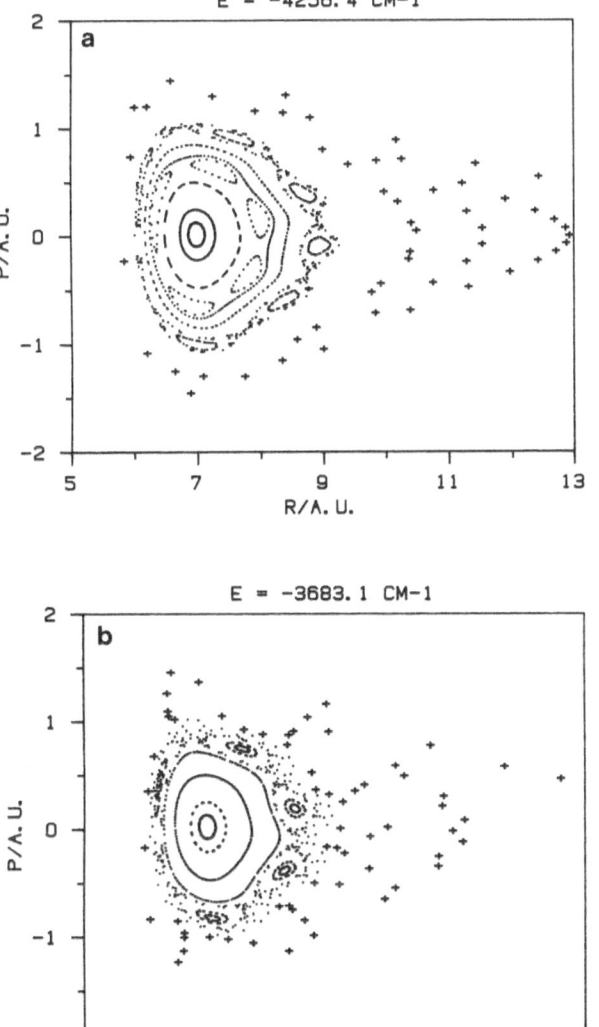

FIG. 1. Phase portrait of the T-shaped model dynamics for I_2He
fragmentation in the van der Waals bond stretching
Poincaré (R,P) surface for several total energies: (a)
$E = -4256.4$ cm^{-1} (v = 5); (b) $E = -3683.1$ cm^{-1} (v = 10);
(c) $E = -2661.6$ cm^{-1} (v = 20); (d) $E = -1807.0$ cm^{-1} (v =
30). v is the vibrational quantum number of I_2. From
Ref. 33.   (continued)

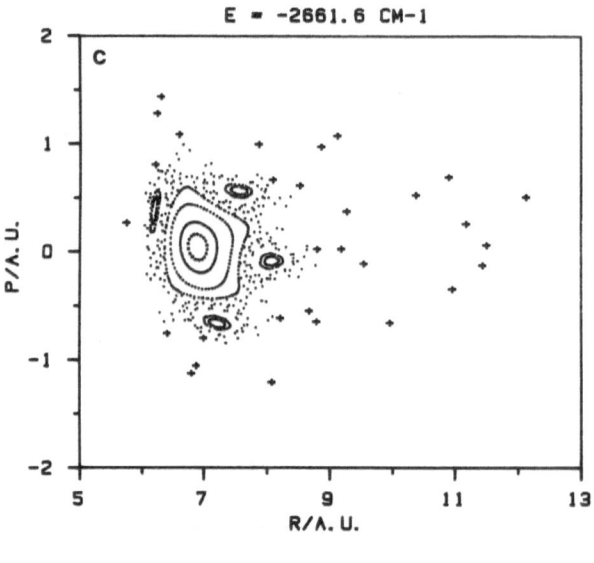

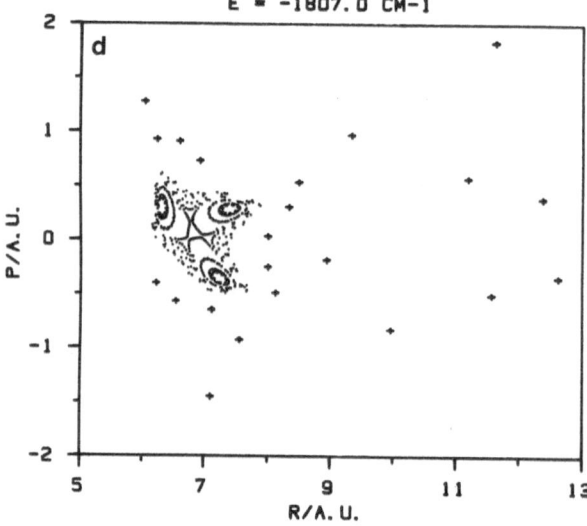

Fig. 1 (continued)

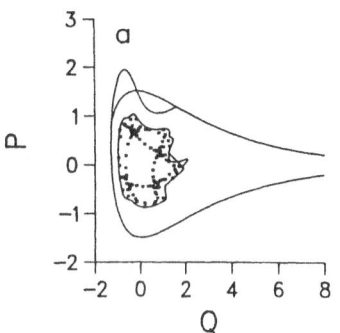

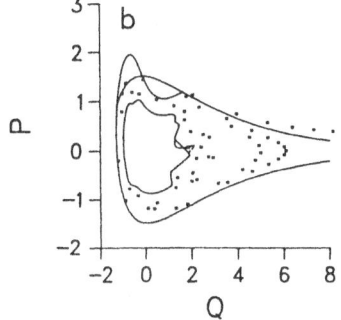

FIG. 2. An example of a trajectory which undergoes
predissociation at $E = -2661.6$ cm^{-1} ($v = 20$) for the
same system as in Fig. 1. In (a) the trajectory is
trapped inside the cantorus of winding number 4.618….
In (b) the trajectory crosses the cantorus and after a
while escapes from inside the partial separatrix. From
Ref. 34.

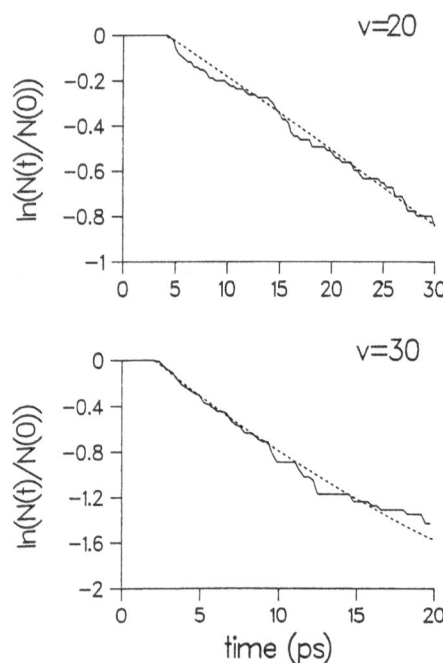

FIG. 3. Decay of an ensemble of I_2He systems. Exact calculations from classical trajectories (____); theoretical analysis (----). The theoretical model includes the partial separatrix as intermolecular bottleneck and one cantorus as intramolecular bottleneck. From Ref. 34.

inclusion of rotational motion has been made by Gray, Rice, and Davis,[38] and Tersigni and Rice[41] have shown that coupling the two degree of freedom model system to a third oscillator has little effect on the calculated fragmentation rate for a surprisingly large range of frequency of, and coupling strength to, the third oscillator.

Application of the same theoretical scheme to the reaction $C_6H_6He \rightarrow C_6H_6 + He$ leads to an examination of the influence of molecular vibration symmetry on the fragmentation rate.[42] Both out of plane vibrations (modes 11 and 16) and in plane vibrations (modes 1 and 6) of benzene were modelled. The results lead to the conclusion, for all the initial states studied, that the reaction rate is more influenced by the rate of ring mode to van der Waals bond stretch vibrational redistribution than by the rate of crossing the separatrix. In one sense this result is to be expected. That is, the ratio of frequencies of the two vibrational degrees of freedom of I_2He is about 5:1, but is much larger for all choices of one of the C_6H_6 vibrations and the C_6H_6He van der Waals stretching vibration, implying more difficult energy redistribution to the van der Waals bond in the latter case. Models of C_6H_6He which include the van der Waals bond stretching motion and as many as six ring vibrations (in plane and out of plane) and cubic and quartic anharmonic coupling between the latter, also imply that ring mode-to-van der Waals bond stretch intramolecular energy transfer is the most important rate limiting step in the unimolecular fragmentation. Thus, despite the existence of Arnold diffusion[43] in any nonlinear system with three or more degrees of freedom, on the time scale defined by the fragmentation reaction this model of C_6H_6He behaves as if the degrees of freedom are nearly quasi-periodic. We will return to this observation in Section IV.

It must be noted that the quantum mechanical version of the alternative statistical theory of unimolecular reaction rate remains to be developed. The difficulties to be surmounted are: (i) the alternative rate theory makes extensive use of the detailed characteristics of trajectories in the nonlinear system, and there is no good quantum mechanical analogue of a classical trajectory; (ii) there is very little understanding of how the quantum mechanical description of a system is influenced by the existence of chaos in the classical mechanical description of the same system. To address (ii) we have studied, for a simple model system, the signatures of classical chaos in quantum mechanics; this model system is described in Section V.

IV. EFFECTIVE DECOUPLING OF SLOW FROM RAPID VIBRATIONAL MOTIONS

We now examine intramolecular vibrational redistribution and its coupling to fragmentation from a different point of view. Our approach attempts to develop an understanding of the conditions under which a subset $n < N$ of the degrees of freedom of a nonlinear system are effectively decoupled from the remainder.

We find it convenient to take advantage of the observation that the representation of the evolution of a classical mechanical system as a flow of points in phase space can be thought of as a Hamiltonian mapping of the phase space into itself. The features of this mapping have been explored in great detail for bounded systems with two degrees of freedom. Indeed, many of the important advances in our understanding of the relationship between regular and stochastic flow, in particular the discovery of the role of the cantorus as a bottleneck to the flow of phase points,[35,36] have emerged from the study of Hamiltonian mappings in two-dimensional bounded systems. In contrast, our knowledge of the properties of Hamiltonian mappings in unbounded systems is much less well developed than for the case of bounded systems. Of course, any discussion of unimolecular fragmentation must allow for separated fragment trajectories that go to infinity, i.e., be based on the study of the Hamiltonian for an unbounded system.

Gaspard and Rice have shown that, just as in the case of bounded systems, insights into the dynamics of molecular fragmentation can be gleaned from a study of model Hamiltonian mappings in unbounded systems.[44] In particular, they have proposed several simple model mappings designed for the study of unbounded systems; they use these mappings as surrogates in the study of the Hamiltonian dynamics of molecular fragmentation. There are two principle motivations for this study. First, the properties of a mapping can often be determined from analytical considerations, thereby providing invaluable guidance to understanding the mechanics of the system. Second, the numerical calculation of the properties of a mapping is often several orders of magnitude more efficient than is integration of the equations of motion for which the mapping is surrogate. This feature allows the exploration of complexities in the mechanics of the system which are more difficult to discover and analyze when direct integration of the equations of motion is employed.

Mappings can be constructed in several different ways. In the case of autonomous systems with two degrees of freedom, with the representative Hamiltonian

$$H = H (X_1, P_1, X_2, P_2),$$
(IV.1)

a mapping is generated by the flow in a Poincaré surface of section

$$S (X_1, P_1, X_2, P_2) = \text{Constant},$$
(IV.2)

in the constant energy hypersurface $H = E$. The mapping is represented by

$$A (q, q') = \int_0^{T(q,\, q')} dt L (X_1, \dot{X}_1, X_2, \dot{X}_2),$$
(IV.3)

which is the action associated with the unique trajectory from q to q', where q is the intrinsic position of the Poincaré surface of section, and L is the Lagrangian corresponding to the Hamiltonian (IV.1). The mapping in the (p, q) plane is obtained from

$$p = - \partial A / \partial q,$$
(IV.4)

$$p' = \partial A / \partial q',$$
(IV.5)

where p and q are the canonical momentum and coordinate used to define the Poincaré surface of section. In principle, a two-dimensional mapping can be generated in this fashion. However, in practice, development of the generating function (IV.3) usually requires the introduction of analytical approximations. Thus, a mapping corresponding to (IV.3) is usually generated numerically.

The situation is simpler for the case of periodically perturbed ("kicked") systems, which can be represented by the generic Hamiltonian

$$H = H_0(X,P) + TG(X) \sum_{n=-\infty}^{\infty} \delta(\tau - nT). \qquad (IV.6)$$

In (IV.6), $G(X)$ is the amplitude of the perturbation (kick) at position X and T is the time between kicks. This Hamiltonian models, for example, the dynamics of a material system with Hamiltonian H_0 interacting with a pulsed laser; the laser pulses are presumed to have a duration which is small with respect to the time between pulses. Of course, the material system evolves under the Hamiltonisn during the time between pulses. If this "zero order" Hamiltonian system is integrable, the equations describing the mapping can be written out explicitly. In general, the presence of the periodic external perturbation converts the regular motion of the "zero order" system into irregular motion. Gaspard and Rice[44] show that the dynamics of these symplectic mappings capture the qualitative features of Hamiltonian flows such as (IV.1).

In the more general case of a system with N degrees of freedom, the 2N-dimensional phase space has intermingled regions where the resonance layers near the separatrices are not isolated from each other, since the N-dimensional surfaces that support quasiperiodic motion do not separate the (2N-1)-dimensional surface of constant energy into distinct regions, as they do when N = 2. For N > 2 all the stochastic regions of the phase space are connected in a single complicated web, the Arnold web.[43] For an initial condition within the web the trajectory will eventually intersect every finite region of the energy surface, a process called Arnold diffusion. In

principle, Arnold diffusion implies that randomization of an
initial energy distribution will always occur in a system with N
> 2 degrees of freedom. In practice, the few numerical studies
reported indicate that in the model systems studied Arnold
diffusion is sufficiently slow that many of the qualitative
features of regular and stochastic motion characteristic of an
isolated two degree of freedom system survice for a very long
time in a two degree of freedom subsystem of an N degree of
freedom system. Gaspard and Rice show that a similar inference
can be drawn from the properties of a model four-dimensional
Hamiltonian mapping. These observations provide motivation for
further examination of the dynamics of systems with only two
degrees of freedom.

Gaspard and Rice report the results of studies of the
following model Hamiltonians:

$$H = \frac{P^2}{2m} + TG(X) \sum_{n=-\infty}^{\infty} \delta(t-nT),$$ (IV.7)

$$H = \frac{L^2}{2I} + \frac{P^2}{2m} + TG(\theta, X) \sum_{n=-\infty}^{\infty} \delta(t-nT).$$ (IV.8)

Eq. (IV.7) describes a free particle subject to periodic kicks.
When $G(X)$ is chosen to have the form

$$G(X) = D(1 - e^{-aX})^2,$$ (IV.9)

the scaling $p_n = (aT/m) P_n, q_n = aX_n, d = 2 a^2 T^2 D/m$, leads to the mapping

$$p_{n+1} = p_n + d[\exp(-2q_n) - \exp(-q_n)],$$ (IV.10)
$$q_{n+1} = q_n + p_{n+1},$$

which is characterized by the single parameter d. For positive q
the perturbation vanishes, while for negative q it becomes very
large. The kicks are then repulsive (i.e., transfer momentum
opposite to that of the free particle) if we assume the
parameter d positive. If d were not chosen to be positive, the
particle would be indefinitely accelerated for negative q. There
is another reason for choosing $d > 0$. If we think of $G(X)$ as a

potential well in the limit where T goes to zero, then the mapping is equivalent to a flow and $D > 0$ defines the depth of the well relative to the energy zero at $X = 0, P = 0$. Defining $\omega \equiv a(2D/m)^{1/2}$ to be the frequency associated with $G(X)$ near $X = 0$, we see that $d = \omega^2 T^2$.

The mapping (IV.10) has two fixed points, $q = p = 0$, and $p = 0, q = +\infty$, whose stabilities are determined by the eigenvalues Λ of the linearized mapping obtained from (V.10).

If the perturbation amplitude in Eq. (IV.7) is constructed to have several minima and maxima, one obtains mappings with several bottlenecks, each associated with fixed points. The existence of several bottlenecks allows this model to mimic some properties of a molecule with intramolecular restrictions to energy flow. For instance, the following kick amplitude function,

$$\frac{dg}{dq} = -d\,e^{-q}\,(e^{-q} - r)\,(e^{-q} - s)\,(e^{-q} - t), \tag{IV.11}$$

(with $d > 0$) generates a mapping with 4 fixed points,

$$q = +\infty, \ \ln\frac{1}{t}, \ \ln\frac{1}{s} \text{ and } \ln\frac{1}{r}, \tag{IV.12}$$

all at $p = 0$; these fixed points are centers if $0 < d^2g/dq^2 < 4$, and saddles otherwise.

Fig.4 shows the global invariant set of the mapping

$$p_{n+} = p_n + d\,e^{-q_n}\,(e^{-q_n} - 1)\,(e^{q_n} - 1/2)\,(e^{q_n} - 1/4), \tag{IV.13}$$
$$q_{n+1} = q_n + p_{n+1},$$

with two centers at $p = 0, q = 0$ and $p = 0, q = \ln 4$, and an intermediate saddle at $p = 0, q = \ln 2$, representing an intramolecular bottleneck. We show in Fig. 5, for comparison, the global invariant set for the mapping (IV.10); the plot shows only those trajectories which remain at finite distance for a long time.

The Hamiltonian (IV.8) represents a rotator in free motion

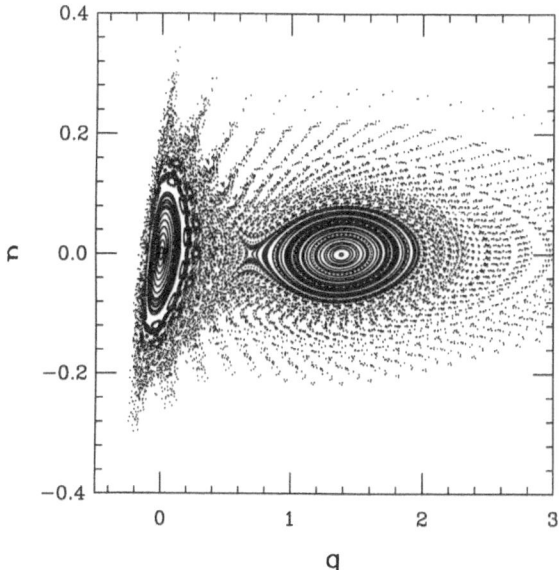

FIG. 4. Global invariant set of the mapping (IV. 13) for d = 2
in the (q, p) plane. 113 trajectories of 400
iterations are plotted. From Ref. 44.

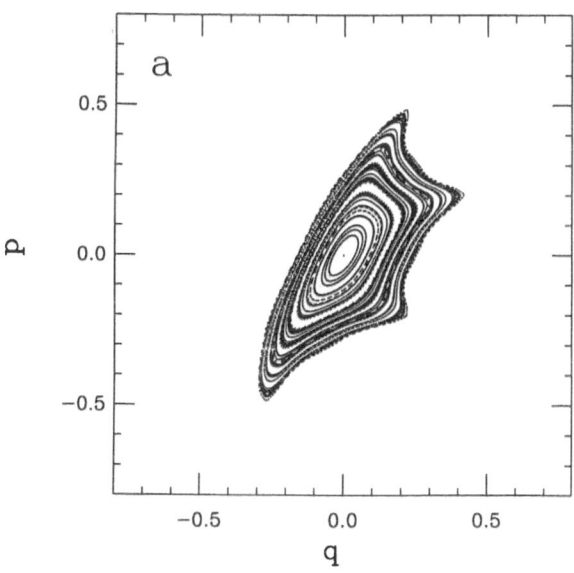

FIG. 5. Global invariant set of the mapping (IV. 10) in the (q,
P) plane for: (a) d = 1.8; (b) d = 2; (c) d = 2.2.
The collapse of the global invariant set at d = 2 is
caused by a period 4 resonance. The global invariant
set rapidly recovers from this collapse as d increases
past the resonance at d = 2. From Ref. 44. (continued)

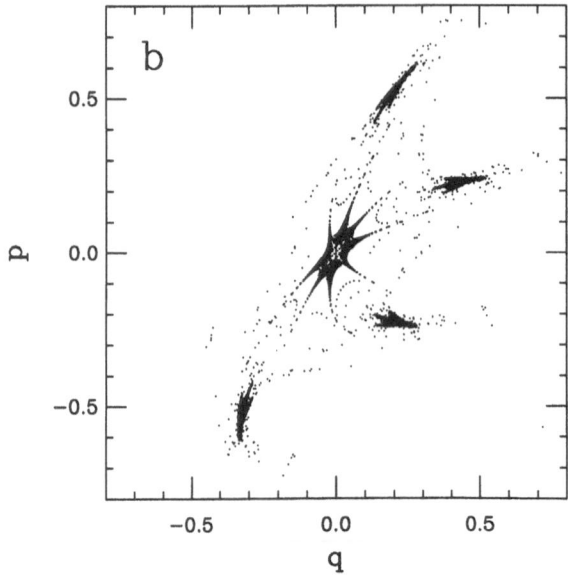

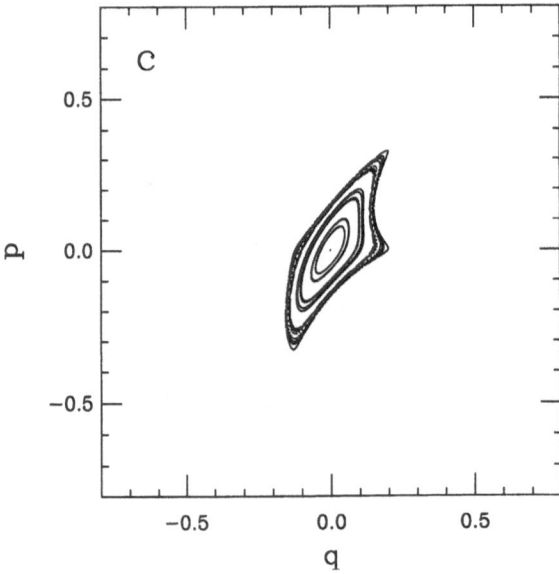

Fig. 5 (continued)

between kicks of amplitude $G(\theta, X)$; the rotator can escape from the perturbing field. Choosing

$$G(\theta, X) = D\,[(1 + g\cos\theta)\,e^{-2aX} - 2e^{-aX}] \qquad\qquad (IV.14)$$

generates a four-dimensional mapping which is pseudosymplectic. Gaspard and Rice have analyzed the escape dynamics for the system defined by (IV.8) and (IV.14) using the escape time function to investigate whether the quasi-invariant set will slowly disappear after a very long time. The mapping can be written in terms of the following rescaled variables and parameters:

$$q = aX,\ p = (aT/m)\,P$$

$$\ell = (T/I)\,L,\ c = g\,T^2\,D/I,\ d = 2a^2T^2D/m.$$

Fig. 6 shows the escape time function at $d = 2$ for $\theta = 1$ and varying initial angular momentum ℓ over 10^3 iterations. As for two-dimensional mappings (see Fig. 7), the escape time function is complex due to the presence of a fractal repellor. Fig. 8 depicts the same function over the period corresponding to 10^5 iterations. Some parts of the 10^3 – quasi-invariant set are depleted after 10^5 iterations but some other parts (around $\ell = 0$ and $\ell = 0.35$) remain, showing that the depletion dynamics is extremely slow.

An important difference with respect to two-dimensional maps is that escape of a trajectory may be rapid between two regions inside of which it is extremely slow. This novel feature occurs because of the four-dimensional topology and is a general phenomenon in many degree of freedom systems.

Gaspard and Rice have also calculated the decay of an ensemble of particles for $d = 2$. As for two-dimensional mappings (see Fig. 9), the decay occurs over two different time scales and it may be approximated here by a biexponential curve for intermediate times. However, an extremely slow decay still occurs after a very long time, as shown in Fig. 10, which is due to the slow depletion of the quasi-invariant set. As a consequence, we may not define either a global invariant set nor its volume for this system.

As for the decay in two-dimensional mappings, it is very difficult to describe any characteristic features of the long time behavior of the system. Over a very long time, periods of fast decay alternate with periods of slow decay. The origin of this phenomenon can, presumably, be traced back to Arnold diffusion. In molecular fragmentation, only the short time dynamics of these classical models need be considered because the long time dynamics will be governed by quantum mechanics, emission of radiation or collisions.

This study of Hamiltonian mapping models of fragmentation displays the differences and similarities between several systems.

The two-dimensional mappings of periodically perturbed systems have Poincaré portraits which are comparable to the Poincaré portraits of two degree of freedom Hamiltonian flows. Integrability is achieved when the time interval between the kicks vanishes.[42] This property is quite unexpected and it should not be confused with the trivial limit where the amplitude of the kicks goes to zero, which also leads to integrability. The kicked Morse system is an example where the two limits are different. In this case, the limit where the kicking period goes to zero is equivalent to the limit where the ratio between the Morse oscillator intrinsic frequency divided by the kicking freguqncy goes to zero. That integrability is observed in this limit has been inferred recently by Tersigni, Gaspard and Rice in numerical studies of Hamiltonian flows.[42] The adiabatic hypothesis is valid in this limit.

In two-dimensional mappings, a global invariant set with a positive area is preserved by the escape dynamics. However, in four-dimensional mappings Arnold diffusion precludes the existence of complete barriers formed by invariant tori. Accordingly, no invariant set of positive Legesgue measure is expected to exist. Nevertheless, numerical integration shows that a quasi-invariant set persists for a very long time. This quasi-invariant set shows a property similar to the invariant

232

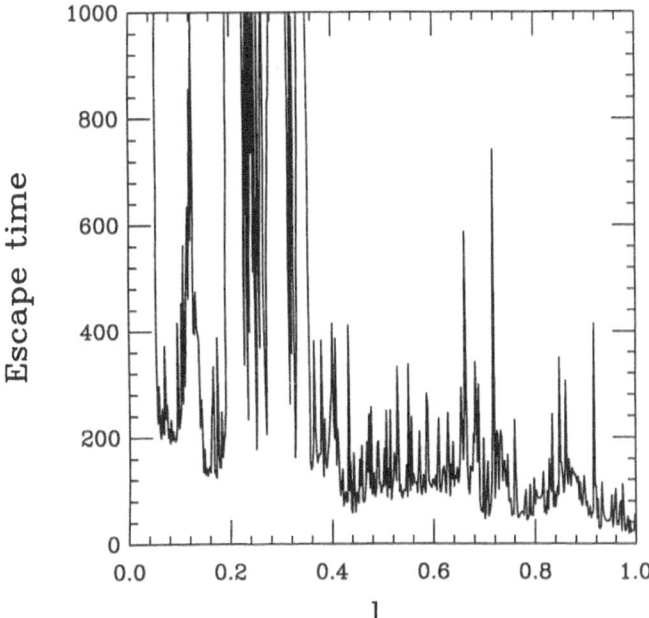

FIG. 6. Escape time function of the four dimensional mapping
generated by (IV. 8) and (IV. 14) for d = 2, c = 0.1,
q = 0.1. The initial conditions were p = 0, q = 0,
variable l. The 500 trajectories were each followed
for 10^3 iterations. From Ref. 44.

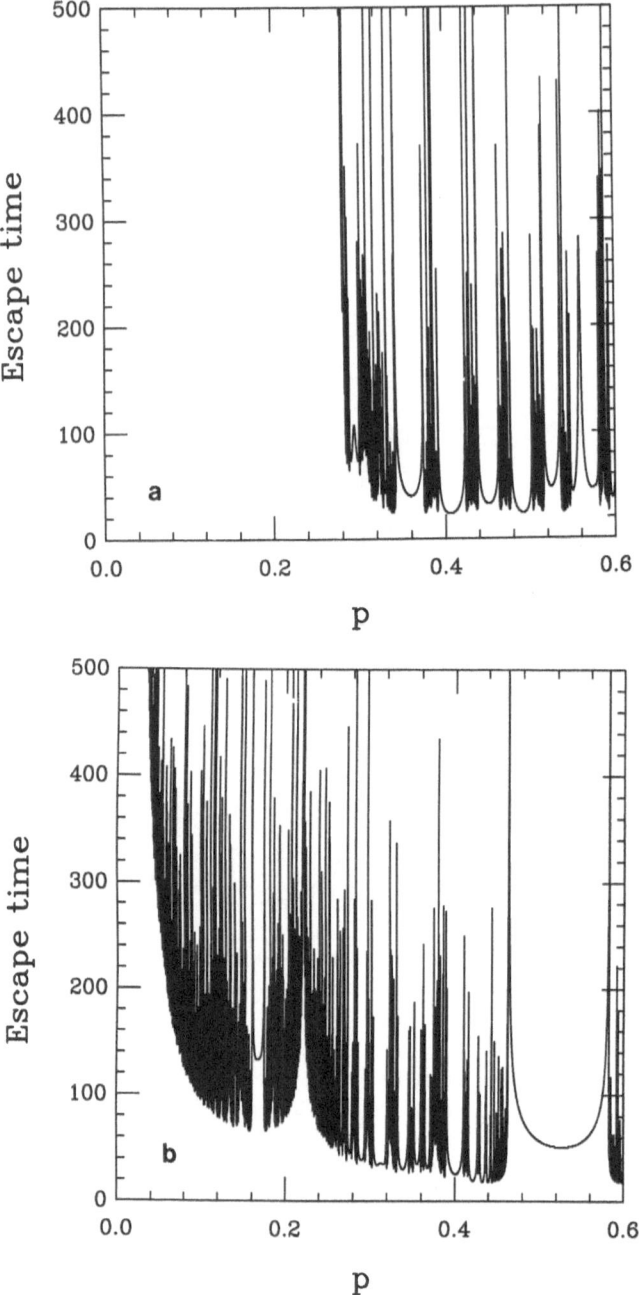

FIG. 7. Escape time function of the two dimensional mapping
(IV. 10) for (a) d = 1.8 and (b) d = 2. The function
is based on 6000 trajectories of 500 iterations each,
with initial conditions q = 0, variable p. Escape is
assured when q > 50. The 4-resonance at d = 2 has a
dramatic effect on the escape time function, which
becomes finite close to the q = 0, p = 0 center. From
Ref. 44.

set of two-dimensional mappings: it collapses near low-order resonances.[44] Furthermore, an ensemble of particles decays with two widely different time scales at intermediate times in four-dimensional as well as two-dimensional mappings. This important observation implies that a two degree of freedom subsystem of a larger system retains many of the properties of an isolated two degree of freedom system for a time long enough for physical processes of interest to occur. An obvious corollary of this observation is that two degree of freedom models of more complex systems can incorporate the major features of the larger system dynamics and are, therefore, worthy of detailed study.

V. SIGNATURES OF CLASSICAL CHAOS IN QUANTUM MECHANICS

Given that our goal is to understand better the role of molecular dynamics in determining the rate and selectivity of a unimolecular reaction, we are naturally led to investigate the character of the quantum mechanical behavior of model systems such as described in Sections III and IV. There are many difficulties in such an investigation, a number of which are related to how underlying classical mechanical chaos influences the quantum dynamics of a system. The approach sketched in this Section takes the view that qualitative aspects of the signatures of classical chaos in quantum mechanics can be established from studies of model systems and then, hopefully, used to characterize complex systems resembling real molecules.

Chaotic behavior, already referred to in Section III, is a general feature of the classical mechanics of nonlinear systems. As in the preceding Sections, we focus attention on unbounded Hamiltonian systems. For trajectories with infinitesimally different initial conditions, an unbounded Hamiltonian system with two or more degrees of freedom can be focussing (the trajectories come closer together after passing through a region of changing potential energy), or defocussing (the trajectories diverge after passing through a region of changing potential energy), or both. Consequently, the long time structures generated in the system phase space are typically composed of: (i) regular bounded stable motions forming quasiperiodic islands; (ii) irregular bounded unstable motion forming a chaotic repellor which governs the escape dynamics; and (iii)

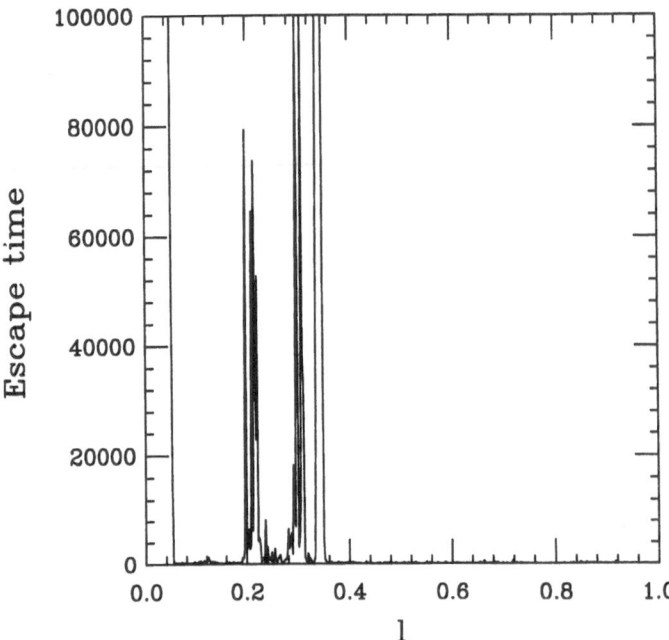

FIG. 8. The same as Fig. 6, but the trajectories are now followed over 10^5 iterations. From Ref. 44.

unbounded trajectories coming from and going to infinity. The entire structure is hierarchical with scaling properties that have only recently been investigated.[45]

The study of the signatures of classical chaos in the quantum mechanical description of a general system is too complex for us to undertake at present. However, the phase space structure of a classical system that is exclusively defocussing is simpler than that of a general system. In particular, in an exclusively defocussing system the quasiperiodic motions of type (i) are absent. Examples of exclusively defocussing systems are the elastic collisions of a point particle with an assembly of hard discs or hard spheres or, indeed, any hard objects with smooth convex boundaries.

Gaspard and Rice[46] have studied the classical, semiclassical and full quantum mechanical dynamics of the scattering of a point particle from three hard discs fixed in a plane (see Fig. 11). We note that the classical motion (which is chaotic) consists of trajectories which are trapped between the discs. If the three discs are labeled with the integers {1,2,3}, each trapped orbit is characterized by the sequence of integers labeling the successive impacts on the discs:[47]

$$\omega = \cdots 1231321 \cdots. \tag{V.1}$$

Each finite string of n integers defines a cell on the repellor and all such cells $\{\omega_0 \omega_1 \cdots \omega_{n-1}\}$ form a partition of the repellor.

Both the classical and the quantum dynamics of this system can be described as a multiple scattering process which is governed by the Perron-Frobenius matrix defined as[48]

$$Q_{ab}(z, \beta, r) = \frac{r \exp(-z \ell ab)}{\Lambda_{ab}^{\beta}}. \tag{V.2}$$

The rows of this matrix are labeled by the strings $a = \omega_0 \omega_1 \cdots \omega_{n-1}$ and the columns by the strings $b = \omega_1 \omega_2 \cdots \omega_n$. In Eq. (V.2), ℓ_{ab} is the length of the path between cells a and b, and $\Lambda_{ab} > 1$ is the defocussing factor of the impact in cell b for a trajectory

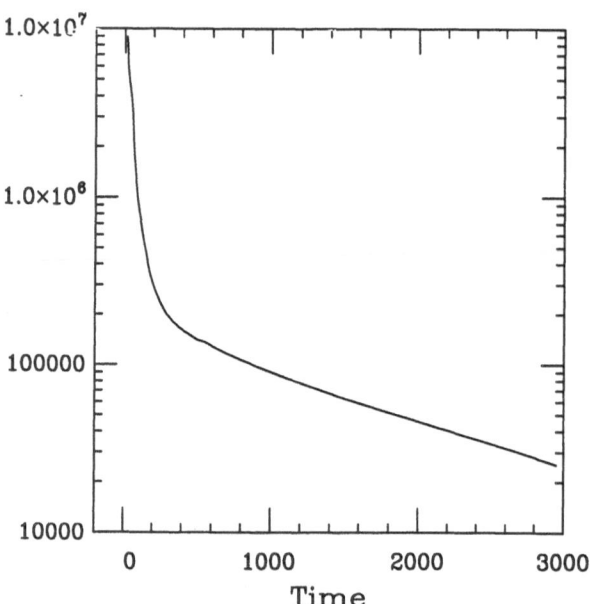

FIG. 9. Decay of an ensemble of 10^7 particles under the mapping
IV. 10 for d = 1.8. The initial ensemble is uniformly
distributed in the rectangle (q_1 = 0, q_2 = 0.01) x (p_1 =
0.275, p_2 = 0.375). N(t) is the number of particles
with q < 10 after t iterations. The function M(t) =
N(t) − N(5000) is plotted, with N(5000) = 1.1 x 10^6.
From Ref. 44.

coming from cell a and going to cell $c = \omega_2\omega_3\cdots\omega_n$, while r is a coefficient modelling reflection or absorption at the impact in cell b.

Consider, first, the classical dynamics of this system. If the discs scatter an ensemble of particles, each of which has the same velocity v, we have to take $\beta = 1, r = 1$ and $z = -(v\tau_{cl})^{-1}$, where τ_{cl} is the classical lifetime for a particle on the repellor. Now consider the quantum dynamics of the same system. If the discs scatter a plane wave of wave vector k, the same matrix describes the scattering processes in the classical and semiclassical descriptions, but in the latter the parameters must be fixed as follows: $z = -ik, r = -1$ for Dirichlet boundary conditions,[49] and $\beta = 1/2$ because the square of the wave function is the density of probability.[50] Therefore, we can analyze both the classical and the quantum dynamics of this model system with the same formalism, keeping in mind the different definitions of the parameters z, β and r. In the following we shall consider only the case $|r| = 1$.

For the Perron–Frobenius matrix to define an invariant measure, it must have 1 as eigenvalue, so that the variable z must be a solution of the equation

$$\det [1 - Q (z,\beta,r)] = 0. \qquad (V.3)$$

This equation has a set of zeroes $\{z_j (\beta, r)\}, j = 1, 2, \cdots$, in the complex plane; these zeroes characterize the kinetic properties of the repellor. An important role is played by the largest real zero of (V.3) when r is replaced by $r = 1$, which defines the function $P (\beta)$ called the topological pressure of Ruelle. This real zero bounds the real parts of all the zeroes of (V.3).[51]

$$\text{Re} [z_j (\beta, r)] \leq P (\beta); j = 1, 2, \cdots. \qquad (V.4)$$

An invariant probability measure μ_β associated with this real zero can now be constructed. The stability of the trajectories

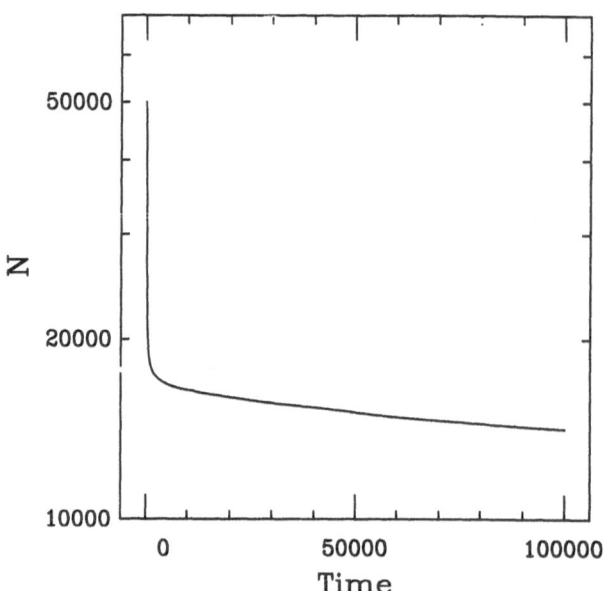

FIG. 10. Decay of an ensemble if 5×10^4 particles under the
four dimensional mapping IV. 14, with $d = 2$, $c = 0.1$,
$q = 0.1$. The initial ensemble is uniformly
distributed in the rectangle ($\theta_1 = -1$, $\theta_2 = 1$) x (ℓ_1
$= 0$, $\ell_2 = 2$) of the plane $q = 0$, $p = 0$. In this case
we plot $N(t)$ directly. From Ref. 44.

is then determined by the mean Lyapunov exponent per unit
length,[52]

$$\lambda(\beta) = - P'(\beta).$$ (V.5)

The dynamical randomness is characterized by the Kolmogorov-
Sinai entropy per unit length. The Kolmogorov-Sinai (KS)
entropy is the rate of accumulation of data necessary and
sufficient to follow unambiguously an orbit on the repellor. It
is given by the relation[51,52]

$$h_{KS}(\beta) = P(\beta) + \beta\,\lambda(\beta).$$ (V.6)

The classical lifetime for an ensemble of particles of velocity
v is calculated from

$$\frac{1}{v\tau_{cl}} = - P(1) = \lambda(1) - h_{KS}(1) \geq 0.$$ (V.7)

For a bounded system, the escape rate is zero and we recover the
well-known Pesin result that the KS entropy is equal to the
Lyapunov exponent. However, when escape occurs from a chaotic
system the rate is given by the difference between the sum of
positive Lyapunov exponents and the KS entropy.[53]

As to the semiclassical quantum mechanical description, the
zeroes $\{z_j\,(1/2, - 1)\}, j = 1, 2, \cdots,$ of Eq. (V.3) generate a family of
scattering resonances which are poles of the S-matrix. This
relationship is established using the Gutzwiller trace formula,[54]
which Gaspard and Rice employ to carry out a semiclassical
quantization based on the classical orbits of the system.
Cvitanovic and Eckhardt[55] recently reported a high precision
calculation of the exact scattering resonances of the point
particle-three hard disk system using related methods. In the
semiclassical regime, the resonances are located only in the
half plane[46]

$$\mathrm{Im}(k_j) \leq P(1/2).$$ (V.8)

If we introduce the standard definition of the complex energy of

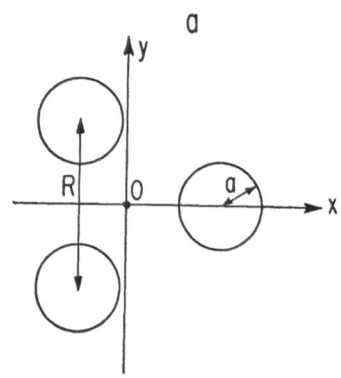

a

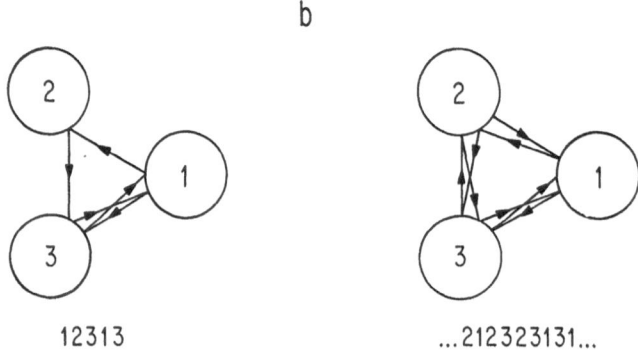

b

12313   ...212323131...

FIG. 11. (a) The geometry of the three hard disc scattering
system. The discs of radius a are fixed in the plane
at the vertices of an equilateral triangle with side
R.(b) Two examples of trapped trajectories: The
periodic trajectory 12313 and the trajectory
...212323131.... From Ref. 46.

a resonance,

$$E_j = \frac{\hbar^2 k_j^2}{2m} = \varepsilon_j - i\frac{\hbar}{2\tau_j},$$

(V.9)

and of the associated velocity

$$v_j = \frac{\hbar}{m} Re(k_j),$$

(V.10)

then the lifetimes $\{\tau_j\}$ of the resonances satisfy[21]

$$\frac{1}{v_j \tau_j} \geq -2P(1/2) = \lambda(1/2) - 2h_{KS}(1/2),$$

(V.11)

in the semiclassical limit. The right-hand member of (V.11) is positive when the discs are not too close to each other.

This semiclassical relation is the analog of the classical relation (V.7). It is known that the function $P(\beta)$ satisfies the general inequality $P(1) \leq 2P(1/2)$.[56] The equality sign holds when the repellor is periodic but not, in general, for a chaotic repellor. Consequently, we expect there will be some resonances with longer lifetimes than the classical lifetime. The difference between the classical and the quantum decay rates is a dramatic effect of underlying classical chaos on the quantum dynamics of the system.

The scattering resonances exist only below the gap in the complex wave number plane defined by (V.8). This gap is a function of the interdisc distance and is shown in Fig. 12 along with calculations based on two successive approximations to $P(1/2)$.

Gaspard and Rice also calculated the distribution function of the $Im(k_j)$ and compared the result with the semiclassical prediction of the same distribution function (Fig. 13). As expected, there are no resonances in the predicted gap and the distribution of resonances becomes denser as one goes deeper in the complex plane. These resonances are overlapping. They do not generate peaks in the scattering cross section with the typical Fano profile,[57] but they do modulate the magnitude of the cross section.

The preceding results suggest that the spectrum of resonances described by the Gaspard-Rice formalism is characteristic of a class of systems which is different and complementary to the class of scattering systems described by R-matrix theory.[58] In this latter case Fano resonances appear in our problem when the three discs are close to each other and when the classical repellor is bulky with a Hausdorff dimension[43] close to three. In contrast, the Gaspard-Rice formalism is applicable when the classical repellor is filamentary with a Hausdorff dimension close to one, as when the three discs are far from each other. A typical example of the distribution of scattering resonances when there is a filamentary repellor is given in Fig. 13. On the other hand, for a bulky repellor with long-lived resonances, the distribution of scattering resonances can be derived from random matrix theory, and is given by the density[58]

$$(2\pi\xi x)^{-1/2}\exp(-x^2/2\xi)dx, \quad 0<x<\infty \qquad (V.12)$$

where $x = -\mathrm{Im}(k)$ and $\xi = -\langle\mathrm{Im}(k)\rangle$. This distribution function is sketched in Fig. 14. Note that when V.12 is valid the resonances accumulate near the real axis and their density decreases deep in the complex plane. The distribution of scattering resonances corresponding to a bulky repellor is thus very different from that corresponding to a filamentary repellor, and a distinction must be established between these classes of scattering systems.

The Gaspard-Rice results are relevant to the description of rapid fragmentation in systems which are strongly defocussing. To treat real molecules the theory must be extended to typical unbounded Hamiltonian systems, for which we expect there to be long-lived resonances associated with the quasi periodic islands in the classical description.

The conceptual relationship between classical chaos and scattering resonances we described above has been established in the semiclassical domain where the classical trajectories still

244

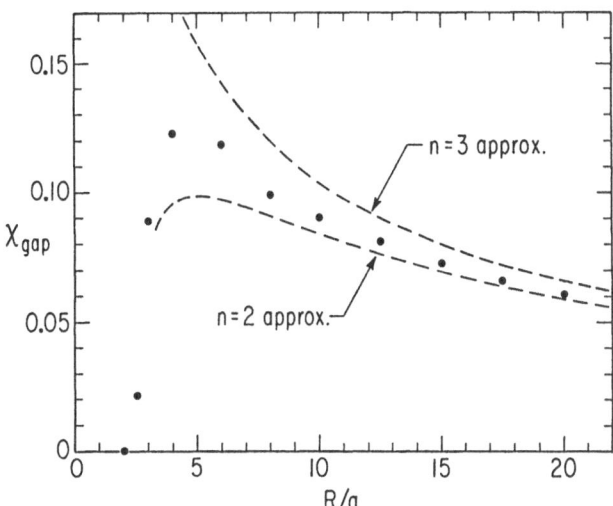

FIG. 12. The gap V.8 in the distribution of Im(k) versus the interdisc distance R. The radius of each disc is taken to be a = 1. The dots are the results from a full (numerical) quantum mechanical calculation. The dashed lines are the gaps − $P^{(2)}(1/2)$ and − $P^{(3)}(1/2)$, for two successive semiclassical approximations to the gap calculated from V.3.

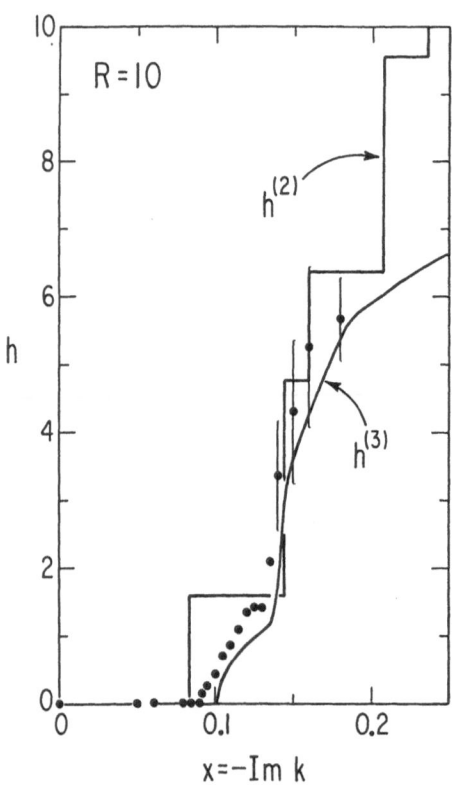

FIG. 13. Distribution function of $x = -\text{Im}(k)$ of the
scattering resonances when the disc radius $a = 1$ and
the interdisc distance $R = 10$. The density is given
by the derivative of $h(x)$. The dots are the values
of $h(x)$ obtained from a full (numerical) quantum
mechanical calculation. $h^{(2)}(x)$ and $h^{(3)}(x)$ are the
distribution functions corresponding to the same,
successive, semiclassical approximations as shown in
Fig. 12.

play a dominant role in the description of the dynamics.   In
this sense a relation like (V.11) is a scaling property of the
quantum dynamics with respect to the underlying classical
dynamics.   While the success of the Gaspard-Rice analysis shows
that such a relationship describes unbounded quantum systems
rather than bounded quantum systems, the analysis does not
answer the fundamental and intriguing question of the origin of
dynamical randomness in quantum systems.   Since the work by
Kosloff and Rice[59] it has been known that bounded quantum systems
can not have a positive KS entropy per unit time because of the
discreteness of the energy spectrum.   Accordingly, it is now
clear that dynamical randomness must be sought in the
characteristic properties of unbounded quantum systems, which
have a continuous energy spectrum.

VI.   A REACTION CONTROL PARADIGM

As the final topic in this paper we consider the use of a
van der Waals molecule fragmentation reaction to test ideas
concerning the active control of product formation.   Specifically, we examine how the fundamental idea underlying the
Tannor-Rice[60,61] scheme can be tested.

The Tannor-Rice scheme for controlling the selectivity of a
chemical reaction is based on exploitation of the coherence
properties of evolving wave packets, and uses an excited state
potential energy surface to mediate reaction on the ground state
potential energy surface.   Briefly, it is assumed that the
ground electronic state Born-Oppenheimer potential energy
surface has two or more exit channels, corresponding to the
formation of two or more distinct chemical species.   It is also
assumed that the system possesses an excited electronic state
Born-Oppenheimer potential energy surface whose minima are
displaced from those of the ground state surface, and whose
normal coordinates are rotated with respect to those of the
ground state surface.   In broad outline, Tannor and Rice[60]
suggest that an ultrashort pulse be used to excite the system.
This pulse generates a wave packet, consisting of a
superposition of rovibronic states built on the excited state
potential energy surface, which is not stationary with respect

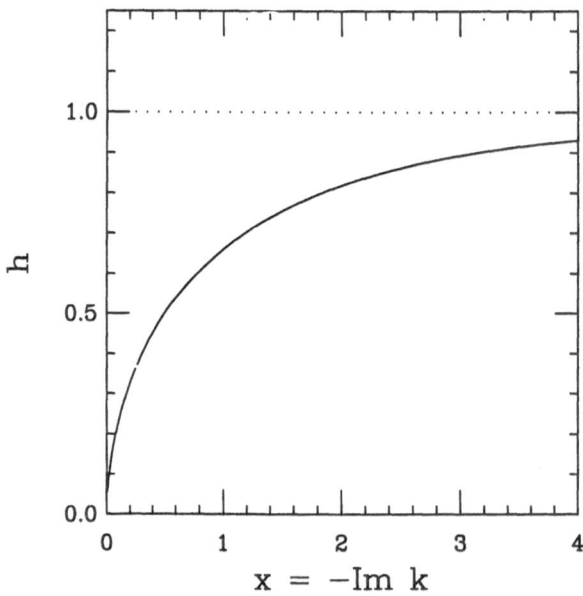

FIG. 14. Distribution of the scattering resonances, obtained
from random matrix theory combined with R−matrix
theory. The distribution function h(x) is plotted
versus x = − Im(k). The derivative of h(x) is the
density given by V.12 with ξ = 1. Note the absence
of a gap as compared with the results in Fig.13, and
the saturation for large x.

to the Hamiltonian for the excited state. As the wave packet evolves, it spreads over the excited state potential surface. After some interval, suitably chosen, an ultrashort "dump" pulse is used to stimulate population of the ground state; the wave function amplitude is unchanged in this vertical transition to the ground state surface. If the time interval for propagation on the excited state surface is chosen properly, a significant part of the amplitude of the wave packet dumped to the ground state surface can be placed in a particular product channel with momentum such as to generate that product. Tannor, Kosloff and Rice[61] have shown, for a model two degree of freedom system with the ground and excited state potential energy surfaces modelled on (but not identical to) those for the collinear reaction HHD -> H_2 + D or HD + H that the selectivity scheme outlined above can be used to alter the ratio of products formed in a reaction, i.e., varying the interval between the pump and dump pulses varies the ratio of products formed, so a favorable interval can be selected (Fig. 15).

The language used in the preceding paragraph emphasizes an interpretation which is useful when the initially prepared wave packet is well localized, hence has many components. Then the motion of the wave packet is easily visualized and the position of its centroid with respect to geometric features of the potential energy surface can easily be identified and used to signal the timeliness of a dump pulse. Of course, the wave packet spreading is a consequence of interference between the wave packet components, and the time interval between pump and dump pulses is the control variable for the phase difference between the initial wave packet and the evolved wave packet to be dumped to the ground state potential energy surface. The cited interference effects also exist in delocalized wave packets consisting of only a few components, and control of the ratio of products formed can still be achieved by varying the phase via the time delay between pump and dump pulses, without any direct reference to geometric features of the excited state and ground state potential energy surfaces.

Once the conceptual features of the Tannor-Rice scheme are grasped, modifications of the methodology that enhance product selectivity are readily suggested. One such modification takes

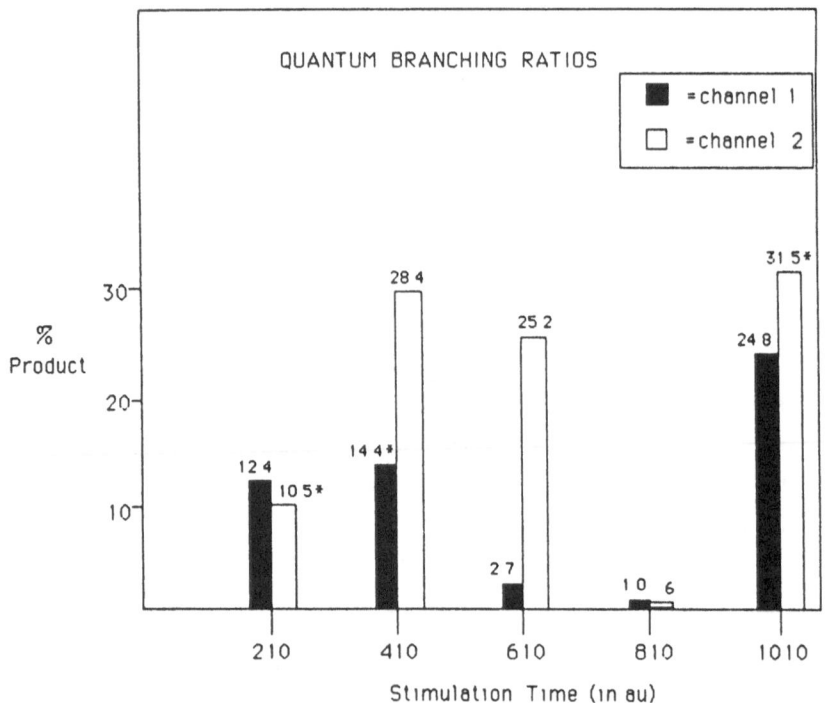

FIG. 15. Results of quantum mechanical calculations of the branching to products 1 and 2 in the Tannor-Rice pump-dump pulse scheme for controlling product formation, as a function of the time difference between pump and dump pulses. Note the dramatic differences between the yields of products 1 and 2 for different pump-dump pulse separations. The yields are quoted in relative units, including unreacted parent molecules.

advantage of the interference effects inherent in wave packet propagation. Suppose we ask what wave packet must be initially excited in order that a "target" wave packet, whose projection onto the ground state surface is particularly favorable, be in position for dumping to the ground state surface. The brute force approach to answering this question is to generate the special initial wave packet by integration of the Schrodinger equation backwards in time using the target wave packet as initial state. If the form of the special initial wave packet thus found is simple enough to be excited, say by the method introduced by Heritage,[62] enhanced selectivity is achieved. Calculations by Tersigni, Kosloff, Tannor and Rice[63] show that, for the surface used for the HHD-> H+HD and HH + D. reactions, back integration of the motion of a simple Gaussian wave packet, initially placed so as to have good projection onto an exit channel of the ground state, yields an exceedingly complex special initial wave packet, one not amenable to excitation.

A more sophisticated approach is based on the use of optimal control theory to choose the shapes of the pump and dump pulses so as to achieve a particular chemical outcome. A modification of the Tannor-Rice scheme of this type has been proposed by Kosloff, Rice, Gaspard, Tersigni and Tannor.[64] When applied to a similar model to that used in the earlier calculations (same ground state potential energy surface, a more anharmonic excited state potential energy surface), optimization of the dump pulse improved the selectivity for D versus H from 5:4 with a total yield of 10^{-4} (gaussian pump and dump pulses, not optimized) to 13:3 with a total yield of 0.132 (gaussian pump pulse, optimized dump pulse). The Wigner distribution function for the optimized pulse, defined by

$$W(t,\omega) = \frac{1}{\pi} \int_{0}^{tf} ds \, [E(t+s) \, E^*(t-s)] \exp{(2i\omega s)}, \qquad (VI.1)$$

is shown in Figure 16. Note that the optimized pulse does not have a simple shape. The behavior of the wave function under this pulse is shown in Fig. 17.

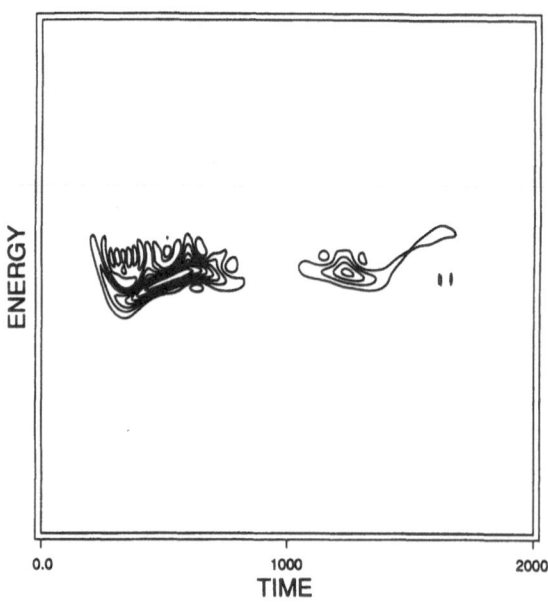

FIG. 16. The Wigner distribution function of the dump pulse that leads to optimal yield of product 1, calculated from the global optimization theory described in Ref. 64.

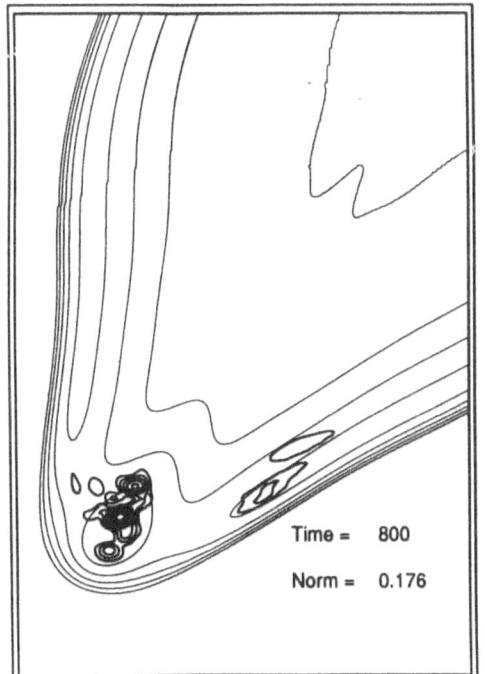 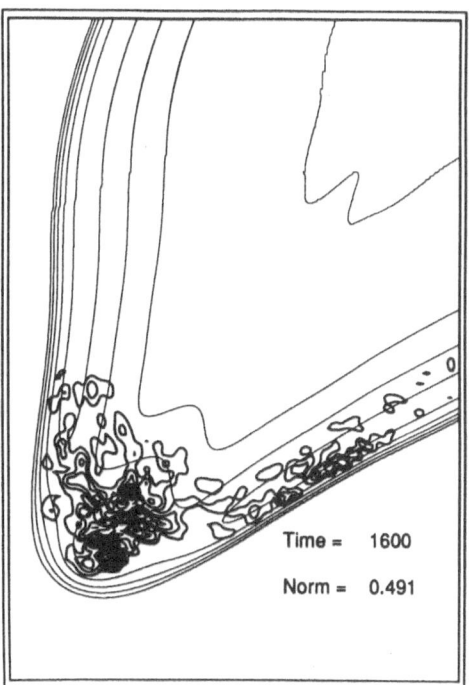

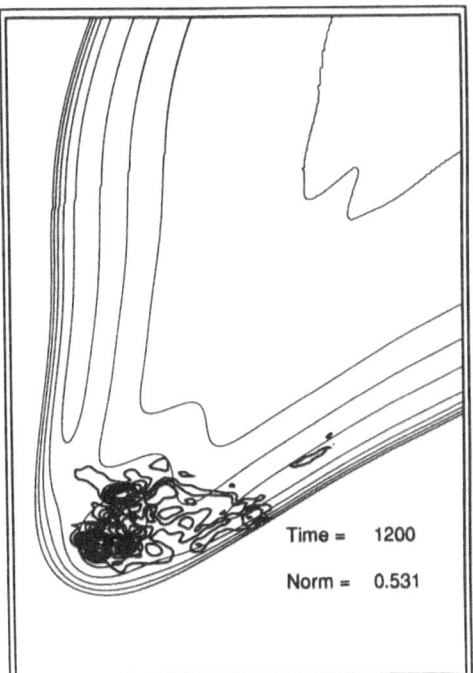 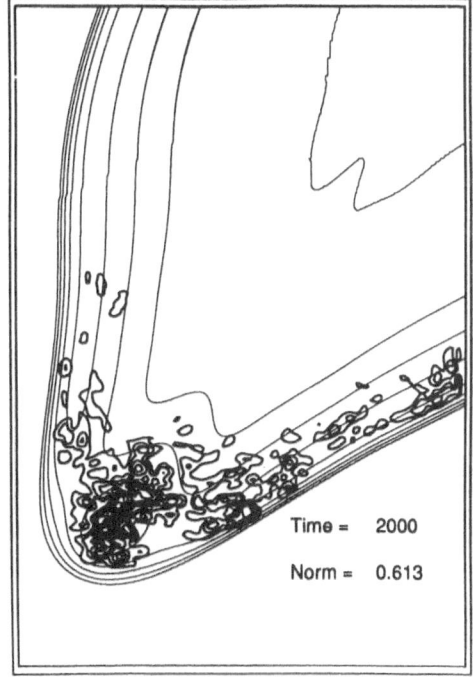

FIG. 17. The evolution of the wave function under the influence
of the pulse shown in Fig. 16. The time units are
a.u.

To demonstrate the principles involved in the Tannor-Rice selectivity scheme we have examined the use of van der Waals trimers such as Ar-tetrazine-Xe,[65] since these have low frequency bond vibrations which ease the technical requirements for the duration of the pump and dump pulses. Moreover, these molecules absorb light in a convenient spectral region. However, the shift in equilibrium geometry following electronic excitation of Ar-tetrazine-Xe and related molecules is too small to generate much evolution of the wave packet on the excited state potential energy surface, hence no opportunity to dump the wave packet so as to enhance formation of Xe or Ar. Finding a suitable molecule to test the Tannor-Rice scheme has turned out to be a nontrivial exercise when the constraints of available technology are applied.

Dr. Roger Carlson has suggested an approach to testing the key concepts of the the Tannor-Rice scheme which is simpler, both conceptually and experimentally, than the alteration of product yields in the photofragmentation of a triatomic molecule. In the experiment suggested,[66] a diatomic molecule is subjected to a femtosecond duration pump-dump pulse sequence. Since there is only one reaction coordinate, product selectivity can not be achieved. However, the delay between pulses can still be used to control the kinetic energy of the final wavepacket; one should be able to switch the dissociation on and off as a function of delay.

A suitable diatomic molecule is a van der Waals complex of mercury and argon. The spectroscopy of HgAr has been extensively studied by single[67,68,69] and double[70,71] resonance laser induced fluorescence. These measurements have yielded accurate potential surfaces for the complex,[72] some of which are sketched in Figure 18. Most of the excited states show large negative offsets from the ground state, suggesting that sufficient kinetic energy might be acquired on one of them to induce dissociation on the ground state surface. Since this complex can be made at relatively high concentrations in a molecular beam and the electronic transition moments are large, very high signal levels can be expected. We shall use this system as an example of the diatomic analog of the Tannor-Rice scheme.

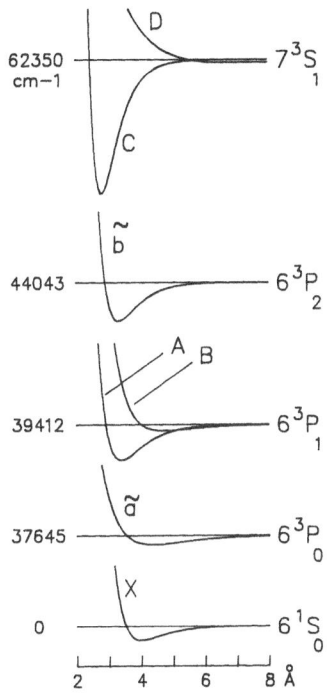

FIG. 18. Potential energy surfaces for HgAr. Each curve is a
Morse potential with the parameters given by
Breckenridge et al.[72] Two additional sublevels
corresponding to the Hg $^3P_2$ state have been omitted.

In order to avoid the problem of a large background of dissociation products, Carlson, Schmidt and Rice[66] propose using the excited A (6^3P_1) state as the initial potential surface upon which the dissociation will occur. The Hg resulting from dissociation on this surface will be in the excited 6^3P_1 state, which should have zero population in the absence of the field that excites it. Emission from 6^3P_1 to 6^1S_0 can be monitored to directly determine the reaction yield.

Imagine that the HgAr molecule is excited with a strong nanosecond pump pulse to the zeroth vibrational level of state A, which state will persist for many nanoseconds (lifetime= 120ns).[72] The femtosecond pump-dump pulse pair is then applied, the excitation pulse being to the C state. This scheme is sketched in Figure 19. The transition moment for the A-C transition is extremely large, as has been demonstrated by the reported double resonance experiments, and it is at a wavelength (~~450 nm) which is easily reached with visible dye lasers.

Figure 20 shows a classical Wigner swarm calculation of dissociation windows for the HgAr A-C transition. A period of 406.5 femtoseconds is clear, corresponding to one vibrational period of the excited state. The swarm calculation indicates that the contrast of the window pattern will be appreciably degraded within a few periods.

Figure 21 shows the results of some typical exact quantum mechanical wavepacket calculations for the HgAr system. The excitation energy (22451.4 cm^{-1}) was chosen to match the potential energy difference between the C and A surfaces. The deexcitation energy (20432.7 cm^{-1}) was chosen to match the classical potential energy difference for a delay of 203 femtoseconds; that is, a delay corresponding to a dissociation window. The pulse width (FWHM on an intensity scale) was chosen to be 50 femtoseconds and the pulse peak power density was approximately 10^{10} W/cm^2. The spectral width (also FWHM on an intensity scale) of such pulses is approximately 290 cm^{-1}. Note that the deexcitation step is to vibrational energy levels greater than 290 cm^{-1} above the dissociation limit (D_e=369 cm^{-1}). Thus, any wavepacket transferred to the lower state will inevitably lead to dissociation. Dissociation windows are still

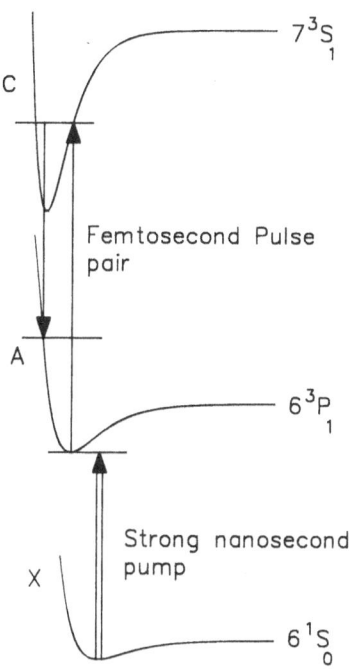

FIG. 19. Proposed transition scheme for HgAr.

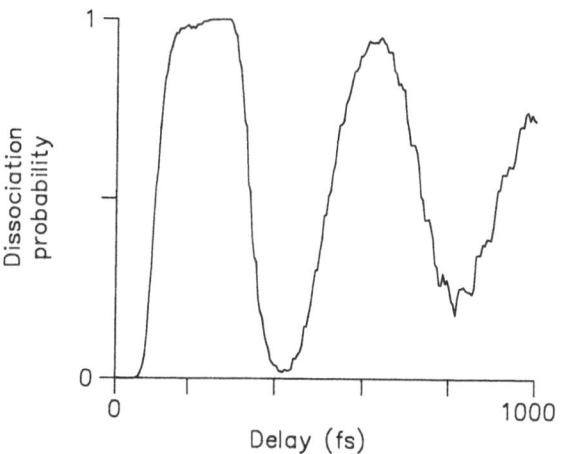

FIG. 20. Wigner swarm calculation of the dissociation
probability of HgAr in the A state. Excitation is to
the C state.

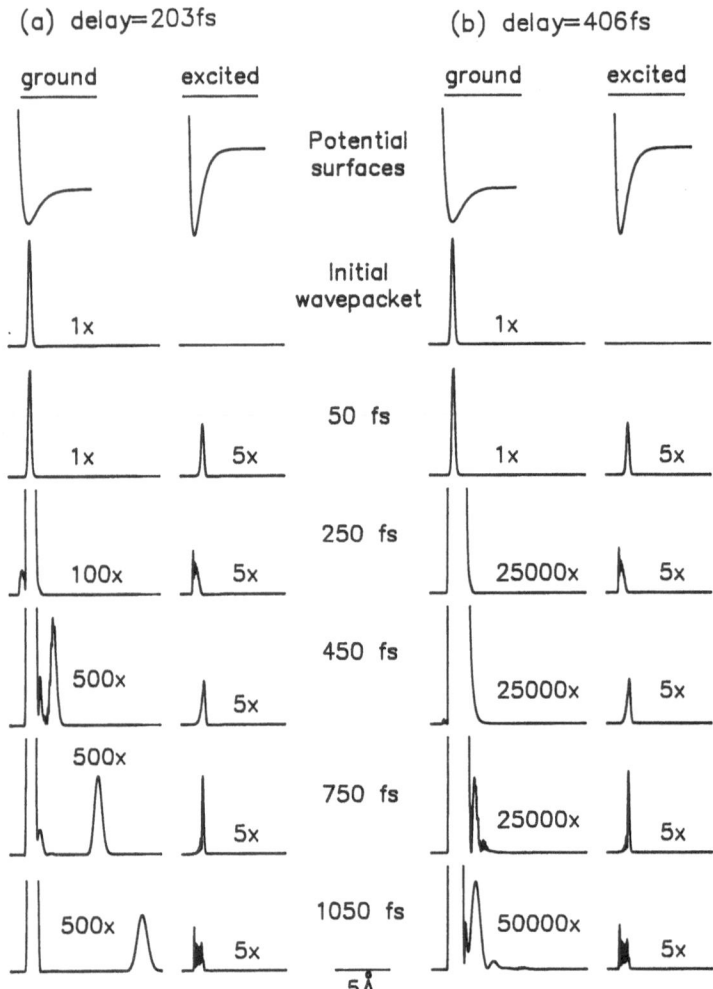

FIG. 21. Results of an exact quantum mechanical wavepacket calculation for HgAr. The ground and excited states are the A and C states respectively. The time scale is defined such that t=0 corresponds to the peak intensity of the first laser pulse.

expected, however, because of the variation of the Franck-Condon factors with the vibrational coordinate. This effect is the quantum mechanical analog of momentum conservation in the classical picture. The wavepacket transferred in the dissociation window Figure 21a represents 0.33% of the total population.

Figure 21b shows a wavepacket picture for a delay of 406 femtoseconds, for which the classical description predicts little dissociation. A very small amount of dissociation (0.0001%) is observed due to the nonzero laser pulsewidths. These results imply a contrast ratio of the yield in a dissociation window to that in a nondissociation window of 3300. Provided such short pulses can be achieved and the absolute detected signal is sufficiently large, this window pattern should be experimentally observable. The choice of 10^{10} W/cm for the pulse intensity is realistic. For a focal area of 0.0001 cm^2 (100μm x 100μm), this power corresponds to a 50 nJ pulse, which can be achieved by laser systems currently available.

Experimentally, 50 femtosecond pulses may not be easily obtainable. Table X therefore gives the results of quantum mechanical wavepacket calculations for pulses of various temporal widths, but with a constant pulse energy. Note that as the pulse width increases, the yield in the dissociation window only decreases slightly. However, the yield in the nondissociation window increases dramatically, resulting in a reduced contrast ratio. For 120 femtosecond pulses, the contrast has dropped to 17, which probably represents the limit of detection for the window pattern. Our net conclusion, then, is that pulses shorter than 120 femtoseconds should be adequate to resolve the windows, and thereby demonstrate the principle underlying the Tannor-Rice scheme for controlling the product formation in a photoinduced reaction.

VII. CONCLUDING REMARKS

This paper has presented a personally biased overview of the use of van der Waals molecule fragmentation as a tool for the study of unimolecular reactions. The common themes in the diverse studies sampled are the ubiquity of phenomena rooted in

259

TABLE X.   VARIATION OF PRODUCT YIELD IN DISSOCIATION OF HgAr WITH PUMP-DUMP PULSE SEPARATION.

Pulsewidth (fs)*	(%) Yield at 203 fs	(%) Yield at 406 fs	Contrast Ratio	Peak Power Density (W/cm^2)[+]
50	0.33	1.0×10^{-4}	3300	1.1×10^{10}
83	0.18	1.5×10^{-3}	120	6.1×10^9
120	0.17	0.010	17	4.4×10^9
170	0.089	0.032	2.8	3.1×10^9

* FWHM on intensity scale assuming a Gaussian profile

[+] assuming 58 nJ pulse energy and a focal area of $1 \times 10^{-4} cm^2$

From Ref. 66.

the nonlinearity of the system dynamics, the importance of interference effects in quantum dynamics, and the subtlety of the signatures of underlying classical mechanical chaos in the quantum mechanical description of a system. Clearly, we are only at the threshold of understanding the dynamics of van der Waals molecule fragmentation, and how to control product selection.

ACKNOWLEDGEMENTS

The research reported has been supported, over a period of years, by the National Science Foundation. It should be clear that I have acted as reporter for a large number of collaborations with very talented scientists who have been referred to in the text. The work described could not have been carried out without them.

REFERENCES

1. F.A. Lindemann, Trans. Faraday Soc. **17**, 598 (1922).

2. See, for example, K.J. Laidler, Chemical Kinetics, McGraw Hill, New York (1965).

3. C.N. Hinshelwood, Proc. Roy. Soc. (London) **A113**, 230 (1927).

4. L.S. Kassel, J. Phys. Chem. **32**, 225 (1928); O.K. Rice and H.C. Ramsberger, J. Am. Chem. Soc. **49**, 1617 (1927); **50**, 617 (1928).

5. See, for example, W. Forst, Theory of Unimolecular Reactions, Academic Press, New York (1973).

6. There are so many reports of measurements of vibrational energy redistribution in, and vibrational predissociation of, vdW molecules that only one review will be cited here; other relevant papers are cited in references 16-23. F.G. Celli and K.C. Janda, Chem. Revs. **86**, 507 (1986).

7. We cite only a few examples from the extensive literature:

    a. J.A. Beswick and J. Jortner, J. Chem. Phys. **74**, 6725 (1981).

    b. E. Segev and M. Shapiro, J. Chem. Phys. **78**, 4969 (1983).

    c. G. Dalgado-Barrio, P. Mareca, P. Villarreal, A.M.

Cortina and S. Miret-Artes, J. Chem. Phys. **84**, 4268 (1986).

    d. J.M. Hutson, C.J. Ashton and R.J. LeRoy, J. Phys. Chem. **87**, 2713 (1983).

    e. J.M. Hutson, J. Chem. Phys. **81**, 2357 (1984).

    f. I.F. Kidd and G.G. Balint-Kurti, J. Chem. Phys. **82**, 93 (1985).

    g. A.C. Peet, D.C. Clary and J.M. Hutson, Faraday Disc. Chem. Soc.**82**, 327 (1986).

    h. G.E. Ewing, J. Phys. Chem. **91**, 4662 (1987).

    i. G.E. Ewing, J. Chem. Phys. **90**, 1790 (1986).

    j. A.R. Tiller, A.C. Peet and D.C. Clary, Chem. Phys. **129**, 125 (1989).

8.   K.E. Johnson, L. Wharton and D.H. Levy, J. Chem. Phys. **69**, 2719 (1978).

9.   K.E. Kenny, K.E. Johnson, W. Sharfin and D.H. Levy, J. Chem. Phys. **72**, 1109 (1980).

10.   K.E. Kenny, T.D. Russell and D.H. Levy, J. Chem. Phys. **73**, 3607 (1980).

11.   K.E. Johnson, W. Sharfin and D.H. Levy, J. Chem. Phys. **74**, 163 (1981).

12.   D.H. Levy, private communication.

13.   J.A. Beswick and G. Delgado-Barrio, J. Chem. Phys. **73**, 3653 (1980).

14.   R.L. Waterland, J.M. Skene and M.I. Lester, J. Chem. Phys. **89**, 7277 (1988).

15.   J.C. Drabits and M.I. Lester, J. Chem. Phys. **89**, 4716 (1988).

16.   a. D.V. Brumbaugh, J.E. Kenney and D.H. Levy, J. Chem. Phys. **78**, 3415 (1983);

    b. J.J.F. Ramaekers, H.K. van Dijk, J. Langerlaar and R.P.H. Rettschnick, Faraday Disc. Chem. Soc. **75**, 183 (1983).

    c. M. Heppener and R.P.H. Rettschnick, in <u>Structure and Dynamics of Weakly Bound Molecular Complexes</u>, ed. A. Weber, Reidel, Dordrecht (1987);

    d. P.M. Weber and S.A. Rice, J. Chem. Phys. **88**, 6107 (1988);

    e. P.M. Weber and S.A. Rice, J. Chem. Phys. **88**, 6120 (1988).

f. P.M. Weber, J.T. Buontempo, F. Novak and S.A. Rice,
J.Chem. Phys. **88**, 6082 (1988).

17. N. Halberstadt and B. Soep, J. Chem. Phys. **80**, 2340 (1984);
Chem. Phys. Lett. **87**, 109 (1982).

18. T.A. Stephenson and S.A. Rice, J. Chem. Phys. **81**, 1083
(1984).

19. R. Rosman and S.A. Rice, J. Chem. Phys. **86**, 3292 (1987).

20. E.R. Bernstein, K. Law and M. Schauer, J. Chem. Phys. **80**,
207 (1984).

21. K.W. Butz, D.L. Catlett, Jr., G.E. Ewing, D. Krajnovich and
C.S. Parmenter, J. Phys. Chem. **90**, 3533 (1986);

22. B.A. Jacobson, S. Humphrey and S.A. Rice, J. Chem. Phys.
89, 5624 (1988).

23. H. Abe, Y. Ohyanagi, M. Ichijo, N. Mikami and M. Ito, J.
Phys. Chem. **89**, 3512 (1985):
N. Mikami, Y. Sugahara and M. Ito, J. Phys. Chem. **90**, 2080
(1986).

24. See, for example, C.A. Haynam, D.V. Brumbaugh and D.H.
Levy, J. Chem. Phys. **80**, 2256 (1984).

25. E.J. Bieske, M.W. Rainbird, I.M. I.M. Atkinson and A.E.W.
Knight, J. Chem. Phys. **91**, 752 (1989).

26. J. Boesiger and S. Leutwyler, Chem. Phys. Lett. **126**, 238
(1986).

27. J.A. Menapace and E.R. Bernstein, J. Phys. Chem. **91**, 2533
(1987).

28. P.M. Weber and S.A. Rice, J. Phys. Chem. **92**, 5470 (1988).

29. C.J. Ballhausen and A.E. Hansen, Ann. Rev. Phys. Chem. **23**,
15 (1972).

30. E.B. Wilson, Jr., .C. Decius and P.C. Gross, <u>Molecular
Vibrations</u>, McGraw Hill, New York (1955).

31. B.A. Jacobson, Ph.D. dissertation, University of Chicago
(1988).

32. G. Herzberg and E. Teller, Z. Phys. Chem. C. Leipzig) **21**,
410 (1933); G. Orlandi and W. Şiebrand, J. Chem. Phys. **58**,
4513 (1973).

33. S.K. Gray, S.A. Rice and D.W. Noid, J. Chem. Phys. **84**, 3745
(1986).

34. M.J. Davis and S.K. Gray, J. Chem. Phys. **84**, 5389 (1986).

35. R.S. MacKay, J. Meiss and I.C. Percival, Physica **D13**, 55
(1984).

36. D. Bensimon and L. Kadanoff, Physica **D13**, 82 (1984).

37. S.R. Channon and J. Lebowitz, Ann. N.Y. Acad. Sci. **357**, 108 (1980).

38. S.K. Gray, S.A. Rice and M.J. Davis, J. Phys. Chem. **90**, 3470 (1986).

39. S.K. Gray and S.A. Rice, Faraday Disc. Chem. Soc. **82**, 307 (1986).

40. S.K. Gray and S.A. Rice, J.Chem. Phys. **86**, 2020 (1987).

41. S.H. Tersigni and S.A. Rice, Ber. Bunsenges. Phys. Chem. **92**, 227 (1988).

42. S.H. Tersigni, P. Gaspard and S.A. Rice, work in progress.

43. A.J. Lichtenberg and M.A. Lieberman, Regular and Stochastic Motion, Springer, New York (1983).

44. P. Gaspard and S.A. Rice, J. Phys. Chem., in press.

45. J.M. Greene, R.S. MacKay, F. Vivaldi and M.J. Feigenbaum, Physica **D3**, 468 (1981); S.J. Shenker and L.P. Kadanoff, J. Stat. Phys. **27**, 631 (1982); R.S. MacKay, Physica **D7**, 283 (1983).

46. P. Gaspard and S.A. Rice, J. Chem. Phys. **90**, 2225 (1989); P. Gaspard and S.A. Rice, J. Chem. Phys. **90**, 2242 (1989); P. Gaspard and S.A. Rice, J. Chem. Phys. **90**, 2255 (1989).

47. B. Eckhardt, J. Phys. **A20**, 5971 (1987).

48. Y.G. Sinai, Russian Mathematical Surveys **27**, 21 (1972); L.P. Kadanoff and C. Tang, Proc. Nat. Acad. Sci. U.S.A. **81**, 1276 (1984).

49. For Neumann boundary conditions $r = 1$; in the case of scattering of electromagnetic waves from imperfect conducting cylinders, $|r| < 1$.

50. The plane wave dynamics also makes use of corrections with $\beta = 3/2, 5/2, …$.

51. D. Ruelle, J. Stat. Phys. **44**, 281 (1986); D. Ruelle, Thermodynamic Formalism Addison-Wesley, Reading, MA (1978).

52. D. Bessis, G. Paladin, G. Turchetti and S. Vaienti, J. Stat. Phys. **51**, 109 (1988).

53. H. Kantz and P. Grassberger, Physica **D17**, 75 (1985); J.-P. Eckmann and D. Ruelle, Rev. Mod. Phys. **57**, 617 (1985).

54. M.G. Gutzwiller, J. Math. Phys. **12**, 343 (1971); M.C. Gutzwiller, J. Phys. Chem. **92**, 3154 (1988).

55. P. Cvitanovic and B. Eckhardt, Phys. Rev. Lett., in press.

56. P. Walters, An Introduction to Ergodic Theory, Springer, Berlin (1981).

57. U. Fano, Nuovo Cimento **12**, 156 (1935); U. Fano, Phys. Rev. **124**, 1866 (1961).

58. C.E. Porter (ed), <u>Statistical Theories of Spectral Fluctuations</u>, Academic Press, New York (1965).

59. R. Kosloff and S.A. Rice, J. Chem. Phys. **74**, 1340 (1981); S.A. Rice and R. Kosloff, J. Phys. Chem. **86**, 2153 (1982); J. Manz, preprint entitled "A simplified Proof of the Kosloff-Rice Theorem: Intramolecular Quantum Dynamics Cannot be Chaotic".

60. D.J. Tannor and S.A. Rice, J. Chem. Phys. **83**, 5013 (1985).

61. D.J. Tannor, R. Kosloff and S.A. Rice, J. Chem. Phys. **85**, 5805 (1986); S.A. Rice, D.J. Tannor and R. Kosloff, J. Chem. Soc. Faraday Trans 2, **82**, 2423 (1986).

62. A.M. Weiner, J.P. Heritage and E.M. Kirschner, J. Opt. Soc. Am. **B5**, 1563 (1988).

63. S.H. Tersigni, R. Koslorr, D.J. Tannor and S.A. Rice, manuscript in preparation.

64. R. Kosloff, S.A. Rice, P. Gaspard, S.H. Tersigni and D.J. Tannor, Chem. Phys., in press.

65. M. Schmidt, R. Carlson and S.A. Rice, unpublished work.

66. R. Carlson, M. Schmidt and S.A. Rice, work in progress.

67. K. Yamanouchi, J. Fukuyama, H. Horiguchi, S. Tsuchiya, K. Fuke, T. Saito and K. Kaya, J. Chem. Phys. **85**, 1806 (1986).

68. K. Fuke, T. Saito and K. Kaya, J. Chem. Phys. **81**, 2591 (1984).

69. K. Fuke, T. Saito and K. Kaya, J. Chem. Phys. **79**, 2487 (1983).

70. W.H. Breckenridge, M. Duval, C. Jouvet and B. Soep, Chem. Phys. Lett. **122**, 181 (1985).

71. M. Duval, O.B. D'Azy, W.H. Breckenridge, C. Jouvet and B. Soep, J. Chem. Phys. **85**, 6324 (1986).

72. W.H. Breckenridge, M. Duval, C. Jouvet and B. Soep, in <u>Structure and Dynamics of Weakly Bound Molecular Complexes</u>, ed A. Weber, Reidel, Boston (1987).

IVR OF VAN DER WAALS AND

HYDROGEN-BONDED COMPLEXES AS

STUDIED BY STIMULATED EMISSION

ION DIP SPECTROSCOPY

Mitsuo Ito,
Toshinori Suzuki,
Mikako Furukawa
and
Takayuki Ebata

Department of Chemistry
Faculty of Science
Tohoku University
Sendai 980
Japan

INTRODUCTION

Intramolecular vibrational redistribution (IVR) is an important nonradiative process in an isolated large molecule, and it is being extensively studied experimentally and theoretically. Especially, IVR in electronically excited state has been studied by various experimental means such as fluorescence excitation, dispersed fluorescence spectroscopies and the measurement of fluorescence life-time.[1] As a result, our information on IVR for electronically excited states is now considerably accumulated. On the other hand, the study on IVR in electronically ground-state is very few. This is due to lack of suitable experimental means. The study by infrared radiation is an orthodox way. However, because of poor time response of the infrared detection and of severe selection rule for infrared absorption, the IVR study of a ground state molecule by infrared light is greatly restricted. Instead of direct vibrational excitation by

Dynamics of Polyatomic Van der Waals Complexes
Edited by N. Halberstadt and K. C. Janda
Plenum Press, New York, 1990

infrared light, the ground-state vibrational level of an
isolated molecule can be populated by stimulated emission
from an electronically excited state, say, S_1 state with
UV/visible laser light (ν_2). One of the stimulated emission
methods is "stimulated emission pumping" in which the
stimulated emission from S_1 to a ground-state vibrational
level is monitored by the dip of the fluorescence from the S_1
level pumped with ν_1.[2-3] The depth of the fluorescence dip
represents the decrease of the S_1 state population which is
determined by the balance of loss of the S_1 state population
by the stimulated emission with ν_2 and the gain of the
population by the reabsorption with same ν_2 from the ground-
state vibrational level. When the life-time of the ground-
state vibrational level to which the stimulated emission
occurs is long, the molecules in the vibrational level have a
great chance to come back to S_1, resulting in a small
fluorescence dip. Conversely, when the life-time (decay
rate) is short (large), we have a large dip. Since the lifetime of the ground-state vibrational level of an isolated
molecule is in most cases determined by IVR process, we can
obtain the IVR rate from the observed depth of the
fluorescence dip. Therefore, the depth of the fluorescence
dip measured in percentage relative to the fluorescence
signal in absence of the stimulated emission provides us with
very useful information on the IVR rate of ground-state
vibrational level. However, the quantitative determination
of the percentage fluorescence dip depth is very difficult
because it requires complete spatial matching of the two
laser beams ν_1 and ν_2. The observed percentage dip depth
sensitively varies by mismatching of the two beams whose
elimination is practically impossible.

Recently, Suzuki et al. developed another version of
stimulated emission method.[4-6] In this method, the first
laser light ν_1 pumps a jet-cooled molecule to a particular
level in S_1, and the second laser beam ν_2 is introduced to
induce the stimulated emission from S_1 to the vibrational
level in S_0. ν_2 is also used for the ionization of the S_1
state molecule. The ν_1 laser power is suppressed as low as
possible, therefore, the ionization by ν_1 alone does not

occur. Under this condition, all the ion signals occur only in the sample space where both ν_1 and ν_2 coexist. When the ν_2 frequency coincides with the energy difference between S_1 and a ground-state vibrational level, the stimulated emission occurs and the population of the S_1 state decreases. The decrease of the population is detected as a dip of the ion signal due to $\nu_1 + \nu_2$ ionization. This method may be called "stimulated emission ion dip (SEID) spectroscopy". The method has a great advantage to obtain quantitative data of the ion dip intensity. If we measure the dip intensity by the percentage depth of the ion dip relative to the ion signal in the off-resonant condition, this percentage dip intensity is not seriously influenced by mismatching of the two laser beams ν_1 and ν_2 because both the ionization and stimulated emission occur only in the space where the two laser beams overlap. The quantitative ion dip intensity leads us to the determination of reliable IVR rate of the ground-state vibrational level which is very difficult by other means. Therefore, we use here SEID spectroscopy for the study of the IVR rate of ground-state vibrational levels.

We applied SEID spectroscopy to determine the IVR rate of the ground-state vibrational levels of weakly bound complexes. Two kinds of complex are chosen here, one is the hydrogen bonded complex of phenol with water (Ph-H_2O) and another the van der Waals complex of m-fluorotoluene with Ar (mFT-Ar). For both complexes, it was found that a great enhancement of the IVR rate occurs by the formation of the complex. The enhancement may be explained as a result of the increase of the density of vibrational levels by the addition of low frequency hydrogen bond or van der Waals modes by the complex formation and also of large vibrational coupling. The results are important to provide some insight to the nonradiative relaxation mechanism occurring in condensed media.

EXPERIMENTAL

The apparatus for the measurement of SEID spectra was described elsewhere.[4-6] Two tunable dye lasers (ν_1 and ν_2)

were pumped simultaneously by an excimer laser. The first laser light ν_1 pumps the jet cooled molecule or complex to the zero point level in the S_1 state. The second tunable laser light ν_2 was used for the ionization of the S_1 state moleucle or complex and also for the stimulated emission to ground-state vibrational levels. The depth of the ion dip (here-after called the dip intensity) caused by the stimulated emission was measured for several main vibronic bands of each species. The dip intensity was expressed by percentage depth of the ion dip relative to the background ion signal obtained when ν_2 is off-resonant to the vibrational level.

The hydrogen bonded complex of $Ph-H_2O$ was prepared by supersonic expansion of gaseous mixture of phenol and water seeded in He carrier gas (1 atm). The van der Waals complex of mFT-Ar was also prepared by supersonic expansion of m-fluorotoluene vapor seeded in Ar carrier gas.

RESULTS AND DISCUSSION

The 0,0 band of the electronic absorption spectrum of bare Ph due to the $S_1(\pi,\pi^*) \leftarrow S_0$ transition is located at 36348 cm^{-1}. The SEID spectrum of bare Ph in a supersonic jet was measured after tuning the first laser light ν_1 to this 0,0 band. The second laser light ν_2 which is used for both the ionization and stimulated emission from the zero point level in S_1 was scanned to cover the ground-state vibrational energy region from 500 to 4000 cm^{-1} ($\nu_1 - \nu_2$). The SEID spectrum obtained is shown in Figure 1a. It is seen that the background ion signal gradually decreases with the increase of $\nu_1 - \nu_2$. This background signal represents the ionization efficiency for the vertical ionization from the zero point level in S_1, which is governed by the Franck-Condon factor. No prominent dip was found in the entire spectral region. Weak dips can be seen in the region above 1500 cm^{-1}, but their dip intensities (measured by percentage dip) were less than 20 %. The result shows that the stimulated emission is not efficient for bare Ph.

The 0,0 band of the $S_1 \leftarrow S_0$ transition of Ph-H_2O complex is located at 35993 cm^{-1},[7] red-shifted by 335 cm^{-1} from the 0,0 band of bare Ph. Figure 1b shows the SEID spectrum of Ph-H_2O complex prepared by the supersonic expansion obtained by tuning the ν_1 frequency to the 0,0 band of the complex. The laser powers of ν_1 and ν_2 were nearly the same as those for bare Ph. In contrast to the case of bare Ph, a number of prominent ion dips appear in the entire spectral region of the complex. Even very weak dips are reproducible and represent the stimulated emission to the ground-state vibrational levels. It is noted that the largest ion dip intensity amounts to as large as 75 %.

In Table I are shown the dip intensities of several main vibronic bands of the complex. The dip intensities given are those obtained under the condition that the ν_2 laser power is strong enough to saturate the stimulated transition.

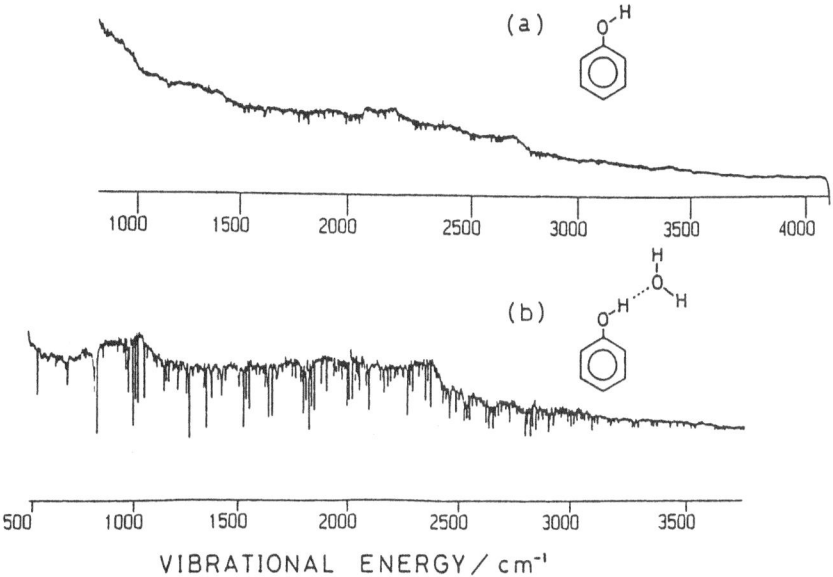

Figure 1. (a) SEID spectrum of bare phenol obtained after exciting the molecule to the zero-point level in S_1 with ν_1. The abscissa is $\nu_1 - \nu_2$ which corresponds to the ground-state vibrational frequency. (b) SEID spectrum of phenol-H_2O complex obtained after exciting the complex to its zero-point level in S_1 with ν_1.

Table I.   Observed dip intensitiy, IVR rate
           and calculated vibrational state
           density of main vibronic bands in
           SEID spectrum of Ph-H_2O complex.

Mode	energy (cm^{-1})	dip int. (%)	IVR rate $k_v (\times 10^9 s^{-1})$	State density $(/cm^{-1})$
6a	527	45	0.1	7
12	825	74	3	44
1	999	67	1	110
7a	1272	74	3	420

Using these dip intensities together with the known life-time
of the S_1 state complex,[8] the decay rates (k_v) of the ground-
state vibrational levels can be evaluated by solving
numerically the rate equations. The decay rates evaluated
are given in the fourth column in Table I. The rate is the
order of $10^8 \sim 10^9$ s^{-1}. It is interesting to compare the
value with that of bare Ph. Unfortunately, the dip intensity
is zero or very small for bare Ph and the IVR rate could not
be evaluated. However, the IVR rate of bare Ph must be very
small because the molecule has no intramolecular vibrational
mode of very low frequency and the vibrational levels are
sparse even at the vibrational level of 3000 cm^{-1}.
Therefore, the value of $10^8 \sim 10^9$ s^{-1} for the complex is
larger by many orders of magnitude than that for bare Ph.

The great increase of the IVR rate by the complex
formation is ascribed to a great increase of the density of
the vibrational states by the addition of low frequency
hydrogen bond modes. There are six hydrogen bond modes in
the complex. One stretching (150 cm^{-1}) and two bending modes
(104 and 70 cm^{-1}) of the hydrogen bond were found from the
analysis of the dispersed fluorescence spectrum. Three
rotational modes of H_2O relative to the Ph molecule were
assumed that the rotational motion around the C_2 axis of H_2O
molecule coinciding with the direction of the hydrogen bond

is free rotation of H_2O molecule and the other two rotational oscillations around the directions perpendicular to the C_2 axis are similar to the two bending modes of the hydrogen bond. Based on the above assumptions for the hydrogen bond modes and by using the known frequencies of the intramolecular modes of Ph, we calculated the densities of the states at the 6a, 12, 1 and 7a levels, which are given in the last column in Table I. It is seen that the density rapidly increases with the increase of the vibrational energy. The increase is parallel to the increase of the IVR rate with an exception for the value of the 12 level. The parallel relation suggests that the IVR rate is qualitatively determined by the density of the states contributed mainly by the hydrogen bond modes. The considerably large IVR rate for the 12 level seems to be a typical example of the mode selectivety for IVR. The mode 12 of Ph involves considerable motion of the OH group. Therefore, this mode will mix easily with the hydrogen bond modes by anharmonic coupling. This will result in a large value for the matrix element of the anharmonic coupling between the 12 level and isoenergetic bath levels which constitute mainly from various combinations and overtones of the hydrogen bond modes.

We also studied the IVR rate of the van der Waals complex of mFT-Ar. The 0,0 bands of the electronic absorption spectra of bare mFT and mFT-Ar due to their $S_1(\pi,\pi^*) \leftarrow S_0$ transition are located at 37385.5 and 37364.0 cm^{-1}, respectively.[6] The SEID spectra of bare mFT and mFT-Ar complex in supersonic jets were measured after exciting them to their zero-point levels in S_1. The observed spectra are shown in Figure 2a and 2b for bare mFT and mFT-Ar complex, respectively. As seen from the figure, the strong dips are observed even for the bare molecule. This is in great contrast to the case of bare phenol, where the dip is absent or very weak even with the same laser powers as those used for mFT (see Figure 1a). The large dip for mFT is probably due to a large IVR rate of the ground-state vibrational level contributed from the low frequency levels of the internal rotation of the CH_3 group.[9] It is seen from the figure that the dip intensity increases by the formation of the van der

Waals complex. For example, the dip intensity of the 6a band
(728 cm^{-1}) is 35 % for mFT, but it increases to 50 % in the
complex. A similar enhancement is also found for all the
bands. The measurements of the ν_2 power dependence upon the
dip intensity have been carried out for several main vibronic
bands to determine the IVR rates of their vibrational levels.
The IVR rates for the 1_1(728 cm^{-1}), 12_1(1004 cm^{-1}) and
$1_1 12_1$(1733 cm^{-1}) levels of bare mFT were evaluated to be 1 $\times$
10^8, 1 $\sim$ 3 $\times$ 10^8 and 3 $\sim$ 7 $\times$ 10^8 s^{-1}, respectively.

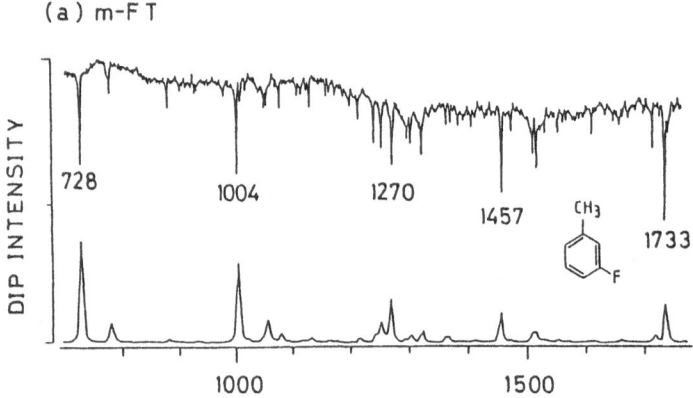

(a) m-F T

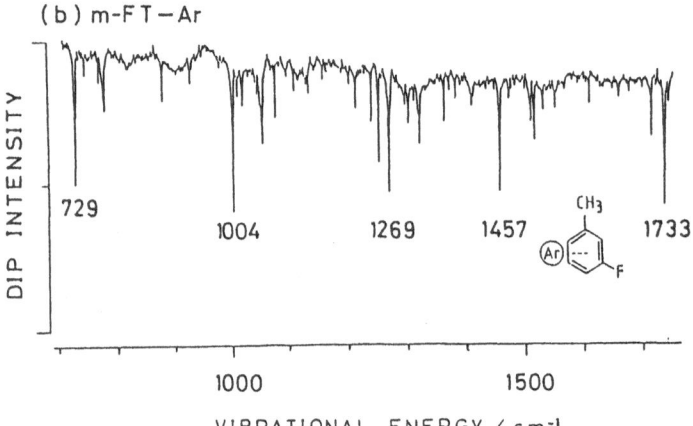

(b) m-F T $-$ Ar

VIBRATIONAL ENERGY / cm^{-1}

Figure 2. (a) SEID (upper) and dispersed fluorescence
(lower) spectra of bare mFT obtained after exciting the
molecule to the zero-point level in S$_1$ with ν_1. (b) SEID
spectrum of mFT-Ar complex obtained after exciting the
complex to its zero-point level in S$_1$ with ν_1.

The rate increases by the complex formation. The enhancement factor was found to be 3 ~ 7 for the 1_1 level and a little smaller for the other levels.

Although the IVR rate is quite different between bare Ph and bare mFT, it is the same order of $10^8 \sim 10^9$ s^{-1} for their complexes. For both the complexes, the increase of the IVR rate by the complex formation is concluded to be about 10^8 s^{-1}. Finally, a comment should be made for the IVR rates of the complexes. Since in most cases the ground-state vibrational levels of the complex studied exceed the dissociation energy of the hydrogen bond or the van der Waals bond, we have a direct dissociation route other than IVR. Then, the rate obtained here should be the sum of the contributions from the direct dissociation and IVR. However, when the dissociation occurs after IVR, the IVR rate given here has its own meaning.

References

1. See for example, (a) C. S. Parmenter, J. Phys. Chem. 86, 1736 (1982), (b) R. E. Smalley, J. Phys. Chem. 86, 3504 (1982), (c) R. E. Smalley, Ann. Rev. Phys. Chem. 34, 129 (1983).

2. C. E. Hamilton, J. L. Kinsey and R. W. Field, Ann. Rev. Phys. Chem. 37, 493 (1986).

3. H. L. Kim, S. Reid and J. D. McDonald, Chem. Phys. Lett. 132, 361 (1986).

4. T. Suzuki, N. Mikami and M. Ito, Chem. Phys. Lett. 120, 333 (1985).

5. T. Suzuki, N. Mikami and M. Ito, J. Phys. Chem. 90, 6431 (1986).

6. T. Suzuki, M. Hiroi and M. Ito, J. Phys. Chem. 92, 3774 (1988).

7. H. Abe, N. Mikami and M. Ito, J. Phys. Chem. 86, 1768 (1982).

8. R. J. Lipert, G. Bermudez and S. D. Colson, J. Phys. Chem. 92, 3801 (1988).

9. K. Okuyama, N. Mikami and M. Ito, J. Phys. Chem. 89, 5617 (1985).

IVR IN A POLYATOMIC VAN DER WAALS COMPLEX

André G.M. Kunst and Rudolf P.H. Rettschnick

Laboratory for Physical Chemistry, University of Amsterdam
Nieuwe Achtergracht 127,   1018 WS Amsterdam
The Netherlands

ABSTRACT

We have studied the vibrational energy flow in photoexcited van der Waals complexes T·Ar of s-tetrazine and argon. The aim of this paper is to present experimental information about the coupling among the van der Waals vibrational modes in T·Ar and about the interaction between ring modes and vdW vibrations.

Time-resolved laser-induced fluorescence experiments have demonstrated that the photodissociation of T·Ar is preceded by intramolecular vibrational redistribution (IVR) processes in which the pumped level C_i develops into levels C_j of the complex which are not accessible by optical excitation because they include highly excited van der Waals vibrations. The isoenergetic levels C_i and C_j of the cluster are characterized by different vibronic states of the tetrazine moiety.

The intensity patterns of the vibrational substructure of the fluorescence spectra indicate that the van der Waals stretch and bend modes are extensively mixed. High resolution experiments performed with a single frequency laser show that the mixing of the pumped level C_i (in the present study $6a^1$) with near-resonant levels C_j (in this case $16a^2$) depends on the rotational state of the complex. The quantum yields of IVR and photodissociation depend on which rotational states of the cluster are initially excited. Almost all the selectivity with respect to the rotational degrees of freedom is in the first step (IVR) and not in the last step (fragmentation).

Dynamics of Polyatomic Van der Waals Complexes
Edited by N. Halberstadt and K. C. Janda
Plenum Press, New York, 1990

INTRODUCTION

Very few experiments providing direct information about the pathways of intramolecular energy flow in van der Waals complexes have been reported to date. In most cases studied so far, distinct channels for IVR and vibrational predissociation (VP) could not be detected separately. Among the polyatomic van der Waals complexes the cluster T·Ar of s-tetrazine and argon is one of the few favourable exceptions. It exhibits several channels for IVR and VP.[1,2] The temporal characteristics of these processes were previously investigated by us.[3,4,5] Pathways for the energy flow within the cluster could be mapped by performing time-resolved experiments on a picosecond time scale. In these experiments the fluorescence from different vibronic states of the complex and from the molecular dissociation fragment were selectively detected. Selective detection of emission bands originating from the complex or the molecular dissociation fragment as well as selective excitation of vibronic levels of T·Ar can be achieved since the transitions in the complex are red shifted with respect to those in the bare tetrazine molecule by $22-24$ cm^{-1} or 8 cm^{-1} dependent on which vibronic state is excited.

Time-resolved experiments revealed that each of the observed VP processes is preceded by at least one IVR step in which the vibrational energy initially present in a ring mode is redistributed between the vibrational modes of tetrazine and the van der Waals vibrations. During such an IVR process the initially prepared state C_i of the complex develops into a state C_j from which the complex may dissociate. The cluster states C_i and C_j are characterized by vibronic states v_i and v_j of the tetrazine molecule. The energy difference $(\varepsilon_i - \varepsilon_j)$ between the molecular states v_i and v_j is taken up by the relative motion of the two constituents of the complex as vdW vibrational energy. Subsequent dissociation of the complex leads to the formation of electronically excited tetrazine in a specific vibronic state v_f.

$$Ar \cdot T(i) \xrightarrow{\text{IVR}} Ar \cdot T(j) \xrightarrow{\text{VP}} T(f) + Ar$$

The total rotational and translational energy of the fragments equals $(\varepsilon_i - \varepsilon_j - D_0')$ where D_0' is the binding energy of the vdW complex in the S_1 electronic state. The photofragmentation pathways $i \longrightarrow f$ observed for T·Ar indicate that the dissociation energy D_0' is ca. 320 ± 40 cm^{-1}. However the upper limit of D_0' might be as low as 306 cm^{-1} because of the rotational energy present in the tetrazine molecule after fragmentation of the complex.[1]

The ring vibration 16a appears to be predominantly involved in the IVR and VP processes occurring in the S_1 state of T·Ar. Apparently, this mode couples exceptionally strong with the vdW vibrations. However, in the electronic ground state of the cluster IVR and VP are at least an order of magnitude slower than in the excited electronic state. Neither the 16a mode nor any other ring vibration seems to affect the intermolecular potential strong enough to cause any significant IVR or VP on a 15 ns time scale.[6] It has been pointed out by Weber and Rice[6,7] that the torsional mode 16a has the same nodal structure as the highest occupied electronic orbital in the S_1 state of s-tetrazine which implies that the modulation of the intermolecular potential by the 16a vibrational motion can bring about a repulsive interaction between S_1-tetrazine and argon. Weber and Rice[6,7] explained the specific activity of the 16a mode as well as other features of the energy conversion process in T·Ar in terms of a perturbation theory treatment.

Kelly and Bernstein[8] looked at the energy conversion in T·Ar in a different way. In their view the rate of the transition $C_i \longrightarrow C_j$ depends only upon the total number of vibrational quanta exchanged during the IVR process. In this approach the VP process is described in terms of a restricted RRKM theory, based on the assumption that IVR among the vdW vibrations is rapid.

Ewing[9] discussed the anharmonic coupling of the zero-order cluster states C_i and C_j and rationalized some of the observed propensities in the vibrational dynamics of T·Ar. He suggested to carry out high-resolution experiments in order to examine the specific properties of the vdW modes involved in the Fermi resonance between C_i and C_j.

In the present paper we report upon the first results of such a study and we pay attention to the mixing of the vdW stretch and bend modes. The high-resolution experiments provide information about the role played by the rotational degrees of freedom in the energy conversion processes.

EXPERIMENTAL

Tetrazine-argon van der Waals complexes were produced by expanding s-tetrazine seeded in argon at a pressure of 1.2 bar through a 100 μm orifice into a vacuum chamber which was evacuated by two roots pumps (500 and 250 m^3h^{-1}) backed by a two stage rotary vane pump (15 m^3h^{-1}).

Rovibronic excitation was achieved by using a CR-599-21 scanning single mode CW linear dyelaser. The dyelaser was pumped by a CR-5 Ar^+ laser operating at 488 nm. The resulting dyelaser output was in the range of 10-60 mW depending on the wavelength and the age of the dye. The bandwidth could be reduced by the insertion of two etalons inside the cavity and active stabilization. The resulting bandwidth was 1 MHz. However, due to Doppler and lifetime broadening of the transitions of the complex the effective width of the rovibronic transitions is 0.02 cm^{-1} fwhm. Scans were made by varying the cavity length with an intracavity Brewster plate at the same time electronically adjusting the etalons. Due to this set-up scans were limited to a range of 1 cm^{-1}. For measuring a whole rotational contour several overlapping scans were needed. The linearity of the scans was checked using a 1.5 GHz etalon. No corrections were necessary. The excitation spectra were corrected for variations in laserpower during a scan.

The laserbeam crossed the gas jet at right angles. The distance between the laser-jet interaction region and the nozzle was adjusted between 5 and 40 nozzle diameters downstream of the nozzle orifice. Fluorescence was collected using a high quality aspherical glass condensor which was imaged onto the slit of a Jobin-Yvon 1.5 m monochromator, equipped with a 2400 lines/mm holographic grating. Doppler broadening could be reduced by limiting the slit height, thereby detecting only the fluorescence originating from the center of the jet.

Fluorescence excitation spectra were generated by collecting the fluorescence from only one emission band and recording the intensity as a function of the laser wavelength.

Light passing the monochromator was detected by an EMI 9558 QA photo-multiplier, connected to a PAR 231 photon counter, whose signals were fed into an EG&G 918I multichannel analyzer used in a multichannel scaling mode. Spectra were stored into an IBM XT personal computer for further processing.

RESULTS AND DISCUSSION

One of the objects of this study is to search for information concerning the interaction between the vdW vibrational modes. The geometry of T·Ar was deduced by Levy and co-workers[10] from the resolved rotational structure in the electronic origin of the fluorescence excitation spectrum. The argon atom is situated on the out-of-plane C_2 axis of tetrazine at a distance of about 0.34 nm. The rotational constants are such that the cluster can be considered as a prolate near-symmetric top. The small amplitude motions of argon with respect to tetrazine can be described in terms of three distinct normal modes. The stretching mode σ belongs to the irreducible representation a_1 of the C_{2v} point group of the cluster and the two bending modes $_x\beta$ and $_y\beta$ belong to b_2 and b_1 respectively. The y-axis coincides with the HCCH axis of tetrazine. The electronic transitions of the cluster may display progressions of the totally symmetric stretching mode whereas the anti symmetric bending modes may participate only in transitions with Δv = even.

Weber et al[11] have investigated van der Waals vibrational transitions in T·Ar and other tetrazine-rare gas complexes. They observed that (i) the frequencies of these transitions in the cluster T·X (X = Ar, Kr, Xe) fit fairly well in with harmonic or near harmonic progressions and (ii) the relative intensities of corresponding transitions of σ and β's are very similar for these clusters. These observations led them to the notion that the bend and stretch vibrational modes are sensibly uncoupled, unless the vibrational mixing in the three different clusters would be very similar.

We have investigated the vdW vibrational mode structure of several vibronic transitions of T·Ar with a higher spectral resolution than used by Weber et al. It turns out that the relative intensities of stretch and bend transitions depend strongly on the vibrational state of the tetrazine moiety. However, the ring vibrations have only little influence on the vdW vibrational frequencies (differences of the order of 1 cm^{-1} and less). The results of this work will be published elsewhere. In the context of this paper we will consider the vdW mode structure of the $\overline{0}_0$ state of T·Ar (a bar on top of the assignment is used to designate the complex).

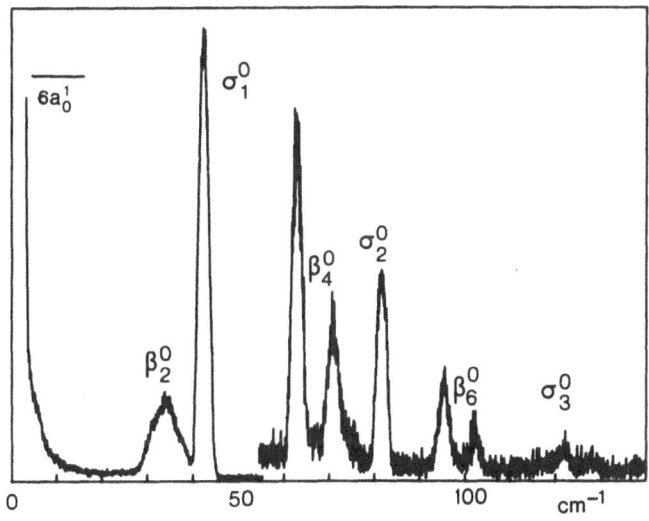

Fig. 1. Dispersed fluorescence obtained while pumping the $\overline{6a}_0^1$ transition (Q branch) of T·Ar. The spectrometer resolution is 2 cm^{-1} fwhm. The intensities beyond 55 cm^{-1} have been enlarged by a factor of 10.

Table 1. Conventional assignments, measured frequencies and intensities of transitions to van der Waals modes in the 0_0 state of T-Ar. Frequencies are relative to the origin. The intensity of σ_1^0 is set to 1000.

conventional assignment	frequency (cm^{-1})	intensity (arb. units)
origin	0.0	(50000)
$x\beta_2^0$	30.5	170
$y\beta_2^0$	34.6	230
σ_1^0	42.9	1000
$x\beta_4^0$	63.0	70
$y\beta_4^0$	71.0	40
σ_2^0	81.5	48
$x\beta_6^0$	95.5	18
$y\beta_6^0$	102.7	7
σ_3^0	120.5	6

Fig. 1 shows part of the dispersed fluorescence after excitation at 18808.7 cm^{-1}, i.e. at the peak position of the Q branch of the $6a_0^1$ transition. In this Figure we have used the conventional notation for the vdW vibrational transitions, neglecting any mixing of the stretch and bend modes. The emission band at about 33 cm^{-1} can be resolved if the spectrometer resolution is improved. Spectra taken with a resolution of 0.7 cm^{-1} fwhm show two distinct transitions at 30.5 and 34.6 cm^{-1} from the origin. The spectral positions and the relative intensities of the vdW bands were found by fitting each peak to a Gaussian profile. During the fit, the background as well as the position, height and width of the profile were allowed to vary. The area under the profile is considered to represent the intensity of the emission band. The spectral positions and the relative intensities as presented in Table 1 are believed to be accurate within 0.2 cm^{-1} and 10% respectively. The intensity of the σ_1^0 band is set to 1000. In this particular case the origin band coincides with stray light from the laser. The relative intensity of 50000 is an estimated value based on the average intensity ratio of the origin band and the σ_1^0 band in several vibronic transitions of T·Ar which are free from any interfering stray light.

The observed frequency of the stretch vibration agrees fairly well with the value of 41 cm^{-1} calculated by Menapace and Bernstein[12], but the experimental values of the bending modes are not in agreement with theoretical values.[12,13] Nevertheless we assign the bending mode with the lowest frequency as $_x\beta$ in conformity with the results of theoretical calculations.[12,13] We could have expected 3 emission bands $_x\beta_2^0$ (i.e. $_x\beta_4^0$, $_x\beta_2^0{}_y\beta_2^0$ and $_y\beta_4^0$) and 4 bands $_x\beta_6^0$ (i.e. $_x\beta_6^0$, $_x\beta_4^0{}_y\beta_2^0$, $_x\beta_2^0{}_y\beta_4^0$ and $_y\beta_6^0$). However, the spectrum displayed in Fig. 1 exhibits only 2 distinct emission bands in each of both spectral regions. In view of the observed frequencies (including those of $_x\beta_2^0$ and $_y\beta_2^0$) and the analysis of the relative intensities, the assignments as given in Table 1 look to be the most appropriate. It should be noted that combination bands like $\sigma_1^0\beta_2^0$ are also lacking in the spectra.

From the fluorescence excitation spectra we have obtained the frequencies 34 and 38 cm^{-1} (with respect to the origin band 0_0^0) for the transitions $_x\beta_0^2$ and $_y\beta_0^2$ respectively. The frequency ratio $\delta = \sqrt{(\omega''/\omega')} =$ = 0.95 for $_x\beta$ and for $_y\beta$ we also find $\delta = 0.95$. The relative intensities of the transitions $_x\beta_n^0$ can be calculated[14] from the distortion parameter δ if it is assumed that the bending modes behave as harmonic oscillators. In this way we find for the intensity of $_x\beta_4^0$ (relative to that of the origin band) a value that is 400 times smaller than the observed intensity. The calculated intensity of $_y\beta_4^0$ appears to be 230 times too low. The discrepancy is even larger for both transitions β_6^0. It was already pointed out by Weber et al[11] that the observed intensity of the transitions β_n^0 is too high to result from just the small frequency difference $\omega' - \omega''$. They suppose that a large part of the observed intensities is to be ascribed to a change of the anharmonicity of the bending mode potential upon electronic excitation.

We have tried to account for the anomalous intensity distributions in terms of anharmonic coupling between the zero-order vdW vibrational levels. We followed the procedure developed by Bieske et al[15] who investigated the interaction between vdW vibrations in complexes of argon with a number of benzene derivatives including s-tetrazine. Bieske et al considered only the lowest three vdW levels; we now have the opportunity to involve the lowest nine levels in the calculations. It is assumed that only the zero-order transitions σ_n^0 carry oscillator strength, the intensities of the zero-order transitions β_n^0 are completely neglected.

The spectra are interpreted in terms of the mixed states Ψ_1 through Ψ_9 which arise from the anharmonic interaction among the nine zero-order states $_x\beta_2$ up to σ_3. The transitions owe their intensity to the σ components of the mixed states Ψ_i. We have assumed that the stretching motion is subject to a

Table 2. Composition of the mixed states Ψ_i in terms of the zero-order vdW vibrational states of $\overline{0}_0$ T·Ar. The zero-order energies are given as ε_0 (in cm^{-1}).

	$_x\beta_2$	$_y\beta_2$	σ_1	$_x\beta_4$	$_y\beta_4$	σ_2	$_x\beta_6$	$_y\beta_6$	σ_3	–
ε_0	31.5	36.4	42.6	63.0	71.8	81.2	94.5	106.8	115.8	–
Ψ_1	0.91	0.22	-0.33	0.05	0.04	-0.01	0.02	0.03	0.00	0.00
Ψ_2	-0.35	0.85	-0.39	0.06	0.05	-0.02	0.03	0.03	-0.01	0.00
Ψ_3	0.20	0.48	0.81	-0.19	-0.13	0.05	-0.08	-0.08	0.01	0.00
Ψ_4	0.02	0.03	0.21	0.95	0.00	-0.22	0.00	-0.01	0.00	0.00
Ψ_5	0.01	0.02	0.14	-0.12	0.91	-0.36	0.04	0.02	-0.01	0.00
Ψ_6	0.00	0.00	0.00	0.19	0.37	0.83	-0.32	-0.16	0.09	-0.01
Ψ_7	0.00	0.01	0.09	0.06	0.08	0.34	0.90	0.03	-0.24	0.04
Ψ_8	0.00	0.00	0.06	0.02	0.03	0.12	-0.22	0.82	-0.50	0.09
Ψ_9	0.00	0.00	0.05	0.01	0.01	0.07	0.16	0.52	0.73	-0.40
Ψ_{10}	0.00	0.00	0.01	0.00	0.00	0.02	0.05	0.14	0.38	0.91

Morse potential with an estimated steepness parameter of 15.1 nm^{-1} in both S_0 and S_1 and potential well depths $D_e'' = 305$ and $D_e' = 327$ cm^{-1}. Since the displacement of the equilibrium position upon electronic excitation (0.003 nm) is known from the literature[10], the zero-order intensities of the transitions σ_n^0 can be calculated. We obtained the following relative intensities: 0.955 (σ_1^0), 0.043 (σ_2^0) and 0.002 (σ_3^0). These values are much more dependent on $D_e'-D_e''$ than on the precise value of D_e.

The intermode coupling has been calculated by constructing a model Hamiltonian based upon the nine zero-order states plus a 10th level that represents the zero-order vdW levels β_8 and σ_4. Our spectra do not show any transitions that terminate in these levels. The levels β_8 and σ_4 do not have any significant effect on the observed intensities, but they do influence the positions of the observed energy levels. The mixing coefficients are slightly dependent on the additional assumptions concerning the positions of the zero-order states β_n. The coefficients of the zero-order states in each of the mixed eigenstates Ψ_i, as listed in Table 2, have been obtained by choosing the zero-order energies in such a way that the bending overtones are as harmonic as possible. Zero-order energies (i.e. the diagonal matrix elements) are included in Table 2 as ε_0 (in cm^{-1}). All the coupling matrix elements are smaller than 10% of the zero-order energies. The frequencies and relative intensities as presented in Table 1 can be reproduced exactly in terms of the eigenstates Ψ_i.

It should be noted that the states σ_1, σ_2 and σ_3 are completely uncoupled in first order and also the six states β_n do not interact in first order. However, in second order these states couple via the anharmonic interactions between β_n and σ_m. As a consequence of the second order coupling, a large part of the intensity of the transition indicated as σ_3^0 is borrowed from the transition σ_1^0. The transitions indicated as β_n^0 derive their oscillator strength completely from the transitions σ_m^0. The data listed in Table 2 demonstrate that the (zero-order) vdW modes in T·Ar are extensively mixed. For instance, the eigenstate Ψ_3 (conventionally assigned σ_1) is composed of 66% σ_1 character, 23% $_y\beta_2$ character, and the zero-order states $_x\beta_2$ and $_x\beta_4$ contribute each for about 4%. As a result of the anharmonic coupling the vdW vibrational levels do not possess any distinct bending or stretching character.

One of the channels for IVR and VP observed by means of time-resolved experiments[5] is

$$Ar{\cdot}T(|6a^1>|0>) \xrightarrow{\text{IVR}} Ar{\cdot}T(|16a^2>|w>) \xrightarrow{\text{VP}} Ar + T(|16a^1>)$$

The quantum number w denotes some vdW vibrational state which is probably not characterized by a completely distinct stretch or bend character. However, adjacent levels $|w>$ in the vdW vibrational manifold can have fairly different identities which may be reflected by their dominant bending or stretching character or by significantly different rotational constants.
The picosecond laser pulse excites coherently a linear combination of cluster eigenstates

$$|e_i> = \alpha_i |6a^1>|0> + \sum_j \beta_{ij} |16a^2>|w_j>$$

The zero-order vibronic state $|6a^1>|0>$ carries all the oscillator strength whereas the zero-order states $|16a^2>|w>$ are Franck-Condon inactive. In these experiments the dephasing of the linear combination of cluster eigenstates into different near iso-energetic states $|16a^2>|w>$ causes an increase of the decay rate of the $\overline{6a^1}$ level compared to that of the $6a^1$ level in the bare molecule.

The energy of the vdW vibrational states $|w>$ is approximately 180 cm^{-1}. The average density of vdW levels with the right symmetry in this energy region is of the order of 1 per cm^{-1} and therefore we think that the number of interacting states is rather low. Nevertheless we could not observe any oscillatory decay of the fluorescence. The reason for the absence of quantum beats must be that so many rovibronic levels $|6a^1>|0>|J,K>$ are excited in the ensemble of clusters because of the relatively large spectral width

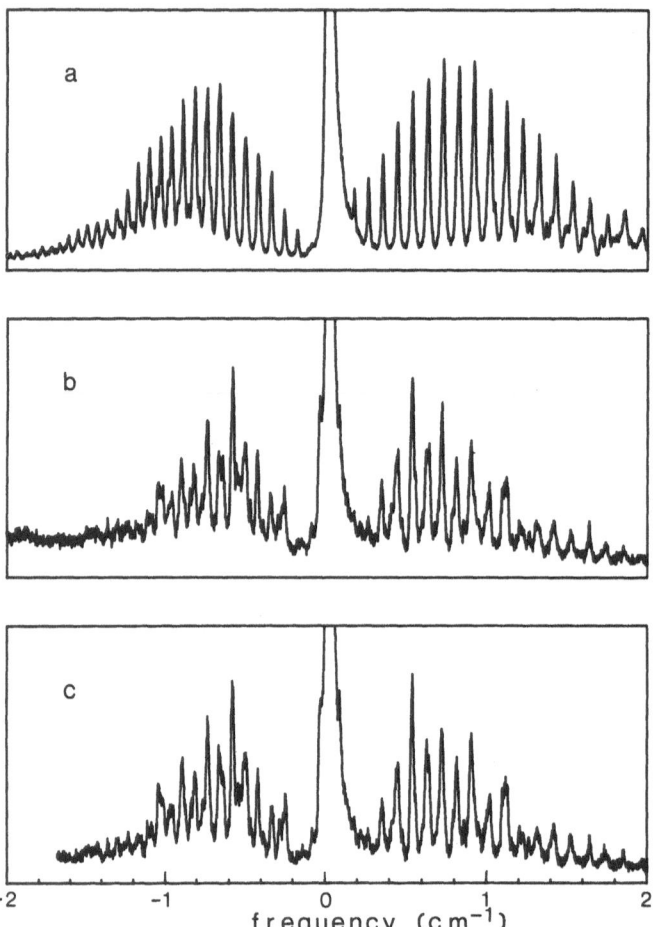

Fig. 2. High resolution fluorescence excitation spectrum of the $\overline{6a}_0^1$ transition of T·Ar. The top trace a of this figure shows the spectrum obtained when detecting the resonant emission band $\underline{6a}_2^1$. Selective detection of the relaxed emission band $\overline{16a}_2^2$ or the fragment emission $16a_1^1$ yields the excitation spectra displayed in traces b and c respectively.

(2 cm^{-1} fwhm) of the laser pulses. These levels interact with near-resonant levels $|16a^2>|w_j>$ with equal rotational quantum numbers. Owing to the floppiness of the cluster, the rotational constants depend strongly on the identity of the levels $|w_j>$ and hence the rotational levels may pile up differently for different vdW vibrational levels $|w_j>$. Therefore, the energy separation between the levels $|6a^1>|0>|J,K>$ and $|16a^2>|w_j,J,K>$ which are in Fermi resonance (and hence the beat frequencies) depend on the rotational states of the individual clusters in the ensemble. Due to the large number of different rovibronic states prepared by the laser pulse any oscillatory behaviour of the decay is washed out.

In order to search for information about the Fermi resonance between the cluster states we have carried out c.w. experiments using a single frequency laser. Fig. 2 displays the $6a^1_0$ band of the fluorescence excitation spectrum. The fluorescence was selectively detected in the resonant $6a^1_2$ band and in the $16a^2_2$ band of the complex (traces a and b respectively), and also in the $16a^1_1$ emission band of the molecular dissociation fragment (trace c). In spite of the narrow spectral width of the laser, it is not possible to excite individual rotational levels J,K. The individual rotational lines overlap because of lifetime broadening (0.013 cm^{-1} fwhm for the $6a^1$ level) and Doppler broadening (0.015 cm^{-1} fwhm). This implies that separate rotational transitions $\Delta K = 0$ can not be resolved. Fig. 2a displays the transitions $\Delta J = 0, \pm 1$ which terminate in the $6a^1$ component of the cluster eigenstates, while spectrum b scans the transitions terminating in rotational states of several different components $|16a^2>|w>$.

A comparison of both spectra shows that the quantum yield of IVR depends on the rotational state of the cluster. Two distinctions between both spectra are conspicuous. The transitions R(5), R(7) and R(9), corresponding to P(7), P(9) and P(11), are stronger than neighbouring transitions R(6) etc. and the relative intensities of the transitions beyond R(11) and P(13) collapse. A close examination of spectrum b shows several conformities between the P and R branches. A detailed analysis of the spectrum shown in Fig. 2a, b is now being performed in connection with an analysis of the intensity distribution in the IVR. emission band $16a^2_2$. If measured with a high spectral resolution (1 cm^{-1} fwhm) this emission band exhibits a marked substructure which is assigned to sequence transitions of the vdW vibrations. Experiments that are carried out for the present, show that the shape of the $16a^2_2$ emission band depends on which rotational levels of the complex are selectively prepared. The results of the analysis will be published elsewhere.

There is a striking similarity between the spectra b and c which implies that the influence of rotation on the vibrational predissociation of T·Ar following the excitation of $6a^1$ is almost completely restricted to the IVR step preceding the actual dissociation process. We think that the major role of the rotational degrees of freedom is to bring the levels $|6a^1\rangle|0\rangle$ and $|16a^2\rangle|w\rangle$ in and out of resonance, while the mixed character of the various vdW levels $|w\rangle$ causes the dissociation process to be fairly insensitive to which particular vdW states $|w_j\rangle$ are involved in the IVR process.

This interpretation is supported by the rotational structure of the emission from the molecular dissociation fragment.

Fig. 3 shows the $16a_1^1 6a_1^0$ emission band of tetrazine recorded after exciting the complex to the $6a^1$ level. In this case the laser was tuned into the Q branch of the $6a_0^1$ transition. We observed that the rotational contour of the product emission hardly depends on which rotational levels of the complex are initially excited. Specific excitation of rovibronic levels of $6a^1$ gives rise to almost identical emission spectra of $16a_1^1$ and $16a_1^1 6a_1^0$. The spectra exhibit a fairly broad intensity distribution which seems to indicate a bimodal state distribution of the dissociation product. A full analysis of the observed intensity distributions has not yet been completed. From computer simulations that are now being performed, we tentatively conclude that the VP process yields tetrazine molecules in a wide variety of rotational states with energies $E(J,K)$ from about 0 up to about 150 cm^{-1} and a preference for the intermediate energy values. The population distribution is probably bimodal.

Halberstadt et al.[16] have discussed the rotational state distribution of Cl_2 produced by vibrational predissociation of chlorine-rare gas complexes.

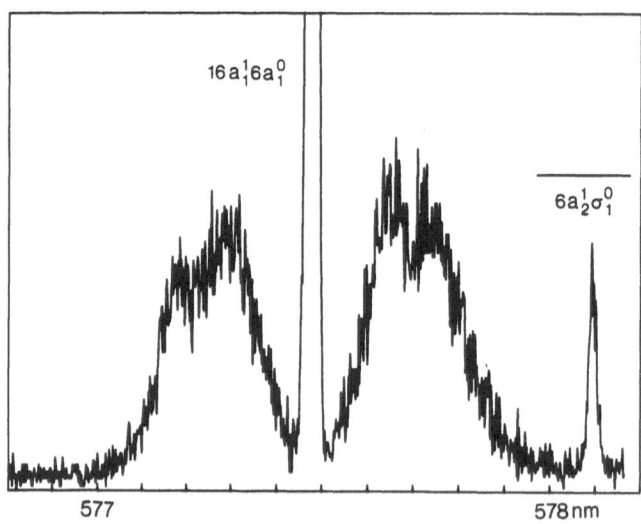

Fig. 3. Dispersed fluorescence spectrum of the fragment emission of tetrazine after excitation of the $6a^1$ level of the cluster.

Using a sudden approximation for the description of the VP process, and assuming a negligible final state interaction, they have shown that the excitation of a tight van der Waals bending mode in the quasi bound state of the complex will yield a wide distribution of rotational quantum numbers in the dissociation product.

The wide distribution of rotational states in tetrazine $16a^1$ is an indication that vdW bending modes are always excited in the metastable (dissociative) levels produced by IVR, whatever rotational states of the $6a^1$ level were initially excited. We consider this as another indication for the mixed character of the van der Waals vibrational levels.

ACKNOWLEDGEMENT

The experimental assistance of ing. D. Bebelaar and Mr. G. Jansen is kindly acknowledged. This work was supported by the Netherlands Foundation for Chemical Research (SON) and it was made possible by financial support from the Netherlands Organization for Scientific Research (NWO).

REFERENCES

[1] D. V. Brumbaugh, J. E. Kenny and D. H. Levy, J. Chem. Phys. **78**, 3415 (1983)

[2] J. J. F. Ramaekers, H. K. van Dijk, J. Langelaar and R. P. H. Rettschnick, Faraday Discuss. Chem. Soc. **75**, 183 (1983)

[3] J. J. F. Ramaekers, L. B. Krijnen, H. J. Lips, J. Langelaar and R. P. H. Rettschnick, Laser Chem. **2**, 125 (1983)

[4] M. Heppener, A. G. M. Kunst, D. Bebelaar and R. P. H. Rettschnick, J. Chem. Phys. **83**, 5341 (1985)

[5] M. Heppener and R. P. H. Rettschnick, in "Structure and Dynamics of Weakly Bound Complexes", A. Weber (ed.) Reidel, Dordrecht (1987), 553.

[6] P. M. Weber and S. A. Rice, J. Chem. Phys. **88**, 6120 (1988)

[7] P. M. Weber and S. A. Rice, J. Phys. Chem. **92**, 5470 (1988)

[8] D. F. Kelley and E. R. Bernstein, J. Phys. Chem. **90**, 5164 (1986)

[9] G. E. Ewing, J. Phys. Chem. **90**, 1790 (1986)

[10] C. A. Haynam, D. V. Brumbaugh and D. H. Levy, J. Chem. Phys. **80**, 2256 (1984)

[11] P. M. Weber, J. T. Buontempo, F. Novak and S. A. Rice, J. Chem. Phys. **88**, 6082 (1988)

[12] J. A. Menapace and E. R. Bernstein, J. Phys. Chem. **91**, 2533 (1987)

[13] G. Brocks and T. Huygen, J. Chem. Phys. **85**, 3411 (1986).

[14] J. R. Henderson, M. Muramoto and R. A. Willett, J. Chem. Phys. **41**, 580 (1964)

[15] E. J. Bieske, M. W. Rainbird, I. M. Atkinson and A. E. W. Knight, J. Chem. Phys. **91**, 752 (1989)

[16] N. Halberstadt, J. A. Beswick and K. C. Janda, J. Chem. Phys. **87**, 3966 (1987)

DISCUSSION OF WEDNESDAY AFTERNOON SESSION
MORE DYNAMICS OF LARGER MOLECULES

Philippe BRECHIGNAC

Laboratoire de Photophysique Moléculaire CNRS

Bâtiment 213 - Université de Paris-Sud

91405 - ORSAY Cedex France

The presentation by E.R. Bernstein of his work on the vibrational dynamics of aniline-Ar and aniline-CH_4 van der Waals complexes has raised two major questions. The first one deals with the actual singificance of the more and more common and widely used expressions : intramolecular vibrational redistribution (IVR) and vibrational predissociation (VP). These terms normally designates well-defined and often competing mechanisms :

1) IVR is the process governing within an isolated molecule the relaxation of excess energy into all the accessible vibrational degrees of freedom. When IVR is completed the system must be at equilibrium. When IVR is not completed, i.e. within a period of time which is such that $k_{IVR}\, \tau \leqslant 1$ (k_{IVR} = rate for IVR to process), the system must be found in a state intermediate between the initially prepared state and equilibrium.

2) VP is a process through which a molecular system, containing an excess vibrational energy larger than the energy required to break the system into two given fragments (dissociation energy of a given channel), evolves from the initially prepared state to a state made of these two fragments, one of which having less vibrational energy. Essentially VP must be accompanied by a change of at least one vibrational quantum in a particular mode. In most cases, VP is a multichannel process since the fragments can be found in various (vibrational or rotational) states. Each channel is characterized by a partial rate k_{VP}^{i}, such that :

$$\sum_i k_{VP}^i = 1/\tau_{VP}$$

where τ_{VP} is the global VP lifetime of the initially prepared state.

Dynamics of Polyatomic Van der Waals Complexes
Edited by N. Halberstadt and K. C. Janda
Plenum Press, New York, 1990

Some confusion may arise in E.R. Bernstein's model of serial IVR/VP processes, due to the fact that it basically divides the vibrational phase space of the weakly-bound cluster into two separate set of modes : the chromophore modes and the WdW modes. It must be realized that the injected vibrational energy into the chromophore modes is not large enough for IVR being efficient within this restricted set of modes. Consequently the IVR which is actually taking place should rather be described in terms of energy transfer from the chromophore set of modes to the WdW modes. Although, strictly speaking, this process is true IVR from the initially prepared state of the cluster, it is incorrect to call VP the second step of this serial model. Indeed calculation of a rate according to statistical theory of unimolecular dissociation in the phase space restricted to WdW modes hides the discrete change of vibrational energy in the chromophore modes which is essential to the VP process from the initially excited cluster state. Alternatively this step could be named "VP from the relaxed cluster state", but again this is confusing since the change of the vibrational energy in the WdW set of modes is statistically distributed rather than discrete. Apart from this language remarks, the serial IVR/VP model, although attractive for the description of the vibrational dynamics in such large molecules clusters, may have a weakness which also derives from the division of the phase space : the IVR step ignores the presence (i.e. the coupling) of the continuum states, and consequently the possible competition and/or intereferences between IVR and actual (direct) VP from the initially excited state.

The second question addressed during the discussion deals with the calculation of the density of VdW states N(E). The major conclusion of the interpretation of the data reported by E.R. Bernstein is to point out this density of states, as being the primary factor controlling the IVR/VP dynamics. These conclusions are based on the comparison between the behaviours of the aniline-Ar complex and the aniline-CH_4 complex. It is clear that, within the framework of the serial IVR/VP model, a proper counting of the VdW modes is essential since it drastically affects N(E) and consequently the predictions of the model. The simplest approach is to remark that, besides the three usual VdW modes (one stretch and two bends) aniline-CH_4 complex offers three hindered rotation degrees of freedom. But the proper way to take them into account may be non trivial, due to selection rules imposed by symmetry and spin statistics. Simultaneously the exact values of associated (vibrational) frequencies may depend very much on the relative magnitude of the barrier to internal rotation compared to the free rotation frequency. These factors are expected to reduce the density of states.

On the other hand the values of the vibrational frequencies evaluated near the minimum of the potential energy surface may not be appropriate for large excess energy. In particular, potential models seem to show that the VdW bending motions must correlate with the overall rotation of the Ar atom around the aromatic plane for energies of about 250 cm^{-1} above the minimum. This effect must produce a sudden increase in the density of states. In connection to the above points it has been suggested during the discussion that the CH_4 fragments should be found rotationally excited if internal rotation does participate in IVR.

In conclusion, although noble gas atoms are intrinsically simple as complexing partners of aromatic molecules, small molecules appear like a complementary and more efficient tool for probing VdW interactions and the associated dynamical behaviour of weakly-bound complexes.

VIBRATION DYNAMICS IN SOLUTE/SOLVENT CLUSTERS: ANILINE (Ar)$_1$ AND ANILINE (CH$_4$)$_1$

E. R. Bernstein
Department of Chemistry
Colorado State University
Fort Collins, CO 80523

ABSTRACT

The first excited electronic state (S$_1$) vibrational dynamics of aniline(Ar)$_1$ and aniline(CH$_4$)$_1$ van der Waals (vdW) clusters have been studied using molecular jet and time resolved emission spectroscopic techniques. The rates of intramolecular vibrational energy redistribution (IVR) and vibrational predissociation (VP) as a function of excess vibrational energy are reported for both clusters. For vibrational energy in excess of the cluster binding energy, both clusters are observed to dissociate. The dispersed emission spectra of these clusters demonstrate that aniline(Ar)$_1$ dissociates to all energetically accessible bare molecule states and that aniline(CH$_4$)$_1$ dissociates selectively to only the bare molecule vibrationless state. The emission kinetics show that in the aniline(Ar)$_1$ case, the initially excited states have nanosecond lifetimes, and intermediate cluster states have very short lifetimes. In contrast, the initially excited aniline(CH$_4$)$_1$ states and other vibrationally excited cluster states are very short lived (<100 ps), and the intermediate cluster 0^0 state is observed. These results can be understood semiquantitatively in terms of an overall serial IVR/VP mechanism which consists of the following elements: 1. the rates of chromophore to vdW mode IVR are given by Fermi's golden rule, and the density of vdW vibrational states is the most important factor in determining the relative [aniline(Ar)$_1$ vs. aniline(CH$_4$)$_1$] rates of IVR; 2. IVR among the vdW modes is rapid; and 3. VP rates can be calculated by a restricted vdW mode phase space RRKM theory. Since the density of vdW states is three orders of magnitude greater for aniline(CH$_4$)$_1$ than

Dynamics of Polyatomic Van der Waals Complexes
Edited by N. Halberstadt and K. C. Janda
Plenum Press, New York, 1990

aniline(Ar)$_1$ at 700 cm^{-1} of excess energy in S$_1$, the model predicts that IVR is slow and rate limiting in aniline(Ar)$_1$, whereas VP is slow and rate limiting in aniline(CH$_4$)$_1$. The agreement of these predictions with the experimental results is very good and is discussed in detail.

I.   INTRODUCTION

Photoinduced unimolecular decomposition reactions are among the simplest reactions which can be studied experimentally and theoretically.   One such reaction which has received considerable attention is the vibrational predissociation of small isolated van der Waals (vdW) clusters for which one molecule is a chromophore and the other is a small "solvent" molecule.   Two dynamical events may transpire in such a system following the initial photoexcitation to S$_1$ vibronic levels: vibrational energy may be redistributed to modes other than the optically accessed zero order chromophore states; and at sufficient energies the cluster may dissociate.   The fundamental theoretical understanding of these two kinetic processes should be accessible in terms of Fermi's golden rule[1] and unimolecular reaction rate[2] concepts.

A.   Theoretical Considerations: T$_1$, T$_2$, IVR

The theoretical discussion of molecular vibrational dynamics is best begun by defining the concepts of small, intermediate and large molecules.[1]   The distinction between these three "cases" of vibrational dynamical behavior is based upon the number of vibrational molecular eigenstates per unit internal energy (density of vibrational states).   The zero order states accessed by optical excitation are composed of a coherent superposition of molecular (cluster) eigenstates.   As is typically the convention, T$_1$ refers to a population relaxation time and a loss of energy from the system.   T$_2$ refers to a dephasing time in which the phase information of the initially excited (zero order) wavefunction(s) is lost.   IVR is basically a redistribution of vibrational energy within a system without loss of total energy.   The concept of IVR arises because the (optically accessed) coherent superposition of molecular eigenstates evolves in time and develops into other zero order, optically active (and inactive) states.   The dephasing process takes place in such a way as to conserve total energy.   This time development or evolution constitutes the IVR "T$_2$ time" or IVR "dynamics".

In the above context a small molecule is one in which the density of molecular eigenstates is low; so low in fact that only one eigenstate is accessed by the exciting laser pulse. In this instance, the molecular eigenstate is stationary, energy is lost only to the radiation field, $T_1 = T_2$ and no "IVR" takes place.

The intermediate molecule case arises for a higher density of molecular eigenstates. The zero order optically accessed state may be decomposed into a coherent superposition of many molecular eigenstates. If VP does not occur, the system can lose energy only by radiative processes: $T_1 = \tau_{rad}$ and $T_2^{-1} =$ "rate of IVR". The notion of an "intermediate case" implies several (not one, not hundreds) molecular eigenstates are coherently excited by the laser pulse. Thus, typically several quantum beat components can be observed in the (wavelength resolved) emission intensity from the intermediate case molecule, as the molecular eigenstates change phase and the zero order states recur in time. One would typically see a recurring beat pattern for the intermediate case molecule.

In the large molecule limit, many (greater than ~20) molecular (cluster) eigenstates are accessed by a laser pulse and therefore the zero order optical state contains many fourier components in its dephasing or quantum beat spectrum. The summation of these many fourier components leads to an exponential time dependence - an "IVR decay" or "dissipative IVR". IVR in this case can be treated as a relaxation process and rate constants for the "decays" can be measured by characterizing the rise and fall times of zero order molecular chromophore vibronic state emission. If VP does not take place then $T_2 = $ (IVR rate)$^{-1}$ and $T_1 = \tau_{rad}$.

If the vibrational (vdW) modes which lead to VP are coupled to the optically accessed zeroth order state, then above the VP threshold recursion times are infinite, and quantum beats do not occur. This is the typical situation in van der Waals clusters, and we expect that above the VP threshold a cluster is within the large molecule limit.

The density of states for a cluster can be estimated by the Marcus-Rice semiclassical approximation.[2] A chromophore(Ar)$_1$ cluster at ca. 250 cm^{-1} above the S_1 origin would fall within the intermediate molecule boundaries. Due to the low VP thresholds, low vibrational frequencies of vdW modes, and coupling between the zero order chromophore modes and the vdW modes, we conclude that at most accessible excess vibrational energies ($\geq$ 400 cm^{-1}) the clusters of interest to us in this work will probably fall within the large molecule limit. IVR and VP dynamics may be therefore described in general by phenomenological rate constants.

B. Theory of IVR and VP

Theoretical studies of vibrational redistribution and dynamics including VP have been quite numerous over the last 15 years.[1-3] While we will not review the theoretical developments in detail in this report, we will present some of the important conclusions that can be drawn from the theoretical studies referenced.

Two theoretical models can be put forward for the dynamics of IVR and VP processes, and most of the available data has been interpreted in terms of one or the other of these models. These models treat IVR and VP to be either "parallel" or "serial" processes. The Beswick-Jortner[3] model for VP considers dissociation to be a process which occurs in parallel with IVR. This treatment considers the direct coupling between the chromophore vibrational states of the bound complex and the plane wave states of the dissociated complex, and is most appropriate when the amount of energy put in vdW modes by a one quanta change in the chromophore is large compared to the binding energy. Under these high energy, weak binding conditions one chromophore quantum of energy in the vdW modes must dissociate the cluster in one-half of a vibrational period. As a result, IVR/VP occurs "directly" into the "dissociative continuum". This parallel model is appropriate to diatomic/He type systems. The binding energy in the polyatomic clusters of interest in this work is typically large compared to the separations of chromophore vibrational energy levels, however. Thus, for polyatomic clusters, the Beswick-Jortner treatment seems inappropriate and is directly at odds with the more conventional serial model based on RRKM unimolecular reaction rate theory.[2]

In the serial model,[2b] the vibrational phase space of the cluster can be divided into two regions: that of the chromophore modes and that of the vdW modes. The rationale behind this partitioning of phase space is two fold: 1) an energy mismatch exists between the vdW modes (typically less than ~50 cm^{-1}) and the chromophore modes (typically greater than ~200 cm^{-1}); and 2) the coupling between these two sets of modes is small. Optical excitation typically puts most or all of the vibrational energy into the chromophore region. The amount of energy in the vdW modes then increases as chromophore to vdW mode IVR proceeds. With the approximation that IVR among the van der Waals modes is very rapid, the VP rate can be calculated by a restricted RRKM theory in which the rate constant depends only upon the total amount of energy in vdW modes. Therefore, in this latter model, VP can occur only after chromophore to the vdW mode IVR has occurred; the rate depends essentially upon the amount of energy in vdW modes and this energy varies with time, due to chromophore to vdW mode IVR.

The two theories therefore predict much different appearance kinetics for individual vibronic states of solute/solvent clusters generated by IVR and the bare chromophore molecule generated by VP. While the interpretation of wavelength and time resolved measurements does not depend on the theoretical model imposed or envisioned, the interpretation of the cw experiments is indeed highly model dependent. Thus, in the absence of temporal resolution, the assumption of only parallel or only serial relaxation processes is important for the data interpretation. The question of serial vs. parallel processes for vibronic dynamics can in any cases be uniquely answered by time resolved studies. Indeed, we have shown zpreviously for tetrazine(Ar)$_1$,[2b] and herein for aniline(Ar)$_1$, (N$_2$)$_1$, (CH$_4$)$_1$, that the serial IVR/VP process is the appropriate one.

C   Previous Experimental Studies of IVR and VP

Studies of IVR and VP in molecular clusters available in the literature come from the laboratories of Levy,[4,5] Ito,[6] Soep,[7] Rice,[8] Parameter,[9] Rettschnick,[10] and Bernstein.[11] In most of these instances cw (pulse width greater than 5 ns) experiments are performed on the clusters and dynamical behavior is inferred from the dispersed emission. Some time resolved dynamical studies have recently appeared on tetrazine/argon,[8,10] p-difluorobenzene/argon, the dimethyltetrazine dimer,[5,12] and the benzene/phenol dimer[13] which will be briefly discussed below.

The studies of references 4-9 involve the measurement of intensities of various dispersed emission peaks as a function of different accessed vibronic transitions in clusters. The systems studied are tetrazine(Ar)$_1$, glyoxal(Ar)$_1$, (Kr)$_1$, (N$_2$)$_1$, p-difluorobenzene(Ar)$_1$, pyrimidine(Ar)$_1$, (N$_2$)$_1$, respectively. Other work is referenced in these papers. While we will not discuss the extensive detail in these papers, we will present a general discussion of their major findings and approach to their data.

Each of the groups analyzes the cluster "dynamical" data with a particular prejudice which is often not stated. In most cases, dynamical processes are assumed to occur in parallel. Thus, IVR and VP are competitive "channels" and the various IVR "pathways" are also competitive. This point of view with regard to VP is implicit in the theoretical work of Beswick and Jortner.[3] A cluster can thereby undergo VP or IVR into several different lower modes depending on the excitation energy. In this approach the intensity of various cluster and free molecule emissions following single vibronic level excitation of the cluster yields "branching ratios" for IVR and for VP channels. The branching ratios then lead to characterization of "propensity" rules with molecular chromophore modes assigned as having a "special" relationship to VP. With the exception of the reported studies for the benzene(Ar)$_1$ cluster[8a,b] this conceptualization has led to the reporting of individual rate constants for IVR and VP for the clusters. Rate constants of the order of 10^8-10^9 s^{-1} seem to be typical. In the study of benzene(Ar)$_1$, Rice, et al.,[8a,b] have reported only "vibrational redistribution/ predissociation" rate constants because the distinction between these two processes based solely on the presented cw measurements is not possible. We would agree with this latter position. The reporting of branching ratios and channels for IVR and VP is an outgrowth of the highly parallel conceptualization of the vibrational dynamics and cw measurements.

Bernstein, et al.,[11] have done similar cw experiments on aniline(He)$_n$, aniline(CH$_4$)$_1$ and, to a lesser extent, toluene(CH$_4$)$_n$ and have augmented these results with 2-color time of flight mass spectroscopy.

Several picosecond cluster studies on tetrazine(Ar)$_1$ have recently appeared in the literature.[10, 8c,d] The tetrazine(Ar)$_1$ studies are concerned with decay times of various vibrational states above and below the binding energy of the complex. Both excited (S$_1$) and ground state picosecond results have been published. The excited state studies are an appropriate preliminary effort but firm conclusions are difficult to reach based on the presented data because spectroscopic resolution could not always separate cluster and molecular emission and some of the reported lifetimes are somewhat inconsistent. Nonetheless, these studies show that the excited state IVR/VP dynamics take place on the ≤ 2 ns timescale.

The ground state data for tetrazine/argon,[8c,d] obtained by a three photon resonant fluorescence technique, indicate that little or no IVR occurs on the 15 ns timescale for S$_0$. The ground state IVR relaxation times are in sharp contrast to the excited state times: this difference has been explained in terms of different chromophore-vdW mode interactions in each electronic state. These authors postulate that the extent of vibrational coupling is indicated by the spectral bandshifts which occur upon clustering.

In a later paper, these authors[8c,d] also use the bandshift idea, along with our ideas of serial relaxation dictated by energy gap laws and density of states considerations, to interpret the entire relaxation/dissociation process.

Picosecond studies of the dimethyltetrazine dimer,[12] and the benzene/phenol dimer[13] have also appeared. The dimethyltetrazine dimer study simply reports an IVR time of ~35 ps for the $6a^1$ vibronic state. This result is in reasonable agreement with the linewidth estimated from spectroscopic data. The benzene/phenol study reports the rate of complex disappearance following single vibronic level excitation. This rate is found to increase with excitation energy above 1275 cm^{-1}.

II. SUMMARY OF ANILINE(X)$_1$ (X = Ar, CH$_4$) RESULTS

We have recently performed time resolved and static emission studies on aniline(X)$_1$, X = Ar, and CH$_4$, clusters. These time correlated single photon counting dispersed emission studies are still in progress and much more work needs to be done; however, the experimental and calculational results obtained thus far have elucidated the basic IVR and VP mechanisms. A summary of these results and calculations is given here.

The spectroscopy of the aniline molecule has been extensively studied,[14,15] and the most intense vibronic features have been assigned. The assignments of some of these peaks are indicated in Figure 1. The corresponding aniline(Ar)$_1$ spectrum has its origin shifted about 40 cm^{-1} to the red of the corresponding bare molecule origin. The other (vibronic) peaks are shifted by approximately the same amount. These shifts can be understood in terms of an approximately 40 cm^{-1} increase in the aniline(Ar)$_1$ binding energy upon electronic excitation. The value of the excited state binding energy is important in the interpretation of the spectroscopic and kinetic data presented below.

The dispersed emission (DE) spectrum of the aniline(Ar)$_1$ cluster following $\overline{0^0_0}$ excitation is shown in Figure 2a. (The bar over a spectral assignment indicates a cluster, rather than a bare molecule, transition.) This spectrum is very simple, and as expected, all peaks can be assigned to the vibrationless cluster. Figure 2b shows the DE spectrum following cluster excitation at 442 cm^{-1} above the $\overline{0^0_0}$. The assignment of the pumped feature is not clear and is unimportant in the discussion here. Emission occurs from the initially pumped cluster state, and the $\overline{0^0}$ (cluster) state. Only IVR is apparent in these results, suggesting that the cluster binding energy

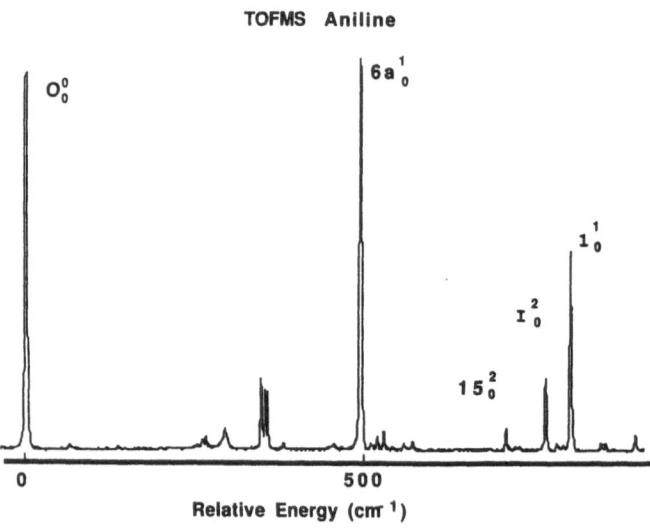

Figure 1. Time of flight mass selected excitation spectrum of bare aniline, in a molecular jet. Several of the more intense vibronic transitions are assigned.

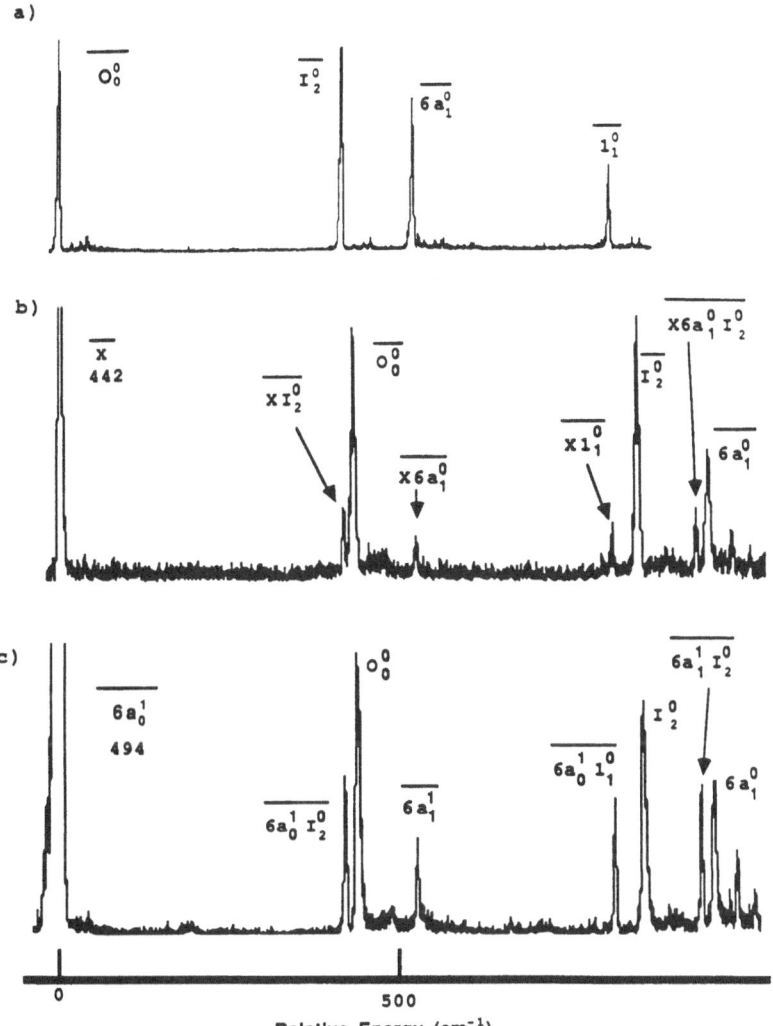

Figure 2. Dispersed emission spectra of aniline(Ar)₁ following excitation at a) 0_0^0, b) 442 cm^{-1} and c) $6a_0^1$ (494 cm^{-1}). Assignments of the more intense features are indicated.

is greater than 442 cm^{-1}. Figure 2c shows that following $\overline{6a_0^1}$ excitation (494 cm^{-1}) emission appears from both the $\overline{6a^1}$ (cluster) and 0^0 (bare molecule) states. Thus, an upper limit of 494 cm^{-1} can be put on the excited state binding energy. Interestingly, no detectable emission is observed from the lower cluster states. This observation will be discussed below.

The spectra become more complicated as the excitation energy is increased. Figure 3 presents the emission spectra arising from $\overline{15_0^2}$, $\overline{I_0^2}$, and $\overline{I_0^1}$ cluster excitation. As the cluster excitation energy increases, more bare molecule (dissociated) vibronic states become energetically accessible and are observed. In the case of $\overline{I_0^1}$ excitation, emission is observed from the $10b^1$, $16a^1$, I^1, and 0^0 states, as well as the initially pumped state. Figure 3c shows that all energetically possible final states are populated in approximately equal amounts (within a factor of 2 or 3).

We have also performed time resolved studies (time correlated single photon counting dispersed emission spectroscopy) on all of the above emission spectral features. An example of the data typically obtained is presented in Figure 4. This figure shows that the decay of the $\overline{6a^1}$ state is matched by the rise of the 0^0 state. Similar kinetics are obtained following $\overline{15_0^2}$, $\overline{I_0^2}$, and $\overline{I_0^1}$ excitation. In all cases the rise times of the final (dissociated) bare molecule states are within the experimental uncertainty of the initially populated state decay time.

Figure 5a presents the DE spectrum following $\overline{0_0^0}$ excitation of the aniline(CH$_4$)$_1$ cluster. As expected, this spectrum is very similar to the corresponding aniline(Ar)$_1$ spectrum. Very different spectra and kinetics are observed in the aniline(CH$_4$)$_1$ cluster when vibrational excitation is present, however. Figure 5b shows the DE spectrum following $\overline{6a_0^1}$ excitation of the aniline(CH$_4$)$_1$ cluster. Two types of spectral features are present: sharp peaks which originate from the vibrationless bare molecule (0^0), and a broad emission assigned to the vdW transitions built on the $\overline{0_0^0}$ band. This latter feature is very clearly due to residual energy transferred from the $\overline{6a^1}$ mode to the vdW modes (i.e., $\overline{6a^1} \rightarrow \ \rightarrow \ \rightarrow \overline{0^0}$ + 494 cm^{-1} vdW). $\overline{6a_0^1}$ excitation for aniline(CH$_4$)$_1$ thus leads to both bare molecule 0^0 state emission and cluster $\overline{0^0}$ emission. Absent from these DE spectra

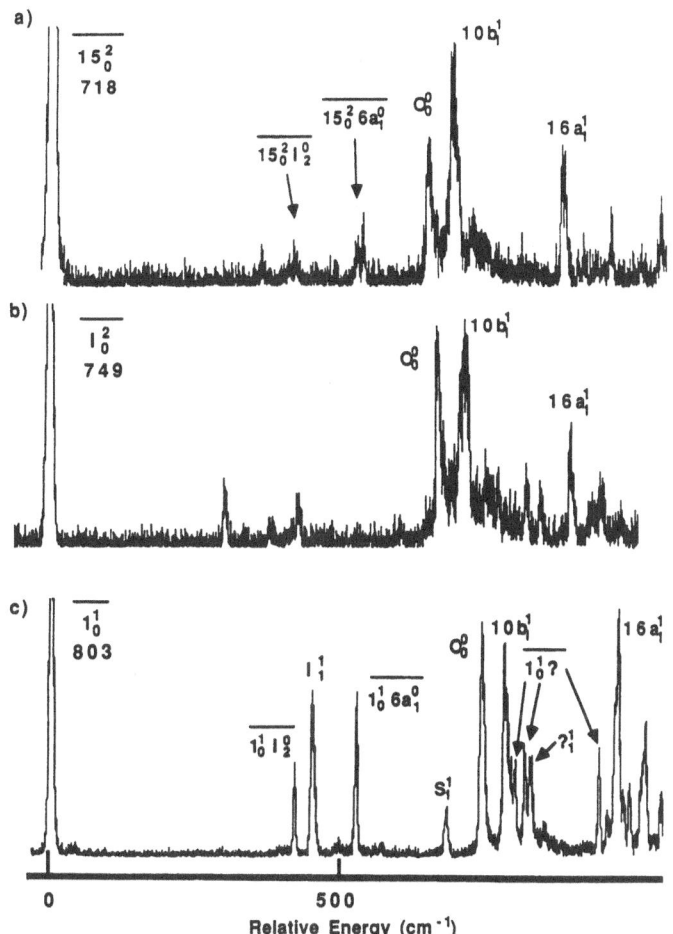

Figure 3. Dispersed emission spectra of aniline(Ar)$_1$ following excitation at a) $\overline{15}_0^2$ (718 cm^{-1}), b) $\overline{I}_0^2$ (749 cm^{-1}) and $\overline{1}_0^1$ (803 cm^{-1}). Assignments of the more intense features are indicated.

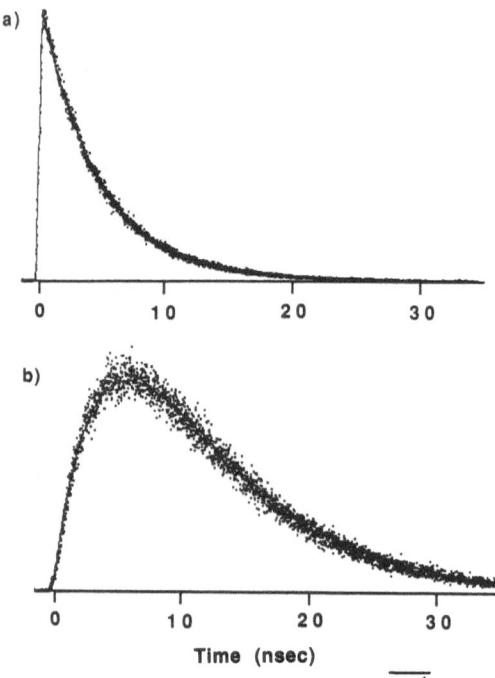

Figure 4. Emission kinetics following $\overline{6a_0^1}$ excitation of aniline(Ar)$_1$. The kinetics of the $\overline{6a_0^1 I_2^0}$ and $\overline{0_0^0}$ transitions are shown. Also shown are calculated curves corresponding to a) a 4.7 ns decay, and b) a 4.7 ns rise and a 7.6 ns decay.

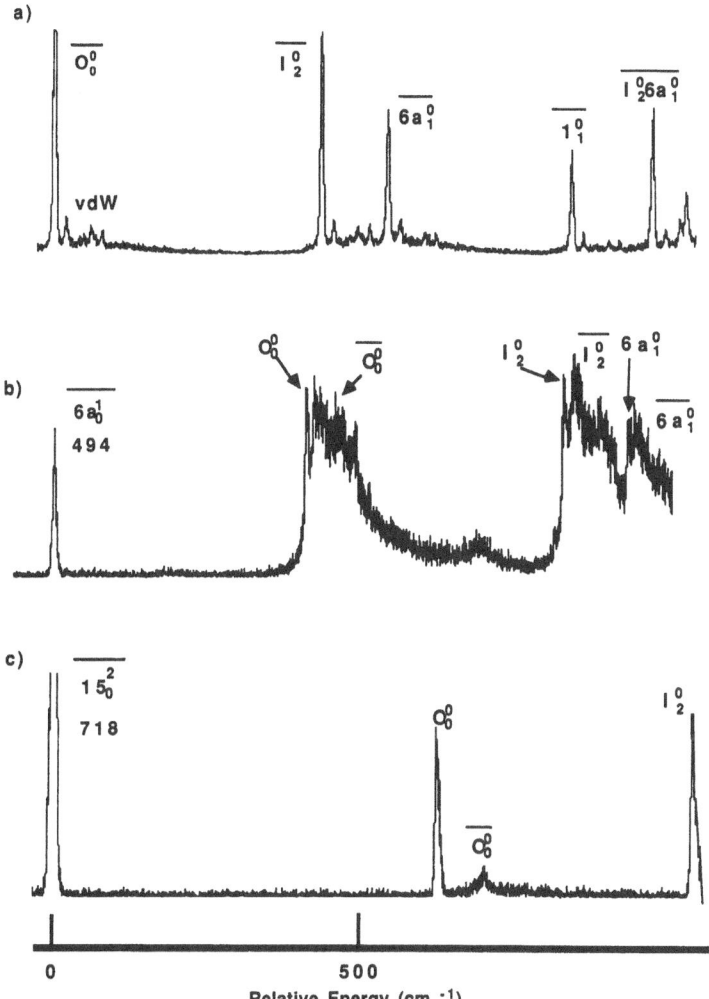

Figure 5. Dispersed emission spectra of aniline(CH_4)$_1$ following excitation at a) $\overline{0_0^0}$, b) $\overline{6a_0^1}$ (494 cm^{-1}), and c) $\overline{15_0^2}$ (718 cm^{-1}). Assignments of the more intense features are indicated.

for aniline$(CH_4)_1$ $\overline{6a^1}$ excitation is any initially excited state (i.e., $\overline{6a^1_1}$, etc.) emission. These results are in stark contrast to those found for the aniline$(Ar)_1$ cluster.

Figure 5c presents the DE spectrum which results following $\overline{15^2_0}$ excitation of the aniline$(CH_4)_1$ cluster. Two types of spectral features can be seen: 0^0 and much weaker $\overline{0^0}$ bands. The relative intensity of $\overline{0^0_0}$ band is found to increase upon increasing the CH_4/He ratio (i.e., from 0.2% to 2.0%) in the expansion gas mixture: much of this $\overline{0^0}$ emission is a result of excitation of aniline$(CH_4)_n$ $(n>1)$ clusters. From the DE spectra observed following $\overline{6a^1}$ excitation we conclude that the aniline$(CH_4)_1$ excited state binding energy is less than 494 cm^{-1} and that the bare molecule $10b^1$ and $16a^1$ states are energetically accessible upon $\overline{15^2_0}$ excitation. Conspicuously absent from the DE spectra is any emission from $10b^1$ (177 cm^{-1}), $16a^1$ (187 cm^{-1}), or the initially excited $(\overline{15^2})$ state. [Again we emphasize the difference between the aniline$(Ar)_1$ and aniline$(CH_4)_1$ spectra.]

This latter observation, in conjunction with the very high signal to noise ratio, indicates that the initially excited state has a lifetime of less than about 100 ps. Time resolved measurements indicate that the rise time of the 0^0 state is 240 ps (See Figure 6). The disparity in these rise and decay times suggests an important question: what intermediate states can be responsible for this kinetic difference? The most likely candidate for an intervening state in the kinetic IVR/VP process is $\overline{0^0}$. We note that when a considerable amount of energy is in the vdW modes the DE peaks are very broad, as in the case of the $\overline{0^0}$ emission in Figure 5b. Therefore, we would expect that any $\overline{0^0}$ emission in Figure 5c would also be quite broad. This breadth, in combination with the relatively short lifetime (ca. 240 ps), would make intermediate (cluster) state emission difficult to detect. The $\overline{0^0}$ cluster state must therefore be populated by IVR from the $\overline{15^2}$ state in less than 100 ps, and must live for about 240 ps prior to VP.

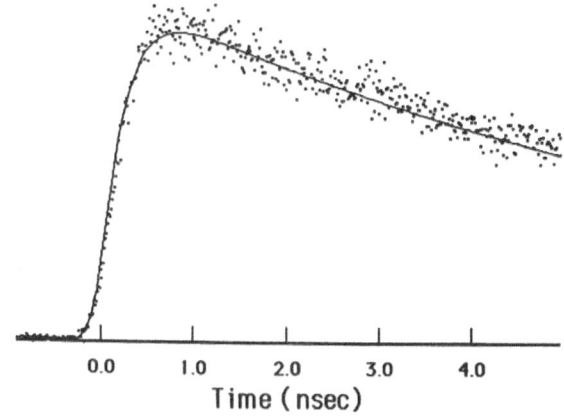

Figure 6. Emission kinetics following 15_0^2 excitation of aniline(CH$_4$)$_1$. The kinetics of the 0_0^0 transition are shown. Also shown is a calculated curve corresponding to the convolution of the instrument response function with a 240 ps rise and a 7.6 ns decay.

III. DISCUSSION

A. The General Model

The above aniline(Ar)$_1$ and aniline(CH$_4$)$_1$ results are summarized in Tables I and II, respectively: these *individual* data sets cannot be explained in terms of a parallel IVR/VP mechanism. As stated above, aniline(Ar)$_1$ dissociates into all energetically accessible vibrational levels of the bare molecule, whereas aniline(CH$_4$)$_1$ dissociates only to the vibrationless level of the bare molecule. Moreover, no plausible explanation for this *difference* in dynamical behavior can be found within the framework of a parallel IVR/VP mechanism.

These data sets can be understood in terms of a serial IVR/VP model.[2b] In this model chromophore to vdW mode IVR precedes VP. Once the amount of energy in the vdW modes exceeds the binding energy, either of two processes can occur: further chromophore to vdW mode IVR, or VP. The VP rate is determined by the amount of energy in the vdW modes; the VP rate is therefore time dependent and varies with the extent of IVR. The observed spectra and kinetics for both aniline(Ar)$_1$ and aniline(CH$_4$)$_1$ clusters systems will be the result of a competition between these two processes. The general theme for rationalizing the aniline(Ar)$_1$ and aniline(CH$_4$)$_1$ data is as follows: for the aniline(Ar)$_1$ cluster, IVR is very slow, due in part to the low density of vdW mode receiving states, and VP is very fast, because of the small (three mode) vdW phase space. For the aniline(CH$_4$)$_1$ cluster, IVR is very fast, due in part to the high density of vdW mode receiving states, and VP is relatively slow, because of the larger (six mode) vdW phase space.

We consider now each individual component of the overall redistribution/dissociation (cluster to bare molecule) process, and propose a simple model to understand the observed results.

The chromophore to vdW mode IVR in this model is simply given by the Fermi Golden rule transition probability expression: the product of a matrix element squared times a density of final receiving states. The rate of chromophore to vdW mode IVR may then be given by a product of three terms: a coupling coefficient between chromophore and vdW mode (initial and final) states; a vibrational wave function overlap term for initial and final vdW mode states; and the density of final vdW vibrational receiving states. While the above ideas have not been previously applied to the problem of chromophore to vdW mode IVR, they are not without precedent. The first two terms lead to a general "energy gap law" for

IVR in which the exchange of many quanta of energy between the chromophore and vdW modes is not favored.

The qualitative results for IVR can be readily calculated: the model generates differences in the predicted behavior of aniline(Ar)$_1$ and aniline(CH$_4$)$_1$ that are at the same time both striking and in agreement with the observed results. Aniline(CH$_4$)$_1$ has six degrees of freedom in the vdW modes: one stretch, two bends and three torsions. In contrast, aniline(Ar)$_1$ has only a stretch and two bend degrees of freedom. This difference in the number of vdW vibrational modes results in very different densities of states for these two complexes. The density of vibrational states at any given energy $N(E)$ can be estimated. For aniline(CH$_4$)$_1$ at 700 cm^{-1} of energy, $N(E) \simeq 3 \times 10^4$ states/cm^{-1}, whereas for aniline(Ar)$_1$ at the same energy, $N(E) \simeq 15$ states/cm^{-1}. Thus, the density of vdW vibrational states is about 10^3 greater for aniline(CH$_4$)$_1$ than for aniline(Ar)$_1$ at 700 cm^{-1} of energy.

Consider now how this density of states difference can affect the IVR rates for aniline clusters: IVR should be much faster in the aniline(CH$_4$)$_1$ system than in the aniline(Ar)$_1$ system. IVR rates are measured to be on the order at a few nanoseconds in the aniline(Ar)$_1$ system and if the coupling constants are comparable for both systems, IVR rates for aniline(CH$_4$)$_1$ will be three orders of magnitude faster than those measured for aniline(Ar)$_1$ - a few picoseconds.

The VP rate constants can also be estimated for any given amount of energy placed in the vdW modes by IVR.[2] The calculations are based upon a restricted (to the vdW mode phase space) RRKM theory.[2b] For aniline(Ar)$_1$ with 700 cm^{-1} of energy in the vdW modes and a binding energy of 450 cm^{-1}, $k = (5 \text{ ps})^{-1}$ Thus, in the case of aniline(Ar)$_1$ chromophore to vdW IVR is slow and rate limiting, and subsequent VP is very fast. Any IVR process which puts energy in excess of the binding energy into the vdW modes is immediately followed by VP. All of the lower chromophore levels are populated by IVR, and as a result the bare molecule is formed in all of the energetically accessible states. This model predicts that IVR will be the overall rate controlling process for aniline(Ar)$_1$ dissociation.

Very different VP kinetics are predicted for the aniline(CH$_4$)$_1$ case. Here we find (assuming a tight transition state) that with 700 cm^{-1} of energy in the vdW modes and a 480 cm^{-1} binding energy, $k = (200 \text{ ps})^{-1}$. Thus, in sharp contrast to the aniline(Ar)$_1$ case, aniline(CH$_4$)$_1$ IVR is very fast compared to VP. IVR is expected to populate all the lower chromophore levels just as in the aniline(Ar)$_1$

case; however, in the aniline$(CH_4)_1$ case subsequent IVR to the $\overline{0^0}$ level is complete before VP can occur. Finally VP occurs slowly from $\overline{0^0}$ and is the rate limiting step in the entire process. This model predicts that formation of the bare vibrationless molecule will be limited by the rate of VP for aniline$(CH_4)_1$, and will be on the hundreds of picoseconds time scale.

Thus, the observed spectra and kinetics for cluster systems in general will be the result of a competition between the two rates k_{IVR} and k_{VP}. All of the above qualitative predictions of this model are borne out by the experimental result.

B.   Numerical Simulations

We have performed detailed quantitative numerical simulations based on the above qualitative ideas.[14]   These simulations which closely follow those published by us for the tetrazine/argon system,[2b] generate quantitative predictions of the spectral quantum yields and the kinetic behavior of all observed features.   The calculated and observed intensities agree quite well - within a factor of two.   The calculated and observed kinetics are also in good agreement - again within a factor of two in nearly all instances.

In these calculations, the probability of an IVR transition from chromophore state j to state i, P_{ij}, is given by $P_{ij} = (C_{ij})$ x (energy gap term) x (ρ_i), in which ρ_i is the density of vdW states at the energy $(E_{exc} - E_i)$.   The quantity $(E_{exc} - E_i)$ is the total vibrational energy minus the vibrational energy remaining in the chromophore modes, and hence is the amount of vibrational energy in the vdW modes following the IVR transition.   The coupling coefficients C_{ij} for any set of initial and final chromophore states and the exact functional form of the energy gap law are difficult to determine.   Both depend upon the details of the potential surface.   Nonetheless, some reasonable approximations can be made which result in a phenomenological theory of IVR.   The notion of partioning the vibrational phase space into chromophore and vdW regions is based on the observation that the vdW mode frequencies are much lower than the chromophore mode frequencies.   The comparatively slow chromophore to vdW energy transfer is due in part to this frequency mismatch.   This suggests that the coupling constant C_{ij} should be inversely proportional to v_i and v_j, the chromophore vibrational frequencies.

Energy gap laws, based on wave function overlaps, have been worked out for a variety of situations.[16]   The functional form of the energy gap term depends on the nature of the final states. For example, small molecules in rare gas matrices undergo vibrational relaxation to the hindered rotational degrees of freedom. In this case, the energy gap term has the functional form $\exp(-\sqrt{\Delta E})$, in which ΔE is the amount of energy transferred.[1,17] A similar form can be adopted for chromophore to vdW mode energy transfer.

The density of the vdW vibrational states is easily evaluated, if the vdW mode energies are either known or if reasonable guesses concerning their energies and anharmonicities can be made. Thus the matrix of IVR rates can be constructed with a single adjustable parameter which scales all the IVR rates.

With the above assumptions and approximations, numerical simulations of the IVR/VP process can be performed. Two population vectors, N_c and N_{nc} are defined. $N_c(i)$ is the cluster population with the chromophore in the ith vibrational level and similarly, N_{nc} refers to the bare molecule population vector. To calculate the time evolution of the population vectors three matrices must be defined: **P** is the IVR transition probability matrix with matrix elements P_{ij} for the chromophore transition to level i from level j in a time interval Δt; **K** is the VP matrix; and **R** is the radiationless transition matrix. Then

$$R_{ij} = (1 - k_{isc+ic}\Delta t)\delta_{ij},$$

$$K_{ij} = (1 - k_1\Delta t)\delta_{ij},$$

$$P_{ij} = \frac{A}{v_j} \exp(-\sqrt{\Delta E_{ij}}) \rho_i$$

in which k_{isc+ic} is the rate constant for internal conversion ($S_1 \to S_0$) and intersystem crossing ($S_1 \to T_1$) for aniline, and k_1 is the RRKM unimolecular reaction (dissociation) rate constant characteristic of an energy ($E_{exc} - E_i$) in the vdW modes (see below). A is the only adjustable parameter of the model, which temporally scales the entire IVR process. The temporal evolution of the population vectors is given by

$$N_c(t+\Delta t) = \mathbf{R}\ \mathbf{K}\ \mathbf{P}\ N_c(t)$$

and

$$N_{nc}(t+\Delta t) = \mathbf{R}\ N_{nc}(t) + (\mathbf{1}\text{-}\mathbf{K})\ N_c(t).$$

The RRKM rate constant $k_1 = k(E)$ can easily be calculated using the Marcus Rice approximation,[2a]

$$k(E) = \frac{1}{h} \frac{\sum P(E^+)}{N(E)}$$

with $N(E) \simeq \frac{E^{S-1}}{(S-1)! \prod hv_i}$

and $\sum P(E^+) \simeq \frac{(E^+)^{S-1}}{(S-1)! \prod' hv_i}$,

in which $\sum P(E^+)$ is the sum of vibrational states above the dissociation energy excluding the reaction coordinate, $E^+ = E-E_0$ with E_0 the cluster binding energy, $N(E)$ is the total vibrational density of states at energy E, v_i is the vdW mode energy and S is the number of degrees of freedom of the vdW modes. The prime on the product sign indicates exclusion of the reaction coordinate. The assumption made in these calculations is that the cluster transition state is a "tight binding" one.[17] The calculation of RRKM rates is explained in detail in ref. 2.

The IVR rates in the aniline(Ar)$_1$ case are rather sensitive to the densities of states. In this case, the values of ρ_i (=$N(E)$) used in the construction of the IVR transition probability matrix are calculated with a direct count method.[18] Vibrational frequencies of 45 cm^{-1} (stretch), and 15 cm^{-1} (both bends) and anharmonicities of 3% (i.e., $\Delta v_{i,\, i+1} / \Delta v_{i-1,i} = .97$) are assumed.

Quantitative predictions of the model can be made from the simulation procedure described above. The results of these simulations, in terms of kinetics and spectral quantum yields, are compared with the experimental results in Tables I and II. In all cases the agreement is quite good; within a factor of about two.

Several comments and observations about these calculations and their comparison to the experimental data can be made. The experimental results show that the extent of vdW vibrational overlap tends to decrease with increasing energy change in an IVR transition. This can be seen from the results in Table I. For example, following aniline(Ar)$_1$ 1_0^1 excitation, the 0^0, $10b^1$, $16a^1$ and I^1 final product states are formed in approximately the same amounts. If the vibrational overlap were energy independent, then the 0^0 state would be somewhat favored due to the higher density of vdW states

Table I. Aniline (Ar)$_1$ relative spectral intensities, rise times, and decay times. The intensities are normalized to that of the 0_0^0 transition. Calculated values are in parentheses.[a]

Cluster excitation	Initially excited state decay times (ns)	Product state rise times[b] (ns)	Initially excited state intensity	Product state intensities			
				1^1	$16a^1$	$10b^1$	0^0
$\overline{6a_0^1}$	4.7 (4.3)	4.7 (5.2)	1.6 (1.2)				1.0 (1.0)
$\overline{15_0^2}$	1.5 (1.9)	1.5 (2.0)	4.7 (1.3)		1.0 (.59)	1.6 (.83)	1.0 (1.0)
$\overline{1_0^2}$	c (1.7)	1.0 (1.8)	c (.72)	(.59)	.57 (.59)	1.0 (1.0)	1.0 (1.0)
$\overline{1_0^1}$	2.7 (2.3)	2.7 (2.5)	1.9 (1.1)	.76 (.11)	1.0 (.62)	.9 (.72)	1.0 (1.0)

(a) Franck-Condon factors of the 1_1^1, $16a_1^1$, and $10b_1^1$ sequence transitions are assumed to be unity.
(b) Rise curves generated from the simulations are slightly non-exponential, but are similar to the fitted exponential rises.
(c) Not observed.

Table II. Aniline (CH$_4$)$_1$ relative spectral intensities, rise times, and decay times. The intensities are normalized to that of the 0_0^0 transition. Calculated values are in parentheses.

Cluster excitation	Initially excited decay times (ns)	Product state rise times (ns)	Initially excited state intensity	Product state intensity[a]	
				$\overline{0^0}$	0^0
$\overline{6a_0^1}$	<0.1 (10^{-3})	(160)	0 (0)	20 (20)	1.0 (1.0)
$\overline{15_0^2}$	<0.1 (10^{-3})	.24 (.18)	0 (0)	0 (0)	1.0 (1.0)

(a) States not identified have zero calculated and observed intensity.

corresponding to that level. Energy gap laws are simply generalizations of how vibrational overlaps vary with the energy difference between states. The functional form of the energy gap law used is chosen rather arbitrarily - basically large energy chromophore transitions are not favored.

The assumed energy gap law, in combination with the calculated density of vdW states, determines the calculated energy dependence of IVR transition rates. The calculations predict that for aniline(Ar)$_1$ IVR from the initially excited state is far slower than subsequent IVR transitions. Thus, the calculated final state rise times closely match the initially excited state decay times, in agreement with the experimental results.

VP is predicted to be the rate limiting process for aniline(CH$_4$)$_1$. The calculated VP rate for $\overline{6a}_0^1$ excitation depends strongly on the assumed binding energy, E_0. With an assumed E_0 of 480 cm^{-1} the calculated $\overline{6a}^1$ VP rate is (160 ns)$^{-1}$. The vast majority of emission would be from the cluster $\overline{0}^0$ state and not the bare molecule 0^0 state, in agreement with the observations. The calculated VP rate for $\overline{15}_0^2$ excitation is 100 ps, again in good agreement with the experimental results.

The most severe approximation in this calculation is that the IVR transition matrix P is scaled by a single constant coupling parameter for all modes. This approximation is surely not completely correct; however, the agreement between the experimental and calculated results suggests that this approximation is good to within a factor of about two.

C   Comment on Techniques and Modeling

A word of caution concerning the general experimental techniques for this research is in order. We have consistently found that the most difficult and time consuming portion of this research is *definitely not* the picosecond time resolved measurements as one might naively expect. Time correlated single photon counting dispersed emission measurements are "simply" calibrated by obtaining an accurate instrument response function on low level scattered light at regular intervals and by keeping the overall photon arrival rate below 1000 counts/s. The two most difficult aspects of these studies are the old spectroscopic problems of sample characterization and transition assignments. System characterization is especially important for these kinetic studies because the IVR/VP kinetics are sensitively dependent on the size of the cluster. Thus, if IVR/VP rates in aniline$(CH_4)_1$ are desired, the expansion system must be arranged to keep the concentration of aniline$(CH_4)_n$ (n>1) clusters below the limit of detection. We find typically that for an expansion pressure of 2500 torr of helium carrier gas, both the aniline and methane pressure should be ≤10 torr. If the methane mixing pressure is increased by a factor of five or so, significant distortion in both the emission spectrum and the rise and decay times of the various features is quite pronounced. Moreover, new cluster features appear in the emission spectrum of high methane concentration expansions that can be misinterpreted as arising from aniline$(CH_4)_1$ clusters. If these methane concentration dependent features are incorrectly assumed to arise from the n = 1 cluster, the entire IVR/VP scheme will be misunderstood and the assigned IVR/VP lifetimes will appear to be much different than they actually are. We suspect, for example, that much of the reported tetrazine$(Ar)_1$ kinetic data, which has often been obtained for tetrazine in pure argon expansions, is so distorted.[8,10]

Finally, we comment on the overall description of the dissociation process. In the aniline$(CH_4)_1$ case, chromophore to vdW mode IVR is very fast and VP is the rate controlling process. Thus, both restricted (considering only the vdW mode phase space) and unrestricted or simple (considering both vdW and chromophore mode combined phase space) RRKM theories give the same correct dissociation rate. In the aniline$(Ar)_1$ case, however, IVR is quite slow and it becomes the rate controlling process rather than VP. As we have pointed out above, IVR rates depend on coupling coefficients, density of final receiving states, the amount of energy transferred between chromophore and vdW modes, etc. Consequently, the overall dissociation rate for aniline$(Ar)_1$ is much slower than the simple unrestricted RRKM calculation; in fact, the overall dissociation time (the time between photon absorption and cluster dissociation) for this cluster is unrelated to any RRKM or other statistical unimolecular reaction rate mechanism. Indeed, Table I demonstrates that the total aniline$(Ar)_1$ dissociation rates do not increase monotonically with excitation energy. The kinetics for aniline$(Ar)_1$ dissociation could thereby be called an example of "non-RRKM" or "mode specific" behavior. We believe that such a description is misleading and obscures the physical picture because once energy has been transferred from the chromophore modes to the vdW modes, the dissociation process is indeed a statistical one. That is, IVR amongst the vdW modes is rapid compared to any other kinetics and VP in both clusters can be described by a restricted statistical RRKM model.

317

IV. CONCLUSIONS

The main conclusions of this work can be summarized as follows:

(1) Aniline(Ar)$_1$ and aniline(CH$_4$)$_1$ clusters exhibit much different S$_1$ vibrational dynamics. Aniline(Ar)$_1$ undergoes IVR relatively slowly (nanoseconds), and all energetically accessible bare molecule states are populated by VP. In contrast, aniline(CH$_4$)$_1$ undergoes rapid IVR (<100 ps) and only the 0^0 level is populated by VP from $\overline{0^0}$.

(2) The dynamical differences between aniline(Ar)$_1$ and aniline(CH$_4$)$_1$ can be understood in terms of a serial IVR/VP model. In this model, the rate of IVR is given by Fermi's golden rule, and the rate of VP is given by a restricted RRKM theory with regard to both IVR and VP processes.

(3) The density of vdW vibrational states is the single most important factor in determining the dynamical differences between aniline(Ar)$_1$ and aniline(CH$_4$)$_1$.

ACKNOWLEDGMENT

This work was supported in part by grants from NSF and ONR. This research has been a collaborative effort with Prof. David F. Kelley and postdoctoral fellows M. R. Nimlos and M. A. Young.

REFERENCES

1.a. P. Avouris, W. M. Gelbart and M. A. El-Sayed, Chem Rev. 77, 793 (1977).

b. S. Mukamel, J. Phys. Chem. 89, 1077 (1985).

c. S. Mukamel and J. Jortner, Excited States, Ed. E. C. Lim, (Academic Press, 1977), Vol. III, p. 57.

2.a. D. J. Robinson and K. A. Holbrook, Unimolecular Reactions, (Wiley, 1972).

b. D. F. Kelley and E. R. Bernstein, J. Phys. Chem. 90 5164 (1986).

c. J. I. Steinfeld, J. S. Francisco and W. L. Hase, Chemical Kinetics and Dynamics, (Prentice Hall, 1989).

d. R. D. Levine and R. B. Bernstein, Molecular Reaction Dynamics and Chemical Reactivity, (Oxford, 1987).

e. D. M. Wardlaw and R. A. Marcus, Chem. Phys. Lett. 110, 230 (1984), Adv. Chem. Phys. 70, 231 (1987).

3.a. J. A. Beswick and J. Jortner, Adv. Chem. Phys. 47(pt.1), 363 (1981) and references therein.

b. S. H. Lin, Radiationless Transitions, (Academic, 1980).

4.a. D. V. Brumbaugh, J. E. Kenny and D. H. Levy, J. Chem. Phys. 78, 3415 (1983) and references therein.

b. C. A. Haynam, D. V. Brumbaugh and D. H. Levy, J. Chem. Phys. 80, 2256 (1984).

c. Y. D. Park and D. H. Levy, J. Chem. Phys. 81, 5527 (1984).

5. C. A. Haynam, L. Young, C. Morter and D. H. Levy, J. Chem. Phys. 81, 5216 (1984).

6. N. Mikami, Y. Scigobora and M. Ito, J. Phys. Chem. 90, 2080 (1986) and H. Abe, Y. Okyanagi, M. Iokijo, N. Mikami, and M. Ito J. Phys. Chem. 89, 3512 (1985).

7. N. Halberstadt and B. Soep, Chem. Phys. Lett. 87, 109 (1982) and J. Chem. Phys. 80, 2340 (1984).

8.a T. A. Stephensen, P. L. Radloff and S. A. Rice, J. Chem. Phys. 81, 1060 (1984).

b. T. A. Stephensen and S. A. Rice, J. Chem. Phys. 81, 1083 (1984).

c. P. M. Weber and S. A. Rice, J. Phys. Chem. 92 5470 (1988) and J. Chem. Phys. 88, 1082, 6107, 6120 (1988).

d. B. A. Jacobsen, S. Humphrey and S. A. Rice, J. Chem. Phys. 89, 5624 (1988).

9.a. C. S. Parmenter, J. Phys. Chem. 86 1735 (1982).

b. K. W. Butz, D. L. Catlett, G. E. Ewing, D. Krajnovich and C. S. Parmenter, J. Chem. Phys. 90, 3533 (1986).

c. H. K. O, C. S. Parmenter, and M. C. Su, Ber. Bunsenges Phys. Chem. 92, 253 (1988).

10.a. J. J. F. Ramackers, J. Langelaar and R. P. H. Rettschnick, Picosecond Phenomena III, Ed. K. Eisenthal, R. M. Hochstrasser, W. Kaiser and A. Laubereau (Springer, 1982), p. 264.

b.   M. Heppener, A. G. M. Kunst, D. Bebelaar and R. P. H. Rettschnick, J. Chem. Phys. <u>83</u> 5314 (1985).

c.   M. Heppener, R. P. H. Rettschnick, Structure and Dynamics of Weakly Bound Molecular Complexes, Ed. A. Weber, (Reidel, 1987) p. 553.

d.   J. J. F. Ramackers, L. B. Krijnen, H. J. Lips, J. Langelaar, R. P. H. Rettschnick, Laser Chem. <u>2</u>, 125 (1983).

e.   J. J. F. Ramackers, H. K. van Dijk, J. Langelaar, R. P. H. Rettschnick, Faraday Disc. Chem. Soc. <u>75</u>, 183 (1983).

11.a.   E. R. Bernstein, K. Law and M. Schauer, J. Chem. Phys. <u>80</u>, 107, 634 (1984).

b.   M. Schauer, K. Law and E. R. Bernstein, J. Chem. Phys. <u>81</u>, 49 (1984).

c.   M. Schauer, K. Law and E. R. Bernstein, J. Chem. Phys., <u>82</u>, 726, 736 (1985).

12.   D. D. Smith, A. Lorincz, J. Siemion and S. A. Rice, J. Chem. Phys. <u>81</u>, 2295 (1984).

13.   J. L. Knee, L. R. Khundkar, and A. H. Zewail, J. Chem. Phys. <u>82</u>, 4715 (1985).

14.   M. R. Nimlos, M. A. Young, E. R. Bernstein and D. F. Kelley, J. Chem. Phys. to be published.

15.   D. A. Chernoff and S. A. Rice, J. Chem. Phys. <u>70</u>, 2521 (1979) and references therein.

16.   G. E. Ewing, J. Phys. Chem. <u>91</u>, 4662 (1987).

17.   V. E. Bondybey and L. E. Brus, Adv. Chem. Phys. <u>41</u>, 269 (1980).

18.   S. E. Stein and B. S. Rabinovitch, J. Chem. Phys. <u>58</u>, 2438 (1973).

State to state in van der Waals complexes

C. Jouvet

laboratoire de photophysique moléculaire du C.N.R.S.
batiment 213, Université de Paris Sud, Orsay

1) Non reactive systems

α)Rotational predissociation

$$(AB(J_1)...C)(E,v,J) \quad + \quad h\nu \quad ---> \quad AB(E,v,J_1') \quad + \quad C$$

The dissociation induced by the excitation of high hindered rotation levels above the dissociation limit leads to the slow predissociation of the complex on a single energy surface. This has been observed in two systems $Ar-N_2$, and $Hg-N_2$. No measurement and calculations on test system have been done yet in order to determine the product state distribution or the time evolution. They would be very interesting in order to test the quality of the potential energy surface and of the approximations used to treat the very high vibrational levels.

β)Vibrational predissociation

a) atom diatom system

i) $AB(E,v)...C \quad ---> \quad AB(E,v',J) \quad + \quad C$

The problem is quite well understood and experimentally well characterized. Still, the quality of the potential energy surface is the key to match theory and experiments. The limitations are how to obtain from the experimental data the potential energy surface. The whole problem is still open when the interaction is strong, and when the intramolecular Vibrational Relaxation is the first step of the dissociation $(Ar-Cl_2, \quad Xe-Cl_2)$.

Dynamics of Polyatomic Van der Waals Complexes
Edited by N. Halberstadt and K. C. Janda
Plenum Press, New York, 1990

ii) $AB(E > E_{diss})...C$ ---> $AB(E,v,J')$ +   C

The Cage effect is the object of theoretical investigation. But there is still the need for experimental data on a single energy surface in a one to one complex, since all the present observations cannot exclude an electronic predissociation or the presence of larger clusters (I_2-Rare gas).

b) polyatomic-rare gas atom

$A...M(E,v_i)$ ---> A + $M(E,v_j,j)$

The role of the I.V.R to explain the selection rules within the product in the dissociation of such complexes is the subject of large discussions and need further experimental and theoretical work. It is yet not clear when and how statistical theory should be used.

c) diatom diatom: AB-CD

Here the HF dimer is the key system and the quality of the potential energy surface used is the limiting step to clearly reproduce the experimental data. In the calculation of the vibrational predissociation one can expect some new progress by applying in this field the wave packet propagation approach.

Γ) Electronic predissociation.

It seems that this field is not as well studied experimentally as well as theoretically as the previous ones. Some experimental evidences of such processes (ICl, Glyoxal-rare gas) has been observed but no theory can explain them quite clearly yet.

2) Reactive system

$(A...BC)(E,v,J)$ ---> $AB(E',v',J')$ + C

In chemical reactions the van der Waals systems are very promising to orient the energy distribution ($Xe-CL_2$, $HI-CO_2$) of the products or even the nature of the products (fluorobenzene-NH_3) due to orientation effects in the entrance channel of the reaction. One can expect also some selective chemistry using a precise and selective optical excitation.

THE SOLUTION OF THE BOUND STATE NUCLEAR MOTION PROBLEM FOR

POLYATOMIC CLUSTERS

Jonathan Tennyson

Department of Physics and Astronomy
University College London
Gower Street, London WC1E 6BT, U.K.

INTRODUCTION

The last decade has seen significant advances in our understanding of small, in particular triatomic, molecules and clusters. It is now possible to use high-resolution spectroscopic data to obtain detailed and accurate potential energy surfaces of atom-diatom Van der Waals systems by successive refinement of a parameterised potential function. The chapter by Hutson in this book gives a discussion of these methods.

A crucial step between any proposed potential and the experimentally-observed properties of a system is the characterisation of the nuclear motion. For clusters this step is particularly demanding because the concepts of equilibrium geometry and small amplitude motion, which have proved so fruitful in the analysis of the spectra of chemically-bound molecules, are almost always invalid. This means that methods based solely on perturbation theory are also unlikely to be successful unless a new and reliable zeroth order model for such systems can be developed.

Rather than using perturbation theory, the methods of choice for triatomic clusters are based on the application of the variational principle using either basis sets or a mixture of basis sets and direct numerical integration. Documented computer packages[1,2] are available for such calculations. However, the increased number of degrees of freedom available to polyatomic Van der Waals clusters makes it doubtful whether the same methodologies will be directly applicable to these larger systems with current computer technology.

There are really two problems associated with directly transferring current triatomic methods. Firstly the increased dimensionality of any numerical quadrature required. For example going from a 3 atom to 4 atom cluster in principle increases the numerical quadrature from 3 to 6 dimensional or from 10^3 to 10^6 points if a modest 10 points per mode is assumed. In traditional basis set methods, potential energy integrals are required over all dimensions of the problem. Integrals over the potential have proved to be a severe limitation on variational calculations of even very low-lying vibrational levels of chemically-bound tetratomic systems[3].

Besides difficulties with integrals, the final matrix diagonalisation

Dynamics of Polyatomic Van der Waals Complexes
Edited by N. Halberstadt and K. C. Janda
Plenum Press, New York, 1990

or set of coupled equations also suffers from scaling of order N^3. Thus for coupled-channels approaches, the number of channels that need to be considered rises too rapidly for these to be directly applicable to all but the simplest problems[4]. Similarly the size of the secular matrix in conventional basis set approaches has so far limited the application of these to larger systems - in the case of Van der Waals clusters almost entirely to systems with high symmetry[5].

Recently there has been a proliferation of new methods proposed for the study of nuclear dynamics. While undoubtedly some of these will not prove practicable, many show great promise. I have taken this opportunity to discuss those which I believe to be the most likely to be useful for the problem of polyatomic clusters.

VIBRATIONAL METHODS

The following selection of techniques is not designed to be comprehensive and they are introduced in no particular order of priority. Rather I have attempted to consider several classes of methods together.

Finite element methods

A number of novel techniques based on finite elements have come to forefront recently. The characteristic of these methods is that they work in coordinate rather than function space. These methods have the advantage that the easy correlation between the potential and points used in the calculation can yield considerable physical insight. This insight can be used both for interpretation and for adapting the method to a particular problem of interest.

Currently, the most widely used finite element method for nuclear motion calculations is the Discrete Variable Representation (DVR) of Light and co-workers[6]. This method relies formally on the use of polynomial basis functions which are used to define Gaussian quadrature grids upon which the calculation is performed. At some levels formal equivalences can be shown between basis set and DVR calculations, but the power of the DVR method comes from the hierarchy of diagonalisations and truncations which can be rigorously defined. These lead to final secular matrices of greatly reduced dimensions. The DVR method has proved very powerful for obtaining large numbers of states for a particular problem,[6-8] but it remains to be seen how exactly the method performs on problems of higher dimensionality.

A related method is the collocation method of Peet and Yang[9]. This method circumvents the integration problem and has the attraction of great simplicity. Indeed Peet solved the Ar-HCl vibrational problem in 150 lines of FORTRAN! The method, however, appears intrinsically less accurate than the DVR method.

Other finite element methods are available, including ones which employ fast Fourier Transform techniques. However these have been applied more to time-dependent problems and to my knowledge these have yet to be applied to multi-dimensional vibrational calculations.

Novel basis functions

Obviously any basis set method is heavily reliant on the choice of appropriate expansion functions. Conventional vibrational basis set have usually been constructed from products of one-dimensional expansions of orthogonal polynomials. In particular Hermite or associated Laguerre polynomials have often been used to represent stretching coordinates, and (associated) Legendre polynomials or spherical harmonics used to carry the bending motion. In the spirit of these, the use of hyperspherical coordinates has led to the use of hyperspherical harmonics, Chebyshev polynomials and series of trigometric functions as suitable expansion functions[8].

Recently Hamilton and Light[10] used a distributed Gaussian basis (DGB) to represent vibrational motion. These functions have been widely used in electronic structure calculations and have the advantage of very simple integration properties, and are amenable to very low-order Gauss-Hermite quadrature. Their spatial localisation means that DGBs can be distributed as dictated by the physics of a particular problem. The disadvantage of DGBs is that they are non-orthogonal and that slightly more of them are usually required than orthogonal polynomials which are well adapted to the problem[11]. Of course well adapted polynomials may not exist for a particular problem. DGBs and polynomial basis sets are both currently widely used.

Co-ordinate systems and separability

For triatomic molecules many coordinate systems have been used to represent the vibrational motions, see ref. 12 for example. For polyatomic clusters it is possible to imagine a large number of possible coordinate systems and a similar proliferation of Hamiltonians. In this context it should be noted that the derivation and application of Hamiltonians in arbitrary coordinates is far from simple. The choice of an objectively inferior coordinate system for technical reasons is thus a common occurrance. For polyatomic Van der Waals dimers however, so-called scattering coordinates based upon the interaction coordinate of the two monomers and associated angles of orientation would appear a natural choice. A general, body-fixed Hamiltonian for these coordinates has already been derived[13].

Although the success of all methods is to some extent dependent on the suitable choice of coordinates, methods which depend in some fashion on mode separability rely particularly heavily on the choice of an optimal coordinate system. For example, methods based on the self-consistent field (SCF) or Mean Field approximation can yield highly accurate results with the right coordinates which are rapidly degraded by a poor choice. Further discussion of this problem can be found in the chapter by Gerber.

Similarly adiabatic or Born-Oppenheimer separation schemes, such as the BOARS method[14], rely heavily on the choice of good coordinates. Such schemes are also adversely affected by the presence of Fermi resonances which are quite common in Van der Waals systems. In both SCF and BOARS-type calculations it is possible to account for the inter-mode correlated motions by performing configuration interaction calculations using the original solutions as the "orbital" basis.

The BOARS methods work by solving for the angular motions and hence defining adiabatic radial potentials. A complementary approach, which approximates the radial motions to yield effective angular potentials, has been widely and successfuly used by Bunker and co-workers[14] to invert spectroscopic data on semi-rigid molecules with one large amplitude coordinate. Although the method allows the large amplitude coordinate to be defined in an arbitrary fashion, it seems unlikely that this method will be applicable to Van der Waals complexes without explicitly coupling further degrees of freedom.

An alternative way of using co-ordinates separately is to develop compact basis sets by piece-wise diagonalisation of split (usually 1D) problems. This method has been widely used by Carter and Handy[16] for chemically-bound system. It can be seen that the efficiency of this method is obtained in a manner similar to that of the DVR approach described above.

(Semi)-classical methods.

Although the problems under discussion here are essentially quantal in nature, classical mechanics can still be used to give insight into a particular problem, see the chapter by Gray in this book for example.

Apart from direct classical trajectory calculations, there are two other ways in which classical mechanics can be used to drive quantum calculations. The first is the use of mixed mechanics which is particularly appropriate for systems with a clear quantum/classical partitioning, for example on grounds of mass. The second is the use of classical trajectories in distributing localised basis functions (such as distributed Gaussians) or in determining finite element grids. Such a proposal has been recently made by Huber and Heller[17]. Although this method has yet to be widely used (see ref. 18 for an example) it would appear to have great promise in guiding quantum calculations to regions of importance for spectroscopy or other processes.

OTHER CONSIDERATIONS

Rotational motion

In traditional quantum mechanical calculations the inclusion of rotational motion, particularly to high orders, has proved difficult because of the consequent increase in the size of the secular matrix. For Van der Waals systems, the so-called helicity approximation has often been invoked. This approximation assumes that the projection, k, of the total angular momentum, J, onto some body-fixed z-axis is a constant of motion for the system. The approximation is often good for energy levels but is less reliable for properties such as transition intensities.

For triatomic systems Tennyson and Sutcliffe[2,19] have developed an accurate two-step variational procedure which goes beyond, and removes the need for, the helicity approximation. This method uses the solutions of the J + 1 unique fixed-k Hamiltonians as basis functions to expand the full Hamiltonian for the problem. For systems where k is indeed nearly conserved this second step to the calculation is rapidly convergent and can be performed at very little extra cost.

This two-step formalism has recently been adapted to DVR-based methods[8,11]. No difficulty with its extension to polyatomic clusters is anticipated provided that one works in a body-fixed frame and makes a sensible choice for the orientation of the body-fixed z-axis.

Novel computer architectures

Many of the codes discussed above have been adapted to vector processing machines. Both secular-equation and coupled-channels techniques are well adapted to vectorisation because of the ease with which the problems can be expressed in matrices.

The advent of parallel processing raises other interesting possibilities in the dynamics of Van der Waals clusters. In particular, the multidimensional integration bottleneck alluded to in the introduction comprises largely of evaluation of the potential energy at a very large number of quadrature points. Such calculations are readily performed in parallel. Similarly, many finite element-based methods can naturally be placed on machines with one node performing all the operations associated with one or a batch of grid points.

The potential

Above the possible applicability of a number of techniques to the challenging problem of the ro-vibrational bound states of polyatomic Van der Waals clusters have been considered. Of course to do such calculations it is necessary to have an appropriate potential function.

The generation of potential energy surfaces is a demanding task. For example the characterisation of a full tetratomic surface may require several thousand grid points if this is to be done ab initio or a large set of parameters to be adjusted if an empirical surface is to be constructed. Methods for the generation of potential energy functions are discussed elsewhere in this volume. I would simply like to comment that it is my belief that techniques including the ones discussed above are already available for performing nuclear motion calculations of an accuracy appropriate to the potentials currently available.

CONCLUSIONS

These are exciting times for those of us interested in the dynamics of nuclei. For many years theory and much cluster spectroscopy have been chiefly concerned with elucidating the structural parameters of complexes. This situation no longer exists and a wealth of data is being generated, leading to a much greater insight into potential energy functions.

These improved potentials have provided a stimulus to the dynamicists and have led directly to the rapid increase in methods available to study the nuclear motion problem. Above I have considered some of these methods with particular reference to those appropriate for the large amplitude motion found in Van der Waals complexes and problems of increased dimensionality encountered in polyatomic clusters. It is hoped that the proponents of methods that I described as promising prove me right and the proponents of methods which I have expressed doubts about prove me wrong.

Acknowledgements

I would like to thank the organisers for inviting me to participate in their excellent workshop and Jeremy Hutson for helpful discussions on the techniques described here.

REFERENCES

1. J.M. Hutson, BOUND computer code.
2. J. Tennyson, Computer Phys. Comms. 42:257 (1986); J. Tennyson, Computer Phys. Rep. 4:1 (1986); J. Tennyson and S. Miller, Computer Phys. Comms. in press (1989).
3. S. Carter and N.C. Handy, Chem. Phys. Lett., 79:118 (1981).
4. G. Danby, J. Phys. B: At. Mol. Phys. 16:3393 (1983); G. Danby, J. Phys. B: At. Mol. Opt. Phys. 22:1785 (1989).
5. J. Tennyson and A. van der Avoird, J. Chem. Phys. 77:5664 (1982); G. Brocks and A. van der Avoird, Mol. Phys. 55:11 (1985).
6. J.C. Light, I.P. Hamilton and J.V. Lill, J. Chem. Phys. 82:1400 (1985); Z. Bacic and J.C. Light, Ann. Rev. Phys. Chem. 40: (1989).
7. R.M. Whitnell and J.C. Light, J. Chem. Phys. 90:1774 (1989).

8.   J. Tennyson and J.R. Henderson, J. Chem. Phys. in press (1989).
9.   A.C. Peet and W. Yang, Chem. Phys. Lett., 153:98 (1988); A.C. Peet
       and W. Yang, J. Chem. Phys. 90:1746 (1989).
10. I.P. Hamilton and J.C. Light, J. Chem. Phys. 82:1400 (1985).
11. S.E. Choi and J.C. Light, J. Chem. Phys. in press (1989).
12. B.T. Sutcliffe and J. Tennyson, Mol. Phys. 58:1053 (1986).
13. G. Brocks, A. van der Avoird, B.T. Sutcliffe and J. Tennyson, Mol. Phys.
       50:1025 (1988).
14. J.M. Hutson and B.J. Howard, Mol. Phys. 41:1123 (1982); R.J. Le Roy,
       G.C. Corey and J.M. Hutson, Faraday Disc. Chem. Soc. 73:339 (1982).
15. P. Jensen and P.R. Bunker, J. Mol. Spectros. 118:18 (1986); P. Jensen,
       Computer Phys. Rep. 1:1 (1983) and references therein.
16. S. Carter and N.C. Handy, Computer Phys. Comms. 51:49 (1988).
17. D. Huber and E.J. Heller, J. Chem. Phys. 89:4752 (1988).
18. J.M. Gomez Llorrente, J. Zakrzewski, H.S. Taylor and K.C. Kulander,
       J. Chem. Phys. 90:1505 (1989).
19. J. Tennyson and B.T. Sutcliffe, Mol. Phys. 58:1067 (1986).

THE CALCULATION OF INTERMOLECULAR

POTENTIAL ENERGY SURFACES

A. J. STONE

University Chemical Laboratory
Lensfield Road
Cambridge CB2 1EW
England

ABSTRACT: It is now possible to carry out accurate *ab initio* calculations on molecular complexes by a variety of techniques. The supermolecule approach is widely used, and is capable of high absolute accuracy, but it is subject to Basis Set Superposition Error, especially when electron correlation is taken into account, and this is a difficulty when accurate calculations of small interaction energies are required. Perturbation theory is not subject to BSSE, but perturbation methods as currently implemented are 'uncoupled'; that is, the response of the electrons to the perturbation is not treated self-consistently. Nevertheless this method gives a more detailed description of the interaction than the supermolecule approach, and consequently provides more physical insight into the nature of the interaction. Both of these methods require calculations to be carried out at a wide range of dimer geometries if a full description of the potential energy surface is needed, and this is extremely time-consuming.

A useful alternative approach is to isolate the components of the perturbation expansion, namely the repulsion, electrostatic interaction, induction, and dispersion terms, and to calculate each of them independently by the most appropriate technique. Thus the electrostatic interaction can be calculated accurately from distributed multipole descriptions of the individual molecules, while the induction and dispersion contributions may be derived from molecular polarizabilities. This approach has the advantage that the properties of the monomers have to be calculated only once, after which the interactions may be evaluated easily and efficiently at as many dimer geometries as required. The repulsion is not so amenable, but it can be fitted by suitable analytic functions much more satisfactorily than the complete potential. The result is a model of the intermolecular potential that is capable of describing properties to a high level of accuracy.

Dynamics of Polyatomic Van der Waals Complexes
Edited by N. Halberstadt and K. C. Janda
Plenum Press, New York, 1990

A natural way to calculate the interaction energy U_{AB} between two molecules A and B is to compare the energy of the complex $A \cdots B$ with the energies of the separated molecules:

$$U_{AB} = W_{AB} - W_A - W_B. \tag{1}$$

This is the *supermolecule method*. It has a number of obvious attractions: it is very easy to understand conceptually, and it is easy to apply, using standard *ab initio* computational techniques. Unfortunately it has some less obvious limitations. The small interaction energy U_{AB} is calculated as a difference of much larger quantities, and this is a serious source of inaccuracy. We are all aware of the need to avoid calculating a small quantity as a difference of large ones for reasons of numerical accuracy, but the problem here is not a numerical one; it is that all of the energies involved are subject to error, because of the approximations made in the calculation, and small relative errors in any of the energies lead to very large percentage errors in the interaction energy. Moreover the errors in W_{AB} do not cancel with the errors in W_A and W_B. Consequently the calculations must be carried out at a high level of accuracy if the results are to be of any value. However this means that a large basis set is needed, and since the computer time required for *ab initio* calculations increases roughly with the fourth power of the number of basis functions, these calculations are very expensive.

Furthermore, the calculation has to be repeated at a large number of relative configurations of the two molecules if the potential surface is to be explored adequately. Six coordinates are required to specify the position and orientation of molecule B relative to molecule A, if both are non-linear. If the calculation is repeated for only four values of each coordinate (a very inadequate number) a total of $4^6 = 4096$ points will be required. There are many examples in the literature of calculations of intermolecular potential energy surfaces which use a wholly inadequate strategy for choosing points on the surface. It is common, for example, even in calculations on pairs of diatomics, where there are only three angular coordinates, to perform calculations at only a few relative orientations: often the linear, rectangular, T-shaped and crossed configurations only. Even if calculations are done for a large number of intermolecular distances, it is impossible to characterise the potential energy surface adequately from such a limited number of orientations. Fortunately it is possible to explore the surface much more efficiently[1], but a large number of points is still required.

Electron correlation and size consistency

It is also necessary to include electron correlation in any accurate calculation of intermolecular interactions, since the dispersion interaction is a correlation effect, and is usually an important component of the interaction energy. Moreover the dispersion interaction describes correlated fluctuations of the charge distributions in the two molecules, and this calls for an accurate description of the molecular polarizability, which in turn requires a large basis set containing a large number of diffuse polarization functions.

In the commonly-used singles-and-doubles CI method (abbreviated as CISD), the wavefunction is a linear combination of configurations which includes every configuration derived by one-electron or two-electron excitation from the 'root configuration' Ψ_0. Suppose that this method is used to obtain the energies in eq. (1). The energy W_{AB} then takes into account all single and double excitations. However the energy of W_{AB} at infinite separation is the sum of W_A and W_B, and if each of these is obtained by CISD, the wavefunction for infinitely-separated $A \cdots B$ includes configurations in which both A and B are doubly excited, so that $A \cdots B$ is quadruply excited. Thus $A \cdots B$ is described at short range by a wavefunction that contains only single and double excitations, but at long range by a wavefunction that includes higher excitations. The long-range energy is then variationally better than the energy at short range, and the strength of the interaction will be under-estimated.

To avoid this problem, it is necessary to ensure that the CI method is *size consistent* or *size extensive:* These two terms have slightly different meanings, but both require that the energy calculated for an assembly of molecules at infinite separation be the same as the sum of the energies of the individual molecules calculated separately. We have seen that the CISD method does not satisfy this requirement, but there are methods which do; they include Møller-Plesset perturbation theory and a variety of 'coupled-cluster' methods[2,3]. CEPA (the Coupled Electron Pair Approximation) is an approximation to the latter.

Basis set superposition error

The need for size-consistency is an illustration of a fundamental difficulty of the supermolecule method. For a single system the variation principle ensures that the better the calculation, the better the energy. When we are interested in energy *differences*, however, the variational principle does not apply. In fact, there is a very serious source of error that can be attributed to the variational principle itself.

We choose a basis for the calculation of molecule A that seems suitable for our purpose. Inevitably it is incomplete, so our wavefunction for A is not exact. The same is true of the calculation for the isolated molecule B. Now we carry out a calculation on the complex $A \cdots B$, and obtain a new energy that includes the interaction between the molecules. But in this calculation there are some new basis functions, belonging to molecule B, that were not present when we calculated the energy of the isolated A molecule, and they allow the wavefunction of molecule A to be improved variationally, so that its energy falls. This is quite separate from the true physical interaction; it is a spurious effect that occurs because the basis set that we initially chose for A was not good enough to describe the wavefunction exactly. This is *Basis Set Superposition Error*, or BSSE, and it occurs in all supermolecule calculations. The effect is to make the interaction seem more attractive than it really is.

This is an extremely troublesome problem. The standard way to deal with it is the *counterpoise correction* proposed by Boys & Bernardi[4]. The reference energy for A is calculated in the presence of the basis functions (but not the electrons or nuclei) of molecule B, and similarly the reference energy for B is calculated in the presence of the A basis functions. In this way, the variational improvement that arises from the presence of the 'foreign' basis functions is included in the reference calculation as well as in the $A \cdots B$ calculation. (The basis functions of the other molecule are sometimes called 'ghost' orbitals when used like this.) An obvious problem is that if we change the relative positions of the two molecules in the $A \cdots B$ complex, then we change the position of the B basis functions relative to A (and *vice versa*) so that the counterpoise-corrected calculation of the reference energies has to be repeated for every point on the surface. Clearly this increases the expense of the potential surface calculation by a factor of about 3.

It has been pointed out that in the $A \cdots B$ complex some of the orbitals of B are not available to A, because they are occupied by the electrons of B, and it has been argued that in the counterpoise calculation these orbitals should be excluded. There has been a great deal of controversy about this point[5].

It is also evident that if a basis is used that is not capable of describing the isolated molecules correctly, then *a fortiori* it will give the wrong interaction energy, even if the effects of BSSE could be corrected precisely. It follows that the ideal solution to the problem is not to find the best procedure for correcting for BSSE, but to perform the calculation in such a way that it does not arise in the first place. This requires that a sufficiently good basis set be used for the individual molecules, so that the basis functions of the other molecule cannot provide any variational improvement. Unfortunately this is not realistic even for the smallest systems at present. However it is currently possible to reduce the BSSE in SCF calculations on molecules containing first-row atoms to about $10\,cm^{-1}$ by using a very good basis[6]. Unfortunately the limit in correlated calculations appears to be an order of magnitude larger[5], in the region of $100\text{--}200\,cm^{-1}$. One might hope that the residual error after correcting for BSSE might be smaller than this, but Szalewicz *et al.*[7] estimate the uncertainty in the total interaction energy of the water dimer, including correlation effects and allowing for BSSE, to be about $\pm 120\,\mathrm{cm}^{-1}$. In that case the well depth is in the region of $1750\,\mathrm{cm}^{-1}$, but many van der Waals molecules are more weakly bound, so that the uncertainty arising from BSSE may be as large as the well depth.

There is a further difficulty that currently limits the accuracy of intermolecular potential calculations, even on small systems where good basis sets can be used and the BSSE correction is relatively small. If the basis set on each molecule is very good, then the basis for the complex may become nearly over-complete; that is, the basis functions are not all independent. Numerical instabilities then ensue. The difficulty seems to arise from the use of atom-centred basis functions. In order to describe the complicated distortions that may result from the presence of another molecule, it is necessary to use basis functions that are very diffuse and have high angular momentum quantum numbers, and they overlap strongly with the corresponding functions on the other molecule, especially at very short range. A better description might result from the use of basis functions centred in the region between the interacting molecules, but no way has been found so far to do this consistently.

INTERMOLECULAR PERTURBATION THEORY

If supermolecule calculations are so troublesome, perhaps perturbation theory can provide a more satisfactory approach. It is often assumed that perturbation theory is a relatively crude approach to the calculation of intermolecular potentials, but there is no reason to believe that it is intrinsically any less accurate than the supermolecule approach. The attraction of perturbation theory is that it leads directly to an expression for the interaction energy itself. Moreover it is possible to analyse the expression into separate terms that can be correlated with distinct physical effects. We can then refine each of these terms, expressing each of them in terms of properties of the individual molecules that can be calculated much more accurately than properties of the complex.

Long-range perturbation theory—first order

For two molecules, sufficiently far apart, the Hamiltonian for the system can be written as

$$\mathcal{H} = \mathcal{H}_A + \mathcal{H}_B + V, \tag{2}$$

where $\mathcal{H}_A$ is the Hamiltonian for the isolated molecule A, $\mathcal{H}_B$ the Hamiltonian for molecule B, and V is the interaction. V arises simply from the electrostatic interaction between the particles of A and those of B:

$$V = \sum_{i\in A, j\in B} \frac{e_i e_j}{4\pi\epsilon_0 r_{ij}}, \tag{3}$$

where e_i and e_j are the charges on particles i and j, and r_{ij} is the distance between them.

If the eigenfunctions of $\mathcal{H}_A$ are ψ_m^A and those of $\mathcal{H}_B$ are ψ_n^B, and if there is no overlap between the two sets of functions, then the unperturbed wavefunctions for $A \cdots B$ are $|mn\rangle = \psi_m^A \psi_n^B$. Ordinary Rayleigh-Schrödinger perturbation theory then gives the first-order energy of interaction as

$$W^{(1)} = \langle 00|V|00\rangle \tag{4}$$

$$= \int \frac{\rho^A(\mathbf{r}_1)\rho^B(\mathbf{r}_2)}{r_{12}}\, d^3 r_1\, d^3 r_2, \tag{5}$$

which is just the classical electrostatic energy of interaction between the charge distributions of A and B. It is easily evaluated if the ground state wavefunctions are known, but the calculation requires a knowledge of all the intermolecular electron repulsion integrals and is fairly time-consuming. If we want to map out the potential energy surface, then it is necessary to repeat the calculation at a large number of relative configurations, just as in the supermolecule method, and this would require an inordinate amount of computer time if it were to be done accurately. The electrostatic interaction is quite sensitive to the effects of electron correlation, because the individual molecular charge distributions are modified quite significantly when electron correlation is taken into account.

It is often useful to replace the operator (3) by its *multipole expansion*:

$$V = \sum_{tu} \tilde{Q}_t^A T_{t\,u}^{AB} \tilde{Q}_u^B, \tag{6}$$

where $\tilde{Q}_t^A$ is an operator for one of the multipole moments of molecule A (charge, dipole, quadrupole, etc.) and $T_{t\,u}^{AB}$ is an interaction function. The index t is an abbreviation for the angular momentum labels lk, and the moment operator $\tilde{Q}_{lk}^A$ is defined by

$$\tilde{Q}_{lk}^A = \sum_{i \in A} e_i r_i^l C_{lk}(\theta_i, \phi_i), \tag{7}$$

where r_i, θ_i and ϕ_i are the spherical polar coordinates of particle i in a local coordinate system for molecule A, and C_{lk} is a modified spherical harmonic:

$$C_{lk}(\theta, \phi) = \left(4\pi/(2l+1)\right)^{\frac{1}{2}} Y_{lk}(\theta, \phi).$$

The derivation of (6) is straightforward[8,9], and is not given here.

If now we use the multipole expansion of the perturbation in the calculation of the first-order energy, eq. (4) becomes simply

$$U_{es}^{AB} = \sum_{tu} Q_t^A T_{t\,u}^{AB} Q_u^B, \tag{8}$$

in which Q_t^A is the expectation value of one of the multipole moments of molecule A, referred to local axes, and the interaction function $T_{t\,u}^{AB}$ takes into account the orientation dependence of the interaction as well as the distance between the molecules. The interaction is now described in terms of monomer properties (the multipole moments Q_t^A and Q_u^B) which need to be calculated only once, and so can be calculated at a much higher level of accuracy than we could contemplate for calculations on the complete complex. To obtain the interaction energy at any arbitrary configuration it is now necessary only to calculate the interaction functions $T_{t\,u}^{AB}$ for that orientation, which is a trivial computation.

We note in passing that it is possible, and much more useful for practical calculations, to use an alternative approach in which the indices t and u refer to real multipole moments, rather than the complex ones defined by (7). The moments are now denoted Q_{lkc} and Q_{lks}, defined, for $k > 0$, by

$$Q_{lkc} = \sqrt{\tfrac{1}{2}}\left[(-1)^k Q_{lk} + Q_{l,-k}\right],$$
$$iQ_{lks} = \sqrt{\tfrac{1}{2}}\left[(-1)^k Q_{lk} - Q_{l,-k}\right]. \tag{9}$$

No transformation is needed for Q_{l0}, which is always real. The notation reflects the fact that Q_{lkc} transforms like $\cos k\phi$ and Q_{lks} like $\sin k\phi$. The factors of $\sqrt{1/2}$ ensure that a rotation of axes induces an orthogonal transformation of the moments. The first few of these moments coincide precisely with the Cartesian charge and dipole moment:

$$Q_{00} = \sum_i e_i,$$
$$Q_{10} = \sum_i e_i z_i = \mu_z, \qquad Q_{11c} = \sum_i e_i x_i = \mu_x, \qquad Q_{11s} = \sum_i e_i y_i = \mu_y, \tag{10}$$

and later ones describe the quadrupole, octopole and so on. A complete list is given in ref. 8 for moments up to hexadecapole. The functions $T^{AB}_{t\,u}$ for this formulation have been tabulated[8,10] for all multipole-multipole interactions up to terms in R^{-5}.

Although the expression (8) is accurate at sufficiently large separation, it converges only if the separation between the molecules is large compared with their size. This is not the case for any pair of molecules at the separations found in condensed phases or in weakly bound complexes. Fortunately this problem is easily overcome, provided that the molecular charge distributions do not overlap, by using a *Distributed Multipole Expansion*, which assigns charges, dipole moments, etc., to *regions* comprising single atoms or (in larger molecules) small groups of atoms. For small molecules it is helpful to treat the bonding regions separately. The electrostatic perturbation then takes the form

$$V = \sum_{ab} \sum_{tu} \widehat{Q}^a_t T^{ab}_{t\,u} \widehat{Q}^b_u, \tag{11}$$

where $\widehat{Q}^a_t$ is the operator for one of the multipole moments of region a of molecule A. The electrostatic energy becomes

$$U^{AB}_{es} = \sum_{ab} \sum_{tu} Q^a_t T^{ab}_{t\,u} Q^b_u. \tag{12}$$

In defining these regional moments it is necessary to specify an origin for each region, and the origin for region a is known as 'site a'. In the Distributed Multipole approach we replace the extended charge distribution of region a by a set of point multipoles at site a, rather than replacing the charge distribution of the entire molecule by a set of point multipoles at the molecular origin.

The Distributed Multipole description of a molecular charge distribution is not unique; it depends on the coice of sites and on the precise definition of the region boundaries. (The conventional single-site multipole description is not unique either, since it depends on the choice of origin[11].) There are many ways of determining these distributed multipole moments[12-17]; many authors have used distributed charges alone, but it is now widely accepted that an accurate and efficient description requires multipoles up to at least quadrupole.

In this way we reduce the elaborate and time-consuming evaluation of the expression (5) to the very much simpler calculation of (8). This is only valid, though, if the molecular wavefunctions do not overlap; if they do, the problem becomes more difficult.

Perturbation Theory at Short Range

The trouble with perturbation theory when the intermolecular distance is short is not so much that the perturbation becomes too large; for configurations that are accessible at thermal energies the interaction is usually still very small compared with the separation between electronic energy levels, so perturbation theory should still converge rapidly. The problem arises from the overlap of the wavefunctions, and the consequences for perturbation theory are profound. In the first place, it becomes impossible to distinguish between electrons that 'belong' to molecule A and electrons that 'belong' to B. It is therefore no longer possible to separate the Hamiltonian for the entire system in the manner of eq. (2), and there is no satisfactory way to define a perturbation operator representing the interaction between the molecules. In the second place, the wavefunctions for the unperturbed system cannot be taken to be simple products of the form $\psi^A_m \psi^B_n$; they have to be antisymmetrized with respect to all electron permutations. A more serious difficulty is that whether antisymmetrized or not, the functions $\psi^A_m \psi^B_n$ are not orthogonal to each other. Ordinary Rayleigh-Schrödinger perturbation theory assumes that the unperturbed Hamiltonian has a complete set of eigenfunctions, and that they are orthogonal.

Many versions of perturbation theory have been proposed to overcome these problems. A large number of them rely on an expansion of the perturbation equations in powers of the overlap between the functions on A and those on B. This approach appears to work when small basis sets are used, but as the basis is improved, the overlap between the functions on the two molecules becomes larger, and the expansion ceases to converge. This failure of the overlap expansion occurs with quite modest basis sets. Accordingly it is necessary to use a method that deals explicitly with the natural non-orthogonal basis functions for the problem.

If we have to use non-orthogonal wavefunctions, then the natural one-electron orbitals in which to express them are the SCF molecular orbitals of the non-interacting molecules. From these we can construct antisymmetrized (determinantal) wavefunctions in which some orbitals of each molecule are occupied. Because of the non-orthogonality of the orbitals, these determinantal wavefunctions will also be non-orthogonal. It is possible to construct a perturbation theory in which the wavefunction is expanded in terms of these determinants. Fortunately it is possible to formulate it in such a way that the separation of the Hamiltonian into an unperturbed part and a perturbation is unnecessary. The resulting Intermolecular Perturbation Theory (IMPT)[18] has been incorporated into the Cambridge Analytical Derivatives Package (CADPAC)[19].

The first-order perturbation energy in the short-range theory can be separated into two parts: the electrostatic energy and the exchange-repulsion energy. The formal expression for the electrostatic interaction is still eq. (5) at short range, but because the charge densities overlap, it can no longer be expressed completely in terms of a multipole or distributed-multipole expansion. There are now additional terms that describe the effects arising from the interpenetration of the charge distributions. This *penetration* effect is illustrated by the simple case of a He^+ ion interacting with a proton. The electrostatic interaction can be evaluated explicitly in this case, and is

$$U_{es}(He^+ \cdots H^+) = \frac{1}{R} + \left(Z + \frac{1}{R} \right) e^{-2ZR},$$

where R is the separation. Here we can distinguish two terms: the 'multipolar part' $1/R$ which is the classical repulsion between two unit charges at distance R, and the 'penetration term' $(Z + 1/R) \exp(-2ZR)$ (where $Z = 2$ is the helium nuclear charge) which describes the modification to the multipolar expression that arises from the penetration of the proton within the electronic charge distribution of the He^+. We see that the latter decays exponentially with separation. It is often said that the multipole expansion of the complete potential is an *asymptotic* expansion in $1/R$, because it is impossible to find a convergent series in $1/R$ for an exponential e^{-aR}. However it is much more satisfactory to separate the multipolar part of the interaction, which converges under well-defined and easily attainable conditions[17,20], from the exponential penetration part, for which any attempt at an expansion in powers of $1/R$ is pointless, and which is much better regarded as part of the short-range interaction.

Unfortunately we have no way at present of evaluating the penetration part of the electrostatic interaction other than by taking the difference between the exact electrostatic energy (5), which includes the penetration effects, and the multipolar approximation (8), which does not. This is clearly no help at all. However Hall has found[21] that a good account of the molecular electrostatic potential, including penetration effects, is given by a 'current bun' model comprising point charges together with a sum of a small number of spherical Gaussian charge distributions. The point charges in such a model yield a form of distributed multipole expansion, while the spherical Gaussians yield both multipolar and penetration contributions. Here we are dealing with the electrostatic interaction between a molecule and a formal test charge, but a similar idea may be helpful in describing the interaction between two molecules. Whether or not this particular approach does prove fruitful, there is a good prospect that we can find an efficient way to describe the penetration effects as well as the long-range effects in terms of monomer properties.

The second additional term that appears in the first-order energy at short range is the exchange-repulsion energy, which is much more difficult to deal with. It comprises two effects: an attractive exchange term which arises because the electrons can exchange between the molecules, and a repulsive term which occurs because the electrons cannot occupy the same region of space if they have the same spin (Pauli repulsion). However it is not usually helpful to separate these two parts. Because the exchange-repulsion is a first-order perturbation term, it is still only necessary to know the unperturbed wavefunctions to evaluate it. Nevertheless it is time-consuming to calculate, because all the intermolecular electron-repulsion integrals are needed, and they are different for each configuration of the complex. Moreover many of the usual simplifications that arise in the evaluation of two-electron integrals do not apply, because the wavefunctions of the two molecules are not orthogonal. The labour is even greater if correlated wavefunctions are used for the monomers, though Rijks *et al.*[22] have suggested how the computational effort may be reduced.

Unfortunately there is as yet no known way to obtain the repulsion energy from properties of the separate molecules. An attempt has been made to characterise the repulsive surface of a molecule by performing IMPT calculations between the molecule and a suitable test particle, such as a helium atom. Because the helium atom has only one molecular orbital and is spherically symmetrical, such calculations can be done much more easily than calculations involving two ordinary molecules. From the data for the repulsion between molecule A and the test particle, and between B and the test particle, it may be possible to construct a repulsive potential between A and B. Some limited progress has been made with this idea[23]. An alternative approach[24] has been based on the suggestion[25-27] that the repulsion energy is closely correlated with the overlap between the molecular wavefunctions, but this seems likely to be more useful as a guide to the form of analytic models than as a direct route to accurate potential functions.

Long-range perturbation theory—second-order effects

A similar philosophy can be applied to the higher-order terms in the perturbation expansion. The second-order energy is, according to Rayleigh-Schrödinger perturbation theory,

$$W^{(2)} = -\sum_{mn}{}' \frac{\langle 00|V|mn\rangle\langle mn|V|00\rangle}{W_m^A + W_n^B - W_0^A - W_0^B}. \tag{13}$$

This is conventionally separated into three terms: those in which $n \neq 0$ but $m = 0$, those in which $m \neq 0$ but $n = 0$, and those in which neither m nor n is 0. These three terms are

$$U_{\text{ind}}^A = -\sum_m{}' \frac{\langle 00|V|m0\rangle\langle m0|V|00\rangle}{W_m^A - W_0^A}, \tag{14}$$

$$U_{\text{ind}}^B = -\sum_n{}' \frac{\langle 00|V|0n\rangle\langle 0n|V|00\rangle}{W_n^B - W_0^B}, \tag{15}$$

$$U_{\text{disp}} = -\sum_m{}'\sum_n{}' \frac{\langle 00|V|mn\rangle\langle mn|V|00\rangle}{W_m^A + W_n^B - W_0^A - W_0^B}. \tag{16}$$

In the first of these, we may perform the integrations over the ground-state wavefunction of B to obtain

$$U_{\text{ind}}^A = -\sum_m{}' \frac{\langle 0|V^B|m\rangle\langle m|V^B|0\rangle}{W_m^A - W_0^A}, \tag{17}$$

where $V^B = \langle \psi_0^B|V|\psi_0^B\rangle$ is the potential at A due to the unperturbed charge distribution of B. Eq. (17) then describes the response of molecule A to this potential, and is the induction energy of molecule A in the field of B. Similarly eq. (15) is the induction energy of B in the field of A.

These induction energy expressions can be reformulated so as to depend only on properties of the individual molecules. To do this, we start from the expression (17) for the induction energy of molecule A, and use the expression (6) for the perturbation V. We evaluate the integral over the coordinates of molecule B, and arrive at the expression

$$
\begin{aligned}
U_{\text{ind}}^A &= -\sideset{}{'}\sum_m \sum_{aa'bb'} \sum_{tt'uu'} Q_u^b T_{ut}^{ba} \frac{\langle 0|\widehat{Q}_t^a|m\rangle\langle m|\widehat{Q}_{t'}^{a'}|0\rangle}{W_m^A - W_0^A} T_{t'u'}^{a'b'} Q_{u'}^{b'} \\
&= -\frac{1}{2} \sum_{aa'bb'} \sum_{tt'uu'} Q_u^b T_{ut}^{ba} \alpha_{tt'}^{aa'} T_{t'u'}^{a'b'} Q_{u'}^{b'},
\end{aligned}
\tag{18}
$$

where

$$
\alpha_{tt'}^{aa'} = \sideset{}{'}\sum_m \frac{\langle 0|\widehat{Q}_t^a|m\rangle\langle m|\widehat{Q}_{t'}^{a'}|0\rangle + \langle 0|\widehat{Q}_{t'}^{a'}|m\rangle\langle m|\widehat{Q}_t^a|0\rangle}{W_m^A - W_0^A}.
\tag{19}
$$

Here $\alpha_{tt'}^{aa'}$ is a polarizability that describes the response of the moment t at site a to a perturbation (a change in potential, field, field gradient, etc.) at site a'[28].

Now the expression (19) is an uncoupled formulation of the polarizability. We can replace it by a polarizability derived from coupled Hartree-Fock perturbation theory, which is more accurate, because it takes account of the reorganisation of the electron distribution in a self-consistent manner. Better still would be to evaluate the monomer polarizability by a method that takes account of electron correlation as well[29]. But whatever the level of calculation, we can once again perform a much better calculation of the monomer property than is possible for the dimer. In this way we arrive at a description of the induction energy that is far more accurate than we can obtain through either intermolecular perturbation theory, where the perturbation is treated in an uncoupled fashion, or from a supermolecule calculation, where the size of the basis is limited by the need to perform calculations at a large number of points on the potential energy surface.

There is another way in which the expression (18) might be improved upon. It gives the induction energy of molecule A in the field arising from the multipole moments of molecule B. However molecule B is also polarizable, and its moments will be modified by the presence of molecule A. If this effect is taken into account, we arrive at an expression for the induction energy that is a power series in the molecular polarizabilities[30]. In practice, the effects of molecular polarization are usually calculated in an iterative fashion; the polarized moments of each molecule are evaluated in the field due to the other molecules, and the calculation is repeated until the polarized moments are self-consistent. This however is equivalent[30] to taking some, but not all, of the terms in a perturbation series to infinite order, and moreover it is known[31] that the perturbation series for the induction energy is asymptotic, i.e. divergent. This means that the conventional iterative procedure is highly questionable, and indeed it is known to lead to singularities at short range[32]. Numerical investigation[33] suggests that the simple expression (18) is more satisfactory, provided that distributed polarizabilities are used and provided that polarizabilities up to at least quadrupole rank are included.

As in the case of the first-order energy, there are two modifications that have to be made to this description when the wavefunctions overlap. In the first place, there must be a penetration effect that arises from the overlap of the charge densities. This has not been examined in detail, but it appears to be small[33]. The second modification arises because when two molecules overlap, it becomes possible for electron density from either molecule to flow onto the other. This effect is called charge transfer. In perturbation theory terms, it can be described by excitations from the occupied orbitals of one molecule to the virtual orbitals of the other. As such, it incorporates not only the genuine physical effect of charge transfer, but the BSSE. One of the virtues of IMPT is that the effects of BSSE do not arise in most of the energy terms, and the charge-transfer interaction is the only term involving single excitations in which they do occur. It is possible to correct the calculated charge-transfer energy for these effects by a procedure similar to that used in the supermolecule method, but as in that case some uncertainties remain.

There is a further problem with charge transfer, however. If we could perform the calculation with a complete basis set on molecule A, then it would be possible to describe any virtual orbital of B in terms of the basis set for A. In this case, the charge transfer effects would be included completely in the induction energy for molecule A. If we were then to calculate the effects of excitations from occupied orbitals of A to virtual orbitals of B we would be counting the same effects again. In other words, the charge-transfer energy is formally spurious. In practice, the basis sets used today are too small for this to happen to any great extent, but we should be aware that in principle the charge-transfer contribution is subject to this kind of double-counting error.

The dispersion energy

The expression (16) for the dispersion energy can also be replaced by a more accurate expression in terms of monomer properties. Eq. (16) involves one-electron excitations on both molecules, and cannot appear in an SCF calculation on the supersystem; accordingly the dispersion energy is a manifestation of electron correlation. Nevertheless, because there is a single excitation on each molecule, the dispersion energy can be reformulated in terms of monomer polarizabilities, which are one-electron properties and can be calculated reasonably accurately at the SCF level.

The key is the replacement of the energy denominator via the Casimir-Polder identity[34], which can be established by a simple contour integration:

$$\frac{1}{A+B} = \frac{2}{\pi} \int_0^\infty \frac{AB}{(A^2+u^2)(B^2+u^2)} du, \qquad \text{for } A > 0,\ B > 0. \tag{20}$$

In eq. (16) we put $A = W_m^A - W_0^A$ and $B = W_m^B - W_0^B$, and then the use of eq. (20), together with the expression (6) for the perturbation, leads to the following expression for the dispersion energy:

$$U_{\text{disp}} = -\frac{\hbar}{2\pi} T_{t\,u}^{a\,b} T_{t'\,u'}^{a'\,b'} \int_0^\infty \alpha_{t\,t'}^{a\,a'}(iu) \alpha_{u\,u'}^{b\,b'}(iu)\, du, \tag{21}$$

where the polarizability at imaginary frequency $\alpha_{t\,t'}^{a\,a'}(iu)$ is given by

$$\alpha_{t\,t'}^{a\,a'}(iu) = \sum_m{}' \frac{\Delta W_m(\langle 0|\hat{Q}_t^a|m\rangle\langle m|\hat{Q}_{t'}^{a'}|0\rangle + \langle 0|\hat{Q}_{t'}^{a'}|m\rangle\langle m|\hat{Q}_t^a|0\rangle)}{(\Delta W_m)^2 + \hbar^2 u^2}, \tag{22}$$

with $\Delta W_m = W_m - W_0$. This is a much more tractable and useful formulation than it appears to be at first sight. Once again the multipole moment operators $\hat{Q}_t^a$ are referred to *local* axes in the molecule, so the polarizabilities defined by (22) are also referred to local axes. Consequently no information about the relative orientation of the molecules is required to evaluate the dispersion integrals in (21), and they can be obtained once and for all for any pair of molecules, using accurate coupled Hartree-Fock calculations on the monomer to obtain the polarizabilities. The need to evaluate polarizabilities at imaginary frequency is not a problem; they can be calculated just as easily as the static polarizabilities. Moreover eq. (22) shows that $\alpha_{t\,t'}^{a\,a'}(iu)$ is very well-behaved as a function of frequency (it tends monotonically to zero as $u \to \infty$) so the dispersion integrals can be evaluated accurately by numerical quadrature using the values of $\alpha_{t\,t'}^{a\,a'}(iu)$ at a dozen or so frequencies[29].

The expression (21) for the dispersion energy is rather cumbersome, since it involves non-local polarizabilities $\alpha_{t\,t'}^{a\,a'}(iu)$. Many of these are small, and it is possible to transform the expression for the dispersion energy is such a way that their effects are represented exactly, at sufficiently large distances, by local terms[35]. However this cannot always be done for the dispersion integrals involving charge-flow polarizabilities on one or both molecules, and these dispersion integrals contribute significantly to the overall dispersion energy. It seems likely that the conventional site-site picture of the dispersion interaction is invalid, or at least incomplete, for large conjugated molecules.

It should eventually be possible to obtain accurate polarizabilities and dispersion integrals using correlated wavefunctions. Unfortunately there are some difficulties[36], but these are being overcome[29].

Once again there are corrections to be made in the short-range region. Since the dispersion energy is part of the correlation energy for the supersystem, it must remain finite at short range, while the terms in the multipole expansion (21) diverge like some power of $1/R$. It is usual to multiply the dispersion expression by a 'damping function' to cancel this singularity. Several authors have suggested suitable damping functions[37-39].

SUMMARY

The calculation of accurate intermolecular potentials is a difficult task. Supermolecule calculations are subject to inaccuracies that are difficult to avoid or correct. An approach based on perturbation theory principles holds some promise. According to this approach, the intermolecular potential is obtained as a sum of several terms. Many of these terms can be calculated in terms of properties of the separate molecules, which can be obtained much more accurately than is possible when calculations have to be carried out on the whole complex.

The *electrostatic energy* is obtained in terms of distributed multipoles, and is corrected at short range for the effects of penetration. The multipole moments can be calculated without undue difficulty from accurate wavefunctions for the individual molecules. It is necessary to use multipole moments at least as high as quadrupole if multipole sites are taken on each atom.

The *dispersion energy* can be expressed in terms of polarizabilities at imaginary frequency for the individual molecules, and the use of distributed polarizabilities makes it possible to describe the dispersion interaction accurately at short range. Damping functions must be included to correct for the effects of overlap[37-39]. The effects of electron correlation should be included in the polarizabilities; this is currently difficult to do, but the principles are understood[29].

The *short-range repulsion* is the most difficult part of the potential to calculate. Perturbation theory provides an expression for the repulsion interms of the unperturbed wavefunctions of the individual molecules, but it is time-consuming to evaluate, especially when correlated wavefunctions are used. Unfortunately it appears that the effects of correlation are significant[22]. At present it is necessary to perform the calculation at a small number of points and to fit a suitable model function to the results. A successful form of model repulsion potential is the following *anisotropic atom-atom repulsion*:

$$U_{\text{rep}} = \sum_{ab} \exp\left[\alpha(\Omega_{ab})\left(R_{ab} - \rho(\Omega_{ab})\right)\right],$$

in which the shape parameter ρ_{ab} and the hardness parameter α_{ab} depend on the relative orientation Ω_{ab} of atoms a and b. (In the general case, Ω_{ab} is a short-hand for five orientation variables.) There is now overwhelming evidence[20] of the need to introduce this kind of anisotropy into atom-atom repulsions. More elaborate forms for the radial dependence may be necessary for very accurate potentials.

Other contributions to the interaction energy are usually less important. The *induction energy* can be calculated using distributed multipoles and polarizabilities for the individual molecules; as for the electrostatic energy it is necesary to include contibutions at least to quadrupole rank. The *charge transfer* energy is a difficult case. It is formally spurious, since its effects are included in the induction energy. In practice, calculations of the induction energy do not recover the effects of charge transfer, which may therefore need to be included separately, but as usually formulated they are heavily contaminated by basis set superposition error.

One of the difficulties in performing calculations of intermolecular potentials is that it is very difficult to evaluate their accuracy. The calculation of measurable properties from a potential often involves approximations, and generally requires an integration over a large region of the potential surface, so that inaccuracies are smeared out. Errors may cancel, so an inaccurate potential may give better predictions than it deserves to. When there is disagreement between a calculated property and the experimental measurements it is often difficult to know what feature of the potential is at fault. It has often been assumed that supermolecule calculations provide a benchmark for model potentials, but we have seen that for weak interactions this assumption is untenable.

However it is now becoming possible to calculate the infra-red spectra of van der Waals complexes in considerable detail[40], and this will provide a much more detailed source of information about the validity of the intermolecular potential surface than has been available in the past. Some progress has also been made with the RKR inversion of spectroscopic data to provide potential-energy surfaces for polyatomic molecules[41]. With the help of this information it should be possible to improve the calculation of intermolecular potential energy surfaces over the next few years.

References

1. Price, S. L.; Stone, A. J. *Molec. Phys.* **40**:805 (1980).
2. Bartlett, R. J. *J. Phys. Chem* **93**:1697 (1989).
3. Pople, J. A.; Head-Gordon, M.; Raghavachari, K. *J. Chem. Phys.* **87**:5968 (1988).
4. Boys, S. F.; Bernardi, F. *Molec. Phys.* **19**:553 (1970).
5. van Lenthe, J. H.; van Duijneveldt-van der Rijdt, J. G. C. M.; van Duijneveldt, F. B. *Adv. Chem. Phys.* **69**:521 (1987).
6. Knowles, P. J. private communication.
7. Szalewicz, K.; Cole, S. J.; Kołos, W.; Bartlett, R. J. *J. Chem. Phys.* **89**:3662 (1988).
8. Stone, A. J. in *Theoretical Models of Chemical Bonding, vol. 4*, Z. B. Maksić, ed., Springer (1989).
9. Stone, A. J.; Tough, R. J. A. *Chem. Phys. Lett.* **110**:123 (1984).
10. Price, S. L.; Stone, A. J.; Alderton, M. *Molec. Phys.* **52**:987 (1984).
11. Buckingham, A. D. *Adv. Chem. Phys.* **12**:107 (1967).
12. Stone, A. J. *Chem. Phys. Lett.* **83**:233 (1981); Stone, A. J.; Alderton, M. *Molec. Phys.* **56**:1047 (1985).
13. Pullman, A.; Perahia, D. *Theor. Chim. Acta* **48**:29 (1978).
14. Rico, J. F.; Alvarez-Collado, J. R.; Paniagua, M. *Molec. Phys.* **56**:1145 (1985).
15. Cooper, D. L.; Stutchbury, N. C. *J. Chem. Phys. Lett.* **120**:167 (1985).
16. Sokalski, W. A.; Sawaryn, A. *J. Chem. Phys.* **87**:526 (1987).
17. Vigné-Maeder, F.; Claverie, P. *J. Chem. Phys.* **88**:4934 (1988).
18. Hayes, I. C.; Stone, A. J. *Molec. Phys.* **53**:69 (1984); Hayes, I. C.; Stone, A. J. *Molec. Phys.* **53**:83 (1984); Hurst, G. J. B.; Hayes, I. C.; Stone, A. J. *Molec. Phys.* **53**:107 (1984).
19. Amos, R. D.; Rice, J. E. CADPAC: The Cambridge Analytical Derivatives Package, issue 4.0, Cambridge, 1987.
20. Stone, A. J.; Price, S. L. *J. Phys. Chem* **92**:3325 (1988).
21. Hall, G. G.; Tsujinaga, K. *Theor. Chim. Acta* **69**:425 (1986); Tsujinaga, K.; Hall, G. G. *Theor. Chim. Acta* **70**:257 (1986).
22. Rijks, W.; Gerritsen, M.; Wormer, P. E. S. *Molec. Phys.* **66**:929 (1989).
23. Stone, A. J.; Tong, C.-S. in preparation.

24. Wheatley, R. J.; Price, S. L. , submitted for publication.

25. Kita, S.; Noda, K.; Inouye, H. *J. Chem. Phys.* **64**:3346 (1976).

26. Kim, Y. S.; Kim, S. K.; Lee, W. D. *Chem. Phys. Lett.* **80**:574 (1981).

27. Gellert, P. D.Phil. thesis, University of Oxford.

28. Stone, A. J. *Molec. Phys.* **56**:1065 (1985).

29. Wormer, P. E. S.; Rijks, W. *Phys. Rev.* **A33**:2928 (1986); Rijks, W.; Wormer, P. E. S. *J. Chem. Phys.* **88**:5704 (1988).

30. Stone, A. J. *Chem. Phys. Lett.* **155**:102 (1989).

31. Dalgarno, A.; Stewart, A. L. *Proc. Roy. Soc.* **A 238**:276 (1956); Dalgarno, A.; Lynn, N. *Proc. Phys. Soc. London, A* **70**:223 (1957).

32. Buckingham, A. D.; Pople, J. A. *Trans. Faraday Soc.* **51**:1173 (1955).

33. Stone, A. J. *Chem. Phys. Lett.* **155**:111 (1989).

34. Casimir, H. B. G.; Polder, D. *Phys. Rev.* **73**:360 (1948).

35. Stone, A. J.; Tong, C.-S. *Chem. Phys.*, in press.

36. Visser, F.; Wormer, P. E. S.; Stam, P. *J. Chem. Phys.* **79**:4973 (1983); Visser, F.; Wormer, P. E. S.; Jacobs, W. P. J. H. *J. Chem. Phys.* **82**:3753 (1984); Visser, F.; Wormer, P. E. S. *Molec. Phys.* **52**:723 (1984).

37. Douketis, C.; Scoles, G.; Marchetti, S.; Thakkar, A. J. *J. Chem. Phys.* **76**:3057 (1982).

38. Tang, K. T.; Toennies, J. P. *Chem. Phys.* **80**:3276 (1984).

39. Knowles, P. J.; Meath, W. J. *Chem. Phys. Lett.* **124**:164 (1986); *Molec. Phys.* **59**:965 (1986); *Molec. Phys.* **60**:1143 (1987).

40. Clary, D. C.; Lovejoy, C, M.; ONeil, S. V.; Nesbitt, D. J. *Phys. Rev. Letters* **61**:1576 (1988); Clary, D. C.; Nesbitt, D. J. *J. Chem. Phys.* **90**:7000 (1989); Nesbitt, D. J.; Lovejoy, C. M.; Lindeman, T. G.; ONeil, S. V.; Clary, D. C. *J. Chem. Phys.* **91**:722 (1989); ONeil, S. V.; Nesbitt, D. J.; Rosmus, P; Werner, H.-J.; Clary, D. C. *J. Chem. Phys.* **91**:711 (1989); Clary, D. C. this volume.

41. Nesbitt, D. J.; Child, M. S.; Clary, D. C. *J. Chem. Phys.* **90**:4855 (1989).

VIBRATIONAL STATES OF VAN DER WAALS AND

HYDROGEN-BONDED CLUSTERS: A SELF CONSISTENT FIELD APPROACH

R.B. Gerber and T.R. Horn
Department of Physical Chemistry and
The Fritz Haber Research Center
The Hebrew University, Jerusalem 91904, Israel and
Department of Chemistry, University of California, Irvine, CA 92717

C.J. Williams and M.A. Ratner
Chemistry Department, Northwestern University, Evanston, IL 60201

ABSTRACT

The Self-Consistent Field (SCF) approximation is applied to the vibrational states of several floppy clusters. This method is essentially a separable-mode approximation (each mode treated as moving in the average field of the other modes), and its quality depends on making a judicious choice of good coordinates for mutual separability. The SCF states can be used as a basic for a numerically exact configuration interaction (CI) calculation. The convergence of the CI (but not the results!) depends on the coordinates chosen. For cases where "good coordinates" are unknown, we improve the SCF by including an explicit correlation factor between the modes (Jastrow factor) in the wavefunction.

In this study we apply SCF and CI calculations to the vibrational states of the clusters Xe^4He_2, Xe^3He_2, $I_2 \cdot He$, $ArCO_2$ and $(HCN)_2$ and Jastrow-SCF calculations to the $XeHe_2$ clusters. Some of the interesting aspects of vibrational dynamics found for these systems are: (1) The $XeHe_2$ clusters behave as a "hyperatom" even in the ground vib-rotational state. The He atoms are highly delocalized within the cluster, resembling "electrons" within an "atom". (2) There are large differences between the vibrational spectra of (nuclear) spin polarized Xe^3He_2 and that of the corresponding, spin-paired cluster. Likewise, there are substantial differences between the vibrational spectra of the Bose cluster Xe^4He_2 and the Fermi Cluster Xe^3He_2. (3) There is a transition in $ArCO_2$ from large amplitude bending vibrations of the Ar at low levels of excitation ($v = 5$) to full rotation around the CO_2 (for $v=6$).

The results illustrate the power of SCF and ts extensions as a tool for studying the vibrational properties of floppy clusters.

I. INTRODUCTION

A virtually unique, and most interesting feature of van der Waals clusters is that such systems may exhibit strong anharmonic coupling between vibrational modes and large amplitude motions even in the ground state, or for the lowest excited states.[1-3] These aspects are hard to treat quantitatively, and pose a challenge to theory in this field. Basically, two levels of treatment are required. It is clearly desirable to have a simple "zero order" approximation, analogous to the normal mode harmonic model for rigid molecules for semiquantitative understanding. It is likewise essential to have rigorous, efficient numerical methods for quantitative analysis and comparison with spectroscopy. It is proposed in the present article that the vibrational Self-Consistent

Dynamics of Polyatomic Van der Waals Complexes
Edited by N. Halberstadt and K. C. Janda
Plenum Press, New York, 1990

Field approximation may provide a framework for both the simple, first insight treatment and the numerically exact calculations. It is our objective to demonstrate here, that at its simplest level the vibrational SCF approximation provides a very useful and semiquantitatively valid picture of the structure of the levels, the underlying pattern of motion, the coordinates along which they occur, etc. Using methods based on SCF, such as the Configuration Interaction (CI) approach, one can, however, also carry out rigorous, numerically exact calculations that are relatively efficient, and which take advantage of the simple insights offered by the lowest-level SCF description.

The SCF and CI treatments of vibrational states of polyatomic systems were first introduced some ten years ago[4-8] and a substantial literature exists on the methods, and on their applications to the vibrational states of molecules such as CO_2, H_2O and its isotopic variants, H_2O etc.[9a-9c]. Recently the potential advantages of the SCF (and related) methods as a tool for studying the vibrational dynamics of van der Waals clusters have been recognized and several realistic, all-mode treatments of systems such as Xe^4He_2 and I_2 ^4He have been reported.[10-12]

The present article describes recent studies on the vibrational and rotational bound states of van der Waals and of hydrogen-bonded clusters, using SCF, CI and related methods. The purpose of this study is twofold: First to demonstrate the power of SCF-type methods in describing the vibrational dynamics of floppy clusters, both on a quantitative footing and as a convenient tool for interpretation and physical insight. Secondly, using the methods as a tool we aim at finding interesting effects in the vibrational dynamics of various clusters. That is, we use SCF and CI simulations as a means for unravelling new patterns of dynamic behavior in clusters.

The structure of this article as as follows. The SCF, CI, and Jastrow-SCF methods are briefly described in the next section. A summary of recent calculations using these methods for several clusters is given thereafter, with focus on interesting aspects of dynamics behavior and properties revealed by the calculations. Concluding remarks are brought in the last section.

II. SCF AND SCF-BASED METHODS

a. SCF Approximation for Vibrational States

The SCF method is extensively described in several review articles[9] as well as in the original papers[4-8]. We give here a brief summary for convenience.

Consider a system of N vibrational modes whose Hamiltonian in a set of chosen coordinates $q_1,...,q_N$ is given by

$$H = - \sum_{i=1}^{N} \frac{\hbar^2}{2m_i} \frac{\partial^2}{\partial q_i^2} + V(q_1, \ldots, q_N) \tag{1}$$

where m_i is the mass associated with the coordinate q_i, and V is the potential energy function of the system. In the SCF method the eigenfunctions are approximated by the ansatz:

$$\Psi(q_1, \cdots, q_N) = \prod_{i=1}^{N} \Phi^{(i)}(q_i) \tag{2}$$

Application of the variational principle to this ansantz yields the single-mode SCF (Hartree) equations:

$$h_i^{SCF}(q_i)\Phi_n^{(i)}(q_i) = E_n^{(i)}\Phi_n^{(i)}(q_i) \tag{3}$$

where

$$h_i^{SCF}(q_i) = \frac{\hbar^2}{2m_i} \frac{\partial^2}{\partial q_i^2} + V_i(q_i) \tag{4}$$

$$V_i(q_i) = < \prod_{j\neq i}\Phi^{(j)}(q_j)V(q_1, \ldots, q_N)\prod_{j\neq i}\Phi^{(j)}(q_j)> \tag{5}$$

Eqs. (3)-(6) are solved self-consistently for single-mode wave functions $\Phi^{(j)}(q_j)$ and the single-mode SCF potentials $V_i(q_i)$, for all the modes $q_1,...,q_N$. The total energy is then given by:

$$E_{n(1),...,n(N)} + \sum_{i=1}^{N} E_n^{(i)} + (1-N)< \prod_{j=1}^{N} \Phi_{n(j)}^{(j)}(q_j)\,|\,V\,|\,\prod_{j=1}^{N} \Phi_{n(j)}^{(j)}(q_j)> \qquad (6)$$

where $n(i)$ is the quantum number in the i mode. SCF is computationally simple, since only the evaluation of single-mode wavefunctions is involved. Also the fact that the theory reduced the description of the dynamics to single-mode potentials and single-mode wavefunctions makes it easy to interpret in this framework.

b. The Importance of Using "Good Modes"

The SCF method involves, as the above description shows, a kind of separation of the modes. Therefore the results of this approximation depend in general on the choice of the modes that are being mutually separated (9b) (9c).[10,11,13] It is obviously of importance to search for "good modes", that can yield optimal SCF results, and examples demonstrating this were given by Bacic et al. for the HCN molecule[13] and by Gerber et al.[10] and Horn et al.[11] for several van der Waals clusters. If restriction to Cartesian coordinates is made, then the search for optimal modes can be put out on a systematic variational footing.[14–16] Variational refinement of the coordinate system is also possible in other cases, when geometric or dynamical considerations suggest a particular family of coordinate systems within which the search is carried out.[15] However, in most realistic applications so far, "good coordinates" were chosen from physical arguments, without a following optimization. Thus, in recent studies of the clusters I_2 He and $XeHe_2$, physical considerations were used in successfully choosing good modes for SCF calculations on the two systems.[10,11] Several types of physical considerations can be used in making a well-motivated choice of modes for SCF calculations, and Ref. 13, 10, 11 provide good examples in this respect: (1) One may attempt to choose modes which are mutually well-separated in frequency. This reduces the correlation between the modes and leads to good SCF results. (2) Choices of collective coordinates can be made which eliminate the problem of strong short-range repulsive correlations between the individual atoms (basically because suitable collective coordinates, or even bond coordinates "absorb" the pairwise interatomic correlation into the single-mode effective SCF potentials). On the other hand, using the individual atomic coordinates as the SCF modes would be very poor for the approximation, due to the strong correlation associated with the short-range mutual repulsions. (3) For states involving large amplitude motions, choices of coordinates that naturally associate with the form of the minimum energy path should be advantageous, e.g. ellipsoidal modes for the bending excitations in HCN[13] and I_2He[10]. The potential for exploring new and imaginative coordinate choices in this context seems to us an outstanding one. Needless to say, well-motivated coordinate choices can be tremendously useful also in other approaches to vibrational dynamics, as shown e.g. in an elegant recent work by Sibert[17].

c. The Vibrational Configuration Interaction (CI) Method

This method is, in principle, numerically exact:[8,2,9] It is tantamount to expanding the wavefunctions in a basis of SCF states:

$$\Psi(q_1,q_2, \cdots q_N) = \sum_{n(i)} C_{(n(1), \cdots ,n(N))} \prod_i \Phi_{n(i)}^{(i)}(q_i) \qquad (7)$$

where the $C_{(n(1), \cdots ,n(N))}$ are constant coefficients. The expansion can be carried out either in the "true" SCF states as defined above, or in "virtual" SCF states[7] which have the advantage of being rigorously orthonormal. The results of CI calculations, if pursued to correct convergence, are of course independent of the choice of coordinates. However, SCF states in good coordinates provide a more efficient basis, and yield convergence for a smaller number of states. CI calculations are far more demanding computationally than SCF ones, so working with a "good modes" SCF basis could be crucially important.

d. The Jastrow-SCF approach

We introduce here a method aimed at eliminating the need to search for good modes, while still retaining a computational simplicity approaching that of SCF. We have in mind especially future applications to larger clusters, where "good modes" may be hard to guess or calculate. The Jastrow method is familiar from electronic structure, condensed matter physics, nuclear theory etc. For simplicity, we introduce it here for a model system involving one infinitely heavy atom, and two atoms of finite mass having position vectors r_1, r_2 (measured from the heavy atom). As was pointed out above, the main problem in using r_1, r_2 as SCF coordinates is the existence of "hard wall" short range repulsion between the modes, that gives rise to very large correlations between the modes. Indeed, a major difficulty with the SCF is that within this approximation, when r_1, r_2 are used as modes, the light atoms can penetrate the mutual repulsive potential wall between them. This suggests a Jastrow-SCF ansatz, as an improvement to primitive SCF:

$$\Psi(r_1, r_2) = J(r_1 - r_2)\Phi_1(r_1)\Phi_2(r_2) \tag{8}$$

where the Jastrow correlation factor can be, for instance, assumed to be

$$J(r_1 - r_2) = \exp(-\alpha V_{rep}^{(1,2)}(r_1 - r_2)) \tag{9}$$

where $V_{rep}^{(1,2)}$ is the repulsive part of the potential between atoms 1 and 2, and α, a parameter of dimension (energy)$^{-1}$. Eq. (8) reflects the intuitive assumption that when the atoms are relatively far apart, the mutual SCF averaging of the interaction between them should work. When the atoms get near, the Jastrow factor allows for the steep repulsive correlation between them, and prevents them from artificially penetrating the mutual repulsive wall. A detailed presentation of the Jastrow-SCF method for cluster vibrations will be published elsewhere[18]. We note here merely that by the preliminary results obtained the method seems very promising.

e. Hartree-Fock SCF for Clusters Having Two or More ^{3}He or ^{4}He Atoms

^{3}He atoms within a van der Waals cluster must be treated as Fermions, while ^{4}He atoms in a cluster are Bosons. Rather than the Hartree ansatz of a product of single-mode wavefunctions (as in Eq. (2)), one should employ a determinantal Hartree-Fock ansatz in the case of ^{3}He atoms. The Hartree-Fock equations for the single-particle orbtials include exchange terms, as in the case of electrons in atoms. Other atoms in the cluster should be treated on a Hartree footing. For ^{4}He clusters, a symmetrical ansatz for the wavefunction should be used, as required for Boson calculations. All these considerations do not affect the Jastrow factor. A formulation of the HF-SCF and Jastrow/HF-SCF for clusters such as $Xe(^3He)_n$, $Xe(^4He)_n$ will be given in Ref. 18. Calculations we carried out for $Xe(^3He)_2$, $Xe(^4He)_2$ will be briefly discussed in the next section.

III. RESULTS AND DISCUSSION

This section briefly describes some of the main findings in recent calculations on the clusters $Xe(^3He)_2$, $Xe(^4He)_2$, I_2He, $ArCO_2$, $(HCN)_2$. The calculations used the SCF and Ci methods, and for each system an attempt was made to select "good coordinates" on physical grounds. Preliminary results with Jastrow/HF-SCF calculations on $Xe(^4He)_2$ were also obtained.

a "Hyperatomic Clusters": $Xe(^4He)_2$

SCF calculations using several choices of coordinates were carried out for this system, and so were converged CI calculations for all bound vibrational states.[10,11] Hyperspherical coordinates appear to be optimal modes for this problem, SCF in these modes being superior to SCF using other coordinates. Physically, these modes are a successful choice for SCF separability, because the three hyperspherical modes are well-separated in frequency: The Xe-He bond is the stiffest in the system, and this results in the hyper-radial coordinate being the mode of highest frequency. Since changing the volume of the cluster is energetically the most expensive, the lowest excitations for this system correspond to excitations of the angular motions on the hypersphere.[11] A very interesting finding was that even in its vib-rotational ground state the Xe^{4}He$_2$ cluster is highly delocalized, the He atoms being broadly distributed in some sphere around the Xe atom. This is a pronounced quantum-mechanical behavior, resembling the states of electrons in an atom: The light He atoms move essentially on a sphere. It is instructive to consider the He-He distance distribution in this cluster when in its vibrational ground state:

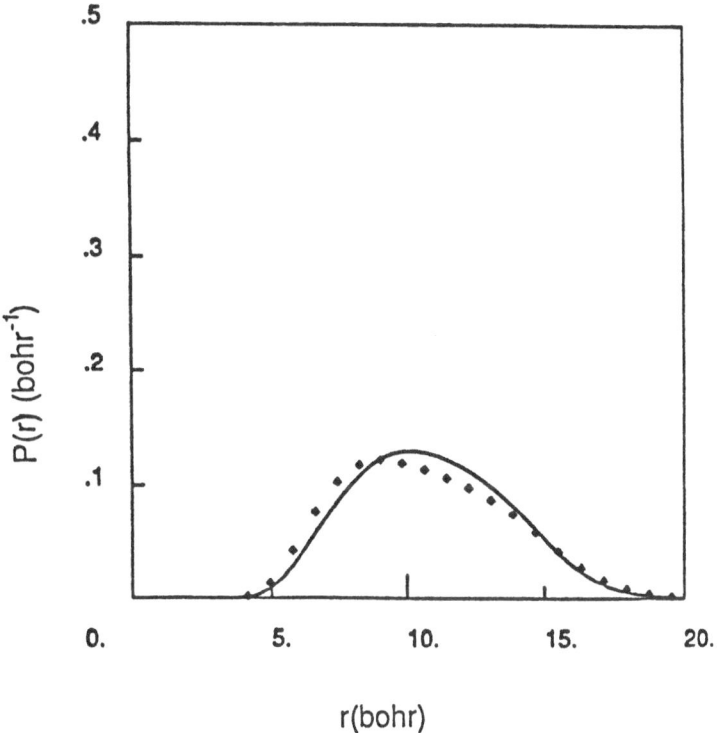

Fig. 1. The probability of the two He atoms in Xe^{4}He$_2$ being at relative distance r. The result shown is for the ground state. The result shows the highly delocalized distribution of the He atoms within the cluster even in its ground state. (An extraordinary behavior!)[11] The solid line in Fig. 1 represents SCF results, the diamonds exact CI results, and the close agreement between the two indicates the high accuracy of SCF in "good coordinates" even for an atom-atom distance distribution, which is basically a wavefunction property.[10,11] This picture is reinforced when one considers the second excited state of Xe^{4}He$_2$. The excited mode is, in the hyperspherical nomenclature, the κ-mode. As before, the copied line is SCF, and the diamond "exact".

The quality of the SCF remains excellent also for the excited state, basically because of the physical validity of the mutual separation of the "good modes", the basis for which was explained above. Note that the excited state (see Fig. 2) is even more delocalized than in the ground state.

(b) Role of Nuclear Spin in XeHe$_2$ Clusters[18]

Preliminary results were obtained on the interesting issue of how does the nuclear spin state of He clusters affect the vibrational dynamics and indeed the (highly nonrigid) "structure" of these species. In these calculations we studied the vib-rotational states of Xe^{4}He$_2$, Xe^{3}He$_2$ in the nuclear S=O spin state, and Xe^{3}He$_2$ in the nuclear S=1 spin state. The method used was Jastrow/HF-SCF, using an antisymmetrized (determinant) wavefunction for Xe^{3}He$_2$, and a corresponding symmetrized wavefunction for the Bose cluster Xe^{4}He$_2$. The result shows that the energy levels, and the He-He distance distributions for the spin-polarized S=1 state of Xe^{3}He$_2$ are very different than for the spin-paired S=O state of the cluster. Basically, the (nuclear spin) S=1 state of the cluster has a more spread out and diffuse He-atom probability distribution. Also, the S=1 state of the cluster has lower excitation energies. Likewise we find large differences between the energy spacings and the atom-atom distance distributions for Xe^{3}He$_2$ and the Xe^{4}He$_2$. Only part of these large differences are due to the masses, and the role of Fermi vs. Bose behavior of the wavefunction is a crucial factor. Finally we note that all the states of all three clusters are, in accordance with our calculations, "hyperatom-like states", and that includes the vibrational ground states of Xe^{3}He$_2$(S=O), Xe^{3}He$_2$(S=1) and Xe^{4}He$_2$: In all cases the He atoms are extensively delocalized within a large sphere that surrounds the Xe atom, and the He-He distance distribution within the cluster are smeared over ranges from say 5 a.u. to 20 a.u., in some of the bound states. In addition to being intrinsically interesting, these unusual clusters may be of relevance to condensed-matter states of He.

(c) The Vibration-Rotation Spectrum of I$_2$He [13]

The bending-stretching-rotation spectrum of I$_2^4$He was calculated by SCF in several coordinate systems and by CI. Since the soft-mode dynamics corresponds basically to a light particle moving in the field of two heavy centers, it is not surprising that elipsoidal coordinates turn out to be the optimal ones for the SCF calculations.[13] T-shaped at the classical equilibrium configuration, the cluster exhibits large amplitude bending oscillations (amplitude $\Theta \approx 20^{\circ}$ in the ground state). Although the bending amplitude increases with energy, there are no bound states for which this bending motion changes into full rotation around the I$_2$ stick. Another interesting soft mode in this system is the procession of the He about the I$_2$ axis, the energy levels for which scale almost perfectly as those of a free rotor[13]. A final point of interest in this system is the "cluster shift" of the I$_2$ vibrational frequency, which is of the order of 0.017 cm^{-1} (to the blue), and depends on the soft-mode stretching-bending state of the cluster. Although this dependence is very weak, it may provide information on the (soft-mode) state of the system if very high resolution spectroscopic data is available.

(d) ArCO$_2$, and the Bending to Free Rotor Transition

ArCO$_2$, a "ball and stick" problem as is I$_2$He, was treated in "all-mode" SCF calculations, and additional detailed computations were done on the soft modes only, with the CO$_2$ vibrations frozen. In its classical equilibrium configuration, the cluster is T shaped, but it exhibits a large zero-point bending amplitude quantum mechanically. Keeping the soft mode stretching at $n_R = O$, and increasing the bending level n_Θ a very interesting transition is observed as one reaches the level $n_\Theta = 6$; The motion in the "bending" mode becomes that of full rotation of the Ar about the CO$_2$ stick. It is useful in visualizing this transition to consider the SCF potentials and wavefunctions for the Ar stretching mode R and "bending" mode, pertaining to the ground state $n_R = n_\Theta = O$, and for the excited state $n_R = O$, $n_\Theta = 6$, as shown in Figs. 3 and 4.

As Fig. 4 shows, the stretching wavefunction is shifted to large distance R for the $n_\Theta = 6$, $n_R = O$ state, compared with the situation for the $n_\Theta = n_R = O$ state. This is

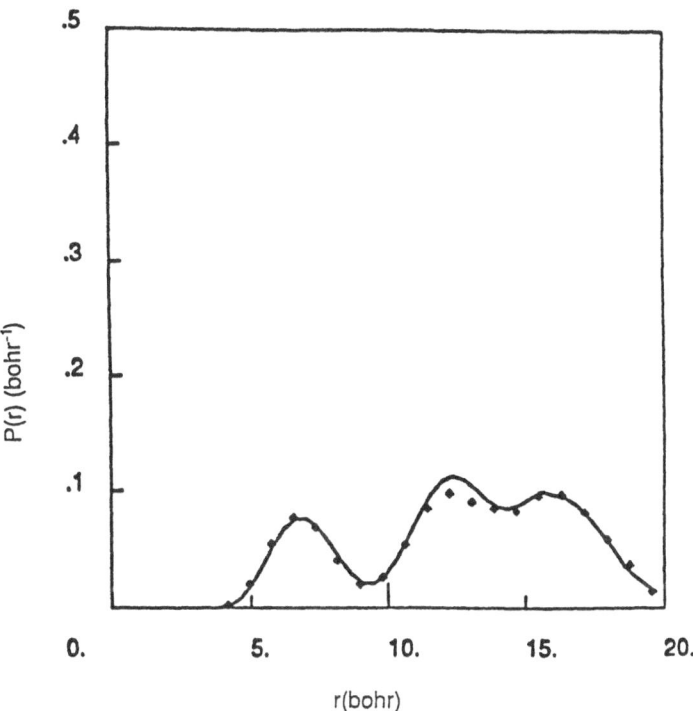

Fig. 2. The He-He distance distribution for the second excited ($k=2$) state of Xe^4He_2 [11].

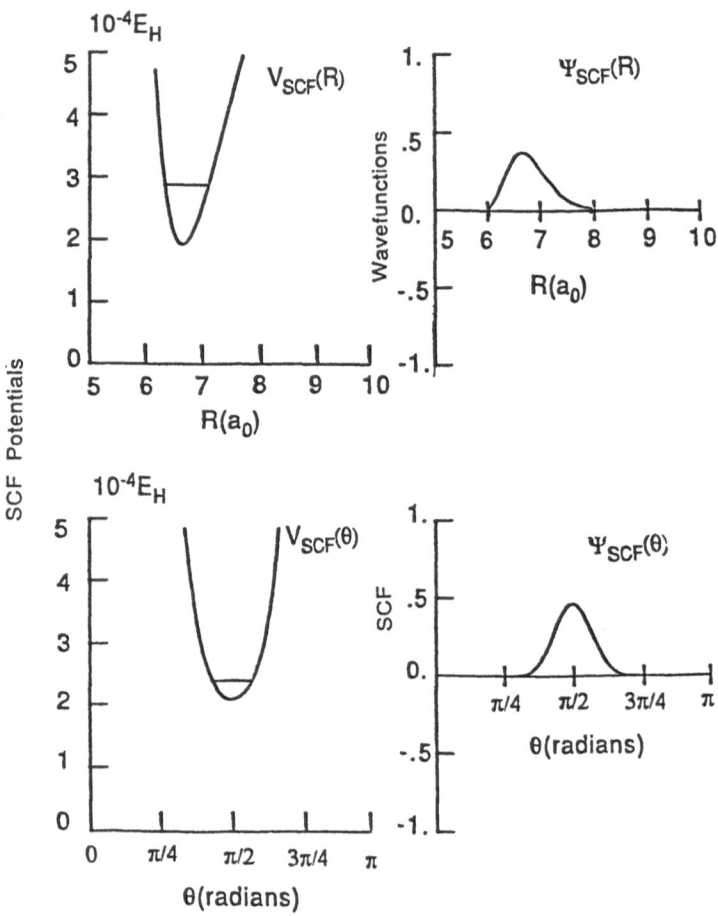

Fig. 3. The SCF potentials and wavefunctions for the stretching mode R and bending mode Θ of Ar vs. CO_2. The result shown is for the ground state $n_\Theta = n_R = O$.

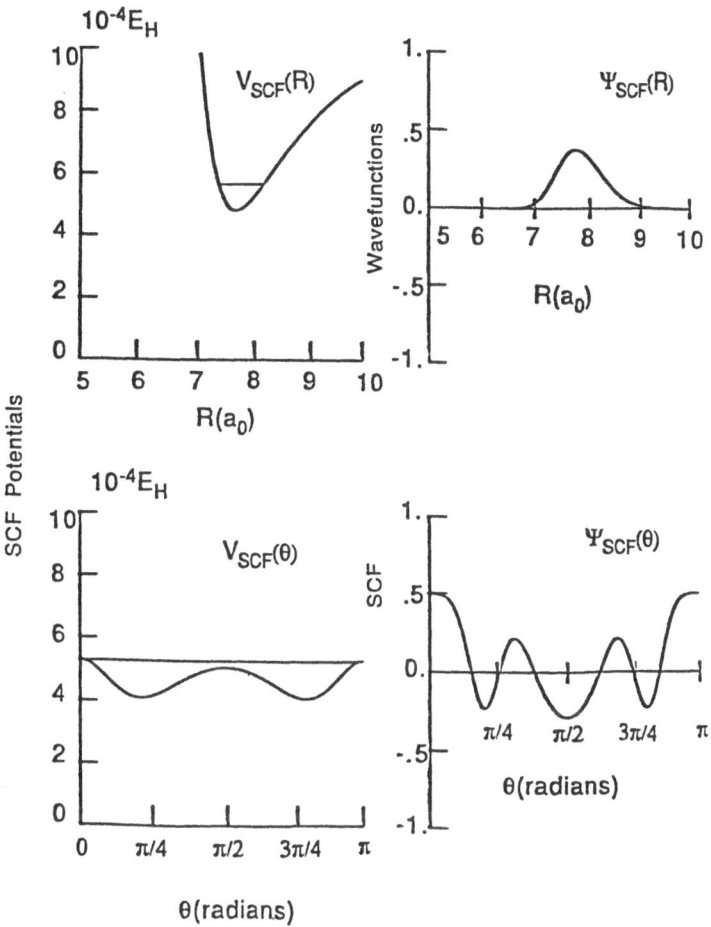

Fig. 4. The SCF potentials and wavefunctions for the stretching mode R and bending mode Θ of R in $ArCO_2$. The result shown is for the state $n_R = 0$, $n_\Theta = 6$.

Table 1. (HCN)$_2$ Vibrational Transition Frequencies
(HCN)$_2$ TRANSITION FREQUENCIES

Transition	SCF(cm^{-1})	CI (cm^{-1})	EXPT.(cm^{-1})
Free H-C Fundamental	3311.80	3311.82	3308.3
1st overtone Free H-C	6519.44	6519.38	6513.2
H-bonded H-C Fund.	3237.98	3237.57	3241.6
1st overtone H-bonded H-C	6343.4	6342.05	?-

For Free HCN (Monomer)	C-H Fundamental	3311.5 cm^{-1}
	C-H 1st overtone	6519.6 cm^{-1}

due to the "centrifugal" coupling between the bending and stretching, represented in an average way by the SCF, and describes an increase in the mean Ar-CO_2 distance, thus making the barrier for the Ar motion around the CO_2 low to the point of becoming classically allowed. Fig. 3 and Fig. 4 can be said to represent "solid like" and "liquid like" quantum levels in this cluster.

(e) The (HCN)$_2$ Cluster[18]

The linearly structure (HCN)$_2$ cluster was studied by SCF and CI calculations. Results available so far are within the framework of a collinear model only, given the fact that a full, reliable potential surface for this system is not available yet. We note that calculations of vibrational predissociation lifetimes that we carried out for this system in the framework of a preliminary collinear treatment have shown that to obtain lifetimes compatible with the experimental ones, the energy released from the excited C-H stretching mode must necessarily be mostly dumped into C-H bending and rotational excitation of the fragments. This means that the collinear treatment is unacceptable even qualitatively for describing the predissociation dynamics. However a collinear treatment may be a reasonable approximation for the spectroscopy of those modes that do not involve bending excitations. Table 1 shows the results of SCF and CI calculations for several of the transition frequencies. Jacobi ("collision") coordinates seemed an intuitively reasonable choice for the SCF calculation, and were employed here.

SCF results are in excellent accord with CI ones, indicating that the SCF "separability" applies in the case of the hydrogen bond, as it did in the earlier examples of the weaker van der Waals bond (and previous examples of strong covalent bonds). This suggests in turn that a reasonable choice of modes for SCF separability was made. The agreement between CI and experiment indicates the quality of the collinear potential surface, and provides inducement to a full all-mode SCF and CI treatments when a complete, reliable surface is at hand.

IV. CONCLUDING REMARK

This study shows the power of the SCF method, and approaches derived from it, in treating the vibrational dynamics and spectroscopy of floppy van der Waals clusters. The SCF was shown a useful tool in providing a reliable first quantitative approximation and qualitative insights on the dynamics and patterns of motion, but with the extension to CI it gives also the practical basis for rigorous, numerically exact calculations. The importance of making a good judicious choice of coordinates was stressed. This is important also for the CI calculations, since CI calculations in the appropriate modes converge faster.

Finally, considerable emphasis was put on extracting and predicting new physical features in the vibrational dynamics of various systems, having employed the SCF and CI calculations as a tool for obtaining the quantum states. It is hoped that the predictions of several interesting features will be tested in future experiments.

Acknowledgement: The Fritz Haber Research Center at the HU is supported by the Minerva Gesellschaft für die Forschung, mbH, Munich, FRG.

REFERENCES

1. An excellent recent review on theoretical treatments of large amplitude motions in molecules and clusters is: Z. Bacic and J.C. Light, Ann. Rev. Phys. Chem. (in press).
2. D.J. Nesbitt, Chem. Rev. 88:843 (1988).
3. R.E. Miller in "Structure and Dynamics of Weakly Bound Molecular Complexes", A. Weber, Ed., Reidel, Dordrecht (1987), p. 131.
4. G.D. Carney, L.I. Sprandel and C.W. Kern, Adv. Chem. Phys. 37:305 (1978).
5. J.M. Bowman, J. Chem. Phys. 68:608 (1978).

6. R.B. Gerber and M.A. Ratner, Chem. Phys. Lett. 68:195 (1979).

7. M.A. Ratner, V. Buch and R.B. Gerber, Chem. Phys. 53:345 (1980).

8. J.M. Bowman, K. Christoffel and F. Tobin, J. Phys. Chem. 83:905 (1979).

9. For reviews, see: (a) J.M. Bowman, Acc. Chem. Res. 19:202 (1986); (b) R.B. Gerber and M.A. Ratner, Adv. Chem. Phys. 70:97 (1988); (c) R.B. Gerber and M.A. Ratner, J. Phys. Chem. 92:3252 (1988).

10. R.B. Gerber, T.R. Horn and M.A. Ratner, in: "The Structure of Small Molecules and Ions", R. Naaman, Ed., Plenum Press, New York (1989).

11. T.R. Horn, R.B. Gerber and M.A. Ratner, J. Chem. Phys. 91:1813 (1989).

12. D.C. Clary and D.J. Nesbitt, J. Chem. Phys. 90:7000 (1989).

13. Z. Bacic, R.B. Gerber and M.A. Ratner, J. Phys. Chem. 90:3606 (1986).

14. B.C. Garrett and D.G. Truhlar, Chem. Phys. Lett. 92:64 (1982).

15. R. Lefebvre, Int. J. Quant. Chem. 23:543 (1983).

16. N. Moiseyev, Chem. Phys. Lett. 98:223 (1983).

17. D.T. Colbert and E.F. Sibert, J. Chem. Phys. 91:350 (1989).

18. T.R. Horn, R.B. Gerber and M.A. Ratner (to be published).

19. C.J. Williams, M.A. Ratner and R.B. Gerber (to be published).

PREDICTIONS OF SPECTRA FOR VAN DER WAALS MOLECULES

David C Clary, Charusita Chakravarty and Andrew R Tiller

Department of Chemistry
University of Cambridge
Lensfield Rd
Cambridge CB2 1EW
UK

ABSTRACT. Predictions are presented of spectra for excitation of the van der Waals rovibrational modes in ArHCl, ArHCN, H_2DF, ArOH and NeC_2H_4. For ArHCN, H_2DF and ArOH the potential energy surfaces used in the spectral computations have been obtained from CEPA calculations with large basis sets. Comparisons with experiment illustrate the power and usefulness of *ab initio* methods in predicting spectra for van der Waals molecules. The results also demonstrate that predictions of spectra can now be made for van der Waals molecules more complicated than the complexes of atoms with closed-shell diatomics.

I. INTRODUCTION

There have been several accurate calculations on the rovibrational energy levels of small van der Waals molecules. For simple systems such as ArH_2 and ArHCl, these calculations have been very useful in enabling high-quality potential energy surfaces to be obtained by comparison with experimental energy level data[1,2]. There have been a much smaller number of computations on the spectra of van der Waals molecules however[3]. Indeed, it is only very recently that the spectrum for ArHCl has been computed[4]. Since the dramatic improvement in high-resolution techniques in recent years has enabled many far-infrared and near-infrared spectra to be observed for simple van der Waals molecules[5], the time is right for

Dynamics of Polyatomic Van der Waals Complexes
Edited by N. Halberstadt and K. C. Janda
Plenum Press, New York, 1990

theoreticians to turn to the calculation not only of line positions but also of line intensities.

The significant improvements and availability of *ab initio* quantum chemistry programs is also of considerable importance for the field of van der Waals molecules. This is because the methods used in these programs can account for electron correlation, and it is now straightforward to perform calculations with large basis sets having high order angular momentum[6]. This enables realistic potential energy surfaces to be predicted for new van der Waals systems. This advance, combined with recent developments in methods for calculating the van der Waals states from these potentials, enables spectra to be predicted from first principles.

The spectra calculated using these procedures will not compete with the accuracy of high-resolution measurements. However, they do point to interesting regions in the spectrum where bands might have unusually large or small intensities, or an unexpected structure that could arise, for example, from Coriolis or Fermi resonance effects. The aim of this paper is to illustrate this useful capability by presenting our own recent predictions of spectra for several van der Waals molecules of considerable interest experimentally. These predictions help to explain the structure in some existing spectra and should also stimulate new experiments.

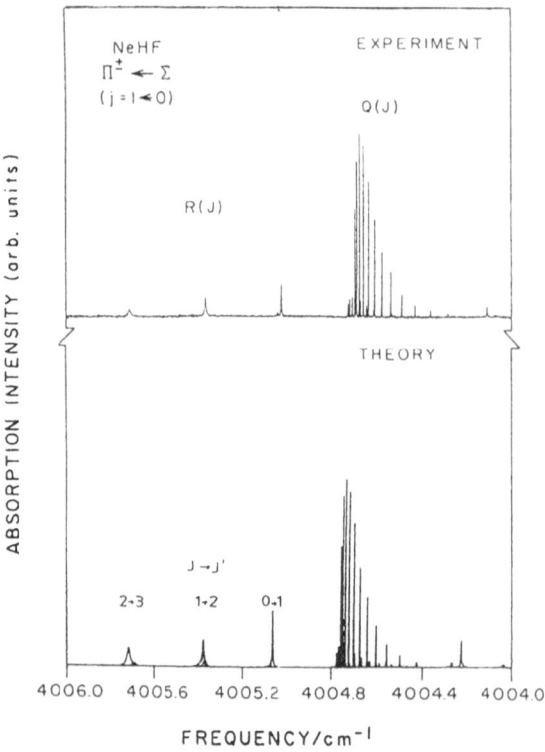

Fig.1. Comparison of experimental and theoretical spectrum for part of the Π–bending mode excitation in NeHF.

Recently published research on the NeHF van der Waals molecule gives an excellent example of the above points. Here, a highly accurate *ab initio* potential energy surface was computed[7] using the Coupled Electron Pair Approximation (CEPA) and this potential was then used in a prediction of the near infrared spectrum for NeHF. The surprisingly large intensities predicted for the fundamental band in NeHF, and the surprisingly narrow linewidths calculated for the excitation of the Π-bending van der Waals mode, led to the first identification of this molecule[8]. Furthermore, good agreement was obtained between the predicted and measured spectra (see Figure 1). Such outstanding agreement cannot be expected for more complicated systems than NeHF, but we still expect this *ab initio* approach to make interesting new predictions. The same CEPA method was used to obtain potential energy surfaces for three of the systems described below: ArHCN, H₂HF and ArOH.

In this paper, we start with the ArHCl system. This has become something of a benchmark system in van der Waals spectroscopy. High resolution far-infrared[9,10] and near-infrared[11] spectra have now been measured and this data has been used to produce an accurate potential energy surface for the van der Waals modes[2]. However, our calculations of the spectrum for this system show that there are still new features to be discovered: we predicted[4] surprisingly large intensities for the doubly-excited bending modes which were subsequently discovered in near infrared measurements[12].

We then go on to predict the spectrum for ArHCN. The rotational structure in the fundamental band in this system has been measured in several microwave experiments[13] and the energy levels are found to display large centrifugal distortion. Furthermore, very recent infrared spectra on ArHCN have been measured by Fraser and Pine[14]. They probed the region of the C-H stretching vibration in HCN and found a band displaced from the ArHCN fundamental by 7.8cm^{-1}. They attributed this to excitation of a bending van der Waals vibration in the ArHCN molecule. However, there have been no reliable determinations of the potential energy surface for ArHCN from this experimental data and little is known about the van der Waals modes in this system.

After this we predict the spectrum for the van der Waals modes of H₂DF. This, and isotopically related systems, have been the subject of high quality infrared experiments by the Nesbitt group[15,16]. These challenging experiments show that these systems have very interesting spectra which are completely different from those observed for NeHF or ArHCl. Furthermore, the measurements are suggestive of van der Waals modes with a very wide amplitude motion and large zero-point energy.

The ArOH van der Waals molecule is particularly interesting as fluorescence excitation spectra have been measured[17,18] in the vicinity of the OH ($^2\Pi \rightarrow {}^2\Sigma$) electronic transition. A summary of the experimental work on this system is given in the paper by Lester in this book[17]. These experimental spectra show a pronounced progression associated with excitation of the van der Waals stretching modes in Ar-OH ($^2\Sigma$). Furthermore, some other non-progression bands are observed that have not been given firm assignments. Our predicted spectrum for this system gives a very promising comparison with the experiment and suggests assignments for some of these non-progression bands.

It is also relevant to examine the van der Waals spectra for more complicated systems involving polyatomic molecules. Experimentalists have been reluctant to attempt to record high-resolution spectra for these systems due to their complexity. Here, we report a predicted infrared spectrum for NeC_2H_4 which has interesting features that should be observable if the experiments can be done with a very cold beam of van der Waals molecules.

Section II gives a simple overview of our approach for calculating the van der Waals spectra. Section III gives brief details of the potential energy surfaces used. In Section IV we present our predicted spectra and compare with experimental data when available. The comparison of the different spectra we have calculated also illustrates many interesting points about how van der Waals spectra vary with the complexity of the system and the structure and anisotropy of the potential energy surface. Conclusions are in Section V.

II. CALCULATION OF VAN DER WAALS SPECTRA

With the exception of NeC_2H_4, we use the same general method for calculating the van der Waals bound states for all the systems discussed above. First, we make the approximation that the bond lengths in the monomers are fixed so that the vibrations in the individual monomers are not considered. This should be a reasonable approximation since the frequencies for this vibrational motion are at least two orders of magnitude larger than the frequencies of the van der Waals modes so that the coupling between these two different types of motion will be very small. Comparison of observed far infrared ($HCl(v=0\rightarrow0)$) and near infrared ($HCl(v=0\rightarrow1)$) spectra for ArHCl confirm the accuracy of this simplification[10-12].

We use a space-fixed system of coordinates that is normally used for the scattering between two monomers. The advantage of this approach is that the appropriate basis functions previously worked out for the scattering of different types of monomers can be directly applied to the calculation of bound states. For a fixed value of the total angular momentum J, the hamiltonian for the vibration-rotation bound states of the van der Waals molecule can be written as a sum of separate hamiltonians that depend on the stretching coordinate R (which joins the centres of mass of the two interacting monomers) and the vibration-rotation bending motion

$$H = H_{stretch} + H_{bend} + V_1 \qquad (1).$$

Here,

$$H_{stretch} = -\hbar^2/(2\mu) \, [\, \partial^2/\partial R^2\,] + V_{stretch}(R) \qquad (2),$$

$$H_{bend} = T_{bend} + V_{bend} \qquad (3)$$

and

$$V_1 = V - V_{stretch}(R) - V_{bend} \qquad (4),$$

where μ is the reduced mass of the system, T_{bend} represents all the angular momentum operators and V is the potential energy surface.

A basis function approach is used to find the eigenvalues of the hamiltonian of equation (1). This is done with the following procedure:

(i) A suitable stretching potential $V_{stretch}(R)$ is chosen. For potentials with weak anisotropy, this could be the isotropic part of the potential V. With larger anisotropies, a cut through V to include the potential minimum is more appropriate.

(ii) The stretching hamiltonian of equation (2) is then diagonalised with a basis set of gaussian functions spaced equally on a grid of R values[19]. The exponent of these gaussian functions is optimised to produce the stretching eigenfunction $\Psi_{stretch}(R)$ with lowest energy.

(iii) This wavefunction $\Psi_{stretch}(R)$ is then used to average the potential energy surface V over R to give a bending potential V_{bend}. The bending hamiltonian of equation (3) is then diagonalised with an appropriate basis set constructed from coupled angular momentum functions that produce eigenfunctions of the total angular momentum for the whole van der Waals complex. For example, in the case of diatom-diatom problems this basis set[20] will couple together the rotational states of both diatomics together with the angular momentum associated with the rotation of the centre-of-mass vector **R** while, for an atom- $^2\Pi$ diatomic interaction, the diatomic eigenfunctions have rotation correctly coupled with the spin and electronic orbital angular momentum[21].

(iv) The bending eigenfunction with lowest energy obtained from (iii) is then used to average V over orientation angles to give a new stretching potential $V_{stretch}(R)$ and the procedure of (ii) is repeated. In turn this enables a new bending potential to be produced so that (iii) can be applied again. This self-consistent procedure is repeated until the stretching and bending eigenvalues do not change.

(v) The above approach produces a simultaneously optimised set of basis functions for the separate bending and stretching motions. These basis functions are then coupled together in a final configuration-interaction expansion that diagonalises the full hamiltonian of equation (1).

(vi) This calculation of the van der Waals bound states is repeated for different values of the total angular momentum J so that line intensities can be calculated. For microwave or infrared transitions, it is assumed that the dipole moment of the complex is directed along the infrared active monomer (eg HCl, HF or HCN). For the Ar-OH problem, a $\Pi \to \Sigma$ electronic transition is of interest, and here the transition dipole is placed at right angles to the OH vector. Since the *ab initio* potential energy surfaces cannot be expected to have a very high accuracy, it does not seem worthwhile to use more accurate dipole functions.

(vii) The spectra can then be predicted for different temperatures of the van der Waals complex by averaging the initial states of the spectral transitions with the appropriate Boltzmann factors and nuclear spin statistical weights.

The above procedures, which have been incorporated into automatic computer programs, are a systematic way of producing spectra that are appropriate for comparison with both far-infrared and near-infrared experiments on van der Waals molecules.

In the calculation on the near-infrared spectrum of Ne-C_2H_4, an approximation[22] is used in which coupling between the bending and stretching vibrations of the van der Waals bond is ignored. Comparison with accurate calculations[23] has shown this to be a good approximation for those low-lying bending modes that do not have similar frequencies to the excited stretching modes. Furthermore, the dipole moment of the Ne-C_2H_4 complex is placed along the direction associated with the v_7 vibration of C_2H_4, which is out of the C_2H_4 plane.

III. POTENTIAL ENERGY SURFACES

CEPA calculations have been carried out at a range of appropriate geometries for H_2HF[24], ArHCN[25] and ArOH[26] with the monomer bond length held fixed. The potential energy surface for these systems is expanded in the general form

$$V = \Sigma_i \, V_i(R) \, A_i \qquad\qquad (5)$$

where the $\{A_i\}$ represent an orthonormal set of appropriate angular functions that depend on the orientation angles of the interacting monomers. By fitting to the CEPA points, the $V_i(R)$ are expanded as simple polynomials multiplied by exponential functions for the short-range region together with the correct (damped) long-range term that has an R^{-n} dependence. For H_2HF, it was found that good convergence was obtained with just six angular terms in the expansion of equation (5) (including the important dipole-quadrupole and quadrupole-quadrupole interactions). In the case of ArHCN, seven Legendre polynomials were needed.

The basis sets used in the CEPA calculations were large. For example with H_2HF, the F atom basis was a (10s,6p) contracted to [6s,4p] augmented with extra diffuse s and p orbitals, and 3d and 4f orbitals optimised to produce the dipole and quadrupole polarisabilities of HF[7]. The results reported here for ArHCN did not include f orbitals in the basis set, although the sensitivity of the results to these functions is currently being examined. The CEPA potential for ArOH was calculated by Esposti and Werner[26]. In all cases, basis set superposition errors were accounted for.

For NeC_2H_4, the potential is much more approximate[27] and is obtained from SCF data for the short-range region with the long-range potential being parameterised from dipole oscillator strength data. For ArHCl, the H6(3) potential of Hutson was used[2].

IV. PREDICTIONS OF SPECTRA

IVa. ArHCl

In figure 2, the spectrum predicted for excitation of the van der Waals modes of ArHCl at 10K is shown. The bands close to $24cm^{-1}$ and $34cm^{-1}$ are the Σ and Π bending modes. The wavefunctions for these states have zero and one unit of angular momentum respectively projected along the intermolecular axis R and correlate roughly with the j=1 levels of HCl. One of the most interesting aspects of this spectrum is the prediction of lines with a fairly strong intensity around $70cm^{-1}$ at a frequency expected for HCl(j=0$\rightarrow$2) transitions. These transitions are, of course, forbidden for pure HCl but the weak anisotropy of the ArHCl potential mixes together different HCl rotational basis functions in the complex, enabling the "j=2" bands to be observed. This prediction[4] stimulated a high-resolution near infrared search for these unusual bands which were subsequently discovered[12]. Indeed, the intensities of all the bands now agree well with experiment, confirming the good accuracy of the potential.

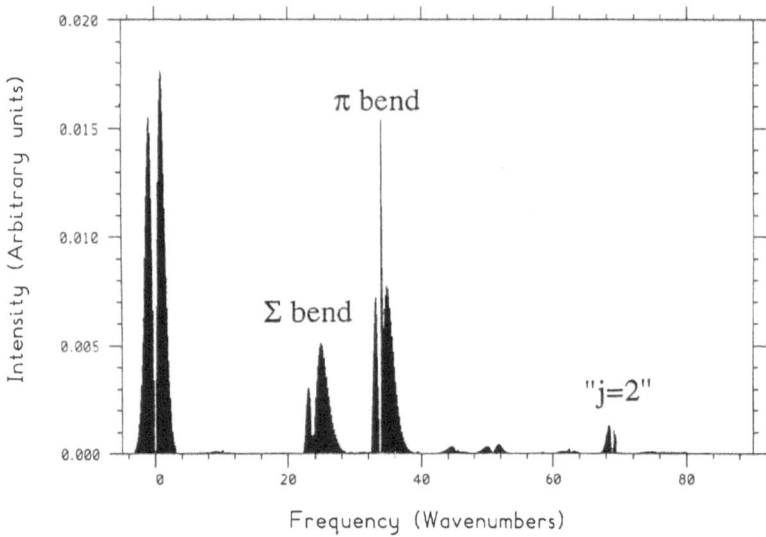

Fig. 2. Calculated spectrum for ArHCl at 10K.

IVb. ArHCN

It is interesting to compare the ArHCN spectrum with that for ArHCl. The CEPA potential[25] for ArHCN is quite similar to that for ArHCl, with the minimum in the potential occurring for the collinear configuration with the Ar atom next to the H atom in the molecule. Furthermore, if we take θ as the angle that defines the orientation of **R** with respect to the vector along the linear molecule, then, if the potential is plotted as a function of θ for fixed R in the region of the minimum, a double-minimum form is obtained with a barrier maximum close to θ=90⁰ for both ArHCl and ArHCN.

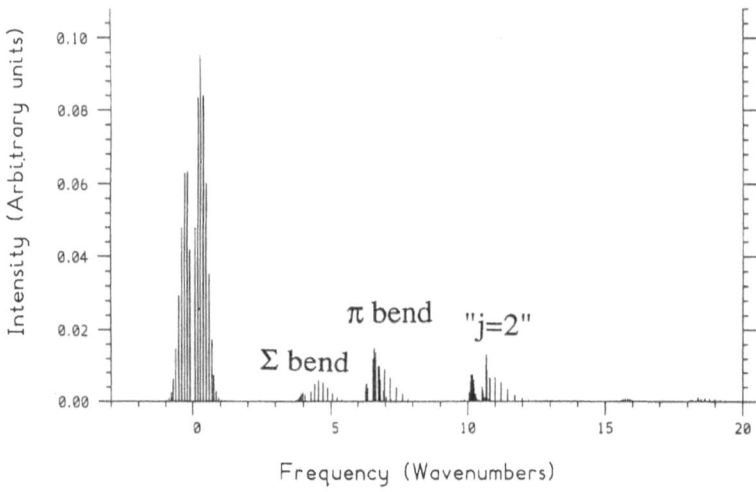

Fig. 3. Calculated spectrum for ArHCN at 1K[25].

The main difference between the two systems is the rotor constant which is seven times larger for HCl than that for HCN. This ensures that the bending modes of ArHCN have a much lower frequency and can perturb the ground state. Furthermore, the Σ and Π bending modes have very similar frequencies so that they strongly interact through Coriolis coupling. This can be seen from the predicted spectrum[25] for ArHCN at 1K shown in figure 3. The Σ bend has a low intensity, around 4.5cm^{-1}, as this vibration involves localisation of the Ar atom at the N end of the ArHCN molecule, while the zero-point vibration is localised at the H end. The Π bend covers mainly the region near $\theta=90^0$ and gives a stronger intensity around 6.5cm^{-1}. Furthermore, the interaction between the Σ and Π bending modes produces an interference effect[11] that results in a low intensity for the P branch, with much larger intensity for the R branch of the Π bend.

These results are in quite good agreement with the experimental findings of Fraser and Pine[14]. They only saw strong intensity in a strongly perturbed R branch of a band which they attributed to the Π bend centred at 7.8cm^{-1}. We are repeating the CEPA calculations with a larger basis set and expect the new potential to give an even better agreement with experiment. Furthermore, we predict remarkably strong intensities for the "j=2" bands of ArHCN around 10.5cm^{-1}. It would be fascinating if the high-resolution spectra could be measured in this region also to see if our predictions are upheld.

IVc. H$_2$DF

The H$_2$HF and isotopically substituted molecules also present an interesting comparison with NeHF and ArHCl. This is because the diatomic nature of H$_2$ produces a very different potential and the light reduced mass results in large zero-point energy effects. The interaction between the permanent quadrupole moment of H$_2$ and the dipole of HF produces a T-shaped complex with the H of the HF pointed towards the midpoint of the H$_2$ bond. This interaction gives a much more anisotropic potential than in NeHF. Combined with the large rotor constants for H$_2$ and HF, this produces a very large bending zero-point energy. Indeed, the *ab initio* CEPA calculations[24] give a minimum in the potential of 305cm^{-1} for H$_2$HF while the lowest energy level for the complex with ortho symmetry of the H$_2$ (ie the H$_2$ rotational states are odd) gives a zero point energy corresponding to 85% of the binding energy. This ensures that, for H$_2$HF, the bending modes correlating with HF(j=1) states will not be bound. However, these bending states are just bound for H$_2$DF due to the smaller rotor constant of the DF monomer. These features are illustrated in the energy level diagram of figure 4.

Another consequence of the T-shaped geometry is that the ground state of the ortho van der Waals level is of the Π type with nearly degenerate odd and even parity energy levels.

This produces a doublet structure, and a Q branch, in the fundamental band. The predicted spectrum also shows $\Pi \to \Sigma$ transitions at frequencies around 30cm^{-1} , and , for H_2DF only, some transitions into $DF(j=1)$ bending levels between 40 and 60cm^{-1}. This latter part of the spectrum is rather complicated due to the many bending levels that arise and the strong interactions between them. However, the transition from the Π to a Δ bending state with two units of angular momentum along the R axis does give the lines with the largest intensity in this region. These features are shown in the predicted spectrum[24] of Figure 5.

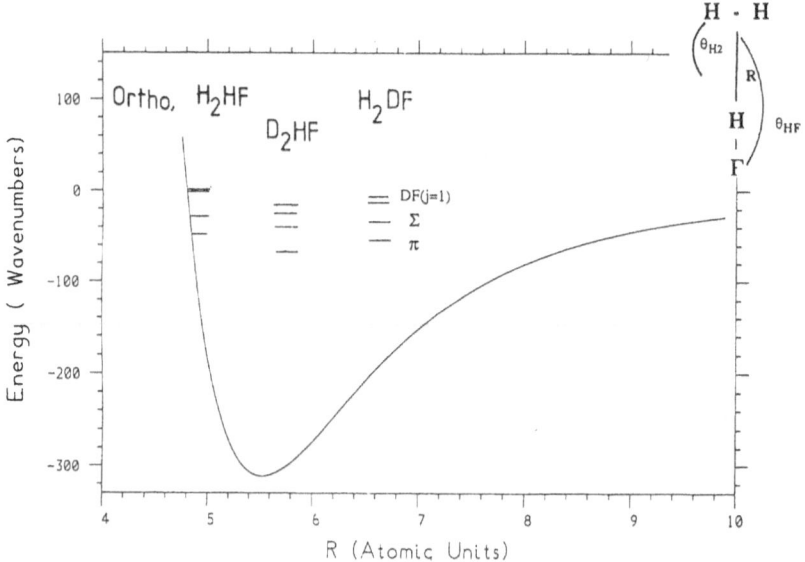

Fig. 4. Energy level diagram for the lowest van der Waals states of H_2HF, D_2HF and H_2DF with ortho symmetry. The levels are plotted on top of the potential calculated for the T shaped approach.

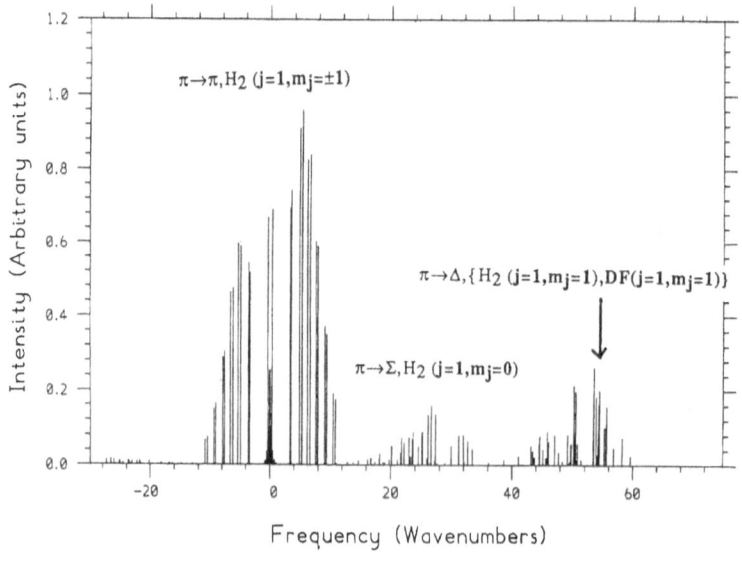

Fig. 5. Calculated spectrum for ortho H_2DF at 16K[24].

The experimental spectrum[15,16] also shows the P,Q and R doublet structure predicted for the fundamental band. The $\Pi \rightarrow \Sigma$ transition at a frequency around 30cm^{-1} has a smaller intensity and has not yet been observed. However, several lines in the region between 40 and 60cm^{-1} have been observed[16] but these are, as yet, unassigned.

IVd. ArOH

The ArOH system presents a fascinating comparison with the systems discussed above as here we are interested in a transition between van der Waals modes of different **electronic** states. The frequency of the transition is around 32000cm^{-1} in the region of the OH ($X\ ^2\Pi \rightarrow A\ ^2\Sigma$) excitation. The ArOH ($^2\Pi$) ground state potential of A' symmetry is similar to that for ArHCl, having a well depth of 104 cm^{-1} and being weakly anisotropic with a double minimum form with respect to θ at the region of the minimum. The potential for the ArOH($^2\Sigma$) state has a much deeper well (D_e=1109 cm^{-1}) and is also of the double minimum form in θ. Furthermore, it is much more anisotropic, having a difference in energy of 2008 cm^{-1} between the collinear and perpendicular configuration for the value of R fixed at the minimum in the potential. This produces a very large zero-point energy in the bending mode (440 cm^{-1}) that is over an order of magnitude larger than that for ArHCl. Figure 6 illustrates the electronic transitions in a schematic diagram in which the full potential has been averaged over the bending wavefunctions for the Ar-OH($^2\Pi$) and Ar-OH($^2\Sigma$) systems.

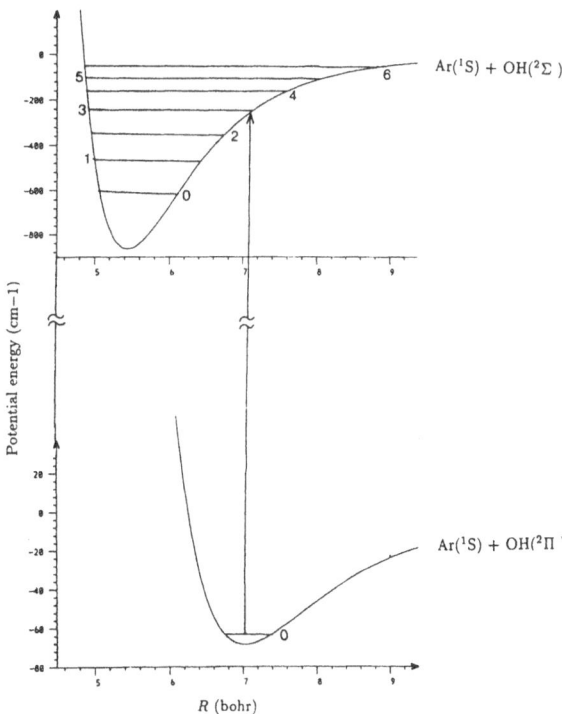

Fig.6 Schematic diagram for the potentials of ArOH averaged over the bending modes.

The predicted spectrum[28] for ArOH is shown in figure 7. The progression in the stretching mode can clearly be seen in this spectrum, with a peak in intensity at v=5. The peaks marked 1,2,3,4 and 5 do not form part of this stretching progression. These peaks involve transitions into bending modes. Peak 1 is the first excited bend, peak 2 is a combination of the first excited bend with the first excited stretch and peaks 3-5 are highly mixed states involving many bending and stretching basis functions.

These theoretical findings present a promising comparison with the experiment[18] (see, in particular, Fig1, in the paper by Lester in this book[17]). The experimentally observed stretching progression has a maximum intensity at v=6. Furthermore, in between v=5 and 6 a non-progression band is observed (marked A in their notation) that corresponds probably to our peak 2 (ie. a bend plus stretch combination band). The first excited bend, peak 1 in our spectrum, is probably too low in intensity to be observed. Beyond the line with maximum intensity in the stretching progression, both the theory and experiment give irregular structure that is typical of interaction between several states near dissociation. Indeed, there are interesting analogies with the H_2DF spectrum which also shows a regular structure for the more strongly bound levels which breaks down as dissociation is approached.

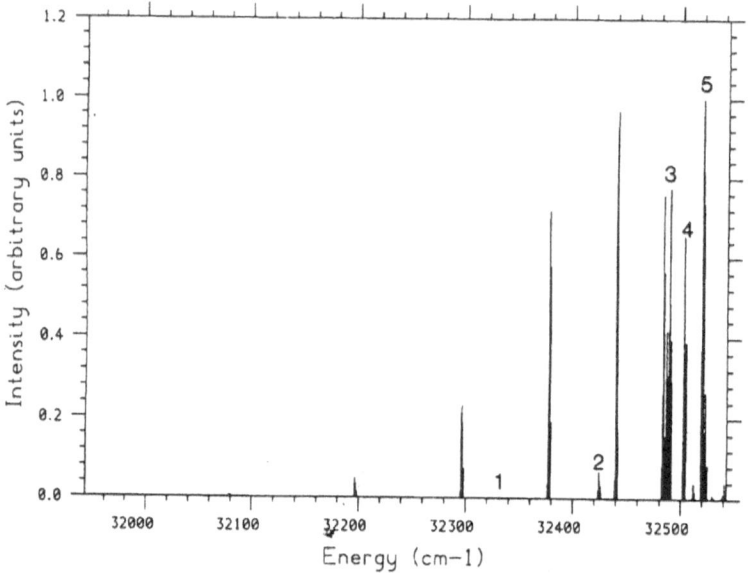

Fig. 7. Calculated spectrum for ArOH ($X\,^2\Pi \rightarrow A\,\,^2\Sigma$)[28].

One of the significant aspects of the experiments[17,18] on ArOH is that the stretching levels are directly probed. These are usually hard to measure in infrared experiments on van der Waals molecules as they normally only gain significant spectral intensities by accidental Fermi resonances with bending modes having high intensity (NeHCl is a good example of this[4]).We are currently calculating the lifetimes for vibrational-rotational predissociation for both the ArOH ($X^2\Pi$,v=1→0) and ArOH (A $^2\Sigma$, v=1→0) systems, as these have also recently been measured[17,29].

IVe. NeC$_2$H$_4$

We have previously reported[22] a calculation of the infrared spectrum for NeC$_2$H$_4$(v=0→v$_7$=1) at 5K, where v$_7$ is an out-of-plane bending mode with a frequency of 949cm^{-1}. The lines in this spectrum showed considerable congestion, the detailed features of which might be hard to measure. However, the spectrum at 1K, shown in figure 8, has a much clearer form. In particular, the excitation of a "θ−bending" mode, displaced 14cm^{-1} from the fundamental band, can be seen. This corresponds to an "isomer" transition in that the configuration for the ground state of the NeC$_2$H$_4$ has the Ne placed perpendicular to the C-C bond , while the θ−bending mode has the Ne atom localised in a direction along the C-C bond between two hydrogen atoms attached to the same carbon atom[23]. Previously calculated vibrational predissociation linewidths for this system[27] have values ~0.002cm^{-1}, which are not broad enough to cause overlap of these spectral lines.

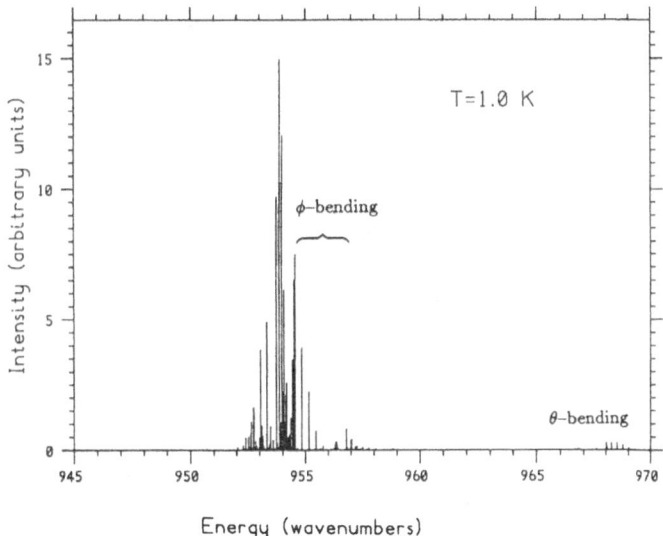

Fig. 8. Calculated spectrum for NeC$_2$H$_4$ at 1K.

Low resolution spectra have been reported for NeC$_2$H$_4$ which do show some structure[30]. It would clearly be interesting if high-resolution infrared experiments could be performed at very low temperatures on NeC$_2$H$_4$ and similar systems.

V. CONCLUSIONS

We have predicted spectra for a variety of van der Waals complexes using potential energy surfaces obtained in several cases from high-quality *ab initio* calculations. The comparisons with available experimental data are promising and the predictions suggest several new high-resolution experiments that will give interesting results. The comparisons of the spectra for the systems of different complexity illustrate the surprising and unusual features that are to be found in the spectra of van der Waals molecules.

ACKNOWLEDGMENTS

The authors would like to acknowledge stimulating interactions and discussions with D. J. Nesbitt, C. Lovejoy, S. V. ONeil, H. -J. Werner, P. J. Knowles, C. E .Dateo, A. C. Peet, J. M. Hutson, M. I. Lester, M. T. Berry, W. Klemperer and G. T. Fraser. This work was supported by the Science and Engineering Research Council.

REFERENCES

1. R. J. Le Roy and J. S. Carley, Adv. Chem. Phys., 42 (1980) 353.
2. J. M. Hutson, J. Chem. Phys., 89 (1988) 4550.
3. See, for example, G. Brocks, J. Chem. Phys., 88 (1988) 578;
    B. P. Reid, K. C. Janda and N. Halberstadt, J. Phys. Chem., 92 (1988) 587;
    I. F. Kidd and G. G. Balint-Kurti, Chem. Phys.Letters, 105 (1984) 91.
4. D. C. Clary and D. J. Nesbitt, J. Chem. Phys., 90 (1989) 7000.
5. D. J. Nesbitt, Chem. Rev., 88 (1988) 843.
6. N. C. Handy in " Supercomputer Algorithms for Reactivity, Dynamics and Kinetics of Small Molecules", Editor A. Lagana, p23 (Kluwer, Dordrecht, 1989).
7. S. V. ONeil , D. J. Nesbitt, P. Rosmus, H.-J. Werner and D. C. Clary, J. Chem. Phys., 91 (1989) 711.
8. D. C. Clary, C. M. Lovejoy, S. V. ONeil and D. J. Nesbitt, Phys. Rev. Lett., 61 (1988) 1576; D. J. Nesbitt, C. M. Lovejoy, T. G. Lindeman, S. V. ONeil and D. C. Clary, J. Chem. Phys., 91 (1989) 722.
9. M. D. Marshall, A. Charo, H. O. Leung, and W. Klemperer, J. Chem. Phys., 83 (1985) 4924.
10. R. L. Robinson, D. -H. Gwo, and R. J. Saykally, J. Chem. Phys., 87 (1987) 5156, and references therein.
11. C. M. Lovejoy and D. J. Nesbitt, Chem. Phys. Lett., 146 (1988) 582.
12. C. M. Lovejoy and D. J. Nesbitt, to be published; see also D. J. Nesbitt and C. M. Lovejoy, Faraday Discuss. Chem. Soc., 86 (1988) 13; ibid 86 (1988) 34.

13. K. R. Leopold, G. T. Fraser, F. J. Lin, D. D. Nelson, Jr., and W. Klemperer, J. Chem. Phys., 81 (1984) 4922; T. D. Klots, C. E. Dykstra and H. S. Gutowsky, J. Chem. Phys., 90 (1989) 30.

14. G. T. Fraser and A. S. Pine, J. Chem. Phys., to be published.

15. C. M. Lovejoy, D. D. Nelson, Jr., and D. J. Nesbitt, J. Chem. Phys., 87 (1987)5621; C. M. Lovejoy, D. D. Nelson, Jr., and D. J. Nesbitt, J. Chem. Phys., 89 (1988)7180.

16. D. J. Nesbitt, C. M. Lovejoy and R. Lascola, private communication.

17. M. I. Lester, in " Dynamics of Polyatomic van der Waals Complexes", Eds. N. Halberstadt and K. C. Janda.

18. M. T. Berry, M. R. Brustein and M. I. Lester, Chem. Phys. Lett., 153 (1988) 17; W. M. Fawzy and M. C. Heaven, J. Chem. Phys., 89 (1988) 7030.

19. I. P. Hamilton and J. C. Light, J. Chem. Phys., 84 (1986) 306.

20. S. Green, J. Chem. Phys., 62 (1975) 2271.

21. M. H. Alexander, J. Chem. Phys., 76 (1982) 3637; M. H. Alexander and S. L. Davis, J. Chem. Phys., 79 (1983) 227.

22. A. R. Tiller, A. C. Peet and D. C. Clary, J. Chem. Phys., 91 (1989) 1079.

23. A. R. Tiller and D. C. Clary, Chem. Phys., in press.

24. D. C. Clary and P. J. Knowles, to be published.

25. D. C. Clary and C. E. Dateo, to be published.

26. A. D. Esposti and H. -J. Werner, to be published.

27. A. C. Peet, D. C. Clary and J. M. Hutson, J. Chem. Soc., Faraday Trans. 2, 83 (1987) 1719.

28. C. Chakravarty , D. C. Clary , A. D. Esposti and H. -J. Werner, to be published.

29. M. T. Berry, M. R. Brustein and M. I. Lester, J. Chem. Phys., 90 (1989) 5878.

30. C. M. Western, M. P. Casassa and K. C. Janda, J. Chem. Phys., 80 (1984) 4781.

STRUCTURES, PHASE TRANSITIONS AND OTHER DYNAMICAL
PROCESSES IN INHOMOGENEOUS VAN DER WAALS CLUSTERS

Robert J. Le Roy, John C. Shelley, Darryl J. Chartrand,
and Mary Ann Kmetic

Guelph-Waterloo Centre for Graduate Work in Chemistry
University of Waterloo
Waterloo, Ontario N2L 3G1, Canada

I. Introduction

This paper is concerned with the properties and dynamical propensities of many-body Van der Waals clusters, species too large to be treated quantum mechanically as a single molecule, yet far too small to take on the properties of a macroscopic sample of matter. The work was motivated by a desire to understand what happens to the familiar macroscopic property known as a "phase transition" when a system becomes "finite", on a molecular scale. More particularly, we wish to learn how the nature of the intermolecular forces between the component particles affect these properties.

Most previous theoretical work in this area has been based on models of clusters of identical spherical particles interacting through pairwise Lennard-Jones potentials. While those studies have proved very instructive,[1] the idealization of the potential energy function prevents comparisons with experiment from providing further insight. The most widely used experimental probe of such homogeneous systems is provided by electron diffraction techniques.[2,3] However, they mainly address the question of the existence and nature of long-range order in such systems, and they cannot readily yield quantitative information on clusters smaller than a hundred or so molecules.

In view of the above, we focus our attention on heterogeneous Van der Waals clusters containing a single chromophore molecule whose spectral perturbation by neighbouring "solvent" molecules can probe the local structure and properties of such systems. The existence of experimental studies of the infrared spectra of clusters formed by SF_6 with Ar atoms,[4–6] as well as of realistic anisotropic potentials for SF_6 interacting with Ar or Kr,[7] has made these systems particularly propitious subjects of study. This stimulated the development of a successful model for predicting the shift of the fundamental v_3 band of SF_6 due to perturbation by surrounding rare gas atoms,[8] and subsequent work has shown that this frequency shift can provide a very perceptive flag for certain types of structural and dynamical changes.[9] The present paper describes our further efforts to use classical simulations of these systems to shed light on the nature of phase transitions in microclusters.

We begin by describing our model for these SF_6-$(Rg)_n$ (Rg = rare gas) systems, and outlining how the simulations were performed. Section III then describes the minimum energy structures and patterns of growth determined for clusters formed by a single SF_6

Dynamics of Polyatomic Van der Waals Complexes
Edited by N. Halberstadt and K. C. Janda
Plenum Press, New York, 1990

molecule with numbers of Ar or Kr atoms, and discusses how these structures depend on features of the overall system potential energy function. The results obtained using molecular dynamics simulations to characterize the "melting" transitions of SF_6-$(Ar)_n$ clusters for $n = 7$–13 are presented and discussed in Section IV, while Section V summarizes the evidence concerning the occurrence and generality of a spontaneous reversing isomerization between liquid-like and solid-like structures at temperatures far above those for the initial melting transitions of these system.

II. Models and Methodology

In the molecular dynamics and Monte Carlo simulations which are the source of the results described herein, the overall potential energy surface was assumed to be a pairwise sum of two-body interactions. The Ar-Ar and Kr-Kr pair potentials were represented by the accurate functions determined by Aziz and Slaman,[10] while the Ar-SF_6 and Kr-SF_6 interactions were represented by the detailed anisotropic functions determined by Pack et al..[7] Although three-body forces may not be negligible in these systems, we doubt that their effects are larger than those of the *uncertainties* in our knowledge of the Rg-SF_6 two-body potentials.

The considerable anisotropy of the Rg-SF_6 pair potential is illustrated by Fig. 1, which compares the shapes of the radial Kr-SF_6 potentials along the two-, three- and four-fold symmetry axes of the octahedral SF_6 molecule (solid curves, see insert). For the sake of comparison, the Kr-Kr potential is also shown (dashed curve). Note that the *anisotropy* of the atom-molecule potential, illustrated by the differences among the solid curves in this figure, is as strong as the atom-atom potential itself. This statement is also true for the Ar-SF_6 system, whose potential, though scaled to slightly smaller energies and distances, is very similar to that of Fig. 1.[7]

Two types of classical simulations were used in the course of the present work. Monte Carlo and molecular dynamics "annealing" calculations were used to determine the equilibrium structures, while isoenergetic molecular dynamics runs at a wide range of energies were used to determine the melting temperatures. Our Monte Carlo calculations used the Pangali et al.[11] force-biased modification of the basic Metropolis et al.[12]

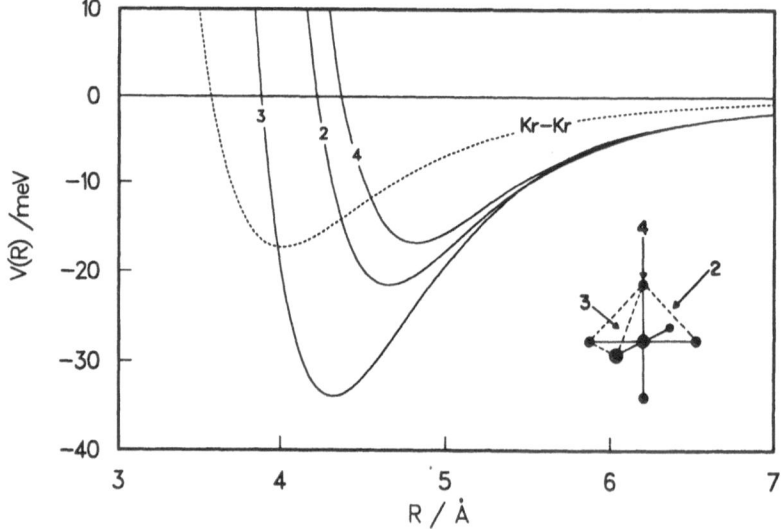

Fig. 1. *Solid curves*. radial behaviour of the empirical Kr-SF_6 potential of Ref. 7 along the two- three- and four-fold symmetry axes of SF_6 (see insert); *dashed curve:* the Kr-Kr pair potential of Ref. 10.

method; for further details regarding our Monte Carlo calculations, see Refs. (8) and (13). The program which performed the molecular dynamics calculations used a fourth-order Adams-Moulton predictor-corrector algorithm, initialized with four steps of Runge-Kutta integration, and an integration stepsize of $2.5\,fs$ ($1\,fs = 10^{-15}$ sec). The numerical integration time step of 2.5×10^{-15} sec was chosen to ensure that in a typical $1500\,ps$ ($1\,ps = 10^{-12}\,sec.$) run, the total energy was conserved to within $ca.$ 1 part in 10^5. This program is described in more detail in Refs. (9) and (13).

As mentioned above, we wish to use the perturbation of the free monomer vibrational frequencies by interactions with its neighbours as a probe of the cluster structure, and if possible, also of its dynamical behaviour. In the SF_6-$(Rg)_n$ systems, the vibrational mode monitored in the experiments is the triply degenerate fundamental v_3 band of SF_6.[4-6] A simple model for predicting its frequency shifts by the surrounding rare gas atoms was developed and tested in Ref. (8), and it is used in the present work as a flag for changes in the clusters.

III. Families of Structures

In their landmark studies of the patterns of growth for clusters of Lennard-Jones atoms, Hoare and Pal[14] showed that for all but the smallest systems, there exist a number of different classically stable structural isomers. The anisotropy of the pair potential for the Rg-SF_6 system will clearly increase this tendency towards the existence of multiple structures. On the other hand, due to the very low temperatures at which such clusters are formed in a molecular beam, only the lowest energy structures will tend to be populated. The emphasis in the present discussion is therefore *not* on providing a comprehensive compilation of the energies and structures of all possible classically stable isomers, but rather on identifying and characterizing the minimum energy form and a few of the other important low energy structures.

For SF_6-$(Ar)_n$ and SF_6-$(Kr)_n$ clusters with $n = 2$-7, the form and relative energies of the lowest energy structures are shown in Fig. 2. In the individual figure segments, the SF_6 molecule is shaded and the cluster drawn with S-F bonds of the SF_6 aligned along the horizontal and vertical directions. Thus, for example, the square arrangement of the atoms in structure-*B* for $n = 4$ shows that the four Rg atoms are arranged symmetrically, roughly centred over the four three-fold "pockets" on one half of the octahedral SF_6 molecule. Similarly, the orientation of the $n = 7$ structures relative to the frame of the figure indicates a rotation of the basic pentagonal pyramid Rg-atom cap away from the symmetric allignment on the underlying SF_6 seen in structure-*A* for $n = 6$. Of course, a Kr atom is distinctly larger than an Ar atom, and effects due to this difference in size may readily be discerned in a visual comparison of analogous structures for these two families.[15] However, for clusters in this size range, the isomer structures are quite similar, so both are represented here by the structures drawn for the Ar species. At the same time, marked effects due to the atomic size difference give rise to the differences in the relative stabilities of analogous alternate structures for the two Rg species.

In view of the strong potential anisotropy seen in Fig. 1, one might have expected that the minimum energy configurations of SF_6-$(Rg)_n$ for $n = 1$-8 would have each of the rare gas atoms lying near a three-fold axis in one of the pockets formed by a set of three adjacent fluorine atoms. For $n = 1$ this is, of course, the case, and when $n = 2$ the second Rg atom preferentially sits in an adjacent pocket (an alternate $n = 2$ configuration in which the second Rg atom sits in a pocket on the *opposite* side of the SF_6 molecule has a much higher energy[15]). For $n = 3$, however, the "pulling" effect of the potential anisotropy is apparently less important than the energy minimization achieved by optimizing the "packing" of the Rg atoms relative to one another. In other words, the third

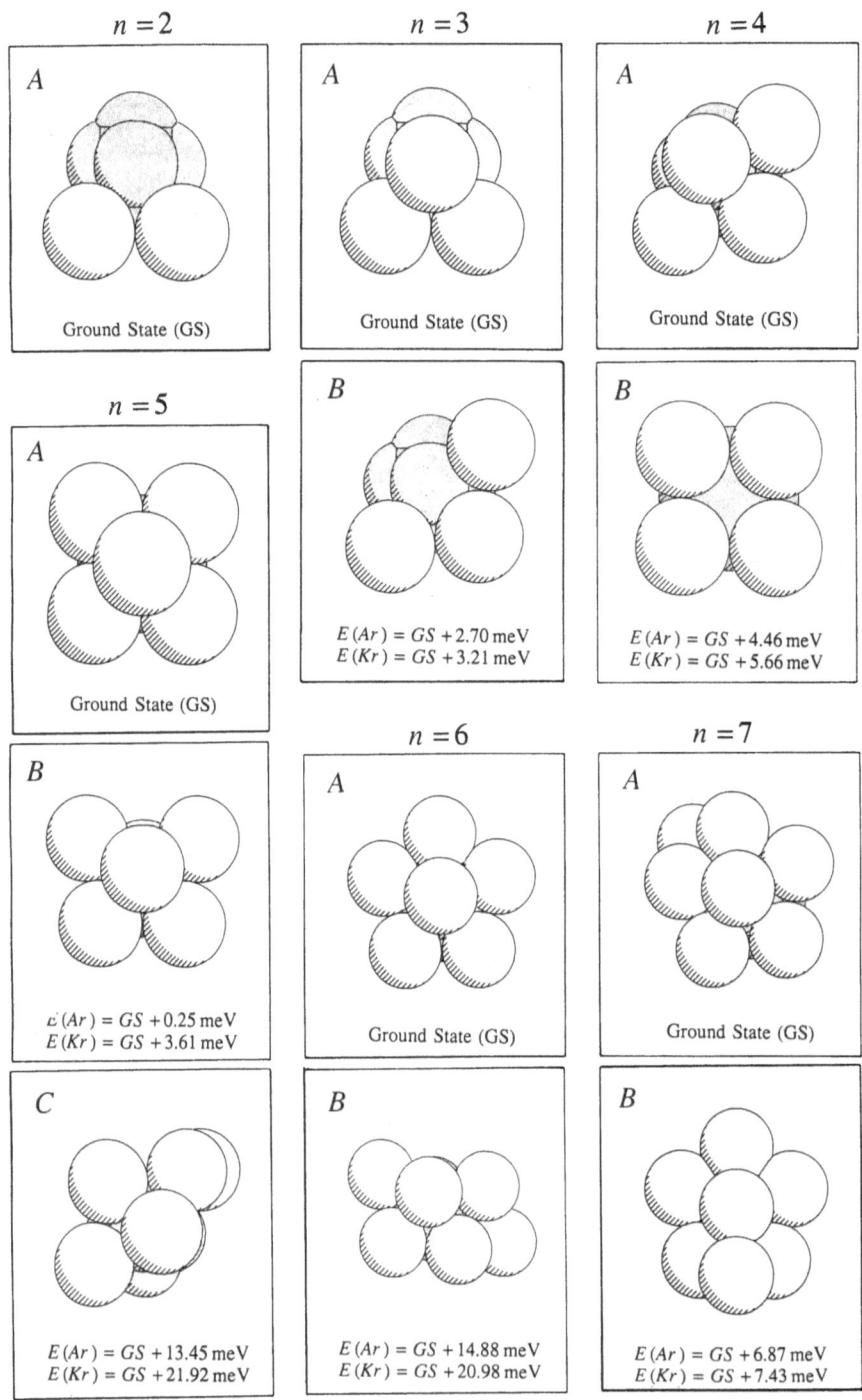

Fig. 2. Structures and relative energies of the lowest energy isomers found for SF_6-$(Ar)_n$ and SF_6-$(Kr)_n$ clusters with $n = 2$-7.

atom added gains more from optimizing the Rg-Rg packing than it loses by being sited on the four-fold symmetry axis at the *least*-favourable orientation relative to the SF_6.

This competition between the "pulling" of the atom-molecule potential and its anisotropy, and "packing" considerations driven by a proclivity for maximizing the number of Rg-Rg contacts, also underlies the patterns of structures for the larger clusters. While perhaps most evident for $n = 3$ and 4, its effects may also be readily discerned for the next larger species. In particular, in the ground state structure for $n = 5$, the first four atoms have been "pulled" by the potential anisotropy into positions near the three-fold axes, while the fifth one sits on top of them in a position which allows contact with the SF_6 along its least-favoured orientation, but maximizes the degree of Rg-Rg packing. In contrast, in structure-*B*, the Rg "packing" has been fully optimized by having two of the atoms displaced towards the distinctly less-favoured two-fold symmetry axes in the SF_6. The much greater relative stability of the ground state (GS) for the $n = 5$ Kr clusters (see Fig. 2) reflects the fact that the size of the Kr atom (i.e., the equilibrium distance on the Kr-Kr pair potential) is more commensurate with the distance between neighbouring pockets. Finally, structure-*C* for $n = 5$ may be interpreted as being the "packed" ground state structure for $n = 4$, plus an additional atom centred over a pocket on the far side of a potential energy "ridge" between two F atoms. Here, the pulling of the anisotropy is the source of the (local) stability of this isomer (it would not exist if the Rg-SF_6 potential was isotropic), while the poor Rg-Rg packing is responsible for its high relative energy.

For $n = 6$, the pentagonal-pyramid rare gas atom "cap" of structure-*A* is far more stable than any other isomer; indeed, for the Ar clusters, the extra stablization energy achieved on adding the sixth atom is greater than the analogous quantity for any other Ar cluster for $n = 1$-11.[13,15] This reflects the fact that the Ar atoms have just the right size for five of them to comfortably form a ring around a central Ar atom on one face of the SF_6, while at the same time the two "lower" (on the figure) Rg atoms are approximately centred on three-fold axes and the two "outer" ones are also partially over pockets. While this same isomer is also the ground state for SF_6-$(Kr)_6$, the larger size of the Kr atom tends to lessen its stability, and very little additional cluster stability is associated with this pentagonal-pyramid cap for that case.[15]

The ground state structure for $n = 7$ clusters is based on that for SF_6-$(Rg)_6$, with an additional atom slipped over the ridge between two neighbouring F atoms into a neighbouring three-fold pocket. However, its stablization in this pocket, combined with the need to optimize its packing with the other Rg atoms leads to a slight rotation of the cap of Rg atoms about the extruding S-F axis. That this also occurs for structure-*B* is a reminder that the effects of the Rg-SF_6 potential anisotropy may also reach into a second layer.

Structure-*B* for $n = 7$ introduces a new class of isomer; species in which one or more of the Rg atoms are "stacked" on top of a close-packed Rg monolayer on the surface of the SF_6. It may also be viewed as introducing a competition between "wetting" and "non-wetting" isomers, or between bulk vs. surface positions for the chromophore impurity. In particular, each subsequent added atom must decide whether to succumb to the Rg-SF_6 pair potential and locate itself in direct contact with the chromophore, or to optimize its packing relative to its more weakly attractive "like" neighbours by becoming stacked in a second layer.

The nature of structure-*B* for $n = 7$ clearly suggests the existence of other $n = 7$ isomers, distinguished from one another by the position of the second-layer atom relative to the underlying SF_6. It also suggests that analogous "stacked" or multilayer structures will occur for all larger clusters. While further studies show that both of these predictions are true, full details are not yet available.[15] For present purposes, therefore, we shall merely summarize certain results regarding the propensity for finding the SF_6 impurity at the surface, rather than maximally solvated in the midst of the Rg atoms.

For SF_6-$(Rg)_n$ clusters with $n = 7$ and 8, the monolayer isomers are the minimum energy form for both Kr and Ar solvents. For $n \geq 9$, however, the stacked structures become the lower energy form for $Rg = Kr$, while this same change occurs for $Rg = Ar$ when $n \geq 11$. In both cases, this crossover to a preference for stacked solvent structures at low temperatures occurs when the first monolayer is approximately half complete (a full shell of Ar atoms consists of some 20 atoms, while for Kr this first layer becomes filled at $n = 16$). At the same time, it is important to realize that the minimum energy isomers are not necessarily those most commonly found in simulations or laboratory experiments. In particular, for SF_6-$(Ar)_n$, independent Monte Carlo and molecular dynamics simulated annealing runs usually converged on close-packed monolayer structures for all cluster sizes up to $n = 20$. This demonstrates that although the stacked structures which have the minimum *potential* energy for $n \geq 11$, they do not necessarily have the minimum *free* energy across much of the temperature range traversed during the simulated annealing process. Since the clusters observed in molecular beam experiments certainly have internal energies corresponding to non-zero temperatures, the minimum energy stacked structures may not dominate the observed population there either.

In summary, therefore, the competition between the *pulling* of the potential anisotropy, a tendency to optimize the *packing* of the more weakly bound solvent atoms, and the likelihood of achieving the latter by *stacking* the Rg atoms in layers rather than wrapping them around the SF_6, seems to underlie the pattern of structures for the species considered here. These considerations should also be characteristic of other heterogeneous clusters involving nonspherical species. However, there are distinct differences between their manifestations for SF_6-$(Kr)_n$ and SF_6-$(Ar)_n$. In particular, trends in this competition depend somewhat on the degree to which the size of the Rg atom is commensurate with the separation between neighbouring anisotropically-favoured sites. Moreover, novel "reverse melting" effects have been found to be associated with isomerization between the monolayer and stacked forms of SF_6-$(Ar)_n$ for $n \geq 7$.[9,13] Some of these results are described below in Section V. First of all, however, we will outline the results of simulation studies directed towards delineating the melting behaviour of these clusters.

IV. Surface vs. Bulk Melting of SF_6-$(Ar)_n$ Clusters

While molecular dynamics simulations are usually performed at a constant total energy E_{tot}, energy partitioning arguments allow one to associate an effective temperature with the average value of the total system kinetic energy $\langle E_{kin} \rangle_t$:

$$T = \langle E_{kin} \rangle_t / (3nk_B / 2) \tag{1}$$

where k_B is Boltzmann's constant, n the number of rare gas atoms in the cluster, and $\langle \ \rangle_t$ the time average over a complete (isoenergetic) molecular dynamics trajectory. For three of the cluster sizes of interest, Fig. 3 shows how the effective temperatures determined in this way depend on the total energy. The slight irregularities in these plots are attributed to the finite lengths of the simulation runs, so this figure demonstrates a smooth monotonic relationship between E_{tot} and this effective temperature. For conceptual convenience, our future discussions will therefore characterize a particular simulation run by this effective temperature, rather than by its E_{tot} value.

If Fig. 3 is rotated by $\pi/2$ and turned upside down, it becomes a plot with the system energy increasing vertically and the effective temperature increasing to the right. In macroscopic physical chemistry, the slope of such a plot is the heat capacity, and discontinuities or abrupt changes in this slope are associated with phase changes. However, any such effects are clearly very muted in Fig. 3, and the results presented below indicate that the changes which do occur are *not* well correlated with changes in the characteristic dynamical behaviour of the system. In particular, the curves for $n = 5$ and 9 can be divided approximately into two linear segments meeting at temperatures around 23 and $28 K$, respectively. However, the examination of system dynamics presented below indicates that these systems achieve the kinds of (relatively) facile re-arrangements associated with liquid-like behaviour at temperatures well below $20 K$. Thus, it appears that the so-called "caloric" curves of Fig. 3 do not provide very useful indicators of phase change in these systems.

In previous work on Van der Waals clusters,[1] it was established that the behaviour of a parameter characterizing the relative root mean square bond length fluctuations,

$$\delta = \frac{2}{m(m-1)} \sum_{i=1} \sum_{j>i} [\langle R_{ij}^2 \rangle_t - \langle R_{ij} \rangle_t^2]^{1/2} / \langle R_{ij} \rangle_t \qquad (2)$$

is an effective flag for the changes in dynamical behaviour associated with "melting". Here, R_{ij} is the instantaneous distance between particles i and j, and $m(m-1)/2$ is the number of such distinct pairs in the cluster. The so called "Lindemann criterion" for melting suggests that having this quantity increase through a value of *ca.* 0.1 is an indication of a solid $\rightarrow$ liquid phase transition.[16] While it is unlikely that any single value of δ will provide a definitive universal criterion, a number of studies have found that its rapid growth through values around 0.1 over a narrow range of energies (or effective temperatures) was typically accompanied by rapid growth in the diffusion rate within the cluster.[1] The behaviour of this property is therefore the criterion used herein to identify "melting".

In studies of homogeneous clusters, the most natural definition of δ would clearly involve inclusion of all particles in the averaging of Eq. (2). In a heterogeneous cluster, however, other definitions are often appropriate. In the present case, the monolayer form

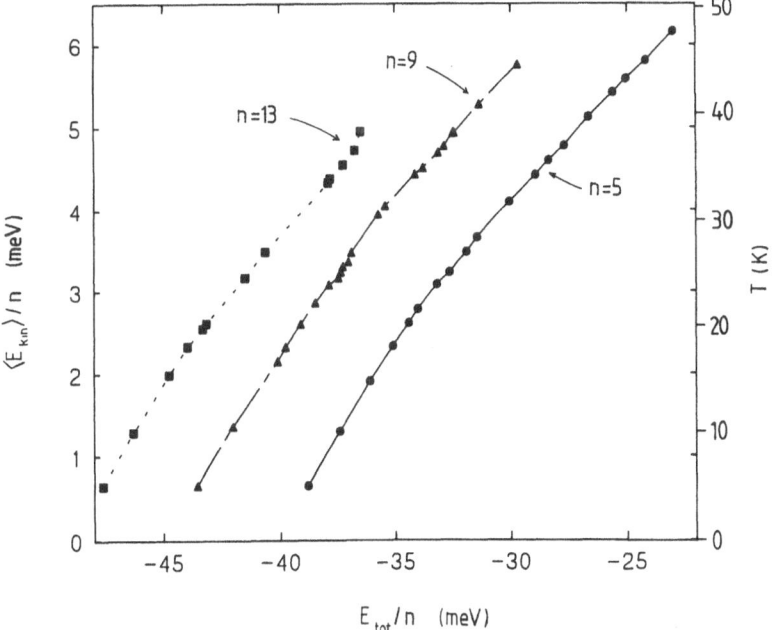

Fig. 3. Plots of the effective temperature calculated from Eq. (1) for SF_6-$(Ar)_n$ clusters with $n = 5$, 9 and 13, as generated from $1.5 ns$ molecular dynamics runs at the indicated total energies.

of the minimum energy structures of Fig. 2 suggests that it might be instructive to examine separately the behaviour of $\delta(Ar-Ar)$, which depends only on the fluctuations in the distances between the various Ar atoms, and $\delta(Ar-SF_6)$, which characterizes the fluctuations in the distances between the Ar atoms and the SF_6.

For $SF_6-(Ar)_n$ with $n = 5$-13, plots of these two types of δ values vs. the effective temperatures associated with a number of 1500 ps simulation runs are shown in Fig. 4.* The triangular points joined by solid lines are the values of $\delta(Ar-Ar)$, and the round points joined by dashed lines the values of $\delta(Ar-SF_6)$ for those same runs. The most prominent feature of these results is the fact that each of the $\delta(Ar-Ar)$ curves rises abruptly on a very narrow temperature range. The range of δ chosen to represent the Lindemann criterion (shaded band on Fig. 4) was therefore based on the fact that all of the $\delta(Ar-Ar)$ curves achieved their maximum slopes within the same narrow range of values. In any case, these results indicate that as far as the Ar atoms' motion *with respect to one another* is concerned, melting occurs at temperatures between 10 and 20 K. On this same interval, however, the much smaller values of $\delta(Ar-SF_6)$ indicate that the Ar atoms are still "frozen" with regard to motion relative to the SF_6. Thus, as was pointed out in Ref. 9, the lowest temperature phase transitions of these heterogeneous clusters appears to be a kind of *two*-dimensional melting in which the Rg atoms become free to move around and rearrange *within* a monolayer wrapped around the SF_6 substrate molecule.

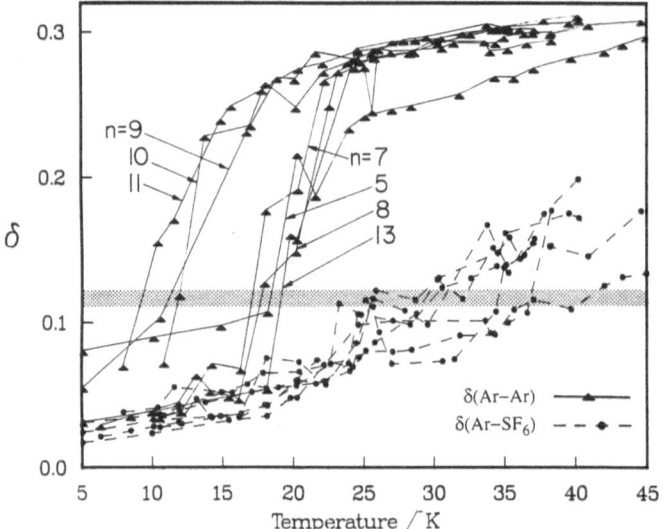

Fig. 4. Averaged relative root mean square bond length fluctuations for the Ar-Ar and Ar-SF$_6$ distances for certain SF$_6$-(Ar)$_n$ clusters. The shaded band identifies the δ values chosen to define the Lindemann criterion for melting.

---

* The simulations for $n = 6$ and 12 are not yet complete, but it appears that the special stability of the pentagonal-pyramid cap structure seen in Fig. 2 makes the $n = 6$ species particularly resistant to melting, and that interesting effects also arise from the stability of a stacked $SF_6-(Ar)_{12}$ isomer which resembles icosahedral Ar_{13} with one of the Ar atoms replaced by SF_6.

Another noteworthy feature of the $\delta(Ar-Ar)$ curves is the grouping of the melting temperatures into two distinct sets. Clusters with $n = 9$-11 Ar atoms melt between 10 and 13 K while those for $n = 5$, 7, 8 and 13 melt at 17-19 K.* A full explanation for this behaviour is not yet available, but it is probably associated with the relative stabilities of the corresponding ground-state structures.

In contrast to the behaviour of $\delta(Ar-Ar)$, the values of $\delta(Ar-SF_6)$ increase relatively gradually with the effective temperature. At temperatures around 30-35 K they too come to satisfy the Lindemann criterion for melting, and in an earlier paper,[9] this was taken as evidence for a distinct *three*-dimensional melting transition. On the other hand, while the slopes of a few of these curves appear to increase sharply as they cross the Lindemann criterion, the overall behaviour of this property suggests that these abrupt changes might be fluctuations associated with the finite length of these simulation runs. Thus, as has been suggested by Garzon et al.,[17] the onset of full *three*-dimensional melting in these clusters may be a much less well define phenomenon than had been suggested in Ref. 9. However, a final conclusion on this point must await further study. In any case, it is clear that in these heterogeneous systems, a two-dimensional "surface" melting of the Rg atoms *within* an incomplete monolayer is the lowest temperature phase transition.

V. Spontaneous Reversing Structural and Dynamical Isomerizations

In simulations of SF_6-$(Ar)_9$ clusters at an effective temperature (26 K) far higher than the 12 K associated with the monolayer melting transition described above, Shelley et al.[9] observed a spontaneous reversing isomerization between the monolayer structure and a stacked isomer in which three of the Ar atoms sit on top of a stable six-atom pentagonal-pyramid cap on the surface of the SF_6. The most remarkable feature of their observation was the fact that for the full *ca.* 0.7 *ns* the system remained in the stacked structure, there occurred *no rearrangements* of the Ar atoms relative to one another. At the same time, the $\delta(Ar-Ar)$ for this portion of the simulation run dropped abruptly to a value much lower than 0.1. Thus, although already far above its initial melting temperature, the system had isomerized from a liquid-like monolayer form to a new solid-like stacked form with a structure quite different from that of the low temperature solid.

This phenomenon appears to differ somewhat from the kind of "phase coexistence" reported by Berry and co-workers[1,18-12] for homogeneous clusters of argon-like Lennard-Jonesium.† In the latter case, the coexistence occurred in the immediate neighbourhood of the first melting transition, and was associated with isomerization between the low-temperature solid-like form and the liquid-like structures formed on melting. Moreover, it often appeared to be characterized by a bimodal kinetic energy distribution, with a distinctly smaller expectation value of the kinetic energy being associated with the liquid-like behaviour. In contrast, Shelley et al.[9,13] found no significant difference between the average kinetic energies of the monolayer and stacked SF_6-$(Ar)_9$ structures, and the latter are quite different that the low-temperature solid monolayer form.

A convenient feature of the isomerization between the monolayer and stacked structures is the fact that the frequency shifts predicted for the fundamental ν_3 band of SF_6 are distinctly different for the two forms.[8,9] In particular, the magnitude of this (negative) frequency shift is approximately proportional to the number of perturber atoms in direct contact with the SF_6, so the magnitude of the shift becomes distinctly smaller when a cluster isomerizes from a monolayer to a stacked form. This provides both a simple property for monitoring the state of the system during a simulation, and a possible means of observing such transformations experimentally.

---

† "Lennard-Jonesium" is defined as a collection of spherical particles held together by pairwise Lennard-Jones (12,6) potentials.

One of the questions raised by the results of Ref. 9 was whether this type of concerted isomerization of both the structural form *and* the dynamical behaviour was specific to SF_6-$(Ar)_9$, or whether it would also occur for a variety of other systems. We have addressed this question, both by examining the SF_6-$(Ar)_9$ case to delineate features which give rise to this behaviour, and by searching for it in simulations of other clusters. With regard to the former, one significant observation is the fact that the pentagonal-pyramid cap which forms the base of this stacked structure is particularly stable relative to neighbouring larger or smaller monolayer forms.[13,15] This suggested that other SF_6-$(Ar)_n$ clusters with $n > 6$ might also have long-lived stacked-structure isomers built on the same first-layer cap. In the following, we report some of our preliminary results for such systems.

To varying degrees, we have now found the novel kind of isomerization behaviour reported by Shelley *et al.*[9] for all SF_6-$(Ar)_n$ clusters in the size range from $n = 7$-12. A first example of this is seen in the the results of the simulation for SF_6-$(Ar)_7$ presented in Fig. 5. The effective temperature for this run was $T = 20.4 K$, which is well above the $17 K$ monolayer melting temperature seen in Fig. 4, and observations of the cluster structures accessed during the run‡ confirmed that the Ar atoms were indeed undergoing rearrangements during this simulation. In the first and last of the structures shown as inserts, the Ar atoms are found to be arranged in the characteristic ground state pentagonal pyramid, with the extra atom in contact with the SF_6 on one side of this cap (and the SF_6 obscured by the Ar atoms). However, the two middle structures associated with the

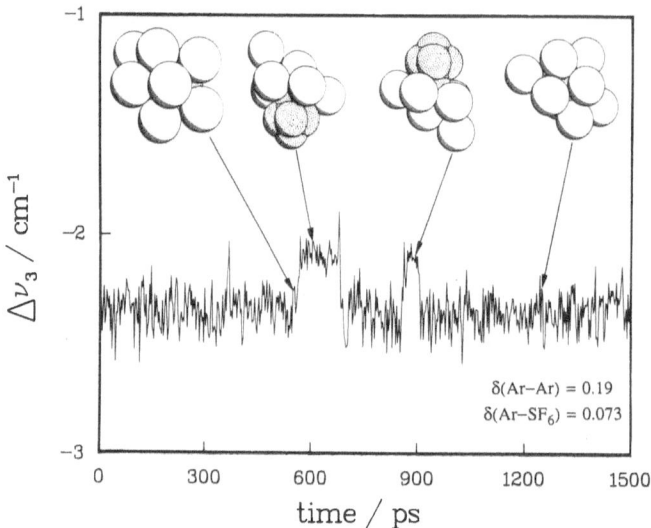

Fig. 5. Predicted frequency shift of the ν_3 band of SF_6 vs. time in a simulation for SF_6-$(Ar)_7$ at an effective temperature of $20.4 K$, together with cluster structures at representative points and the calculated δ values for the entire run.

---

‡ This comment is based on careful examination of a "movie" of the simulation displayed on a Silicon Graphics workstation using software developed by M. Davies.

smaller (in magnitude) frequency shifts are distinctly different, in that one of the Ar atoms has moved up *on top* of this pentagonal-pyramid cap, to form a structure approximately equivalent to that of isomer-*B* for $n = 7$ in Fig. 2. Note that in all of the four structures shown, the absolute orientation of the SF_6 is held fixed, with the same S-F bond protruding towards the viewer. Thus, the changing position of the Ar-atom cap in these pictures indicated a real rotation relative to the SF_6.

While the $n = 7$ cluster spent only relatively short periods of time in the stacked form during this simulation, the considerably damped amplitude of the oscillations in its frequency shifts there attests its more solid-like behaviour. Indeed, during the longer ($100\,ps$) stay of the cluster in the smaller-frequency-shift stacked form, the only rearrangement which occurred was a single displacement of the stacked atom onto a neighbouring site on the pentagonal-pyramid underlayer.‡ This is in accord with the observation of Shelley *et al.*[9] that for SF_6-$(Ar)_9$, a solid-like stacked structure coexisted with a liquid-like monolayer form at a temperature more than twice as high as that for the monolayer melting transition. At the same time, the distinctly shorter lifetimes of the stacked structure for $n = 7$ suggests that it is relatively less stable; this would not be surprising, in view of the fact that its energy minimum lies $6.9\,meV$ above that for the ground state, while for the $n = 9$ case this discrepancy is only $1.4\,meV$.

The second case considered here is SF_6-$(Ar)_{12}$. For this species, the ground state is not the monolayer form, but rather a stacked structure consisting of two staggered pentagonal-pyramids of Ar atoms on one face of the SF_6 (see the upper segment of Fig. 6). It is intriguing to note that while the stacked structure is $22.5\,meV$ more stable than the monolayer form, the latter was found most often in the simulated annealing searches for the minimum energy cluster structure. This may mean that at temperatures significantly above $0\,K$, the monolayer form has a lower *free* energy, and that the annealing was done sufficiently rapidly that the systems tended to get trapped in the form which does not have the lowest potential energy minimum. However, the details of that

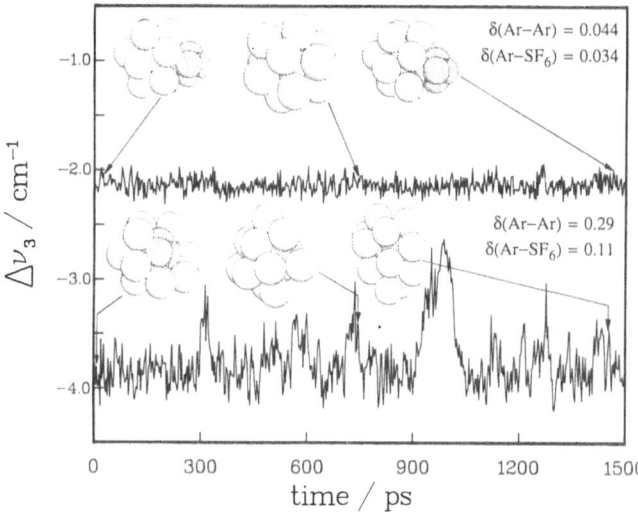

Fig. 6. Representative structures and predicted frequency shift vs. time plots for two separate simulations of SF_6-$(Ar)_{12}$ performed for the same effective temperature, but with different initial structures.

behaviour are not the focus of interest here. We note only that at temperatures above $16K$, δ(Ar–Ar) values for the monolayer form are greater than 0.2 and the system definitely displays liquid-like behaviour.

Fig. 6 presents the frequency shift vs. time behaviour for independent simulations of SF_6-$(Ar)_{12}$ which started from the stacked (upper segment) and monolayer (lower segment) structures, but for which the averaged kinetic energy corresponded to *the same effective temperature* of $30K$. This is far above the melting temperature for the monolayer form. However, as is shown by both δ(Ar–Ar) values and the amplitudes of the oscillations of the frequency shifts with time, the stacked structure is very definitely solid-like and the monolayer form definitely liquid-like. Moreover, isomerization between these forms was remarkably infrequent; in simulations run at a number of temperatures ranging from 20 to $46K$, only a couple of well-defined isomerization events have been seen. Indeed, while the stacked structure remains resolutely solid-like to at least $46K$ (its melting has not yet been observed in our simulations), we expect that Ar atoms will begin to evaporate from the liquid-like monolayer structures at temperatures well below this.

While these studies are far from complete, they definitely confirm the existence of a variety of systems for which coexistence between solid-like and liquid-like forms persists at temperatures far above the lowest-energy melting temperature. However, the conditions which give rise to this type of behaviour are still less than entirely obvious. At the same time, the longevity of at least some of these isomeric forms and the availability of a distinctive spectroscopic signature for a transition from one to the other should facilitate their experimental observation.

Acknowledgments

We are pleased to acknowledge the expert assistance of Mr. M.R. Davies who developed software tools for real-time visualisation of the cluster structure evolution during the course of a simulation. This research was supported by the Natural Sciences and Engineering Research Council of Canada.

References

1. For an excellent review, see, e.g., R.S. Berry, in *The Chemical Physics of Atomic and Molecular Clusters* (Proceedings of Course CVII of the International School of Physics "Enrico Fermi", G. Scoles editor, 1989), Chapters 1 & 2.

2. L.S. Bartell, *Chem. Rev.* **86**, 491 (1986).

3. G. Torchet, in *The Chemical Physics of Atomic and Molecular Clusters* (Proceedings of Course CVII of the International School of Physics "Enrico Fermi", G. Scoles editor, 1989).

4. T.E. Gough, D.G. Knight and G. Scoles, *Chem. Phys. Lett.* **97**, 155 (1983).

5. T.E. Gough, M. Mengel, P.A. Rowntree and G. Scoles, *J. Chem. Phys.* **83**, 4958 (1985).

6. D.J. Levandier, S. Goyal, J. McCombie, B. Pate and G. Scoles, *Faraday Symp. Chem. Soc.* **25**, (1990, in press).

7. R.T. Pack, J.J. Valentini and J.B. Cross, *J. Chem. Phys.* **77**, 5486 (1982).

8. D. Eichenauer and R.J. Le Roy, *J. Chem. Phys.* **88**, 2898 (1988).

9. J.C. Shelley, R.J. Le Roy and F.G. Amar, *Chem. Phys. Lett.* **152**, 14 (1988)

10. R.A. Aziz and M.J. Slaman, *Mol. Phys.* **58**, 679 (1986).

11. C. Pangali, M. Rao and B.J. Berne, *Chem. Phys. Lett.* **55**, 413 (1978).

12. N. Metropolis, A.W. Rosenbluth, M.N. Rosenbluth, A.H. Teller and E. Teller, *J. Chem. Phys.* **21**, 1087 (1953).

13. J.C. Shelley, M.Sc. Thesis, University of Waterloo (1987); also available as *University of Waterloo Chemical Physics Research Report* CP-322 (1987).

14. a) M.R. Hoare and P. Pal, *Adv. Phys.* **20**, 161 (1971); b) *ibid* **24**, 645 (1975); c) *ibid* **32**, 791 (1983).

15. D.J. Chartrand, M.Sc. Thesis, University of Waterloo (1990); D.J. Chartrand, J.C. Shelley and R.J. Le Roy, unpublished work (1989).

16. I.I. Frenkel, *Kinetic Theory of Liquids* (Dover, N.Y., 1955).

17. I.L. Garzon, X.P. Long, R. Kawai and J.H. Weare, *Chem. Phys. Lett.* **158**, 525 (1989).

18. J. Jellinek, T.L. Beck and R.S. Berry, *J. Chem. Phys.* **84**, 2783 (1986).

19. F.G. Amar and R.S. Berry, *J. Chem. Phys.* **85**, 5943 (1986).

20. T.L. Beck, J. Jellinek and R.S. Berry, *J. Chem. Phys.* **87**, 545 (1987).

SPECTROSCOPY AND CLASSICAL SIMULATIONS OF

RIGID AND FLUXIONAL VAN DER WAALS CLUSTERS

Samuel Leutwyler, Thomas Troxler, Jürg Bösiger and
Richard Knochenmuss

Institut für Anorganische, Analytische und Physikalische Chemie, Universität Bern, Freiestr.3, 3000
Bern 9, Switzerland

I. Introduction

In isolated molecules, an increase of internal energy
normally leads to unimolecular rearrangement from the stable
ground-state structure to other shapes and connectivities
(conformers, enantiomers, isomers). Well-studied examples of
compounds undergoing thermal isomerisations are, e.g., (methyl)cycloheptatriene, cyclopropane, and methyl isocyanide
[1-3]. These thermal rearrangements involve the making and
breaking of chemical bonds. In bulk molecular solids, the
increase of thermal energy can lead to formation of new orde-
red or disordered phases and eventually to melting [4]. The
relative forces and motions are intermolecular in nature.
However, there is a close connection between molecular isomerization and bulk phase transitions which becomes obvious in
the case of neat or doped clusters: a range of phenomena occur which are isomerization processes *sensu stricto*, but bear
a close resemblance to bulk phase transitions, due to the
collective and diffusive nature of the transitions [5-16].

We shall focus on various isomerizations and "phase"
transitions in rare gas solvent clusters, describing, on one
hand, new experimental techniques which can provide isomer-
and/or phase-selective [17,18] electronic spectra of van der

Dynamics of Polyatomic Van der Waals Complexes
Edited by N. Halberstadt and K. C. Janda
Plenum Press, New York, 1990

Waals (vdW) complexes and clusters. On the other hand, classical simulations of the same systems are discussed; these yield a deeper understanding of cluster structure and dynamics [9-14], and indicate the connection between molecular and bulk physics which arises in these systems [5-16,41].

II. Isomer- and phase-selective spectroscopy of clusters

Cluster synthesis by supersonic expansion normally yields broad distribution of cluster sizes. The measurement of any size-dependent property depends on the availability of a mass-specific detection scheme [20]. Relatively nondestructive and selective methods for assaying neutral cluster size distributions have been provided by combinations of mass spectrometry with (a) laser techniques (resonant two-photon ionization (R2PI) [21-26], depletion spectroscopy [27]) and (b) helium cross-beam scattering techniques [28].

Isomer selectivity: At a finer level of discrimination, vdW complexes or clusters of a given size may occur as a distribution of structural isomers. Isomers of van der Waals and H-bonded complexes have been detected by rotationally resolved UV or IR spectroscopy [29-34], as have isomers of vdW solvent clusters with aromatic molecules [34-38]. The ionization potentials of vdW isomers can differ substantially [17,18,39], allowing *mass- and isomer-selective* electronic spectroscopy to be performed in a mixture of clusters by selective-ionization (SI) two-color R2PI combined with mass spectrometry in a mixture of clusters [17]. This increase in selectivity is quite general and may also be applied in IR-UV and microwave-UV excitation-ionization schemes.

"Phase" selectivity: We have also observed that rigid (solidlike) Ar_n solvent clusters show sharp steplike photo-ionization efficiency (PIE) curves, while fluxional (liquid-like) clusters exhibit extended tails to lower frequencies. By utilizing SI-R2PI we could discriminate between rigid and fluxional cluster subpopulations, i.e., simultaneously perform *mass- and "phase"-selective* spectroscopy [18].

A. Isomer-selective spectroscopy

We use phenanthrene (Phen) "solute" clustered with Ar as

an example. Microsolvation of Phen was achieved by seeding in pure Ar or 10%Ar/90% Ne supersonic expansions. The two-color R2PI spectra of Phen•Ar_n clusters with n=1-4 are shown in Fig.1. The scans cover a range of 100 cm^{-1} in the spectral vicinity of the Phen $S_1 \leftarrow S_0$ ($^1A_1 \leftarrow ^1A_1$) electronic origin at 29326 cm^{-1}. For n=1 and 3, single strong electronic origins are observed, while for n=2 and 4, two separate features exist, denoted by I and II. These are attributed to origins of two distinct vdW isomers, with various arrangements of the solvent atoms on each side of the Phen "nanosurface", similar to the isomers previously described for perylene [35-38,40]. *Ionization potentials:* The IP of Phen lies above twice the frequency of the $S_1 \leftarrow S_0$ electronic origin, and thus no ion signals were observed when only using the excitation frequency ν_1. The ionization frequency ν_2 was adjusted for minimal excess energy, $h\nu_1 + h\nu_2 - IP < 100$ cm^{-1}. This ensures that ionization takes place without fragmentation. Size-selective sol-

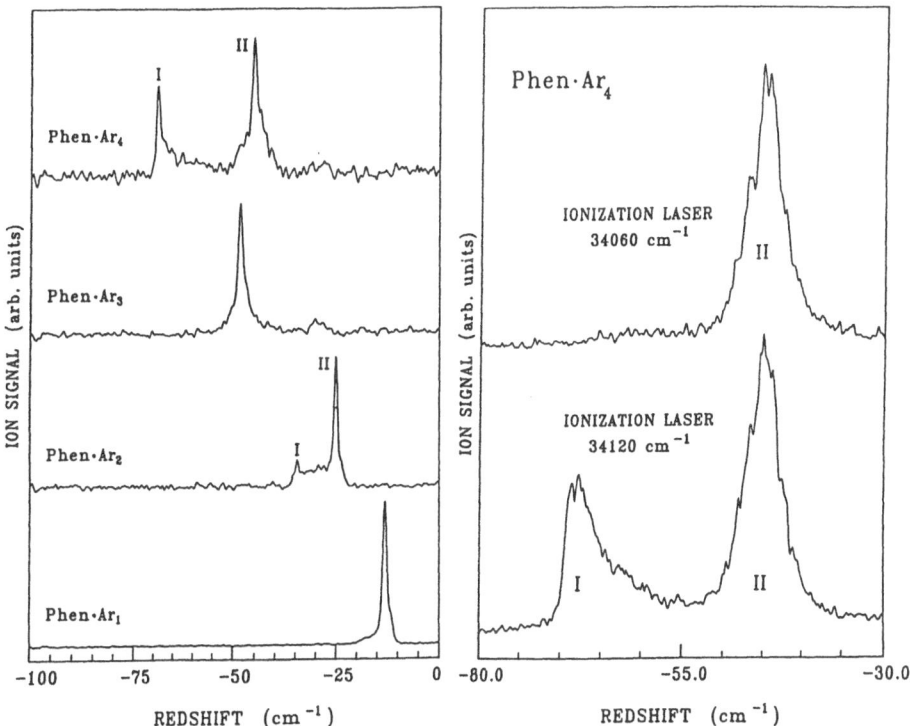

Fig.1. R2PI spectra of Phen•Ar_n (n=1-4) clusters. Note bands of isomers I,II for n=2 and 4.

Fig.2. Size- and isomer-selective two-color R2PI spectra of Phen•Ar_4 isomers.

vent cluster IP's were determined from R2PI PIE curves [23], measured by fixing $h\nu_1$ at the electronic origin of the respective cluster, and scanning $h\nu_2$, which are shown in Fig.3. For Phen the ion signal onset is very sharp, and a linear extrapolation of the threshold to the baseline yields IP=63689 cm^{-1} (7.8965 eV). Fig.3 also shows the PIE curves for Phen·Ar_n (n=2 and 4) when exciting at either band I or II. For both clusters, the IP shift relative to bare Phen depends on the feature which is being excited: for Phen·Ar_2, the IP shifts are ΔIP=-109 cm^{-1} for excitation at band I, and ΔIP=-132 cm^{-1} for band II. This amounts to a difference of 23 cm^{-1} or ~15% of the total solvent shift of the IP, while for Phen·Ar_4, the difference is 56 cm^{-1}, about 20-25% of the total IP shift. From the data in Figs.1-3 we infer that:

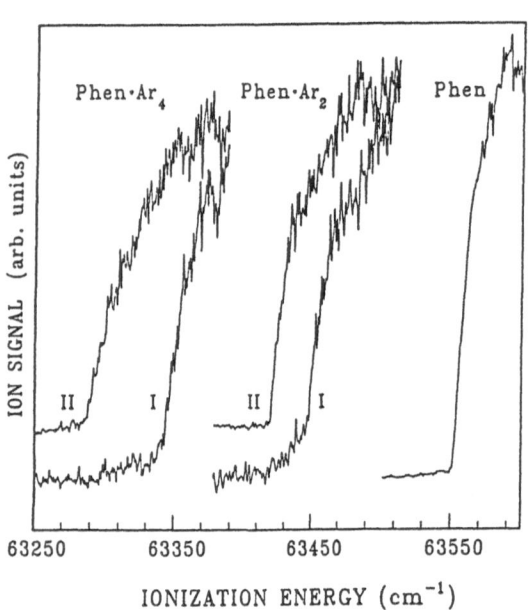

(i) For both n=2 and 4, two different cluster isomers exist (denoted analogously as I and II), since they differ both in the position of the $S_1 \leftarrow S_0$ electronic origin bands, and in the associated adiabatic IP.

(ii) The differences in the spectral and IP shifts must reflect differences in cluster structure. The cluster binding topologies were deduced from a set of additivity relationships of the spectral and IP shifts, which imply that the n=2;II isomer corresponds to the *opposite-sided* (1+1) structure, the n=2;I isomer has the *same-sided* (2+0) structure, and n=3 occurs as the (2+1) structure. For n=4, we assign n=4;I as the (2+2) isomer; the n=4;II structure can not be determined unequivocally by this procedure, but is tentatively assigned as (3+1) on the basis of calculations.

Fig.3. Two-color R2PI PIE curves for Phen·Ar_n [n=0, 2 (I,II),4(I,II)].

(iii) Since both the PIE curves and the two electronic origins signals are completely distinct, no interconversion between the two isomers seems to occur on the ~10 ns time-scale of the excitation/ionization sequence. This is in accord with the calculated side-crossing isomerization barrier of approximately 350 cm^{-1}. However, isomerization should take place at higher internal energies.

(iv) Remarkably, the order of the IP shifts and electronic spectral shifts are reversed for both n=2 and 4: in both cases isomer I exhibits a more red-shifted origin, yet a smaller IP shift relative to isomer II. The former property is related to the transition dipole moment and dipole moment distribution, which for Phen can only be understood on the CI level, while the latter relates to the "hole" density distribution of the frontier orbitals [42].

Isomer-selective spectroscopy: For both n=2 and 4, the difference in IP shifts between the isomers I and II provides a means of selectively ionizing isomer II. We have utilized this to measure size- and isomer-selective spectra in the electronic origin regions. Fig.2 shows the R2PI signal detected in the Phen•Ar$_4$ mass channel when fixing the ionizing laser at ν_2= 34060 cm^{-1}: the spectral feature attributed to isomer II is clearly reproduced, while that of isomer I is completely suppressed. Raising ν_2 by 60 cm^{-1} to 34120 cm^{-1} and repeating the scan results in ionization of both isomers (Fig.2). The spectrum of isomer I can be generated by subtraction of the spectra (not shown). For Phen•Ar$_2$, separation of the spectral features denoted I and II was also possible, establishing them as due to different isomers. Thus, isomer-selective electronic spectroscopy (even of overlapping electronic spectra) can be performed by the selective-ionization R2PI technique. Furthermore, the dependence of IP on cluster structure may make such measurements useful as a structural probe for these and other clusters.

B. Phase-selective spectroscopy

In the second application of selective ionization, we select not different rigid cluster isomers but differentiate between rigid and fluxional cluster subpopulations of carbazole•Ar_n solvent clusters. The carbazole (Car) system differs from phenanthrene in that the cluster buildup from n=1-7 proceeds via a series of same-sided clusters; therefore phenomena such as surface-rotation and surface-decoupling become important at lower cluster size.

The PIE curves for small Car•$(Ar)_n$ clusters excited to their respective $S_1 \leftarrow S_0$ electronic origins vary significantly with size. For the n=0-2 clusters a sharp PI threshold is observed, which reflects the structural similarity of the solvent clusters in the S_1 and ion ground states: when the global minimum region of the intermolecular potential energy (PE) hypersurfaces of the S_1 and ion ground-state electronic states are similar in all coordinates, the Franck-Condon factors for transitions involving no change of vibrational state are large, which for photoionization translates into a sharp ion signal onset at the adiabatic IP. The PIE curves for the larger clusters also show a steplike onset, but in contrast to the smaller clusters, a pronounced tail extending ~100 cm^{-1} (n=4) to ~200 cm^{-1} (n=6) to lower energy is observed.

The PIE curves for n=4-6 can be experimentally decomposed into two separate contributions, as shown in Fig.4 for n=5. When the excitation laser is set to the maximum of the $S_1 \leftarrow S_0$ electronic origin (ν_1=30693 cm^{-1}; see Fig.5a), the solid-line PIE curve is obtained. The dashed curve was obtained when the excitation laser is set to ν_1=30678 cm^{-1}, 15 cm^{-1} below the 0_0^0 band, within the broad absorption (see Fig.5b): the "tail" region of the PIE curve is practically identical to that for excitation at ν_1=30693 cm^{-1}.

We infer (a) that the "tail" and the "step" regions of the solid PIE curve correspond to ionization of two distinct sub-populations of the n=5 clusters; (b) that the subpopulation which contributes to the "tail" part of the PIE curve leads to the spectrally broad $S_1 \leftarrow S_0$ band, while the "step" region of the PIE curve corresponds to the narrow spectral features of the R2PI spectrum. By fixing the ionization frequency and scanning the excitation frequency, selective-ionization R2PI spectra of these two subpopulations were measured. When ionizing at a total energy slightly higher than the adiabatic

IP, the spectrum in Fig.5(a) was measured. Upon reducing the
ionization frequency so that $h\nu_1 + h\nu_2$ lies within the "tail"
region but below the "step" in Fig.4, the spectrum shown in
Fig.5(b) was obtained. This spectrum corresponds to the
subpopulation of the n=5 clusters which exhibit the dashed
PIE curve in Fig.4. Since the spectrum in Fig.5(a) is a com-
posite of both subpopulations, subtraction of Fig.5(b) yields
the spectrum of the subpopulation corresponding to the sharp
step region in Fig.4 [bottom curve (a-b) in Fig.5]. Thus,
Fig.5(a) decomposes cleanly into a spectrum with narrow bands
($\Delta\nu \approx 5$ cm^{-1}) with virtually no residual broad background after
subtraction, and a broad ($\Delta\nu \approx 70$ cm^{-1}) spectrum with

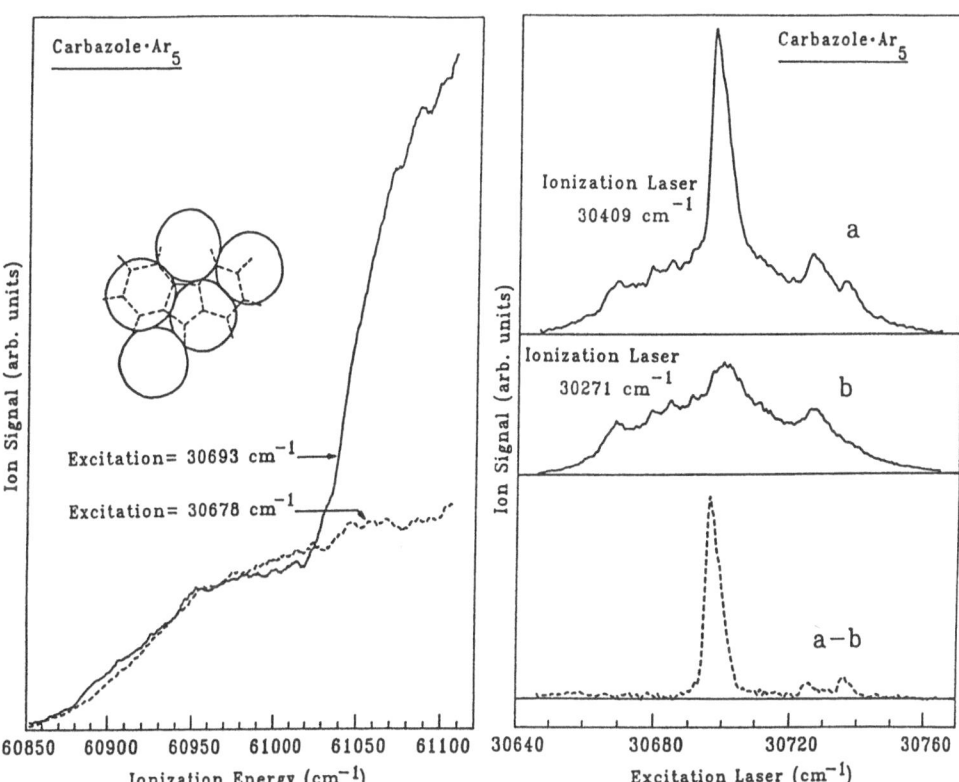

Fig.4. R2PI PIE curves for
Car·Ar$_5$ clusters for two different excitation frequencies.
solid line : on 0_0^0 band.
dashed line: 15 cm^{-1} lower.

Fig.5. SI-R2PI spectra of Car·
Ar$_5$ clusters at different ionization frequencies (see Fig.
4). (a) above step; (b) below
step in tail region of ionization threshold.

weak residual spectral structure. SI-R2PI was also used to measure the spectra of the n=4-6 clusters: in Fig.6 the separate spectra for the two subpopulations are shown. In each case a similar pattern is observed: one spectrum consist of a single strong origin band with short progressions in intermolecular vibrations, very similar to those of the n=1-3 clusters [9-12]. This pattern has been interpreted as due to clusters which exist in a single minimum energy structure (or a few closely related structures) and thus can be designated as structurally rigid [9-12,14]. By extension, we assign the narrow-band spectra of Fig.6 to clusters which do not undergo large-amplitude internal motion, which we denote as *rigid* clusters. The other spectral component of each pair is very wide (50-70 cm^{-1}, ≈10-15 times the width of the narow bands), but exhibits weak residual bands of width comparable to the sharp peaks. These spectra are interpreted as due to clusters which sample a large range of geometries, either on the tiime scale of the experiment or averaged over the ensemble. As a result of the variety of structures (and hence solvent-solute interactions), these clusters should have a broad range of electronic origin shifts and hence broad spectra; in addition, Franck-Condon factors for $\Delta v \neq 0$ may also be large, since the clusters are high up in anharmonic regions of the PE hypersurface with a high associated vibronic level density. These clusters are termed fluxional.

This picture must be extended to explain the relatively narrow features observed atop the broad-band spectra. Some of these are frequency-correlated with bands in the corresponding rigid cluster spectra, but are always slightly blue-shifted. It is known from the asymmetry of the small cluster 0_0^0 bands that the S_1 state potential wells are narrower than those in the ground state, leading to blue shifted sequence bands. From the molecular physics point of view the frequency-correlated features appearing in the broad band spectra can be interpreted as sequence bands of vibrationally highly excited clusters which exist in the global minimum structure. However, these clusters may also visit other minima , i.e., isomeric structures: there are weak features atop the broad spectra which have no counterpart in the rigid cluster spectra, and these are tentatively assigned as due to other cluster isomers which become accessible at higher internal energies. If the lifetime in any given well region is long compared to the relevant vibrational frequencies, and if interwell hopping is infrequent ($<10^{12}s^{-1}$), then relatively sharp spectral features may still be observed. In fluid physics terms, these clusters can be viewed as "inherent structures" in the sense of Stillinger et al. [41].

In principle, SI-R2PI spectra will change with selection of the ionization laser wavelength, if there are distinguishable isomers or fluxional subpopulations with different

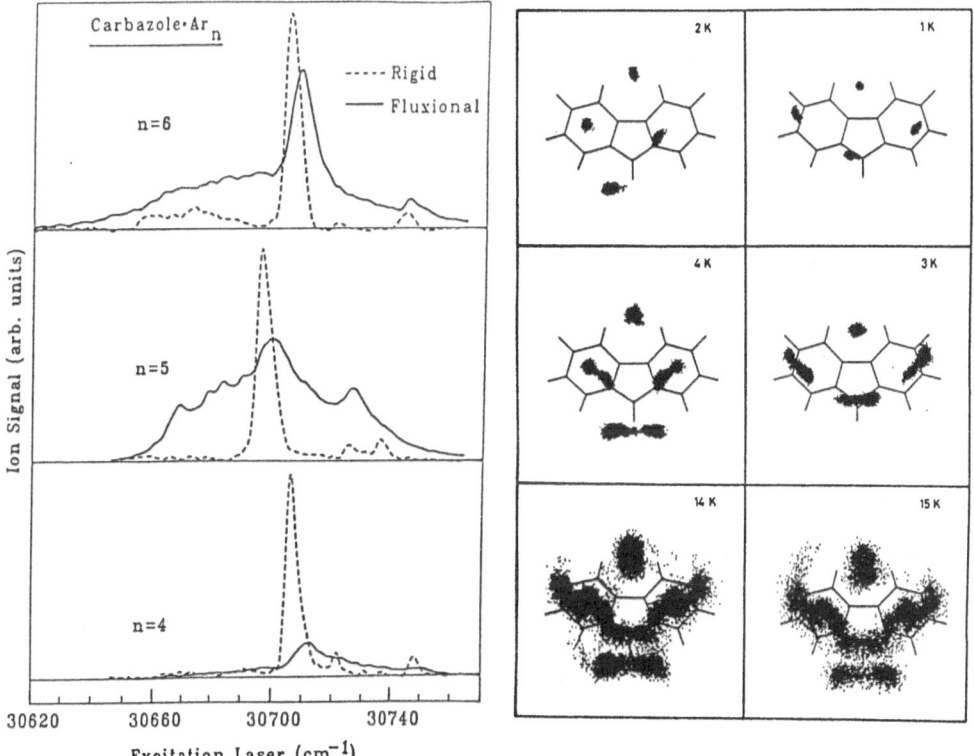

Fig.6. SI-R2PI spectra of rigid and fluxional clusters for Car•Ar_{4-6}. Full lines: "fluxional", dashed lines: "rigid" subpopulations, obtained by subtraction (as in Fig.5).

Fig.7. Spatial probability density distributions obtained by MC simulations of Car•Ar_4. Top: substrate-fixed; middle: racemization transition; bottom: surface-decoupling transition.

IP's; we have not yet exhaustively checked the range of possibilities. For the interpretation of the "phase" transitions
of these clusters which underlie the specific spectroscopic
signatures just discussed, we turn to Monte Carlo simulation
techniques.

III. Classical simulations

Using empirical potentials, solute-solvent and solvent-
solvent interactions can be modeled, allowing many aspects of
solvent clusters to be studied semi-quantitatively by
classical simulation techniques (Monte Carlo, molecular
dynamics). As a specific example we discuss two cluster/sub-
strate-type "phase" transitions of the carbazole•Ar_4 cluster. This is the smallest system that exhibits all types of
two-dimensional transitions observed so far in MC simulations, and, at the same time, allows a direct and intuitive
interpretation of the associated structural changes [9-14].

Simulations were initialized at two isomeric structures
which correspond to the minimum-energy-structure (calculated
intermolecular binding energy $V=2032$ cm^{-1}) and an isomer
which is 12 cm^{-1} higher in energy; these structures were
identified by molecular dynamics slow cooling. Fig.7 displays
the probability density distribution for both isomers at
three characteristic temperatures between 1 and 15 K. At T=1
or 2 K, the clusters are well localized relative to the substrate. The thermal distribution for the Ar atom lying on the
σ_v symmetry plane of the substrate is isotropic, while those
of the three others are distinctly anisotropic, with larger
RMS displacements. This can be interpreted as thermal excitation of low-frequency hindered rotations of the adcluster, with the "axis of rotation" passing through the atom
lying on the σ_v plane. At an internal temperature of 3 or 4 K
the amplitude of this movement increases far enough that
interconversion between the two symmetry-equivalent adcluster
positions are possible, leading to *racemization* between the
low-temperature enantiomers. For both isomers the
characteristic motion is approximately a rigid-body partial
rotation of the entire adcluster by ±15° relative to the
substrate. The RMS atomic displacements $\sigma_i(i=2-4)$ reveal that
this *racemization transition* takes place within a narrow

temperature range, rising by a factor of six within one degree [14].

A next transition at T≥14 K is evident from Fig.7 (middle and bottom panels): the configuration space accessible to either isomer is now the *same*, and is a superposition of the two subspaces of the previously distinct isomers; the probability distribution is weighted slightly in favor of the minimum-energy structure. As the lower panel of Fig.7 shows, the spatial distribution no longer depends on the initial configuration. Below the transition temperature, each Ar atom can reach at most two preferred positions in configuration space, (Fig.7, middle panel). Above the transition temperature, every Ar atom can reach the whole of the available configuration space by cyclic permutation of the Ar_4 as a unit relative to the substrate surface: the adcluster moves in quasicircular surface rotational jumps, on which large-amplitude translational motions are superimposed. We note that the transition range is fairly broad, spanning ≈4 K, which implies that there is a *coexistence temperature region* for this type of transition. This transition type is denoted as a *surface decoupling* transition [9-14].

Since this is the lowest-temperature transition which leads to a considerable "spreading" of cluster atom probability density in configuration space, we interpret the "phase" transitions which were spectroscopically characterized in Sect.II as due to surface-decoupling transitions for n=4-6. This qualitative consideration is borne out by energy- and temperature-selective simulations of the cluster electronic spectra which are currently in progress. These simulations show that the racemization transitions also lead to a broadening of the electronic spectrum, mainly due to the coupling of the "soft" racemization mode to the electronic origin, which amounts, however, only to several cm^{-1}. For clusters containing a larger amount of energy, different other types of rearrangements are possible. The energy (or temperature) thresholds for the different processes tend to be spaced apart, and a temperature sequence of order-disorder or "phase" transitions emerges. Although the sequence of transitions depends on the size of the cluster, an overall hierarchy of transitions can be discerned [12,14].

Acknowledgments

This research was supported by the Schweiz.Nationalfonds (grant no.2.467-87), and by the Wander-Stiftung, Bern.

References

[1] S.H.Luu, K.Glänzer and J.Troe, Ber.Bunsenges.Phys.Chem. **79**, 855 (1975); H.Hippler, K.Luther, and J.Troe, Farad.Soc. Disc. **67**, 173 (1979).
[2] W.Forst, *Theory of Unimolecular Reactions*, (Academic Press, New York, 1973).
[3] H.O.Pritchard, *The Quantum Theory of Unimolecular Reactions*, (Cambridge University Press, Cambridge, 1984).
[4] H.Eugene Stanley, *Introduction to Phase Transitions and Critical Phenomena*, (Oxford University Press, Oxford, 1971).
[5] G.Natanson, F.Amar, and R.S.Berry, J.Chem.Phys **78**, 399 (1983); R.S. Berry, J. Jellinek, and G. Natanson, Phys. Rev. A **30**, 919 (1984); Chem. Phys. Lett. **107**, 227 (1984).
[6] N. Quirke and P. Sheng, Chem. Phys. Lett. **110**, 63 (1984).
[7] J. Jellinek, T.L. Beck, and R.S. Berry, J. Chem. Phys. **84**, 2783 (1986); F.Amar and R.S.Berry, J.Chem.Phys.**85**, 5774 (1986); T.L.Beck, J.Jellinek, and R.S.Berry, J.Chem.Phys. **87**, 545 (1987); H.L.Davis, J.Jellinek, and R.S.Berry, J.Chem.Phys. **86**, 6456 (1987).
[8] J.D.Honeycutt and H.C.Andersen, J.Phys.Chem. **91**,4950 (1987).
[9] S.Leutwyler and J.Bösiger, Zeitschr.Phys.Chemie NF, **154**, 31 (1987).
[10] J.Bösiger and S.Leutwyler, Phys.Rev.Lett., **59**, 1895 (1987).
[11] J.Bösiger and S.Leutwyler, in *Large Finite Systems*, eds. J.Jortner and B.Pullman (D.Reidel, Dordrecht 1987), pp.153-164.
[12] S.Leutwyler and J.Bösiger, Faraday Discuss.Chem.Soc. **86**, 225 (1988); Chem.Rev., in press.
[13] J.Bösiger, R.Knochenmuss and S.Leutwyler, Phys.Rev.Lett., **62**, 3058 (1989).
[14] J. Bösiger and S. Leutwyler, submitted to J.Chem.Phys.
[15] T.E.Gough, D.G.Knight, and G.Scoles, Chem.Phys.Lett. **97**, 155 (1983); T.E.Gough, M.Mengel, P.A.Rowntree, and G.Scoles, J.Chem. Phys. **83**, 4958 (1985).
[16] D. Eichenauer, and R.J.LeRoy, Phys.Rev.Lett. **57**, 2920 (1986); R.J.LeRoy, J.C.Shelley, and D.Eichenauer, in *Large Finite Systems*, eds. J.Jortner and B.Pullman (D.Reidel, Dordrecht 1987), pp.165-172; D.Eichenauer and R.J.LeRoy, J.Chem.Phys. **88**, 2898 (1988); J.C.Shelley, R.J.LeRoy, and F.G.Amar, Chem.Phys. Lett. **152**, 14 (1988).
[17] T.Troxler, R.Knochenmuss, and S.Leutwyler, Chem.Phys.Letters **159**, 554 (1989), and references therein.
[18] R.Knochenmuss and S.Leutwyler, submitted to J.Chem.Phys.
[19] S.Leutwyler and J.Bösiger, Chem.Rev., in press.
[20] see M.Kappes and S.Leutwyler, chap.15 in *Atomic and Molecular Beam Methods*, edited by G.Scoles (Oxford University Press, Oxford,1988) pp.398-406.
[21] A.Herrmann, S.Leutwyler, E.Schumacher, and L.Wöste, Chem.Phys.Lett. **52** (1977) 418; A.Herrmann, S.Leutwyler, E.Schumacher, and L.Wöste, Helv.Chim.Acta **61** (1978) 453.
[22] D.L.Feldman, R.L.Lengel, and R.N.Zare, Chem.Phys.Lett. **52** (1977) 413.

[23] S.Leutwyler, A.Herrmann, L.Wöste and E.Schumacher, Chem.Phys. 48 (1980) 253; S.Leutwyler, M.Hofmann, H.-P.Härri and E.Schumacher, Chem.Phys.Lett. 77 (1981) 257.

[24] J.Hopkins, D.Powers, and R.Smalley, J.Phys.Chem. 85 (1981) 3739.

[25] K.H.Fung, W.E.Henke, T.R.Hays, H.L.Selzle, and E.W.Schlag, J.Phys.Chem. 85 (1981) 3560.

[26] S.Leutwyler, U.Even, and J.Jortner, Chem.Phys.Lett. 86 (1982) 439; S.Leutwyler, U.Even, and J.Jortner, J.Chem.Phys. 79 (1983) 5769.

[27] J.Geraedts, S.Setiadi, S.Stolte and J.Reuss, Chem.Phys.Lett. 78 (1981) 277; J.Geraedts, S.Stolte, and J.Reuss, Z.Phys.A 304 (1982) 167.

[28] U.Buck and H.Meyer, Phys.Rev.Lett. 52 (1984) 109; U.Buck and H.Meyer Surf.Sci. 156 (1985) 275; U.Buck and H.Meyer, J.Chem.Phys. 84 (1986) 4854.

[29] C.A.Haynam, D.V.Brumbaugh, and D.H.Levy, J.Chem.Phys 79 1581, (1983); L.Young, C.A.Haynam, and D.H.Levy, ibid. 79, 1592 (1983)

[30] E.Carrasquillo, T.S.Zwier and D.H.Levy, J.Chem.Phys. 83, 4990 (1985).

[31] R.E.Miller, Science, 240, 447 (1988);

[32] O.Cheshnovsky and S.Leutwyler, J.Chem.Phys. 88 (1988) 4127.

[33] R.Knochenmuss, O.Cheshnovsky and S.Leutwyler, Chem. Phys.Lett. 144 (1988) 317.

[34] see contributions by W.Klemperer and D.Nesbitt in this volume.

[35] M.J. Ondrechen, Z. Berkovich-Yellin, and J. Jortner, J. Am.Chem. Soc. 103, 6586 (1981).

[36] U. Even, A. Amirav, S. Leutwyler, M.J. Ondrechen, Z. Berkovich-Yellin and J. Jortner, Far. Disc. Chem. Soc. 73, 155 (1982).

[37] S. Leutwyler, J. Chem. Phys. 81, 5480 (1984).

[38] M.M. Doxtader, I.M. Gulis, S.A. Schwartz, and M.R. Topp, Chem. Phys. Lett. 112, 483 (1984).

[39] P.D. Dao, S. Morgan, and A.W. Castleman, Jr., Chem. Phys. Lett. 111, 38 (1984).

[40] S. Leutwyler and J. Jortner, J.Phys.Chem. 91, 5558 (1987).

[41] F.H.Stillinger and T.A.Weber, Science 225, 983 (1984); R.A.LaViolette and F.H.Stillinger, J.Chem.Phys. 83, 4079 (1985); F.H.Stillinger and T.A.Weber, ibid.83, 4769 (1985).

[42] T.Troxler and S.Leutwyler, in preparation.

CROSSED MOLECULAR BEAM STUDIES ON

ATOM-MOLECULE VAN DER WAALS COMPLEXES

L.Beneventi, P.Casavecchia, and G.G.Volpi

Dipartimento di Chimica
Università di Perugia
06100 Perugia, Italy

ABSTRACT. Total differential cross section (DCS) measurements
in the thermal energy range have been carried out for $Ne-O_2$,
$Ne-NO$, $Ar-O_2$, $Ar-N_2$, $He-Cl_2$, $Ne-Cl_2$, and $Ar-Cl_2$ under
high-resolution conditions, which have always permitted to
resolve the diffraction and/or rainbow oscillations. Reliable
full anisotropic potential energy surfaces have been derived by
also simultaneously fitting all other available experimental
properties in a multiproperty fashion within the framework of
the infinite-order-sudden approximation, which has been shown
to be accurate under the present experimental conditions.
Information on the anisotropy of the interaction is obtained
from the quenching of the diffraction and rainbow structure in
the total DCS.

INTRODUCTION

The study of potential energy surfaces (PES's) for
atom-molecule van der Waals complexes is a topic of much
current interest.[1] The anisotropy of the PES is responsible for
the rotational energy transfer, which is one of the most
efficient processes in molecular dynamics. Once the PES is
known to a sufficient accuracy, all the necessary physical
observables can be calculated from it. While a hierarchy of
exact and approximate methods[2] is available for calculating
cross sections from a given PES and, in general, for treating
dynamical problems, the calculation of accurate PES's from ab
initio methods, even at the state-of-the-art level, with the
exception of a very few light (few electrons) cases, is still a
big problem because of the weakness of van der Waals
interactions. Furthermore, semiempirical atom-atom interaction
modeling procedures have proven difficult to generalize to
atom-molecule interactions.[3] Most information on atom-molecule
PES's comes from experiments, especially of scattering type,
possibly coupled to theoretical data for the very short-range
repulsion and long-range attraction.

The most detailed information on the anisotropy of the
interaction is generally provided by the measurements of

Dynamics of Polyatomic Van der Waals Complexes
Edited by N. Halberstadt and K. C. Janda
Plenum Press, New York, 1990

399

state-to-state rotationally inelastic differential cross sections.[1,4] However, single rotational transitions can only be resolved experimentally for a very limited number of systems. Differential energy loss spectra still provide a sensitive probe of the anisotropy, especially if rotational rainbows are observed.[1] Rotationally inelastic cross sections, as well as rotational rainbows, give information only on the relative anisotropy of a system, i.e., the difference between potential curves of different orientations. The determination of the absolute scale of the interaction potential relies on additional information, as the precise measurement of diffraction oscillations in the total (elastic + inelastic) differential cross section (DCS). For the determination of the other relevant features of the attractive well of the full PES, measurements of the rainbow structure in the total DCS, and/or the glory structure in the total integral cross section or the second virial coefficient down to low temperatures are required. Transport data can then be used to check the validity of the potential so derived and, if available up to very high temperatures, to extend its validity in the region of the repulsive wall.

In our laboratory we have recently exploited the very detailed information content of diffraction scattering for the determination of interaction potentials for atom-atom[5,6] and atom-molecule[7-9] systems. It has been shown that precise total DCS measurements, carried out under high-resolution conditions and presenting well resolved diffraction oscillations, coupled to absolute total integral cross sections, second virial, diffusion and viscosity coefficients, and semiempirical long-range coefficients from literature, permit deriving reliable PES's for systems as He and Ne interacting with simple diatomic molecules as O_2, N_2, NO, and CO_2.[7-9] In particular, the analysis of the diffraction oscillations, within the framework of the infinite-order-sudden (IOS) approximation, not only gives the absolute scale of the potential, but also provides quantitative reliable information on the anisotropy of the interaction. The reliability of the IOS decoupling scheme in deriving a fully anisotropic PES from the measured scattering dynamics for He-containing systems[10,11] and also for relatively heavy systems, as Ne-N_2,[8] has been examined and demonstrated by performing exact close-coupling calculations.

In this contribution we report experimental results on total DCS for Ne-O_2 and NO, Ar-O_2 and N_2, and He, Ne, and Ar-Cl_2, which represent an extension of previous work on He-N_2, O_2, and NO,[7] Ne-N_2,[8] Ar and Kr-NO,[12] and He-CO_2.[9] The rare gas-chlorine systems have recently attracted much interest in relation to vibrational predissociation studies.[13-16]

In the next section we briefly discuss the experimental set-up to measure differential cross sections. Then, we present some typical results and discuss the procedure of analysis to derive potential energy surfaces from the available experimental data.

EXPERIMENTAL

The experiments were performed in a high-resolution crossed molecular beam apparatus which has been described in detail elsewhere.[5,7,8] Briefly, two well collimated, differentially pumped, supersonic nozzle beams are crossed at 90° in a large scattering chamber and the in-plane scattered lighter particle is detected by a rotating ultra-high-vacuum quadrupole mass spectrometer detector.

Typical beam velocity spreads are about 3% for He, 4.5% for Ne, 6-9% for Ar, and 10-18% for the molecular species. The primary and secondary beam angular divergences are always 0.4° and 1.8°, respectively. The detector angular resolution is 0.5°. These narrow divergences in angle and velocity of the colliding beams, as well as the high angular resolution of the detector, proved to be critical for the observation of a well resolved diffraction structure in the differential cross sections. The beam velocities were measured by absolute time-of-flight analysis to better than 1%, using a high-speed multichannel scaler and a computer controlled CAMAC data acquisition system. The angular locations were determined to better than 0.03°.

The translational temperatures of the molecular species in the beam have been estimated from the velocity distributions to be of the order of 8-20 K. Since the supersonic expansion provides considerable rotational cooling, the assumption of equilibrium between rotational and translational temperatures in the beam leads to the conclusion that almost all the diatomics are in their lowest rotational levels.

The laboratory total angular distributions I(Θ) were obtained by taking at least four scans of 30-90 s counts at each angle. The secondary target beam was modulated at 160 Hz by a tuning fork chopper for background subtraction.

For most of the systems, we have also measured the total DCS at negative angles with respect to the primary beam and have used this in the derivation of the potential surface. This proved to be very useful for deriving an accurate absolute scale of the potential. In fact, shifts between experimental and calculated diffraction oscillations due to a non correct location of the potential on the absolute distance scale appear more pronounced at negative angles in the lab system because of the kinematic c.m.→lab. transformation. In addition, a higher resolution is achieved at negative angles because of a "kinematic cooling" effect.[8]

RESULTS AND DISCUSSION

As an example of the data quality, in Fig. 1 we report the total DCS data, multiplied by $\Theta^{7/3}$, for $Ne-O_2$ measured at positive and negative angles at a collision energy of 74.9 meV. Exemplary error bars are shown, representing ±1 standard deviation. At angles <10° they become smaller than the dots. Five or six clearly resolved diffraction oscillations superimposed on the fall-off of the main rainbow structure can be seen. The oscillatory pattern is significantly quenched with respect to the corresponding isotropic Ne-Ar system[5] measured in the same experimental arrangement: this is a manifestation of the anisotropy of the $Ne-O_2$ potential. In fact, the damping of the diffraction oscillations can be directly related to the anisotropy of the location of the repulsive wall (and then of the minimum position), while the damping of the rainbow is related to the anisotropy of the well depth.[17] Data analysis has been carried out along the lines followed for the He-containing systems[7,9] and Ne-N_2,[8] by simultaneously fitting all other available experimental properties. For Ne-O_2 these properties are: absolute total integral cross sections (also measured in our laboratory), diffusion and viscosity coefficients (from literature), and the Zeeman spectrum (measured in Nijmegen).[18]

Data analysis uses a potential model in which the anisotropy is described by making the size parameters ε and R_m angle dependent, while the reduced form is taken to be the same for all orientations.[7,8] The analytical form of the potential model used is:

$$V(R,\gamma) = \varepsilon(\gamma) \, f(x) \quad , \qquad x = R/R_m(\gamma)$$

where the parameters $\varepsilon(\gamma)$ and $R_m(\gamma)$ are given by:

$$\varepsilon(\gamma) = \bar{\varepsilon}[1 + A_2 P_2(\cos\gamma) + \ldots]$$
$$R_m(\gamma) = \bar{R}_m[1 + B_2 P_2(\cos\gamma) + \ldots] \quad . \tag{1}$$

Here R designates the distance between the center of mass of the molecule and the atomic partner, and γ the angle between R and the internuclear molecular axis. The reduced curve $f(x)$ is generally chosen to be the piecewise analytic exponential--spline-Morse-spline-van der Waals (ESMSV) form.[7,8] Only even terms appear in expansion (1) for homonuclear diatomics, and a P_2 term is sufficient to describe the anisotropy effects experimentally observed for most of the systems. However, preliminary best-fit calculations for rare gas-Cl_2 systems seem

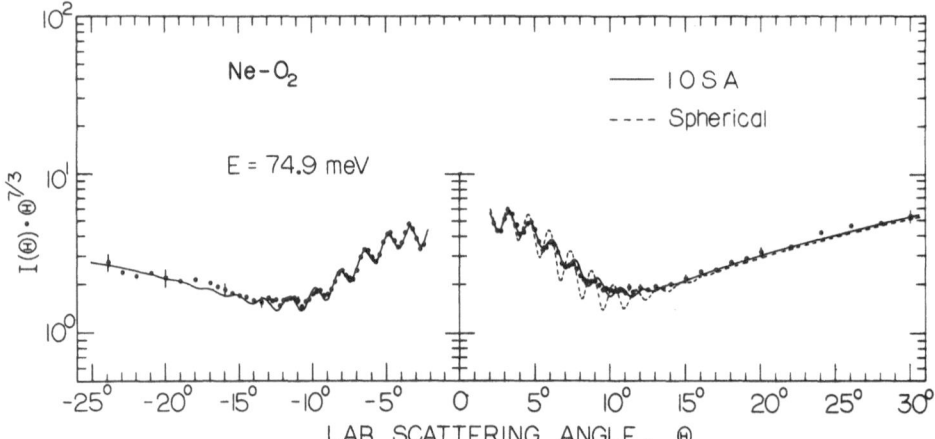

Fig. 1. Total differential cross section data for Ne-O_2 measured at positive and negative angles. Continuous line: IOS calculation with best-fit potential surface (Ref.18). Dashed line: calculation (shown only at positive angles) with the spherical limit potential, obtained from the full potential by setting the anisotropy A_2 and B_2 parameters equal to zero in Equation (1).

to indicate the need also of a P_4 component in equation (1).

Under the present experimental conditions, the scattering data for the He and Ne-diatom systems can be analyzed by using the powerful and simple IOS approximation.[19] In fact, the collision energies of our experiments are sufficiently high and substantially greater than the rotational spacing of the diatomic molecules under study. Therefore, the centrifugal sudden and energy sudden approximations are both expected to be valid. The DCS are calculated for each orientation angle γ by partial wave analysis[20] with JWKB phase shifts.[21] A 32-point Gauss-Legendre quadrature was used to average the c.m. cross sections $\sigma(\theta,\gamma)$ over $\cos\gamma$ according to the IOS formula:

$$\sigma(\theta) = 1/2 \int_{-1}^{+1} \sigma(\theta,\gamma) \, d(\cos\gamma) \ .$$

For comparison with the experimental $I(\Theta)$, the c.m. cross section $\sigma(\theta)$ is transformed into the laboratory frame by using the elastic Jacobian and then averaged over the velocity distributions of the two beams and over the beam/detector geometry.

The best-fit total $I(\Theta)$ for Ne-O_2 is shown as solid line in Fig. 1. A previously reported PES[22] for Ne-O_2 appears unrealistic when compared to the present determination.[18]

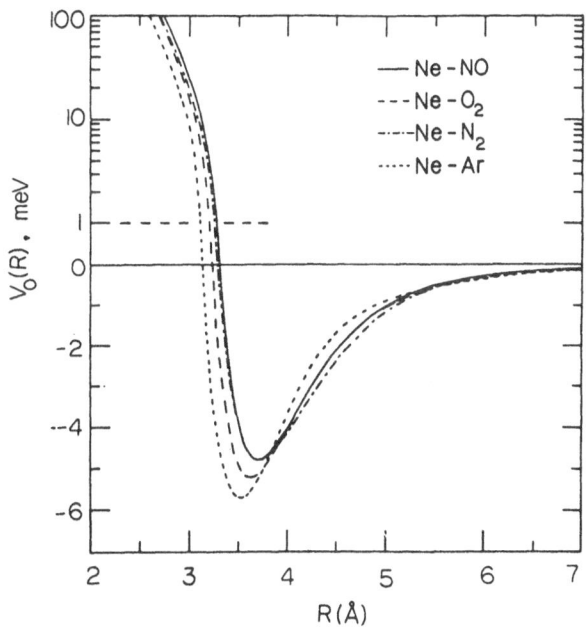

Fig. 2. Comparison between the V_0 terms of the potential surface (spherical average potential) for Ne-NO, Ne-O_2 and Ne-N_2 from this work and the interatomic potential for the corresponding isotropic Ne-Ar system from Ref. 5.

The reliability of the IOS approximation for treating the scattering dynamics in the present experiments involving He and Ne has been investigated and assessed by performing exact close-coupling (CC) calculations of integral and differential rotationally inelastic cross sections at the energy and for the conditions of the He-N_2[10] and Ne-N_2[8] experiments Similar tests of validity have also been reported by others[4,11] for He-N_2 and He-O_2. It was concluded that for such systems the use of IOS to extract the anisotropy of atom-diatom potentials from total DCS data with well resolved diffraction oscillations is certainly reasonable, since IOS fails only for those inelastic transitions which make a negligible contribution to the total DCS. For Ar-containing systems, a comparison between IOS and CC total DCS has been carried out by Bowers et al.[23] on Ar-O_2 at energies similar to our experiments. Although less valid for inelastic transitions, IOS is still satisfactory for the total DCS except at very large scattering angles.[23]

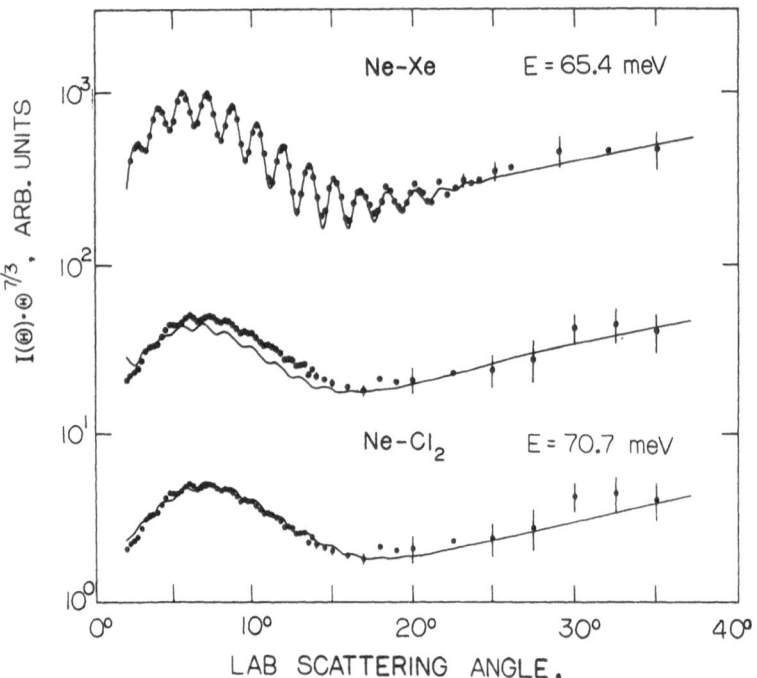

Fig. 3. Total differential cross section data for Ne-Cl_2 compared with the elastic differential cross section for the corresponding isotropic system Ne-Xe (Ref. 28). Continuous lines: spherical calculations with best-fit multiproperty Ne-Xe potential from Ref. 28 (upper plot); IOS calculations with PES from Ref. 14 (middle plot); IOS calculations with preliminary best-fit potential surface from this work (lower plot). The Ne-Cl_2 data are reported twice, displaced by a decade, for clarity.

Scattering results for the open shell species NO($^2\Pi$) interacting with Ne have been fully analyzed[24] and partially reported also in this Volume.[25]

In Fig. 2 we report the spherical average potential $V_0(R)$ for the Ne-N$_2$, O$_2$ and NO systems compared with the corresponding isotropic Ne-Ar case. The $V_0(R)$ term, i.e., the first term in the Legendre expansion of the potential surface, is easily obtained from the full PES by a Gauss-Legendre quadrature, according to the formula:

$$V_0(R) = 1/2 \int_{-1}^{+1} V(R,\gamma) \, d(\cos\gamma) \ .$$

As can be seen from Fig. 2, the position of the $V_0(R)$ repulsive walls (i.e., the σ_0 value, where σ is the distance at which V_0 crosses the zero) increases going from Ne-O$_2$ to Ne-NO and N$_2$, as already noticed in the omologous He-series. The σ_0 values follow the same trend as the molecular polarizabilities (N$_2 \geq$ NO > O$_2$). The Ne-N$_2$ interaction is weaker than Ne-O$_2$, which is not surprising having N$_2$ a larger polarizability and consequently Ne-N$_2$ a longer R$_m$. The location of the repulsive wall could be determined to within ±1% for all three systems, while the well depth can be considered accurate to about ±2-3%.

Data analysis for Ar-O$_2$ and Ar-N$_2$ is currently under way. Comparison of our scattering data with the prediction of previously determined multiproperty PES's[26] shows significant disagreement, indicating a too weak anisotropy and a not sufficiently accurate attractive part of the previous surfaces.

As far as Cl$_2$-rare gas systems are concerned, total DCS's at two collision energies with well resolved diffraction oscillations have been measured for He-Cl$_2$. The diffraction structure superimposed on the main rainbow oscillation has also been resolved for Ne-Cl$_2$. The main rainbow and two supernumerary rainbows have been observed for Ar-Cl$_2$. As example, in Fig. 3 we report the total DCS data for Ne-Cl$_2$ compared with the elastic DCS data for the corresponding isotropic Ne-Xe system measured under the same geometrical arrangement. As can be seen, the diffraction structure is dramatically quenched in the atom-diatom case because of the anisotropy of the repulsive wall of the interaction. The middle plot shows a comparison with the prediction of the PES derived by Janda and coworkers[14] from the best-fit, within a three dimensional quantum mechanical calculation, of the vibrational predissociation dynamics of NeCl$_2$. A simple atom-atom potential was used to reproduce rotational and vibrational product state distributions for the Cl$_2$ fragment of the vibrational predissociation of Ne-Cl$_2$ from several initial quasibound levels. As shown in Fig. 3, although a reasonable agreement is obtained, there is still room left for a considerable improvement. In particular, the minimum position and the well depth of the most stable (T-shaped) Ne-Cl$_2$ configuration from Ref. 14 is, within the error bounds, in agreement with our scattering determination, while the shape of the well and other aspects of the full PES, as the anisotropy of the well depth, need to be varied. In the lower plot we report the results of a preliminary best-fit calculation with a full PES modified as discussed above, using the same potential model employed for Ne-N$_2$ and Ne-O$_2$. A fruitful coupling between scattering measurements and vibrational predissociation data should allow us to achieve an accurate determination of the Ne-Cl$_2$ potential surface.[27] Less good agreement is obtained for He-Cl$_2$ between the prediction of the PES employed by Cline et al. to reproduce their vibrational predissociation data [13] and our scattering results. Work is in progress to derive full PES's for He, Ne and Ar-Cl$_2$, which are consistent with all the available experimental data.

ACKNOWLEDGMENTS

This research is supported by ENEA, the EEC Science program and the CNR bilateral agreements.

REFERENCES

1. U.Buck, Comments At.Mol.Phys. 17, 143 (1986); and references therein.
2. "Atom-Molecule Collision Theory", R.B.Bernstein, ed., Plenum, New York (1979); R.Schinke and J.M.Bowman, in "Molecular Collision Dynamics", J.M.Bowman, ed., Springer, Berlin (1983), Chap. 4.
3. R.R.Fuchs, F.R.W.McCourt, A.J.Thakkar, and F.Grein, J.Phys.Chem. 88, 2036 (1984); P.Habitz, K.T.Tang, and J.P.Toennies, Chem.Phys.Lett. 85, 461 (1982); C.Douketis, J.M.Hutson, B.J.Orr, and G.Scoles, Mol.Phys. 52, 763 (1984); F.R.W.McCourt, F.B. van Duijneveldt, T. van Dam, and R.R.Fuchs, Mol.Phys. 61, 109 (1987).
4. M.Faubel, Adv.At.Mol.Phys. 19, 345 (1983).
5. L.Beneventi, P.Casavecchia, and G.G.Volpi, J.Chem.Phys. 84, 4828 (1986).
6. L.Beneventi, P.Casavecchia, and G.G.Volpi, in "Structure and Dynamics of Weakly Bound Molecular Complexes", A.Weber, ed., NATO ASI Ser. C, Reidel, Dordrecht (1987), Vol. 212, p. 441.
7. L.Beneventi, P.Casavecchia, and G.G.Volpi, J.Chem.Phys. 85, 7011 (1986).
8. L.Beneventi, P.Casavecchia, F.Vecchiocattivi, G.G.Volpi, D.Lemoine, and M.H.Alexander, J.Chem.Phys. 89, 3505 (1988).
9. L.Beneventi, P.Casavecchia, F.Vecchiocattivi, G.G.Volpi, U.Buck, Ch.Lauenstein, and R.Schinke, J.Chem.Phys. 89, 4671 (1988).
10. L.Beneventi, P.Casavecchia, G.G.Volpi, G.C.Corey, and D.Lemoine, to be published.
11. F.A.Gianturco and A.Palma, J.Phys.B 18, L519 (1985).
12. P.Casavecchia, A.Laganà, and G.G.Volpi, Chem.Phys.Lett. 112, 445 (1984).
13. J.I.Cline, B.P.Reid, D.D.Evard, N.Sivakumar, N.Halberstadt, and K.C.Janda, J.Chem.Phys. 89, 3535 (1988).
14. J.I.Cline, N.Sivakumar, D.D.Evard, C.R.Bieler, B.P.Reid, N.Halberstadt, S.R.Hair, and K.C.Janda, J.Chem.Phys. 90, 2605 (1989).
15. D.D.Evard, C.R.Bieler, J.I.Cline, N.Sivakumar, and K.C.Janda, J.Chem.Phys. 89, 2829 (1988).
16. J.I.Cline, D.D.Evard, B.P.Reid, N.Sivakumar, F.Thommen, and K.C.Janda, in "Structure and Dynamics of Weakly Bound

Molecular Complexes" A.Weber, ed., NATO ASI Ser. C, Reidel, Dordrecht (1987), p.533.

17. R.T Pack, Chem.Phys.Lett. 55, 197 (1978).
18. L.Beneventi, P.Casavecchia, F.Pirani, F.Vecchiocattivi, G.G.Volpi, A.van der Avoird, and J.Reuss, to be published.
19. R.T Pack, J.Chem.Phys. 60, 633 (1974); D.Secrest, J.Chem.Phys. 62, 710 (1975); R.Goldflam, S.Green, and D.J.Kouri, J.Chem.Phys. 67, 4149 (1977); G.A.Patker and R.T Pack, J.Chem.Phys. 68, 1585 (1978); D.J.Kouri, in "Atom-Molecule Collision Theory", R.B.Bernstein, ed., Plenum, New York (1979), p. 301.
20. See, for example: R.B.Bernstein, Adv.Chem.Phys. 10, 75 (1966).
21. R.T Pack,J.Chem.Phys. 60, 633 (1974).
22. G.C.Corey and F.R.W.McCourt, J.Chem.Phys. 81, 3892 (1984).
23. M.S.Bowers, M.Faubel, and K.T.Tang 87, 5687 (1987).
24. L.Beneventi, P.Casavecchia, and G.G.Volpi, to be published.
25. P.Casavecchia, this volume.
26. R.Candori, F.Pirani, and F.Vecchiocattivi, Chem.Phys.Lett. 102, 412 (1983).
27. L.Beneventi, P.Casavecchia, G.G.Volpi, K.C.Janda, and N.Halberstadt, work in progress.
28. L.Beneventi, P.Casavecchia, and G.G.Volpi, to be published.

SIMPLE ADDITIVE PAIRWISE POTENTIALS FOR VIBRATIONALLY
PREDISSOCIATING TRIATOMIC VAN DER WAALS COMPLEXES: A
RIOSA MULTIPROPERTY FITTING[1]

G. Delgado-Barrio, J. Campos-Martínez, S. Miret-Artés
and P. Villarreal
Instituto de Estructura de la Materia, C.S.I.C.
Serrano 123, 28006 Madrid, SPAIN

Abstract

Using a Rotational Infinite Order Sudden Approximation ($RIOSA$)
to describe the exiting continua $X + BC(v', j')$ yielded in the vibrational
predissociation of $X \cdots BC$ complexes, we outline a fitting procedure
of the van der Waals interaction to get agreement with the available
experimental data. Assuming an additive pairwise model of Lennard-
Jones or Morse atom–atom functional forms, reasonable parameters
are obtained under a low computational cost for the $He \cdots ICl(B)$
and $Ar \cdots Cl_2(B)$ systems.

1 Introduction

After the pioneering works of Levy's group[1] on vibrational predissociation (VP)
of I_2–(rare gas atoms) complexes, several experimental studies have been lately
conducted on clusters of Cl_2 and ICl with rare gas atoms[2]. By incorporating a
second probe laser it was possible to determine, together with lifetimes and spectral shifts, final state distributions of the diatomic fragment. In this kind of systems, a sum of pairwise atom–atom potentials was able to fit all the experimental
data through accurate three–dimensional quantum mechanical calculations[3]. For
systems like $He-, Ne-, I_2$ we fitted[4] Morse parameters using a $RIOS$ approximation to reproduce the main experimental data, i.e., ground and first excitation
energies of the complexes as well as VP rates depending on the initial vibrational
excitation of I_2. More recently, we have applied this procedure to obtain a potential energy surface (PES) for $Ne - ICl(B)$, where $RIOSA$ has given good results
as compared with those obtained through full $3D$ calculations[5].

---

[1]This work was supported by CYCIT under Grant $N^o P80272$

Dynamics of Polyatomic Van der Waals Complexes
Edited by N. Halberstadt and K. C. Janda
Plenum Press, New York, 1990

409

Here, we study within the $RIOSA$ scheme the VP of $He - ICl(B)$, for which a great deal of experimental data is now available[6] focusing our attention on determining a pairwise additive PES. Beside this interesting example containing an interhalogen subunit, we show a PES for $Ar - Cl_2(B)$, for which valuable data also exist[7]. The paper is organized as follows: Section II deals with the theoretical model applied, and in Section III the results are presented and discussed.

2   Theoretical Background

The Hamiltonian for the nuclear motion of a triatomic $X - BC$ vdW molecule in a given electronic state may be written, after separation of the center of mass motion, as

$$H = h_{BC}(r) + \frac{j^2}{2mr^2} - \frac{\hbar^2}{2\mu} \frac{\partial^2}{\partial R^2} + \frac{l^2}{2\mu R^2} + V_{X-BC}(R, r, \theta) \qquad (1)$$

where m is the BC reduced mass, μ is the reduced mass of $X-BC$, r is the diatomic bond–length, R is the distance between X and the BC center of mass, j and l are angular momentum operators associated to the vectors r and R, respectively, and $h_{BC}(r)$ is the hamiltonian for the isolated BC vibrator, whereas $V_{X-BC}(R, r, \theta)$ represents the vdW interaction with θ being the angle formed by r and R. Because of the fast diatomic vibration within the complex, we can separate this motion diabatically from the rest, and therefore write the total wave function as[8]

$$\Psi(R, r) \simeq \chi_v(r)\varphi(R, \hat{r}) \qquad (2)$$

where $\hat{r} = r/r$ and $\chi_v(r)$ is an eigenfunction of $h_{BC}(r)$ with associated eigenvalue $\varepsilon_{BC}(v)$. The function $\varphi(R, \hat{r})$ satisfies the Schrödinger equation

$$\left[-\frac{\hbar^2}{2\mu} \frac{\partial^2}{\partial R^2} + B_v j^2 + \frac{l^2}{2\mu R^2} + V^{vv}_{X-BC}(R, \theta) \right] \varphi(R, \hat{r}) = E\varphi(R, \hat{r}) \qquad (3)$$

where $V^{vv}_{X-BC}(R, \theta) = <\chi_v(r)|V_{X-BC}(R, r, \theta)|\chi_v(r)>$ and $B_v = \frac{\hbar^2}{2m} <\chi_v(r) | r^{-2} | \chi_v(r) >$.

Eq. (3) has discrete as well as continuum solutions. Assuming a total angular momentum $J = 0$, we express the $k - th$ vdW discrete state through an expansion

$$\varphi^d_k(R, \hat{r}) = \sum_{n,j} a^{(k)}_{nj} \phi_n(R) X^{00}_{j,0}(\hat{R}, \hat{r}) \qquad (4)$$

where ϕ_n are radial eigenfunctions at the equilibrium angle $\theta = \theta_{eq}$

$$\left[-\frac{\hbar^2}{2\mu}\frac{d^2}{dR^2} + V^{vv}_{X-BC}(R, \theta_{eq})\right]\phi_n(R) = E_n\phi_n(R) \qquad (5)$$

and $X^{J=0,M=0}_{j,\Omega=0}$ are body–fixed angular basis functions[9]. The coefficients $a^{(k)}_{nj}$, together with the eigenvalues E_n, are now obtained by diagonalization of the following matrix,

$$h_{nj,n'j'} = \left\{E_n\delta_{nn'} + j(j+1)\left[B_v\delta_{nn'} + \frac{\hbar^2}{2\mu} < \phi_n \mid R^{-2} \mid \phi_{n'} >\right]\right\}\delta_{jj'} +$$

$$+ < X^{00}_{j0} \mid \Delta V_{nn'}(\theta) \mid X^{00}_{j'0} > \qquad (6)$$

where $\Delta V_{nn'}(\theta) = < \phi_n \|[V^{vv}_{X-BC}(R, \theta) - V^{vv}_{X-BC}(R, \theta_{eq})]\| \phi_{n'} >$ and whose matrix elements between angular functions can be easily calculated after they are expanded in terms of Legendre polynomials[9,10]. The special choice of the radial basis functions ϕ_n, Eq. (5), is expected to work correctly for low bending states of the complex, i.e., for small oscillations in the angle θ around its equilibrium value.

Usually, the stretching motion along R is faster than the bending motion associated to θ in the final dissociative channels[5]. Hence, for describing the continuum states we apply $RIOSA$[10,11] and write them as a product of a radial continuum function, that depends on θ as a parameter and is assumed to be *transparent* to the action of the angular momentum operators, times an angular function representing a precise rotational level of the diatomic fragment

$$\varphi^c_{\epsilon,j',j_0}(\underline{R}, \hat{r}) = \phi^{(-)}_{\epsilon,j_0}(R; \theta)X^{00}_{j'0}(\hat{R}, \hat{r}) \qquad (7)$$

where ϵ is the on–shell relative kinetic energy between the fragments and j_0 some averaged final rotational quantum number. Therefore, it is implicity assumed that the total available energy is significantly higher than the rotational energy of the BC fragment, whose states are considered as degenerated. In addition, taking as usual[5] $j_0 = 0$, that will be omited hereafter, the radial function $\phi^{(-)}_\epsilon$ of Eq. (7) is chosen as a scattering solution[12] at each fixed θ value, of the following equation,

$$\left[-\frac{\hbar^2}{2\mu}\frac{\partial^2}{\partial R^2} + V^{v'v'}_{X-BC}(R, \theta) - \epsilon\right]\phi^{(-)}_\epsilon(R; \theta) = 0 \qquad (8)$$

where v' stands for the final vibrational state of the diatomic fragment.

It is important to underline, at this point, the suitability of using this kind of continuum solution that is the product of a standing wave, with sinus behavior at long distances and energy normalization, by an imaginary exponential of the dephase induced by the potential. This choice implies to deal with a complex discret–continuum coupling but avoids any problem concerning continuity of a pure real coupling as function of the angle θ.

Partial half widths for exiting the diatomic fragment in a precise (v', j') vibrotational state may be estimated by the Fermi's Golden Rule,

$$\Gamma_{vk \to v'j'} = \pi \left| \langle \varphi^c_{\epsilon,j'} | < \chi_{v'}(r) | V(R,r,\theta) | \chi_v(r) > | \varphi^d_k \rangle \right|^2 \qquad (9)$$

and the total halfwidth

$$\Gamma^{TOT}_{vk} = \sum_{v'} \sum_{j'} \Gamma_{vk \to v'j'} \qquad (10)$$

allows us to calculate final state diatomic distributions as the ratios

$$P_{v'j'} = \frac{\Gamma_{vk \to v'j'}}{\Gamma^{TOT}_{vk}} \qquad (11)$$

3   Results and Discussion

The VP process is mainly affected by the behavior of the vdW interaction in the region of the well, since the VP rates are essentially determined by the overlap between a continuum wavefunction and a bound state, so that the only nonvanishing contribution originates from this region[8]. So, a simple dumbbell model potential, consisting in a sum of Morse or Lennard-Jones atom-atom interactions is found to be adequate[4]. Then, we write $V_{X-BC}(R,r,\theta)$ in Eq. (1) as

$$V_{X-BC}(R,r,\theta) = V_{X-B}(R_1) + V_{X-C}(R_2) \qquad (12)$$

where R_1 and R_2 are the distances $X-B$ and $X-C$, respectively,

$$R_1 = \qquad [R^2 + \gamma^2 r^2 - 2\gamma r R cos\theta]^{\frac{1}{2}}$$
$$R_2 = [R^2 + (1-\gamma)^2 r^2 + 2(1-\gamma)r R cos\theta]^{\frac{1}{2}} \qquad (13)$$

with

$$\gamma = m_C/(m_B + m_C).$$

3.1 $He - ICl(B)$

To study the VP process $He - ICl(v = 2) \to He + ICl(v' = 1, j')$, we start using Lennard-Jones potentials for describing the two terms in Eq. (12) because of their dependence only on two parameters. Fixing r at its equilibrium value in Eq. (13), we can vary the equlibrium values $\bar{R}_1$ and $\bar{R}_2$, or equivalently $\bar{R}$ and θ_{eq}, in order to see the equilibrium geometry of the complex. For $\theta_{eq} = 140^o$

measured from the I end, that is the configuration obtained for $Ne - ICl$[5], we varied sistematically the equilibrium values $\bar{R}_1 = \bar{R}_{I-He}$ and $\bar{R}_2 = \bar{R}_{Cl-He}$ and, accordingly, the well–depths to get a reasonable dissociation energy. Being the He atom closer to Cl, the $He - Cl$ well–depth has to be significantly larger than the $He - I$ one, as in the $Ne - ICl$ case[5]. Fixing these two parameters, we show in Table I the influence of the variation of $\bar{R}$ through $\bar{R}_1$ and $\bar{R}_2$ on the dissociation energy, D_0, the lifetime, τ, and the final rotational distribution of the ICl fragment. As it can be seen, such variation does not result very important since all these magnitudes remain almost unaltered. In particular, D_0 ranges from 10 to 12 wavenumbers and the rotational distribution shows a bimodal character with maxima placed at $j = 7$ and $j = 11$ or 12 in qualitative agreement with the experiment[6]. However, lifetimes of $\sim 70ps$ seem to be extremely short while the experiment[6(b)] reports an upper limit of about $2ns$. In addition, there is some experimental evidence on the equlibrium geometry, that corresponds to a T–shape configuration. Hence, keeping $\bar{R}$ at $4\mathring{A}$, we investigated the influence of the variation of θ_{eq}, that is shown in Table II. As we get close to $\theta_{eq} = 90^{\circ}$, D_0 shows a moderated increasing but there is a remarkable enhancement of the lifetime, that for $\theta = 110^{\circ}$ is of $\sim 0.8ns$. As regards the rotational distribution, we find an oscillatory behavior that almost loses the bimodal character. Note, at this point, that both sorts of tests done may be, in principle, ambiguous in the following sense: variations in the equilibrium values $\bar{R}_1$ and/or $\bar{R}_2$ imply not only variations in the equilibrium geometry of the complex but also in the steepness of each atom–atom interaction. Hence, we decided to use a more flexible atom–atom interaction like Morse functions, for which the strength, the steepness and the subsequent geometry can be separately analyzed. Repeating the previous calculations but using vdW Morse potentials, conserving the same well–depths and equilibrium distances than the Lennard–Jones ones and estimating the characteristic inverse length parameter α by impossing to both functional forms to get zero at the same distance $\left(\alpha = \left[\bar{R}\left(1 - 2^{-\frac{1}{6}}\right)\right]^{-1} ln2\right)$, the general trends already mentioned are found again: small variations in the relevant magnitudes with the equilibrium value of the R coordinate but dramatical changes depending on the angular equilibrium geometry. On the other hand, Morse functions present an additional convenience, that is, their eigenvalues are analytical. Then, assuming that the lowest vdW level is the simple addition of the ground $He - I$ and $He - Cl$ energies, we dispose of a first estimation of reasonable (D_i, α_i) couples, $i = 1, 2$, leading to energies of dissociation in the neighborhood of 15 wavenumbers[6(a)]. In this way, and taking advantage of the previous results we slightly varied these parameters and moved $\bar{R}_1$ and $\bar{R}_2$ inside the angular region $\theta_{eq} \epsilon [90, 110^{\circ}]$. Final parameters leading to a reasonable agreement with the experiment are listed in Table III. They yield values of $D_0 = 11.3cm^{-1}$, $\tau = 1.050ns$ and a rotational distribution, shown in Table IV, with a maximum placed at $j = 8 - 10$ and a long tail with a secondary maximum at $j = 20$. In fact, these parameters are close to those of Lennard–Jones used by Lester and co.[13] within a semiclassical impulsive model except the steepness of the interactions, that in our case are quite different. Using the same value $\alpha_1 = \alpha_2$, our model in the T–shape region produces very fast oscillations in the rotational distribution for $j \leq 12$. Since the well–depths are rather similar, it can be interpreted as a forced almost–homonuclear behavior of ICl, exception of

Table I

POTENTIAL PARAMETERS			
$\epsilon_{I-He}(cm^{-1})$	12.5	12.5	12.5
$\epsilon_{Cl-He}(cm^{-1})$	24	24	24
$\bar{R}_{I-He}(\mathring{A})$	4.65	4.45	4.27
$\bar{R}_{Cl-He}(\mathring{A})$	2.93	2.75	2.58
$D_0(cm^{-1})$	12.34	11.1	9.9
$\theta_{eq}(degrees)$	140	140	140
$\bar{R}(\mathring{A})$	4.2	4.0	3.8
$\tau(ps)$	63.0	69.0	74.8
ROTATIONAL DISTRIBUTION			
$j = 0$	0.668(-2)	0.717(-2)	0.800(-2)
$j = 1$	0.123(-1)	0.122(-1)	0.127(-1)
$j = 2$	0.739(-2)	0.527(-2)	0.526(-1)
$j = 3$	0.735(-2)	0.505(-2)	0.271(-2)
$j = 4$	0.261(-1)	0.241(-1)	0.210(-1)
$j = 5$	0.585(-1)	0.536(-1)	0.500(-1)
$j = 6$	0.841(-1)	0.752(-1)	0.692(-1)
$j = 7$	0.884(-1)	0.794(-1)	0.730(-1)
$j = 8$	0.755(-1)	0.688(-1)	0.693(-1)
$j = 9$	0.639(-1)	0.594(-1)	0.575(-1)
$j = 10$	0.642(-1)	0.624(-1)	0.599(-1)
$j = 11$	0.685(-1)	0.704(-1)	0.687(-1)
$j = 12$	0.677(-1)	0.726(-1)	0.744(-1)
$j = 13$	0.611(-1)	0.665(-1)	0.716(-1)
$j = 14$	0.516(-1)	0.563(-1)	0.610(-1)
$j = 15$	0.427(-1)	0.466(-1)	0.489(-1)
$j = 16$	0.363(-1)	0.398(-1)	0.407(-1)
$j = 17$	0.323(-1)	0.354(-1)	0.373(-1)
$j = 18$	0.294(-1)	0.323(-1)	0.357(-1)
$j = 19$	0.265(-1)	0.292(-1)	0.320(-1)
$j = 20$	0.230(-1)	0.250(-1)	0.257(-1)

Potential parameters of $L-J$ potentials, and results for the $He-ICl(B, v = 2) \rightarrow He + ICl(B, v = 1)$ process when the configuration is kept constant. D_0 is the bond energy of the lowest vdW level, θ_{eq} the equilibrium angle $\bar{R}$ the equilibrium distance of He to CM of the ICl, and τ the corresponding lifetime.

Table II

POTENTIAL PARAMETERS		
$\epsilon_{I-He}(cm^{-1})$	12.5	12.5
$\epsilon_{Cl-He}(cm^{-1})$	24.0	24.0
$\bar{R}_{I-He}(\mathring{A})$	4.24	4.4
$\bar{R}_{Cl-He}(\mathring{A})$	3.83	3.11
$D_0(cm^{-1})$	15.43	12.85
$\theta_{eq}(degrees)$	110	130
$\bar{R}(cm^{-1})$	4.0	4.0
$\tau(ps)$	794.0	190.2
ROTATIONAL DISTRIBUTION		
$j = 0$	0.456(-1)	0.627(-2)
$j = 1$	0.725(-2)	0.397(-2)
$j = 2$	0.556(-1)	0.177(-2)
$j = 3$	0.767(-1)	0.214(-1)
$j = 4$	0.253(-1)	0.354(-1)
$j = 5$	0.114(0)	0.269(-1)
$j = 6$	0.835(-1)	0.203(-1)
$j = 7$	0.709(-1)	0.304(-1)
$j = 8$	0.788(-1)	0.521(-1)
$j = 9$	0.666(-1)	0.737(-1)
$j = 10$	0.523(-1)	0.767(-1)
$j = 11$	0.258(-1)	0.684(-1)
$j = 12$	0.173(-1)	0.694(-1)
$j = 13$	0.551(-2)	0.762(-1)
$j = 14$	0.321(-3)	0.770(-1)
$j = 15$	0.210(-2)	0.701(-1)
$j = 16$	0.818(-2)	0.616(-1)
$j = 17$	0.145(-1)	0.547(-1)
$j = 18$	0.275(-1)	0.466(-1)
$j = 19$	0.356(-1)	0.362(-1)
$j = 20$	0.361(-1)	0.279(-1)

Same than table I, but now we change the angular configuration of the $He - ICl$ complex, whereas the distance from He to the CM of ICl is fixed.

masses, showing an effective decoupling between even and odd rotational states. Finally, also in Table IV, we list the rotational distribution, obtained with the same parameters, for the VP process $He - ICl(v = 3) \rightarrow He + ICl(v' = 2)$. The two theoretical distributions, $v = 3 \rightarrow 2$ and $v = 2 \rightarrow 1$, are quite similar as can be expected from our model. The difference of energy available to the fragments, inducing a small variation in the continuum $RIOSA$ wavefunctions, is unable to yield great changes if averaged potentials and couplings between the vibrational diatomic states involved remain almost the same. This disagrees with the experiment, which stresses the bimodal character in the $v = 3 \rightarrow 2$ case.

3.2 $Ar - Cl_2(B)$

We have studied the VP process $Ar - Cl_2(v = 6) \longrightarrow Ar + Cl_2(v' = 5, j')$ using Morse functions to describe each $Ar - Cl$ interaction. The equilibrium configuration of this complex is T–shape[7], that is automatically accounted for within this potential model. Starting with the anisotropic Morse potential form of Reid and co[14], we estimated at $\theta = 90^o$ the well–depth and equilibrium distance from Ar to Cl_2 center of mass, getting the values $D = 103.5cm^{-1}$ and $\bar{R}_1 = \bar{R}_2 = 4.2\text{Å}$, and varied the α parameter in the range $1 - 2\text{Å}^{-1}$. For $\alpha = 1.78\text{Å}^{-1}$, the lowest vdW energy results $-174cm^{-1}$, with and associated lifetime of $1.6ns$, while the experiment[7] reports values of $-178cm^{-1}$ (for $v = 7$) and from $83ps$ till $10ns$, respectively. Rotational distribution of the $Cl_2(v = 5)$ fragment is compared with the experiment for j even in TableV. Our results show a highly inverted distribution as we get a monotonous increasing of population with j up to $j = 10$, beyond which the rotational channels become energetically closed. However our result disagrees with the experiment for j even, where a maximum at $j = 4$ followed by a slow decreasing up to $j = 10$ is found, being surprisingly closer to the experimental behavior corresponding to odd j values.

4 Concluding remarks

We have outlined a fitting procedure of potential parameters for $He - ICl(B)$ and $Ar - Cl_2(B)$ complexes resting on an $RIOSA$ treatment of the exiting continua and using a simple additive pairwise atom–atom potential model. These potentials may be considered as starting points to be further improved through more accurate calculations[3,5]. In particular, the influence of continuum–continuum couplings completely neglected in the $RIOS$ approximation, as well as the distorted–wave vibrational approximation, that is expected to fail for high diatomic vibrational excitations, have to be severely examined. Also, long–range terms can play some role and have to be included if a proper description of the whole potential surface is desired.

Table III

	$D(cm^{-1})$	$\alpha(\overset{\circ}{A}{}^{-1})$	$\bar{R}(\overset{\circ}{A})$
$He - I$	14.5	1.2	4.1
$He - Cl$	14.0	1.6	4.1

Potential parameters of Morse potentials used in the final calculations.

Table IV

	$v = 2 \longrightarrow 1$		$v = 3 \longrightarrow 1$	
	Exp	Th	Exp	Th
$j = 0$	0.16(-2)	0.15(-1)	0.15(-1)	0.21(-1)
$j = 1$	0.89(-2)	0.88(-2)	0.18(-1)	0.92(-2)
$j = 2$	0.31(-1)	0.17(-1)	0.43(-1)	0.25(-1)
$j = 3$	0.38(-1)	0.28(-1)	0.19(-1)	0.34(-1)
$j = 4$	0.59(-1)	0.24(-1)	0.65(-1)	0.32(-1)
$j = 5$	0.74(-1)	0.38(-1)	0.82(-1)	0.49(-1)
$j = 6$	0.84(-1)	0.53(-1)	0.96(-1)	0.59(-1)
$j = 7$	0.79(-1)	0.48(-1)	0.12	0.55(-1)
$j = 8$	0.82(-1)	0.76(-1)	0.93(-1)	0.77(-1)
$j = 9$	0.78(-1)	0.61(-1)	0.82(-1)	0.61(01)
$j = 10$	0.66(-1)	0.72(-1)	0.49(-1)	0.67(-1)
$j = 11$	0.61(-1)	0.58(-1)	0.29(-1)	0.55(-1)
$j = 12$	0.49(-1)	0.54(-1)	0.35(-1)	0.48(-1)
$j = 13$	0.50(-1)	0.42(-1)	0.15(-1)	0.40(-1)
$j = 14$	0.42(-1)	0.36(-1)	0.43(-1)	0.34(-1)
$j = 15$	0.41(-1)	0.31(-1)	0.54(-1)	0.31(-1)
$j = 16$	0.43(-1)	0.30(-1)	0.46(-1)	0.31(-1)
$j = 17$	0.33(-1)	0.31(-1)	0.49(-1)	0.32(-1)
$j = 18$	0.30(-1)	0.34(-1)	0.30(-1)	0.33(-1)
$j = 19$	0.24(-1)	0.35(-1)	0.21(-1)	0.34(-1)
$j = 20$	0.16(-1)	0.36(-1)	0.10(-1)	0.33(-1)
$j = 21$	0.58(-2)	0.35(-1)		0.30(-1)
$j = 22$	.	0.31(-1)		0.26(-1)
$j = 23$		0.27(-1)		0.21(-1)
$j = 24$		0.21(-1)		0.17(-1)

Rotational distributions $v = 2 \rightarrow 1$ and $v = 3 \rightarrow 2$ for VP of $He - ICl$. The parameters used are those of Table III. Experimental data from Ref. [6].

Table V

	exp	th
$j = 0$	0.05	0.114
$j = 2$	0.03	0.150
$j = 4$	0.27	0.166
$j = 6$	0.25	0.182
$j = 8$	0.24	0.193
$j = 10$	0.16	0.195

Rotational distributions for the $Ar - Cl_2(B, v = 6) \rightarrow Ar + Cl_2(B, v = 5)$ process. Experimental data are taken from Fig. 3(c) in Ref [6].

References

[1] D.H. Levy, Advan. Chem. Phys., *XLVII*, 323 (1981).

[2] See, for a review, K.C. Janda and C.R. Bieler, in press.

[3] (a) N. Halberstadt, J.A. Beswick, and K.C. Janda, J. Chem. Phys. 87, 3966 (1987).

(b) J.I. Cline, B.P. Reid, D.D. Evard, N. Sivakumar, N. Halberstadt, and K.C. Janda, J. Chem. Phys. 89, 3535 (1988).

(c) J.I. Cline, N. Sivakumar, D.D. Evard, C.R. Bieler, B.I. Reid, N. Halberstadt, S.R. Hair, and K.C. Janda, J. Chem. Phys. 90, 2605 (1989).

(d) N. Halberstadt, O. Roncero, and J.A. Beswick, Chem. Phys. 129, 83 (1989).

[4] (a) E. de Pablo, M.S. Guijarro, P. Villarreal, P. Mareca, and G. Delgado-Barrio, An. Fis. A, 80, 210 (1984).
(b) E. de Pablo, S. Miret–Artés, P. Mareca, P. Villarreal, and G. Delgado-Barrio, J. Mol. Struc., 142, 505 (1986).

[5] O. Roncero, J.A. Beswick, N. Halberstadt, P. Villarreal, and G. Delgado-Barrio, J. Chem. Phys., in press.

[6] (a) J.M. Skene, J.C. Drobits, and M.I. Lester, J. Chem. Phys., 85, 2329 (1986).

(b) J.C. Drobits, and M.I. Lester, J. Chem. Phys., 88, 120 (1988).

[7] D.D. Evard, C.R. Bieler, J.I. Cline, N. Sivakumar, and K.C. Janda, J. Chem. Phys., 89, 2829 (1988).

[8] J.A. Beswick, G. Delgado-Barrio, and J. Jortner, J. Chem. Phys., 70, 3895 (1979).

[9] P. Villarreal, G. Delgado-Barrio, and P. Mareca, J. Chem. Phys., 76, 4445 (1982).

[10] J.A. Beswick, and G. Delgado-Barrio, J. Chem. Phys., 73, 3653 (1980).

[11] (a) D. Secrest, J. Chem. Phys. 62, 710 (1975).

    (b) L.W. Hunter, J. Chem. Phys., 62, 2855 (1975).

    (c) L. Eno and G.G. Balint-Kurti, J. Chem. Phys., 71, 1447 (1979).

[12] G.G. Balint-Kurti and M. Shapiro, Chem. Phys. 61, 137 (1981).

[13] R.L. Waterland, J.M. Skene, and M.I. Lester, J. Chem. Phys., 89, 7277 (1988).

[14] B.P. Reid, K.C. Janda, and N. Halberstadt, J. Phys. Chem., 92, 587 (1988).

R2PI SPECTRA OF THE EXTERNAL VIBRATIONAL MODES OF THE CHLOROBENZENE-,

PHENOL- AND TOLUENE-RARE GAS (Ne, Ar, Kr, Xe) VAN DER WAALS COMPLEXES

M. MONS, J. LE CALVÉ, F. PIUZZI and I. DIMICOLI

Département d'étude des Lasers et de Physico-Chimie
IRDI, DESICP, DLPC, CEN Saclay, Bât 522
91191 GIF sur YVETTE CEDEX, FRANCE

INTRODUCTION

The knowledge of the spectroscopy of the vibrational modes of the van der Waals complexes is a key for the theoretical understanding and modelling of various physical processes occurring in these systems, e.g. intramolecular vibrational relaxation, unimolecular reaction dynamics leading to fragmentation, intersystem crossing, etc...
Extensive studies have been reported on the van der Waals (vdW) complexes formed between helium atom and monocyclic aromatic molecules /1/. Nevertheless the knowledge of the external vibration mode structure of similar complexes involving heavier rare gases is rather limited /2/. Most published works relates to the $S_1 \leftarrow S_0$ electronic transition of the aromatic molecule and report only on the spectral shift of the 0^0_0 transition due to complexation. Some studies report on the vdW stretching vibration but very few experimental data concern the bending modes /3-7/.
Most of the studies yet reported concern aromatic molecules complexes of high symmetry (C_{6v} or C_{2v}) : those of benzene, tetrazine or para-difluorobenzene /4-6/. In these complexes, the rare gas atom is situated on the C_6 or C_2 axis perpendicular to the aromatic ring. The symmetrical vdW stretching vibration (along the Oz direction) is active and well identified. However due to symmetry selection rules, only transitions to even vibrational levels can be observed in the bending progression.
We have chosen to study complexes of monosubstituted benzene (φX) of lower symmetry. Thus we hoped in particular that all bending levels could be observed and that a splitting could occur between the two bending motions which are degenerate in the benzene complexes.
More specifically, it is expected that the presence of the substituent on the benzene ring may produce three main effects :
i) an electronic effect, i. e. a change of the electron density of the benzene π orbitals, with a corresponding change in the strength of the vdW bond in φX-Rg relative to that in φ-Rg.
ii) a mass effect : assuming that no coupling exists between the low frequency vdW modes and the molecular vibrations of φX, the stretching frequency of the vdW bond, generally modelled as that of a diatomic molecule, will be affected by the mass of X.
iii) a dissymmetry effect : as mentionned before, the high symmetry of the complexes of benzene (C_{6v}) and tetrazine (C_{2v}) no longer exists in the

Dynamics of Polyatomic Van der Waals Complexes
Edited by N. Halberstadt and K. C. Janda
Plenum Press, New York, 1990

complexes φX-Rg which belongs at best to the C_s symmetry group, assuming as a first approximation a point mass for the X substituent. The Rg atom will probably be slightly shifted from its symmetrical position on the C_6 axis of the benzene ring, as in aniline-Rg complexes, in which a small angular shift (10°) was obtained from rotational contour simulations /8/. As a result, both stretching (σ) and in-plane bending (β') motions will be of A' symmetry and transitions to each vibronic level should be observed. This is one of the main differences expected from the benzene complexes, in which only the even degenerate bending levels are active /4/. Nevertheless, since the out-of-plane bending (β'') motion belongs to the A'' symmetry class, it still should be active only for the even levels and should exhibit the expected splitting with the corresponding β' levels.

The present paper reports, for the first time with details, the spectroscopic observation of the low frequency external vibrational modes in the S_1 state of the 1:1 vdW complexes formed between rare gas atoms and three monosubstituted benzene molecules : chlorobenzene, phenol and toluene.

RESULTS

The complexes were formed in a supersonic expansion, ionized by a laser induced resonant two-photon ionization process and mass selectively detected in a time-of-flight mass spectrometer. The spectra (Fig. 1) were recorded in the spectral region of the $S_1 \leftarrow S_0$ 0_0^0 transition of each monomer.

Chlorobenzene-rare gas complexes

The $\bar{0}_0^0$ transition in the complex is red-shifted compared to the 0_0^0 transition in the monomer and this shift is found to increase with the polarisability of the rare gas atom (Table 1).

This first series of spectra is characterized by weak vibronic bands involving intermolecular vibrational motion, with a trend to increasing intensity when changing from argon to xenon. Two separate vibrational features can be identified. One of them consists of a single vibronic band, always located near 40 cm^{-1} and with the highest intensity among the vdW modes (between 1/4 and 1/3 of the $\bar{0}_0^0$ intensity). By comparison with spectra obtained on other aromatic-rare gas systems (benzene-argon : band at 40 cm^{-1} /4/;) as well as calculations performed on the the benzene-argon system (40 cm^{-1} /4/), this band is assigned to the first quantum of the stretching motion : σ_0^1.

The second vibrational feature, consisting of a short progression, is assigned to the bending motion of the rare gas. The frequency of the first band (about 15 cm^{-1}) is close to the bending frequency (15 cm^{-1} /4/) of the benzene-argon system (although in this system the first bending band appears at 30 cm^{-1} and corresponds to the β_0^2 transition). For chlorobenzene, the first band is thus assigned to the first quantum (β'_0^1) of the bending motion parallel to the ring-subtituent direction (Ox axis), since the presence of the substituent leads to a lower C_s total symmetry, in which β' (as well as σ) belongs to the A' symmetry class and hence transitions are allowed to each level. An additional justification for this assignment will be given in the discussion section. In spite of the presence of some weak bands in the β'_0^2 region, no clear proof for the occurence of β'' bands can be given.

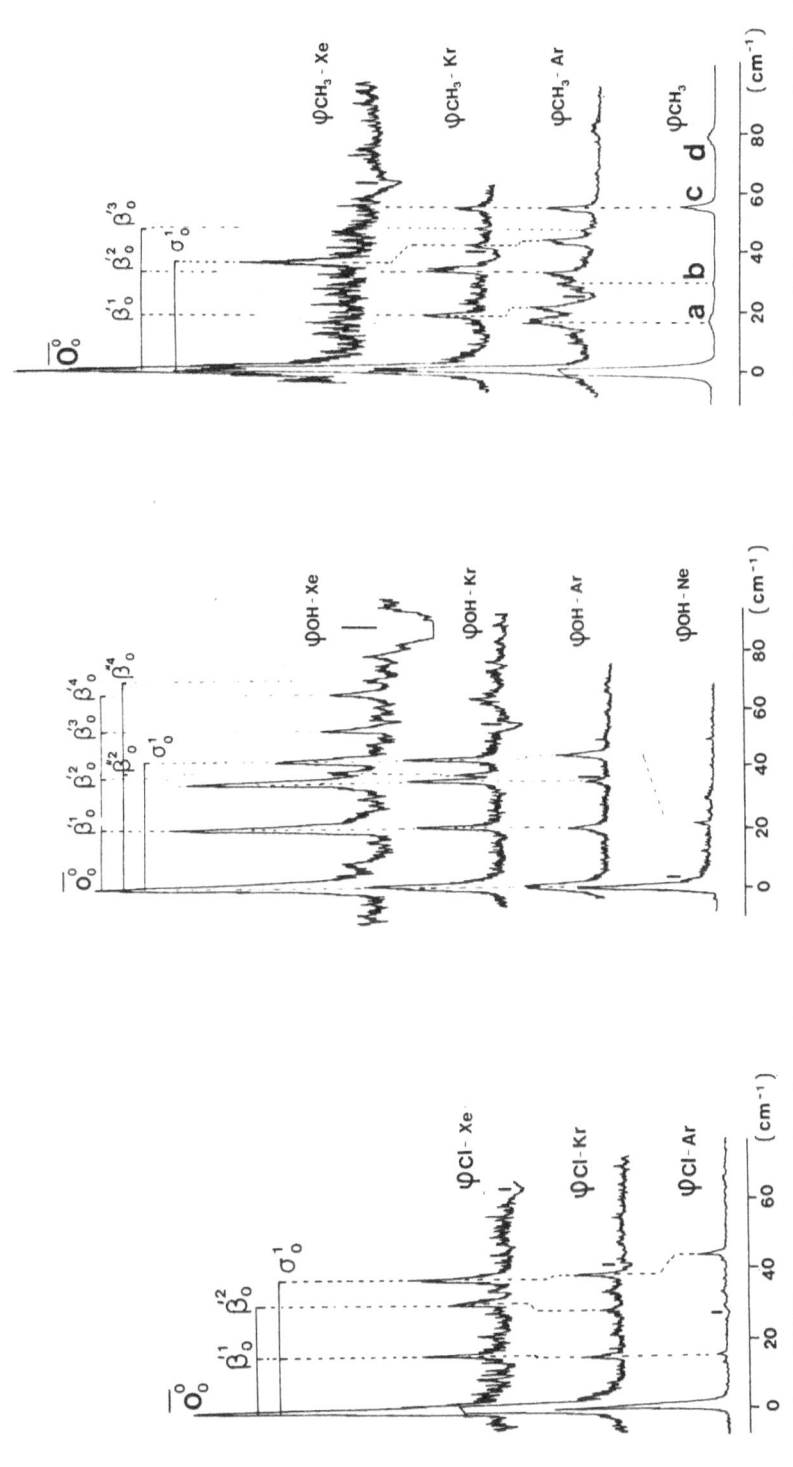

Figure 1 . R2PI spectra of the φCl-, φOH- and φCH₃- complexes. The external vdW band frequencies of the three complexes are given relative to a common origin corresponding to the $\overline{0}^0_0$ transition of each complex. Position of the monomers 0^0_0 origin is indicated by the bars. In the case of the toluene complexes, the methyl group rotational motion gives rise to additional bands noted a, b, c, d.

Phenol-rare gas complexes

The spectra (Fig. 1)exhibit similar features to those of chlorobenzene complexes, with however dramatically increased intensities in the vdW vibronic bands. In the complexes with Ar, Kr and Xe the frequency of the stretching mode (about 43 cm^{-1}) is close to that of the chlorobenzene complexes. The progression in the bending mode extends up to the third overtone. A careful examination of the spectra allows us to detect satellite bands in the vicinity of the first (Kr and Xe) and third (Xe) overtones. According to the symmetry selection rules, the strong bands are assigned to the in-plane bending mode β', whereas the weak bands are assigned to the even harmonics of the out-of-plane β" mode, the odd ones being of A" symmetry and thus forbidden. The anharmonicity analysis of the β' bend progression (Table 1) shows clearly that its first overtone is perturbed in the case of the xenon complex. The experimental frequency is smaller than the calculated one by about 2 cm^{-1}. A Fermi resonance between the σ^1, β'^2 and $\beta"^2$ levels, all of A' symmetry, is probably responsible for this shift. This hypothesis is possibly corroborated by the drastic change of the relative intensity of the σ_0^1 and β'_0^2 bands, when comparing complexes with Ar and Xe. In the phenol-neon complex, the only significant band observed, at 21 cm^{-1} , is assigned to the stretch mode. Further support of this assignment will be given in the discussion.

Table 1 . Spectroscopic data of the studied φX-rare gas atom complexes : red-shift, band frequencies (relative to the $\overline{0}_0^0$ transition) and deduced vibrational constants (ω_e' and $\omega_e' x_e'$) for the bending mode are given in cm^{-1}. Precision is estimated to be 0.5 cm^{-1}.

Complex	φCl-Ar	φCl-Kr	φCl-Xe	φOH-Ne	φOH-Ar	φOH-Kr	φOH-Xe	φCH$_3$-Ar	φCH$_3$-Kr	φCH$_3$-Xe
$\Delta\nu$	26.5	41	63.5	4.5	34	55	89	25.5	39	64
σ_0^1	43.5	38.5	37.5	21	45	43	43	43	42.5	36.5
β'_0^1	14.5	14.5	15.5	-	20	20	20	20.5	18.5	-
$\beta'_0^2 ; \beta"_0^2$	-	28;	30.5;-	-	35.5;-	36; 38	36; 39	33; -	33.5;-	-
β'_0^3	-	-	-	-	(50)	-	53.5	47	48	-
$\beta'_0^4 ; \beta"_0^4$	-	-	-	-	-	63; -	66; 71	-	-	-
ω_e'	-	15.5	16	-	23.5	22.5	22.3	19	21	-
$\omega_e' x_e'$	-	0.5	0.3	-	1.7	1.3	1.2	0.8	1.2	-

Toluene-rare gas complexes

The complexes of toluene with rare gases are an interesting model system for studying the interaction between inter- and intramolecular vibrations. Indeed low frequency transitions between levels of the methyl group pseudorotational motion appear in the same energy region as the vdW external modes of the complex. Fig. 1 presents the spectra of the toluene complexes together with a jet-cooled spectrum of the monomer. On this latter, four bands labelled a,b,c,d are due to the methyl pseudorotation. The vdW stretching vibration is identified at 36.5 cm^{-1} in the xenon complex and slightly above 40 cm^{-1} in the Kr and Ar complexes. By comparison with the phenol complexes, a β' vdW bend mode progression up to the third harmonic level is observed for the krypton complex, in addition to the d band due to the methyl group. With argon, the complexity of the spectrum suggests that Fermi resonances occur between the two first vdW bending levels and toluene levels involved in the a and b transitions. Surprisingly, no vdW bending- nor methyl rotation band appears unambiguously in the toluene-xenon complex.

DISCUSSION

The present results will be compared with similar ones obtained by other groups and correlations will be shown between the presently available spectroscopic data on this series of aromatic complexes and some physical parameters characteristic of the rare gas and the aromatic molecule.

Shift of the 0_0^0 energy

The variation of the $S_1 \leftarrow S_0$ 0_0^0 transition energy from the free aromatic molecule to the complex, i.e. the spectral shift $\Delta\nu$ due to complexation, results from the relative strength of the vdW interaction in

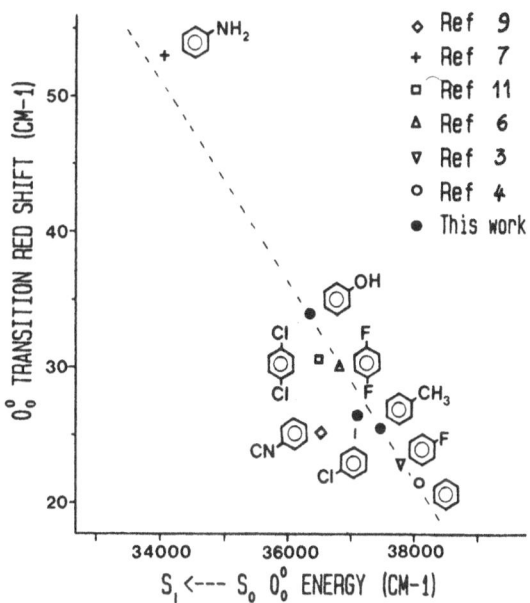

Figure 2 . Plot of the red shift $\Delta\nu$ vs. origin energy of the $S_1 \leftarrow S_0$ monomer transition for the argon - benzene derivative complexes. The dashed line indicates the correlation tendency.

the S_1 and S_0 states of the complex. For each complex studied, complexation induces a red shift of the 0_0^0 transition energy (Table 1), indicating that the vdW interaction is stronger in the S_1 excited state than in the S_0 ground state. Except for helium complexes /1/, this is a general observation for all the 1:1 aromatic molecule-Rg complexes yet studied.

For complexes between a given aromatic molecule and different rare gas atoms, the red shift increases from neon to xenon. A linear correlation exists between this spectral shift and the rare gas atom polarizability as already obtained by other groups for other aromatic-Rg complexes, e.g. the aniline-rare gas complexes /7/, confirming the linear dependence, with the rare gas polarizability α, of the vdW interaction of a mainly dispersive nature.

For a given rare gas atom, one can observe that the red shift increases by substitution in the benzene ring. The relationship existing between the red shift and the nature of the substituent results from numerous physical parameters involved in the complexation phenomenon such as polarizability, transition energies, transition moments, etc... No unique parameter characterizing the solute arises from the theory. However, it is well established that the nature of $S_1 \leftarrow S_0$ electronic transition in most substituted benzenes molecules is the same i.e. a $\pi^* \leftarrow \pi$ transition. Quantitative formulas relating, for instance, intensity and frequency changes of this transition for numerous substituted benzenes have been established from quantum mechanical considerations. Finally one of the most accessible parameters for spectroscopists is the molecular 0_0^0 energy itself. Fig. 2 depicts the relationship between the red shift $\Delta \nu$ and the 0_0^0 energy. A fairly good correlation is obtained but is certainly not expected to be followed by any kind of substitution in the benzene ring. For instance, an eventual steric hindrance between the rare gas and the substituent X should affect the interaction and hence the red shift. Nevertheless, within the series of simple benzene derivatives such as that presented in Fig. 2, the correlation seems reasonable. It still holds for the two para-disubstituted compounds, in agreement with the fact that the effect of X on both $\Delta \nu$ and the 0_0^0 monomer transition is of an electronic nature, without any mass contribution. Only benzonitrile /9/ deviates from the observed correlation. This can be qualitatively explained by the deformation of the complexing area, relative to benzene, caused by the extended delocalization of the benzene π-electrons due to the CN triple bond.

Stretching mode

For nearly all the complexes studied the vdW stretching frequency in the S_1 state is close to 40 cm^{-1}. Comparison of the k_σ force constants of the vdW stretch vibration in the different complexes is obviously more significant than a crude comparison of the measured frequencies. Indeed an increase of the bond strength can be hidden by the mass effect due to a heavier Rg or X substituent. Each complex was approximated to a diatomic molecule and the k_σ force constant of the vdW stretching mode was derived from the experimental frequency, the reduced mass being assumed to be that of the Rg-φX diatomic molecule. This approximation will be valid as long as Rg is not too far distant from a vertical position above the center of mass (c. of m.) of the φX molecule. This is certainly the case for the symmetric (C_{6v} or C_{2v}) complexes like those of benzene or paradisubstituted benzene and accidentally for complexes of monosubtituted benzene if the attraction between Rg and X fairly compensates the c. of m. shift of the φX molecule , relative to benzene. Such a situation seems to occur in the aniline-Rg complexes, according to the available structural information /8/.

As for the dependence of $\Delta \nu$ on the φX molecule (Fig. 2), Fig. 3 shows the plot of k_σ vs. 0_0^0 monomer energy for the series of Ar complexes with several benzenic molecules whose stretch frequency values are available. The unassigned spectrum of the aniline-Ar complex recently reported /7/ is very similar to the present results (Fig. 1). Thus we assign the three vdW bands observed in aniline-Ar at 21, 39, 48.5 cm^{-1} to the β_0^1, β_0^2, σ_0^1 vdW modes. A good correlation is obtained for the symmetric complexes, including benzene, para-difluoro- and para-dichloro-benzene. The aniline-Ar complex fits also well this correlation, suggesting as developed previously that this complex is accidentally symmetric concerning the nature of the stretch vibration. Deviation from this correlation for the other monosubstituted benzene complexes suggests that the diatomic molecule approximation made is then more arbitrary. Nevertheless, the representative points remain in the vicinity of those of the symmetric complexes ; the general tendency being an increase of the force constant with a decreasing 0_0^0 energy.

The correlation existing between the stretching force constant k_σ and the red shift $\Delta\nu$ is presented in Fig. 4. This correlation is general since it allows to eliminate the nature of both rare gas and X substituent. It includes all the available data for complexes with rare gases from helium to xenon. The observed tendency corroborates the intuitive idea that a stronger vdW interaction can be revealed either from the shift due to complexation or from the force constant of the vdW stretching.

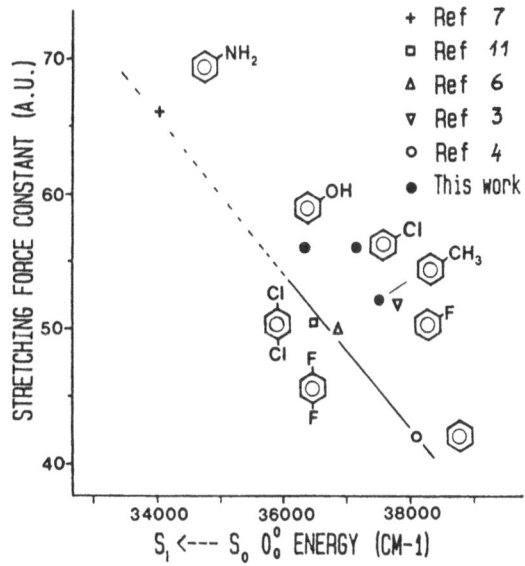

Figure 3 . Plot of the stretching force constant k_σ vs. origin energy of the $S_1 \leftarrow S_0$ monomer transition for the argon - benzene derivative complexes. The dashed line indicates the correlation tendency.

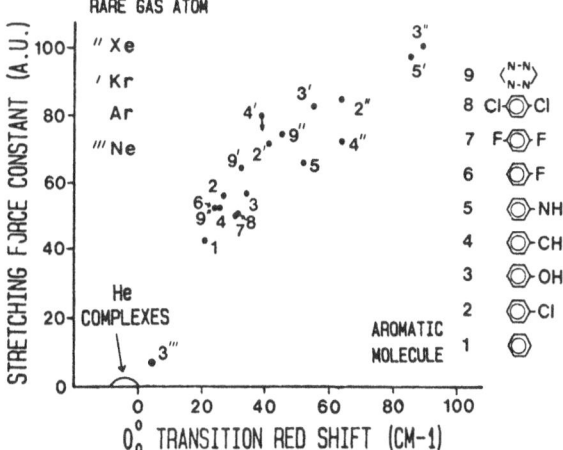

Figure 4 . Plot of the stretching force constant k_σ vs. red shift $\Delta\nu$ for studied rare gas - benzenic molecule complexes. The collected data refer to the present work, to references given in Fig. 2 and 3 and to Ref. /10/ for aniline-Kr. Tetrazine-Rg data, taken from Ref. /12/, have been included

Bending modes

vdW bending modes deserve a particular discussion since only few data are available in the literature. The assignment given in Fig. 1 and Table 1 is proposed under the expectation that, in the three C_s complexes studied, transitions towards any A' bend level are allowed, the out-of-plane bend motion (A") appearing only in the even levels, like the bend vibrations of the symmetrical complexes (C_{6v} and C_{2v}). Another support for the assignment of the first band to the β_0^1 level in the Rg-φX complexes arises from the quantitative comparison with results reported on symmetrical complexes.

In Fig. 5, we have plotted, from the present work and from data available in the literature, the stretching mode frequency vs. the bend frequency in the S_1 state of some Ar-aromatic complexes. The arrow for toluene indicates that the bending frequency is perturbed by a Fermi resonance (see Fig. 1). The tetrazine complex is included in Fig. 9 with its two split bending vibrations /5/.

Fig. 9 shows that points corresponding to Ar complexes of high symmetry (C_{6v} or C_{2v}) with benzene, paradifluorobenzene, tetrazine and that of the "pseudo symmetrical" aniline-Ar complex (see above) are colinear. In these complexes, the "diatomic approximation" is valid for both stretching and bending motions of the Ar atom relative to the solute molecule /4/. Since the reduced mass of these pseudo diatomic molecules is nearly the same, a variation of the frequencies can be directly correlated with a variation in the strength of the vdW interaction. Fig. 5 shows that, in a symmetrical complex, an increase of the stretch frequency, necessarily resulting in a vdW bond length shortening, is accompanied with a simultaneous increase of the bending frequency. A justification can be given when considering the geometrical structure of the interaction in Rg-benzenoid complexes. The vdW interaction results from the sum of atom-atom interactions between rare gas and the aromatic ring atoms (or electron cloud). Each individual interaction is oblique with a vertical (stretching) and lateral (bending) component. A symmetric summing of the

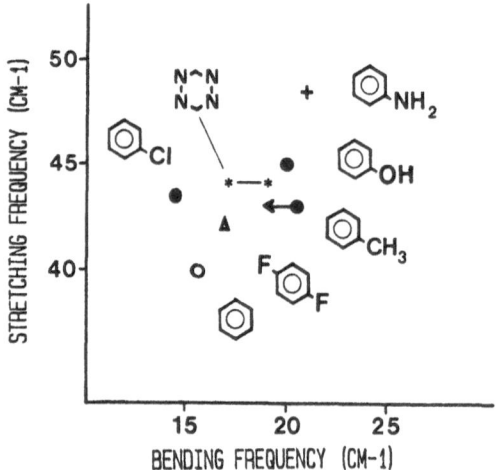

Figure 5 . Comparison between the stretching and bending mode frequencies of the argon - benzenoid molecule complexes. The freqencies are given without anharmonicity correction. References are given in previous figures.

individual bond stretching motions will produce the vdW stretch of the complex, while the vdW bending modes will result from antisymmetrical summings of opposite Rg-ring atom bond stretches. In such a model, the correlation between the frequencies of σ and β is quite straigthforward for symmetrical complexes and is expected to hold also for lower symmetry complexes if the dissymmetry effects can be neglected.

The representative points of the complexes studied in the present work are very close to the "symmetrical" curve. This is a strong argument for our assignment of the bending transitions. This suggests moreover that the dissymmetry effects are not too large within the series studied. The only exception would seem to be the complex with chlorobenzene, with its heavier chlorine substituent.

Anharmonicity and intensity of the bending progessions

In most of the complexes, extended progressions of the β' bending vdW vibration are observed. This allows us to determine the anharmonicity of the β' bending vibration as well as to provide evidence on possible Fermi resonances with stretching levels. The measured anharmonicity is significant and relatively large ($\omega_e' x_e' / \omega_e' \sim 5$ %), in contrast with the results of Bernstein et al. on the benzene-Ar complex /4/. In this latter case the bending potential is symmetric vs. the bending coordinate q_β and contains only even terms. The low value of the anharmonicity $\omega_e x_e$ then means that the contribution of the q_β^4 term is small /4/. Consequently the higher values of $\omega_e' x_e'$ found in the present study are probably related to a significant q_β^3. term in the potential, expressing the dissymmetry induced by the X substituent. No value of the β'' bending anharmonicity could be evaluated since only two bands (β''^2_0 and β''^4_0) are observed with the fisrt one involved in a Fermi resonance. In the phenol complex series, the intensity of the bending progression increases relative to that of the $\bar{0}^0_0$ and stretching bands (Fig. 1). A longer progression could be explained by a significant change between the ground and excited state potentials. The trend observed for the rare gas series might be due to an evolution of this change with the nature of the rare gas, for instance the shift of the Rg equilibrium position along the ring-substituent direction, since the Rg atom - X interaction should increase with the rare gas polarizability.
Concerning the intensities within a bending progression, we expected an alternating intensity law according to a smooth manifestation of the "even level selection rule" of the symmetric complexes. Results on aniline-Ar exhibit this expected alternating intensity distribution, again in accordance with the already noticed "pseudosymmetry" of this complex, which supports our assignment. The monotonic decrease of the intensities observed in our spectra (Fig. 1) suggests a significant shift (toward or away from X) of the rare gas vertical projection relative to the c. of m. of the φX molecule in the complexes we have studied.

REFERENCES

1. M. SCHAUER, K. LAW, and E.R. BERNSTEIN, J. Chem. Phys. 81 (1984) 49 and references therein.
2. M. ITO, T. EBATA, and N. MIKAMI, Ann. Rev. Phys. Chem. 39 (1988) 123 and references therein.
3. K. RADEMANN, B. BRUTSCHY, and H. BAUMGÄRTEL, Chem. Phys. 80 (1983) 129.
4. J.A. MENAPACE and E.R. BERNSTEIN, J. Phys. Chem. 91 (1987) 2533 and references therein.
5. P.M. WEBER, J.T. BUONTEMPO, F. NOVAK, and S.A. RICE, J. Chem. Phys. 88 (1988) 6082.

6. B.A. JACOBSON, S. HUMPHREY, and S. A. RICE, J. Chem. Phys. <u>89</u> (1988) 5624.
7. E.J. BIESKE, M.W. RAINBIRD, and A.E.W. KNIGHT, J. Chem. Phys. <u>90</u> (1989) 2068.
8. K. YAMANOUCHI, S. ISOGAI, S. TSUCHIYA, and K. KUCHITSU, Chem. Phys. <u>116</u> (1987) 123.
9. T. KOBAYASHI, K. HONMA, O. KAJIMOTO, and S. TSUCHIYA, J. Chem. Phys. <u>86</u> (1987) 1111.
10. P. BRECHIGNAC and B. COUTANT, private communication.
11. M. MONS, J. LE CALVÉ, F. PIUZZI, and I. DIMICOLI, subm. to J. Chem. Phys.
12. P.M. WEBER and S.A. RICE, J. Chem. Phys. <u>88</u> (1988) 6121.

THE DIMERS $(HF)_2$ AND $(HCl)_2$: A COMPARISON

OF AB INITIO POTENTIAL ENERGY SURFACES

A. Karpfen[a], H. Lischka[a] and P.R. Bunker[b]

a) Institut für Theoretische Chemie
   und Strahlenchemie der Universität Wien
   A-1090 Wien, Währingerstraße 17, Austria

b) Herzberg Institute of Astrophysics
   National Research Council of Canada
   Ottawa, Ontario K1A 0R6, Canada

ABSTRACT: A new ab initio potential energy suface of $(HCl)_2$ has been computed including electron correlation (ACPF) and applying large, extended basis sets. We present contour plots for selected regions of the in plane intermolecular part of the energy surface and compare with previous calculations on $(HF)_2$. We show that the global minimum energy path for a geared rotation of two molecules in $(HCl)_2$ differs significantly from that found in $(HF)_2$. Energies and structures of the C_s minimum and of the C_{2h} saddle point of $(HCl)_2$ are discussed. Moreover, vibrational spectra and infrared intensities as obtained within the framework of the double harmonic approximation are reported.

INTRODUCTION

The hydrogen bonded dimer $(HF)_2$ was one of the first weakly bound intermolecular complexes investigated by molecular beam electric resonance spectroscopy (MBERS). Dyke et al. /1/ revealed the existence of a large-amplitude hydrogen tunnelling motion between energetically equivalent minima of $(HF)_2$. In the course of this motion "free" and "H-bonded" H-atoms exchange their role. Since then, several other microwave /2-4/ and high resolution infrared /5-12/ studies have appeared which contributed significantly to a more detailed understanding of the spectroscopic properties of the $(HF)_2$ complex.

In case of the related complex $(HCl)_2$ comparatively few accurate spectroscopic investigations have been performed to date. Very recently, Blake et al. /13/ reported a far infrared rotation-tunnelling spectrum. In a related investigation Moazzen-Ahmadi et al. /14/ analyzed the torsional vibration of $(HCl)_2$. Fundamental frequencies of the

Dynamics of Polyatomic Van der Waals Complexes
Edited by N. Halberstadt and K. C. Janda
Plenum Press, New York, 1990

intermolecular Cl-H vibrations in the complex and hence the shifts with respect to the monomer absorption induced by complex formation were accurately determined by Ohashi and Pine /15/.

From the theoretical side a large number of extended ab initio investigations is available for $(HF)_2$. We mention here only those in which significant parts of the energy surface have been scanned or analytic representations of the potential have been suggested /16-23/. More extensive references to earlier $(HF)_2$ work may be found therein. In case of $(HCl)_2$ on the other hand, only few attempts have been undertaken to investigate large portions of the energy surface with ab initio methods. The most extended calculations in this direction presented so far have beeen performed by Ahlrichs and coworkers /24,25/. They performed CEPA calculations applying a pseudopotential for the Cl inner shell electrons and flexible valence shell basis sets. Of the order of 100 points along selected 1D-cuts have been evaluated and were subsequently used to construct an analytic representation of the intermolecular energy surface. Several complete structure optimizations of $(HCl)_2$ were performed previously at the SCF level /26-30/. In some of these investigations harmonic vibrational frequencies /26,29,30/ and infrared intensities /30/ have been calculated as well. To our knowledge comparable calculations on a correlated level have not been published yet.

In this contribution we wish to give a preliminary report of our attempts to obtain a reliable, global 6D representation of the $(HCl)_2$ energy surface. We have performed large scale electron correlation calculations using the ACPF approach /31/. Extended polarized basis sets were applied. We present selected 2D contour plots of the intermolecular energy surface and discuss the minimum energy path for a geared rotation of the two HCl molecules in the complex. We point out significant differences to the case of $(HF)_2$. Planar configurations are considered only.

Apart from these systematic scans we characterize the two most important stationary points, the C_s minimum and the C_{2h} transition state, by their completely optimized structures, harmonic vibrational frequencies and infrared intensities. Again $(HF)_2$ will serve as a reference case. A more detailed discussion of our results will be given elsewhere /32/.

METHOD OF CALCULATION

For all electronic structure calculations reported in this work the COLUMBUS program system was used /33-35/. Starting from a 12s9p basis for Cl and a 6s basis for the H atom /36,37/ we augmented the Cl basis by 1s1p4d1f functions and the H basis by 2p polarization functions. The primitive 13s10p4d1f/6s2p basis was contracted to $|6,5,3,1/4,2|$ with the aid of "generalized contractions" /38,39/. Electron correlation effects were incorporated at the ACPF level /31/. The 2D contour plots have been generated in a 30° mesh in each of the two angular coordi-

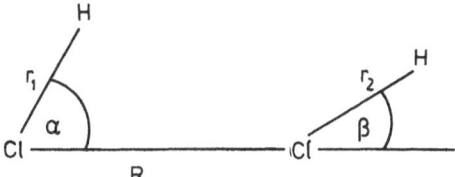

Figure 1. Internal coordinates of $(HCl)_2$

nates α and β (see fig.1). Additionally, about 150 points located in the vicinity of the C_s minimum and of the C_{2h} saddle point were computed. These have been subjected to polynomial fits in order to extract the harmonic force fields at their stationary points.

RESULTS

In fig.2 we compare 2D contour plots of $(HF)_2$ /21/ and of $(HCl)_2$ with α and β as variables at intermolecular distances R close to the respective equilibrium values. The Cl-H distances were frozen at the monomer equilibrium value. In both cases the C_s minima and the C_{2h} saddle points located midway between energetically equivalent C_s minima are clearly visible. The C_s structures of both complexes are characterized by nearly linear (α close to 0°) hydrogen bonds. As is well known the optimal β values differ in these two cases. In $(HF)_2$ the equilibrium value for β is -66° whereas it is close to -90° in $(HCl)_2$ as a consequence of the dominant quadrupole-quadrupole interaction. The main difference in the global minimum energy path for the geared rotation of the two interacting molecules is the presence of a second low lying saddle point in $(HF)_2$ corresponding to the fully linear arrangement ($C_{\infty v}$). In case of $(HCl)_2$ this feature is absent. At R values close to those of the C_s minimum the $C_{\infty v}$ configuration is slightly repulsive in $(HCl)_2$ (see also /24/), again a consequence of the small dipole moment and the large quadrupole moment of the HCl molecule. There exists, however, a further, weakly attractive saddle point in the vicinity of the linear, anti-parallel structure. In the case of $(HF)_2$ there is also a saddle point in the same region, however, already at slightly repulsive energies.

In figs. 3 and 4 we show similar contour plots for $(HCl)_2$ using R values displaced by ± 0.5 and ± 1.0 bohr, respectively. For comparable plots of the $(HF)_2$ surface see fig.3 of ref./21/. As expected the energy surface becomes progressively flatter for larger intermolecular distances (see figs. 3a and 4a). A more interesting behaviour is observed for shorter R values. At $R=6.36$ bohr (fig.4b) the original C_s minima have turned to repulsive saddle points whereas the most stable arrangements correspond now to the C_{2h} structure and a barely attractive valley for

R(FF)=5.2761 bohr

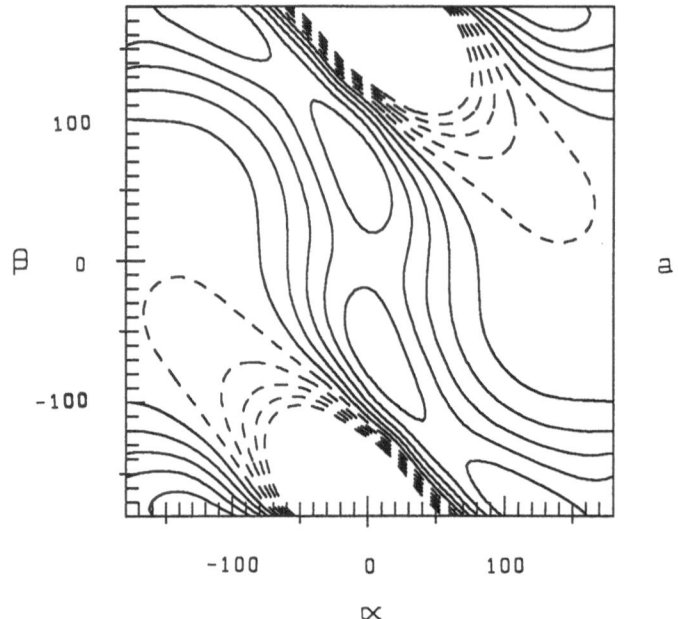

a

R(ClCl)=7.36 bohr

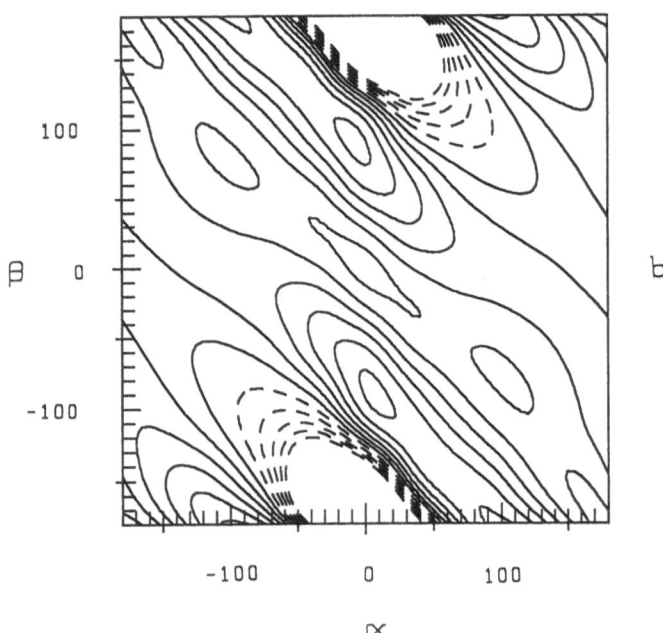

b

Figure 2. 2D planar energy surfaces of (HF)$_2$ (a) and (HCl)$_2$
(b). Successive contour lines differ in energy
by 0.0014 hartree in case of (HF)$_2$ and by 0.0005
hartree in case of (HCl)$_2$. Repulsive regions are
indicated by broken lines.

R(ClCl)=7.86 bohr

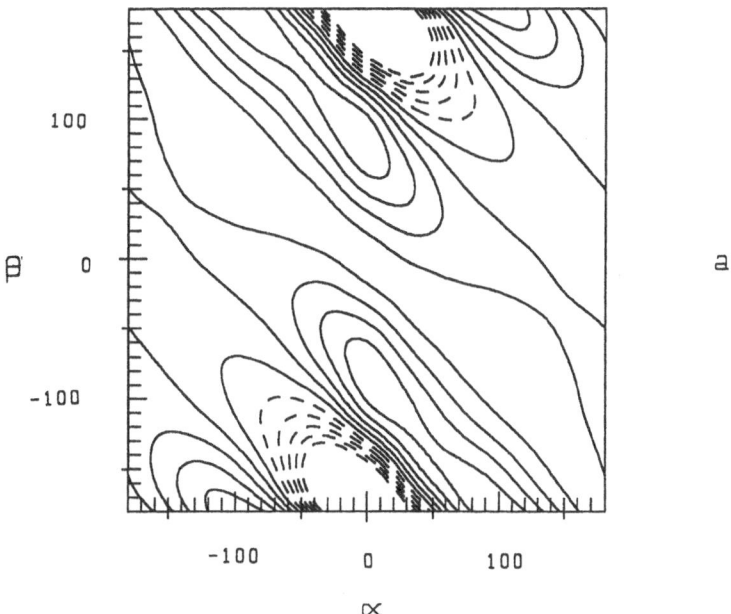

a

R(ClCl)=6.86 bohr

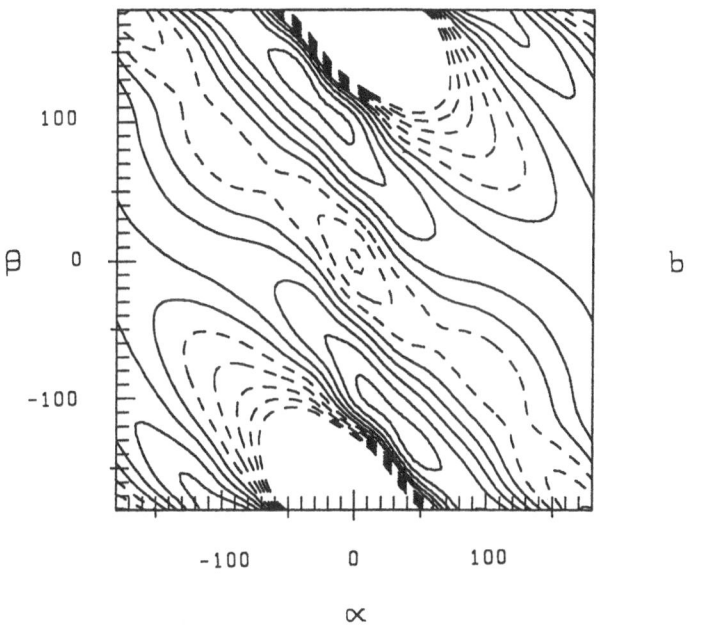

b

Figure 3. 2D planar energy surfaces of $(HCl)_2$ at intermolecular distances displaced from the minimum by
+0.5 (a) and -0.5 bohr (b), respectively. Successive contour lines differ in energy by
0.0005 hartree.

435

R(ClCl)=8.36 bohr

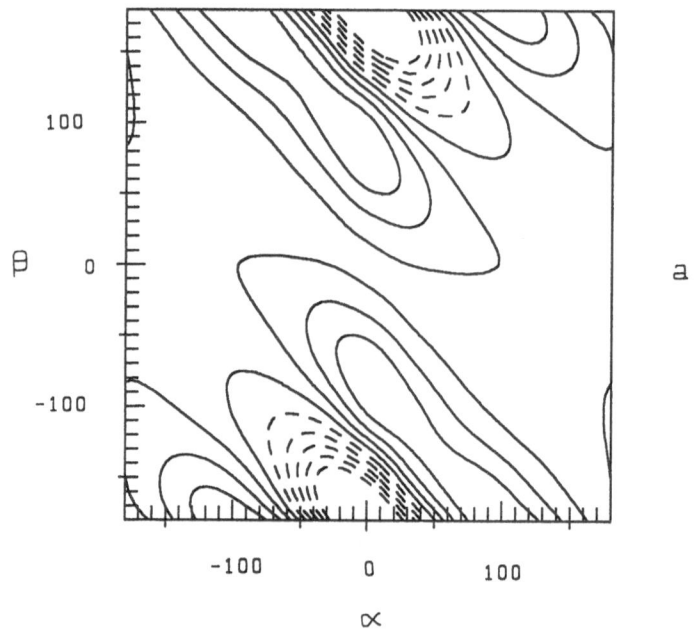

a

R(ClCl)=6.36 bohr

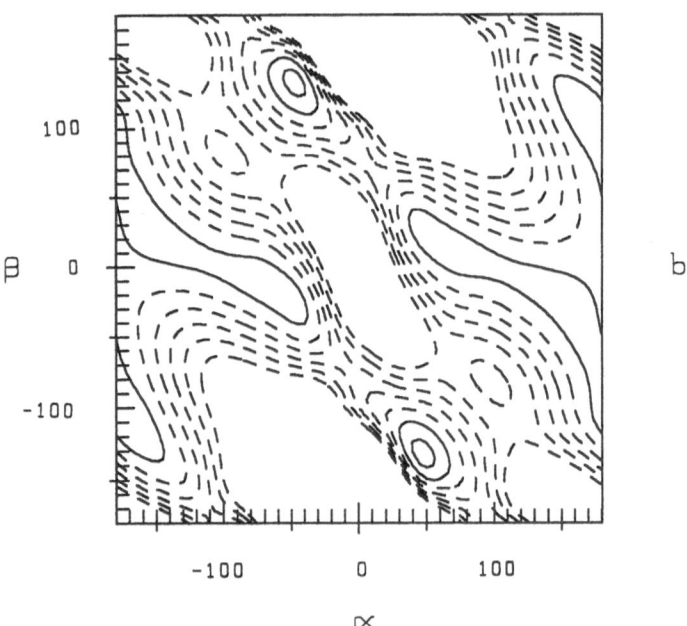

b

Figure 4. 2D planar energy surfaces of $(HCl)_2$ at intermolecular distances displaced from the minimum by
+1.0 (a) and -1.0 bohr (b), respectively. Successive contour lines differ in energy by
0.0005 hartree.

Table 1. Structure and stabilization energies of C_s and C_{2h} stationary points in $(HCl)_2$ and $(HF)_2$. Distances in Å, angles in degrees and energies in kcal/mol.

	$(HCl)_2$		$(HF)_2$	
	C_s	C_{2h}	C_s	C_{2h}
r_1	1.2814	1.2800	0.9236	0.9223
r_2	1.2798	1.2800	0.9220	0.9223
R	3.8871	3.8345	2.7919	2.796
α	6.06	46.85	6.81	54.23
β	-88.60	-133.15	-65.55	-125.77
ΔE	-1.68	-1.44	-4.32	-3.30
ΔH_0	-0.71	-0.97	-2.62	-1.99

configurations where the HCl molecules approach each other with their Cl atoms. The onset of this tendency is already visible in fig.3b. The consequences of this particular shape of the energy surface for dynamics and spectroscopy of the $(HCl)_2$ complex are difficult to estimate at the current stage. Further, extended scans of the non-planar configurations and investigations of the role of intermolecular geometry relaxation along the minimum energy path are necessary to provide a basis for accurate quantum mechanical treatments of the dynamics. Work in this direction is in progress in our group.

Optimized ACPF structures and stabilization energies of the C_s minimum and of the C_{2h} saddle point are shown in table 1. A comparison with our previous $(HF)_2$ results /21/ is also given. For the HCl monomer an equilibrium distance of 1.2784 Å close to the experimental value of 1.2745 Å /40/ was obtained. The ACPF value for the Cl-Cl distance at the C_s minimum of the dimer of 3.887 Å is somewhat larger than the most recent experimental value of 3.81 Å /13/. Our computed interaction energy of -1.68 kcal/mol is in good agreement with the -1.9 kcal/mol reported by Votava and Ahlrichs /24/. From quantitative infrared measurements together with an estimate of the zero point energy contribution of the intermolecular modes Pine and Howard /7/ reported a ΔE of -2.27 ± 0.06 kcal/mol and a ΔH_0 of -1.23 ± 0.25 kcal/mol. Both values are lower than ours by about 0.5 kcal/mol. However, quite recently Blake et al. /13/ reported a distinctly lower estimate of -500 cm^{-1} (-1,42 kcal/mol) for ΔE which would be much closer to our value.

The energy difference between the C_s minimum and the C_{2h} saddle point is much smaller (0.24 kcal/mol) in $(HCl)_2$ than in $(HF)_2$ (-1.02 kcal/mol) /12/. Simple addition of

Table 2. Harmonic vibrational frequencies and infrared intensities of $(HCl)_2$ at C_s and C_{2h} stationary points. Frequencies (ω) in cm^{-1}, intensities (A) in km/mol.

	C_s			C_{2h}		
type of mode	ω	A	type of mode	ω	A
$\omega(r_2)$	2987	45	$\omega(r_1-r_2)$	2982	87
$\omega(r_1)$	2963	188	$\omega(r_1+r_2)$	2979	0
$\omega(\alpha+\beta)$	297	46	$\omega(\alpha+\beta)$	252	0
$\omega(\tau)$	233	87	$\omega(\tau)$	160	105
$\omega(\alpha-\beta)$	129	109	$\omega(R)$	60	0
$\omega(R)$	66	0.29	$\omega(\alpha-\beta)$	77i	-

zero point energy contributions would even revert the stability of C_s and C_{2h} structures. Although we are aware that the application of the harmonic approximation is inappropriate for the various coupled large amplitude intermolecular motions in $(HCl)_2$ we report harmonic frequencies and infrared intensities in table 2. In case of the intermolecular degrees of freedom these serve more to characterize our potential rather than to be reliable predictions for the corresponding fundamentals. For the HCl monomer we obtained a harmonic frequency of 2993 cm^{-1} and an infrared intensity of 36.6 km/mol. This frequency is close to the experimental ω_e of 2991 cm^{-1} /41/. Thus we obtain shifts of -6 and -30 cm^{-1} for "free" and "H-bonded" intramolecular Cl-H vibrations in the complex in close agreement with the high-resolution infrared data of Ohashi and Pine who report shifts of -5.7 and -28.7 cm^{-1}.

In accordance with the weaker intermolecular interaction in $(HCl)_2$ intramolecular structural relaxation and corresponding frequency shifts are much larger in $(HF)_2$ /5,6,21/. Infrared intensity enhancements of X-H stretching vibrations are, however, of comparable magnitude. In case of $(HCl)_2$ we obtain factors of 1.2 and 5.1 for "free" and "H-bonded" stretching frequencies whereas the corresponding factors are 1.1 and 4.2 in case of $(HF)_2$ /21/.

Overall, the two potential surfaces and the C_s minima and C_{2h} saddle points of $(HF)_2$ and $(HCl)_2$ have much in common. The qualitative features of these energy surfaces can be rationalized by simple electrostatic considerations. The dynamical analysis of $(HCl)_2$ will, however, be a formidable task since the role of large amplitude motions is even greater than in $(HF)_2$, a consequence of the smaller energy difference between C_s and C_{2h} structures and the generally flatter energy surface.

ACKNOWLEDGEMENT

This work was supported by the Austrian "Fonds zur Förderung der wissenschaftlichen Forschung", Project Nr. P07174-CH. The calculations have been performed on the IBM 3090-400 VF of the computer center of the University of Vienna within the European Supercomputing Initiative sponsored by IBM, and on the NAS 9160 Computer of the "Interuniversitäres EDV-Zentrum", Vienna. The authors are grateful for ample supply with computer time.

REFERENCES

/ 1/ T.R.Dyke, B.J.Howard and W.Klemperer, J.Chem.Phys. 56, 2442 (1972)

/ 2/ B.J.Howard, T.R.Dyke and W.Klemperer, J.Chem.Phys. 81, 5417 (1984)

/ 3/ H.S.Gutowsky, C.Chuang, J.D.Keen, T.D.Klots and T.Emilsson, J.Chem.Phys. 83, 2070 (1985)

/ 4/ W.J.Lafferty, R.D.Suenram and F.J.Lovas, J.Mol. Spectrosc. 123, 434 (1987)

/ 5/ A.S.Pine and W.J.Lafferty, J.Chem.Phys. 78, 2154 (1983)

/ 6/ A.S.Pine, W.J.Lafferty and B.J.Howard, J.Chem.Phys. 81, 2939 (1984)

/ 7/ A.S.Pine and B.H.Howard, J.Chem.Phys. 84, 590 (1986)

/ 8/ A.S.Pine and G.T.Fraser, J.Chem.Phys. 89, 6636 (1988)

/ 9/ Z.S.Huang, K.W.Jucks and R.E.Miller, J.Chem.Phys. 85, 3338 (1986)

/10/ D.C.Dayton, K.W.Jucks and R.E.Miller, J.Chem.Phys. 90, 2631 (1989)

/11/ K.von Puttkamer and M.Quack, Mol.Phys. 62, 1047 (1987)

/12/ K.von Puttkamer, M.Quack and M.A.Suhm, Mol.Phys. 65, 1025 (1988)

/13/ G.A.Blake, K.L.Busarow, R.C.Cohen, K.B.Laughlin, Y.T.Lee and R.J.Saykally, J.Chem.Phys. 89, 6577 (1988)

/14/ N.Moazzen-Ahmadi, A.R.W.McKellar and J.W.C.Johns, Chem.Phys.Lett. 151, 318 (1988)

/15/ N.Ohashi and A.S.Pine, J.Chem.Phys. 81, 73 (1984)

/16/ D.R.Yarkony, S.V.O'Neil, H.F.Schaefer III, C.P.Baskin and C.F.Bender, J.Chem.Phys. 60, 855 (1974)

/17/ M.H.Alexander and A.E.DePristo, J.Chem.Phys. 65, 6009 (1976)

/18/ A.E.Barton and B.J.Howard, Faraday Discussions 73, 45 (1982)

/19/ D.W.Michael, C.E.Dykstra and J.M.Lisy, J.Chem.Phys. 81, 5998 (1984)

/20/ M.J.Redmon and J.S.Binkley, J.Chem.Phys. 87, 969 (1987)

/21/ M.Kofranek, H.Lischka and A.Karpfen, Chem.Phys. 121, 137 (1988)

/22/ G.E.Hancock, D.G.Truhlar and C.E.Dykstra, J.Chem.Phys. 88, 1786 (1988); D.W.Schwenke and D.G.Truhlar, J.Chem. Phys. 88, 4800 (1088)

/23/ P.R.Bunker, M.Kofranek, H.Lischka and A.Karpfen, J.Chem.Phys. 89, 3002 (1988)

/24/ C.Votava and R.Ahlrichs, in "Intermolecular Forces, Proceedings of the Fourteenth Jerusalem Symposium", ed.B.Pullman, Reidel, Dordrecht (1981)

/25/ C.Votava, R.Ahlrichs and A.Geiger, J.Chem.Phys.
78, 6841 (1983)
/26/ P.Hobza, P.Czarsky and R.Zahradnik, Collect.Czech.
Chem.Comm. 44, 3458 (1979)
/27/ C.Girardet, A.Schriver and D.Maillard, Mol.Phys.
41, 779 (1980)
/28/ M.Allavena, B.Silvi and J.Cipriani, J.Chem.Phys.
76, 4573 (1982)
/29/ M.J.Frisch, J.A.Pople and J.E.Del Bene, J.Phys.Chem.
89, 3664 (1985)
/30/ Z.Latajka and S.Scheiner, Chem.Phys. 122, 413 (1988)
/31/ R.J.Gdanitz and R.Ahlrichs, Chem.Phys.Lett. 143,
413 (1988)
/32/ A.Karpfen, H.Lischka and P.R.Bunker, in preparation
/33/ H.Lischka, R.Shepard, F.B.Brown and I.Shavitt,
Int.J.Quantum Chem. S15, 91 (1981)
/34/ R.Ahlrichs, H.J.Böhm, C.Ehrhardt, P.Scharf, H.Lischka
and H.Schindler, J.Comp.Chem. 6, 200 (1985)
/35/ R.Shepard, I.Shavitt, R.M.Pitzer, D.C.Comeau, M.Pepper,
H.Lischka, P.G.Szalay, R.Ahlrichs, F.B.Brown and
J.-G.Zhao, Int.J.Quantum Chem. S22, 149 (1988)
/36/ S.Huzinaga, J.Chem.Phys. 42, 1293 (1965)
/37/ S.Huzinaga, "Approximate Atomic Functions I",
University of Alberta, Edmonton (1971)
/38/ J.Almlöf and P.R.Taylor, J.Chem.Phys. 86, 4070 (1987)
/39/ J.Almlöf, T.Helgaker and P.R.Taylor, J.Phys.Chem.
92, 3029 (1988)
/40/ D.H.Rank, B.S.Rao and T.A.Wiggens, J.Mol.Spectrosc.
17, 122 (1965)
/41/ K.P.Huber and G.Herzberg "Molecular Spectra and
Molecular Structure IV. Constants of Diatomic
Molecules, Van Nostrand, New York (1979).

AB INITIO STUDIES ON HYDROGEN BONDED TRIMERS:

$(HCN)_x (HF)_{3-x}$, x=0,1,2,3

A. Karpfen, I.J. Kurnig, S-K. Rhee and H. Lischka

Institut für Theoretische Chemie
und Strahlenchemie der Universität Wien
A-1090 Wien, Währingerstraße 17, Austria

ABSTRACT: Ground state properties of several conceivable hydrogen bonded trimers composed of HCN and HF molecules have been evaluated at the SCF level. The most stable ones of these trimeric complexes have subsequently been reinvestigated with electron correlation methods (ACPF). We provide a survey of stabilization energies, dipole moments, selected harmonic vibrational frequencies and corresponding infrared intensities. We also discuss various aspects of the non-additivity of intermolecular interaction taking place in these clusters.

INTRODUCTION

Advances in high-resolution spectroscopic techniques allow currently to study the properties of weakly bound intermolecular complexes with high precision. Dimeric hydrogen bonded complexes have been the favourite objects of these investigations. Very recently, several microwave, infrared and Raman investigations have appeared which describe the detection and partial spectroscopic characterization of hydrogen bonded trimers in the gas phase /1-8/. Among these almost all conceivable trimers composed of HCN and HF molecules have been observed. In case of linear $(HCN)_3$ microwave spectra /1/, CARS and PARS investigations in the C≡N stretching region /5/ and infrared studies of the C-H stretching region /3,4/ are available. A cyclic $(HCN)_3$ complex has also been observed /3/. For the $(HCN)_2HF$ complex microwave data have been reported /2/. The $HCN(HF)_2$ complex was detected in a high-resolution infrared work /6/. Finally, vibrational predissociation spectra of cyclic $(HF)_3$ were obtained too /7,8/.

Despite these experimental successes difficulties often arise in the assignment of observed vibrational bands to individual normal modes of these complexes. This points to the need of accompanying theoretical studies performed

Dynamics of Polyatomic Van der Waals Complexes
Edited by N. Halberstadt and K. C. Janda
Plenum Press, New York, 1990

at a reasonably reliable level. Ab initio quantum chemical investigations may lend a helping hand in these assignment problems and may also assist in structural questions and in the energetics of various stationary points on the energy surfaces. Moreover, predictions for hitherto unobserved vibrational bands and their intensities or of other ground state properties can be made. While at the present stage still being far from quantitative many trends in structural relaxations, shifts of vibrational frequencies, non-additivity effects in stabilization energies and modifications of other ground state properties taking place upon cluster formation may be correctly described.

The aim of this contribution is to give a survey of the computed properties of a large number of trimeric $(HCN)_x$ $(HF)_{3-x}$ complexes. In the first step the ab initio SCF approach is used to find the equilibrium structures (stationary points) and binding energies of these complexes. With this information as a starting point electron correlation calculations were then performed for the more stable trimeric aggregates. Monomers and dimers of HCN and HF molecules were treated at the same level of approximation. Results for the more stable trimers will be reported in more detail elsewhere /9,10/. Here we concentrate on a global overview of some of the relevant energetic, structural and vibrational spectroscopic quantities.

METHOD OF CALCULATION

For all electronic structure calculations reported in this work the COLUMBUS program system was used /11-13/. 10s6p1d basis sets for the atoms C,N and F, and a 6s1p basis for H atoms have been applied /14,15/. In case of the SCF calculations a TZ+P contraction to $|6,4,1/4,1|$ was used throughout. Electron correlation effects were incorporated within the framework of the ACPF /16/ approach. In order to make these calculations feasible for us on all the trimers general contractions /17,18/ to a $|3,2,1/3,1|$ basis were applied in that case. Analytic gradients have been used in the course of SCF structure optimization. Cyclic optimization was performed in the case of ACPF calculations. Force constants on the ACPF level were obtained by fitting polynomials of various degrees to a sufficiently large number of points in the vicinity of the equilibrium structures. The double harmonic approximation was applied in order to evaluate harmonic vibrational frequencies and infrared intensities.

RESULTS

SCF and ACPF stabilization energies (ΔE), ACPF stabilization enthalpies (ΔH_0) obtained by taking zero point energy corrections into account, and intermolecular distances as obtained at the ACPF level for the dimers $(HF)_2$, $(HCN)_2$ and HCN-HF are collected in table 1. The SCF stabilization energy for the less stable HF-HCN complex is included too. Comparison to previous higher level calculations and to experimental data reveals that in case of $(HF)_2$ and HCN-HF our computed ACPF stabilization energies and enthalpies

Table 1. Stabilization energies and enthalpies of and intermolecular distances in dimers composed of HCN
and HF molecules. Energies in kcal/mol, distances in Å.

				stabilization energies			
	SCF	ACPF		best theoretical value		experiment	
	ΔE	ΔE	ΔH_0	ΔE	ΔH_0	ΔE	ΔH_0
$(HF)_2$	-4.4	-5.0	-3.2	-4.32/19/	-2.62/19/	-4.56/20/	-2.97/20,21/
$(HCN)_2$	-4.4	-4.5	-3.7	-4.5/22/	-3.8 /22/	-4.4 /23/	-3.8 /24/
HCN-HF	-6.6	-7.3	-5.7	-6.9/25/	-	-6.2 /26/	-4.5 /26/
HF-HCN	-3.1	-	-	-	-	-	-

		intermolecular distances	
	ACPF	best theoretical value	experiment
$(HF)_2$ r_{F--F}	2.752	2.792/19/	2.79 ±0.05/27,28/
$(HCN)_2$ r_{N--C}	3.335	3.364/22/	3.29 /23/
HCN-HF r_{N--F}	2.835	2.830/25/	2.818 /26/

are too negative by 0.5 to 1.0 kcal/mol, a consequence of the
necessary compromise in basis set size. The computed intermolecular distance is too short in $(HF)_2$ (0.04 Å) and still
somewhat too large in $(HCN)_2$ (0.04 Å). Obviously, these
errors already visible in structure and energetics of dimers
will persist in the trimer calculations.

In table 2 SCF stabilization energies of $(HCN)_x(HF)_{3-x}$
species are compiled. All these correspond to stationary
points on the energy surfaces. We did, however, not perform
a complete vibrational analysis in all cases to prove that
they all correspond indeed to minima. The search for cyclic
mixed trimers was unsuccessful. No stable cyclic configurations could be detected in that case although a number of
different cyclic starting geometries were tried. In the
pure trimers the situation is different. As already reported
previously /29/ linear and cyclic $(HCN)_3$ do not differ very
much ·in stability whereas in case of $(HF)_3$ the cyclic configuration is distinctly more stable than the open chain
arrangements /31-34/. Among the mixed trimeric clusters
the linear HCN-HCN-HF complex and the open chain, cis-bent
HCN-HF-HF complex are the most stable. These findings agree
nicely with the available experimental data. Cyclic $(HCN)_3$
/3/, linear $(HCN)_3$ /1,3-5/, cyclic $(HF)_3$ /7,8/, linear
HCN-HCN-HF /2/ and the open chain HCN-HF-HF /6/ have all
been observed experimentally, whereas the remaining configurations which we predict to be less stable, have not.

Table 2. SCF stabilization energies of pure and mixed trimers composed of HCN and HF molecules with respect to monomers (ΔE_m) and dimers (ΔE_d). All values in kcal/mol.

Trimeric species	ΔE_m	ΔE_d
(HCN)$_3$ cyclic /29/	-8.7	+0.1
(HCN)$_3$ linear /29/	-9.8	-1.0
(HF)$_3$ cyclic	-12.4	-3.7
(HF)$_3$ open	-10.0	-1.3
HCN-HCN-HF /30/	-12.2	-1.2
HCN-HF-HCN	-11.6	-1.9
HF-HCN-HCN	-8.1	-0.5
HCN-HF-HF	-13.4	-2.6
HF-HCN-HF	-10.5	+0.5
HF-HF-HCN	-8.5	-1.1

ACPF stabilization energies and enthalpies of these five more stable trimers are shown in table 3. Corresponding experimental values are not available yet. We note, however, that in the case of (HCN)$_3$ cyclic and linear complexes are now very close in energy, differing by only 0.5 kcal/mol in ΔH_0. In table 4 computed rotational constants are confronted with experimental data. Generally, the computed values are too low by a few percent. Dipole moments of the open chain trimers are given in table 5.

In most vapor phase infrared spectroscopic investigations only the high-lying C-H and F-H stretching vibrations could be observed. Computed monomer, dimer and trimer harmonic vibrational frequencies in this region and the corres-

Table 3. ACPF stabilization energies and enthalpies of trimers composed of HCN and HF molecules. All values in kcal/mol.

trimeric species	ΔE_m	ΔH_{0m}	ΔE_d	ΔH_{0d}
(HCN)$_3$ linear /9/	-10.1	-8.4	-1.0	-1.0
(HCN)$_3$ cyclic /9/	-9.4	-7.9	-0.4	-0.5
(HF)$_3$ cyclic	-15.4	-10.4	-5.4	-4.1
HCN-HCN-HF	-13.2	-10.2	-1.4	-1.8
HCN-HF-HF	-15.3	-11.0	-3.0	-1.7

Table 4. Rotational constants of trimers composed of HCN and HF molecules. All values in MHz.

Trimeric species	rotational constant	ACPF	experiment
$(HCN)_3$ linear /9/	B	458	469 /1,3/
$(HCN)_3$ cyclic /9/	A = B = 2C	2431	2464 /3/
$(HF)_3$ cyclic	A = B = 2C	7517	-
HCN-HCN-HF	B	683	699 /2/
HCN-HF-HF	A	15385	-
	B	1670	1746 /6/
	C	1506	1540 /6/

ponding infrared intensities are compiled in tables 6 and 7 and are compared with experimental stretching fundamentals. Frequency shifts and intensity enhancement factors are indicated as well.

Turning first to the F-H stretching frequencies we note that to date experimental values have been reported for the monomer, the dimers and the cyclic $(HF)_3$ trimer only. The computed frequency shifts for $(HF)_2$ and cyclic $(HF)_3$ are qualitatively correct, although the error is larger in the latter case. With increasing strength of the intermolecular interaction anharmonic contributions to the frequency shifts increase as well. This trend is already visible in the case of the HCN-HF dimer. The hydrogen bond is much stronger in this mixed dimer than in $(HF)_2$. A more extended analysis of the HCN-HF energy surface /26/ has shown that the additional frequency shift caused by anharmonic effects amounts to 77 cm^{-1} just about the difference between our computed shift of -165 cm^{-1} and the experimental -245 cm^{-1} /36/. Since in the mixed trimers the HCN--HF interaction is still larger than in the dimer we expect the additional shifts due to anharmonicity to be more prominent there, of the order of 100 cm^{-1} in $(HCN)_2$HF and probably even larger in HCN$(HF)_2$. Similarly, the predicted intensity enhancement factors, although already quite large in some cases, are probably underestimated.

Table 5. Dipole moments of trimers composed of HCN and HF molecules. All values in Debye.

Trimeric species	ACPF	experiment
$(HCN)_3$ linear /9/	10.4	10.6
HCN-HCN-HF	9.6	-
HCN-HF-HF	7.5	-

Table 6. F-H stretching frequencies (ω), frequency shifts with respect to the HF monomer ($\Delta\omega$), infrared intensities (A) and intensity enhancement factors with respect to the HF monomer (f) in clusters composed of HCN and HF molecules. Frequencies in cm^{-1}, intensities in km/mol.

Species	ACPF				experiment	
	ω	$\Delta\omega$	A	f	ν	$\Delta\nu$
HF	4182		88		3961	/35/
HCN-HF	4017	-165	657	7.5	3716	-245 /36/
(HF)$_2$	4142	-40	88	1.0	3931	-30 /37/
	4099	-83	386	4.4	3868	-93 /38/
(HF)$_3$ cyclic	3964	-218	1098	12.5	3712	-249 /7/
	3863	-319	0	-	-	-
HCN-HCN-HF	3984	-198	791	9.0	-	-
HCN-HF-HF	4027	-155	425	4.8	-	-
	3829	-353	1279	14.5	-	-

Table 7. C-H stretching frequencies (ω), frequency shifts with respect to the HCN monomer ($\Delta\omega$), infrared intensities (A) and intensity enhancement factors with respect to the HCN monomer (f) in clusters composed of HCN and HF molecules. Frequencies in cm^{-1}, intensities in km/mol.

Species	ACPF				experiment	
	ω	$\Delta\omega$	A	f	ν	$\Delta\nu$
HCN	3492		53		3312	/39/
HCN-HF	3480	-12	80	1.5	3310	-2 /36/
(HCN)$_2$ /9/	3488	-4	62	1.2	3308	-4 /3/
	3434	-58	283	5.3	3242	-70 /3,4,40/
(HCN)$_3$ linear	3487	-5	62	1.2	3307	-5 /3/
/9/	3420	-72	0.5	0.01	3231	-81 /3/
	3410	-82	726	13.7	3213	-99 /3/
(HCN)$_3$ cyclic	3471	-21	242	4.6	3274	-38 /3/
/9/	3469	-23	0	-	-	-
HCN-HCN-HF	3485	-7	62	1.2	-	-
	3409	-82	425	8.0	-	-
HCN-HF-HF	3483	-9	96	1.8	3309	-3 /6/

All C-H frequencies of the clusters treated here originate either from "free" C-H bonds or from those hydrogen
bonded to HCN molecules. The C-H frequency shifts are hence
much smaller than the F-H frequency shifts discussed above.
This behaviour is well known from the extremal shifts in
solid HF /41/ and solid HCN /42/ which amount to -920 cm^{-1} in
the former and only to -182 cm^{-1} in the latter case. In
mixed molecular crystals which tend to become more ionic
these shifts can even be considerably larger. With the exception of $(HCN)_2HF$ all C-H frequencies occuring in these
clusters have been determined experimentally and can therefore be compared with our data. We observe quite good agreement between computed and experimental frequency shifts.
In particular the frequencies of the cyclic $(HCN)_3$ are correctly placed higher than the lowest of the linear dimer
while the shifts in the linear trimer originating from the
hydrogen bonded C-H groups are much larger. In case of
$(HCN)_2HF$ we predict that the lower lying C-H frequency
should lie close to the lowest C-H band of the linear $(HCN)_3$
trimer.

ACKNOWLEDGEMENT

This work was supported by the Austrian "Fonds zur
Förderung der wissenschaftlichen Forschung", Project Nr.
P07174-CH and the Korean Ministery of Education. The calculations have been performed on the IBM 3090-400 VF of the
computer center of the University of Vienna within the
European Supercomputing Initiative sponsored by IBM, and
on the NAS 9160 computer of the "Interuniversitäres EDV-
Zentrum", Vienna. The authors are grateful for ample supply
with computer time.

REFERENCES

/1/ R.S.Ruoff, T.Emilsson, T.D.Klots, C.Chuang and
    H.S.Gutowsky, J.Chem.Phys. 89, 138 (1988)
/2/ R.S.Ruoff, T.Emilsson, C.Chuang, T.D.Klots and
    H.S.Gutowsky, J.Chem.Phys. 90, 4069 (1989)
/3/ K.W.Jucks and R.E.Miller, J.Chem.Phys. 88, 2196 (1988)
/4/ D.S.Anex, E.R.Davidson, C.Douketis and G.E.Ewing,
    J.Phys.Chem. 92, 2913 (1988)
/5/ M.Maroncelli, G.A.Hopkins, J.W.Nibler and T.R.Dyke,
    J.Chem.Phys. 83, 2129 (1985)
/6/ D.C.Dayton and R.E.Miller, Chem.Phys.Lett.
    156, 578 (1989)
/7/ D.W.Michael and J.M.Lisy, J.Chem.Phys. 85, 2528 (1986)
/8/ K.D.Kolenbrander, C.E.Dykstra and J.M.Lisy, J.Chem.
    Phys. 88, 5995 (1988)
/9/ I.J.Kurnig, H.Lischka and A.Karpfen, submitted
/10/ A.Karpfen, I.J.Kurnig and H.Lischka, in preparation
/11/ H.Lischka, R.Shepard, F.B.Brown and I.Shavitt,
    Int.J.Quantum Chem. S15, 91 (1981)
/12/ R.Ahlrichs, H.J.Böhm, C.Ehrhardt, P.Scharf, H.Lischka
    and H.Schindler, J.Comp.Chem. 6, 200 (1985)
/13/ R.Shepard, I.Shavitt, R.M.Pitzer, D.C.Comeau, M.Pepper,
    H.Lischka, P.G.Szalay, R.Ahlrichs, F.B.Brown and
    J.-G.Zhao, Int.J.Quantum Chem. S22, 149 (1988)

/14/ S.Huzinaga, J.Chem.Phys. 42, 1293 (1965)
/15/ S.Huzinaga, "Approximate Atomic Functions I",
     University of Alberta, Edmonton (1971)
/16/ R.J.Gdanitz and R.Ahlrichs, Chem.Phy.Lett. 143,
     413 (1988)
/17/ J.Almlöf and P.R.Taylor, J.Chem.Phys. 86, 4070 (1987)
/18/ J.Almlöf, T.Helgaker and P.R.Taylor, J.Phys.Chem.
     92, 3029 (1988)
/19/ M.Kofranek, H.Lischka and A.Karpfen, Chem.Phys.
     121, 137 (1988)
/20/ A.S.Pine and B.J.Howard, J.Chem.Phys. 84, 590 (1986)
/21/ D.C.Dayton, K.W.Jucks and R.E.Miller, J.Chem.Phys.
     90, 2631 (1989)
/22/ M.Kofranek, H.Lischka and A.Karpfen, Mol.Phys.
     61, 1519 (1987)
/23/ L.W.Buxton, E.J.Campbell and W.H.Flygare,
     Chem.Phys. 56, 399 (1981)
/24/ H.D.Mettee, J.Phys.Chem. 77, 1762 (1973)
/25/ P.Botschwina, in: "Structure and Dynamics of Weakly
     Bound Molecular Complexes", A.Weber, ed., Reidel,
     Dordrecht (1987) p.181
/26/ A.C.Legon und D.J.Millen, Chem.Rev. 86, 635 (1986)
/27/ T.R.Dyke, B.J.Howard and W.Klemperer, J.Chem.Phys.
     56, 2442 (1972)
/28/ B.J.Howard, T.R.Dyke and W.Klemperer, J.Chem.Phys.
     81, 5417 (1984)
/29/ M.Kofranek, A.Karpfen and H.Lischka, Chem.Phys.
     113, 53 (1987)
/30/ S.-K.Rhee and A.Karpfen, Chem.Phys. 120, 199 (1988)
/31/ A.Karpfen, A.Beyer and P.Schuster, Chem.Phys.Lett.
     102, 289 (1983)
/32/ J.F.Gaw, Y.Yamaguchi, M.A.Vincent and
     H.F.Schaefer III, J.Amer.Chem.Soc. 106, 3133 (1984)
/33/ G.E.Scuseria and H.F.Schaefer III, Chem.Phys.
     107, 33 (1986)
/34/ S.-Y.Liu, D.W.Michael, C.E.Dykstra and J.M.Lisy,
     J.Chem.Phys. 84, 5032 (1986)
/35/ G.Guelachvili, Opt.Commun. 19, 150 (1976)
/36/ J.W.Bevan, in: "Structure and Dynamics of Weakly
     Bound Molecular Complexes", A.Weber, ed., Reidel,
     Dordrecht (1987), p.149
/37/ A.S.Pine, W.J.Lafferty and B.J.Howard, J.Chem.Phys.
     81, 2939 (1984)
/38/ A.S.Pine and W.J.Lafferty, J.Chem.Phys. 78, 2154 (1983)
/39/ J.Bendtsen and H.G.M.Edwards, J.Raman Spectrosc.
     2, 407 (1974)
/40/ B.A.Wofford, J.W.Bevan, W.B.Olson and W.J.Lafferty,
     J.Chem.Phys. 85, 105 (1986)
/41/ J.S.Kittelsberger and D.F.Hornig, J.Chem.Phys.
     46, 3099 (1967)
/42/ H.B.Friedrich and P.F.Krause, J.Chem.Phys.
     59, 4942 (1973)

RARE GAS-HYDROGEN CHLORIDE COMPLEXES: FAR INFRARED OBSERVATIONS OF

Ar-HCl AND Xe-HCl, AND CALCULATIONS OF EXCITED STATES FOR Xe-HCl

A.R.W. McKellar,[*] J.W.C. Johns,[*] and J.M. Hutson[#]

[*]Herzberg Institute of Astrophysics
National Research Council of Canada
Ottawa, Ontario K1A OR6, Canada

[#]Department of Chemistry
University of Durham
Durham DH1 3LE, England

1. INTRODUCTION

The rare gas - hydrogen chloride (Rg-HCl) systems have become important prototypes for the study of anisotropic intermolecular forces in weakly bound molecular complexes. There has been a large amount of theoretical[1-3] and experimental[4-15] work on these systems, which are now understood in some detail, especially in the case of Ar-HCl. They are known to have linear equilibrium geometries with the rare gas atom located next to the hydrogen; the degree of anisotropy for internal rotation of the HCl increases with the mass of the rare gas, so that Xe-HCl is a more rigid molecule than Ar-HCl. The citations of previous Rg-HCl work given here are not complete, but further references may be found in the papers cited, particularly in Refs. 2 and 3.

Most experiments on these molecules have used supersonic jet sources in order to take advantage of the resulting large abundances of complexes and the spectral simplification afforded by low rotational and vibrational temperatures. However, by working at equilibrium with temperatures just above the condensation points of the constituent gases, it is possible to observe many more rotational and vibrational levels at the expense of spectral simplicity and strength. Thus Howard and Pine have studied mid-infrared spectra of Ar-HCl,[8] Kr-HCl, and Xe-HCl[16] at high resolution, and Boom and van der Elsken[6,7] have studied far infrared (FIR) spectra of Ar-, Kr-, and Xe-HCl at low resolution.

Dynamics of Polyatomic Van der Waals Complexes
Edited by N. Halberstadt and K. C. Janda
Plenum Press, New York, 1990

In the present paper, we present preliminary results of a new far infrared study of Ar-HCl and Xe-HCl made under equilibrium conditions, as well as new energy level calculations for Xe-HCl to aid in the interpretation of these and future spectra. The experiments represent an extension of the work of Boom and van der Elsken to lower pressure, higher resolution, and higher sensitivity. Although the resolution is still not high enough for detailed analyses to be made, the results are of interest because: (1) For Ar-HCl, present knowledge is sufficiently complete that one can begin to envisage[3] a complete rotation-vibration calculation to simulate our results, and (2) For Xe-HCl, there have been very few experiments, so any additional data is valuable. Furthermore, our spectra may be useful as a guide to the appropriate wavelength regions for future studies involving FIR laser/supersonic jet techniques,[14] which have high sensitivity and resolution but are not ideal for spectral searching.

The states of Rg-HCl complexes probed by infrared spectra may be described by six quantum numbers: the HCl vibrational quantum number v; a bending quantum number b, which correlates in the isotropic limit with the HCl rotational quantum number j; the total angular momentum J, and its projection K onto the molecule fixed a-axis; the parity label $p = (-)^{j+\ell+J}$; and the van der Waals stretching quantum number n. J and p are rigorously good quantum numbers, v and K are nearly conserved, and b and n are simply useful labels which qualitatively describe the wavefunctions. The definition of p implies that the spectroscopic parity is $p(-)^J$, but p itself is more commonly used; levels with $p=+1$ and -1 are generally designated either by superscripts $+$ and $-$, or as e and f, respectively. The space-fixed orbital angular momentum quantum number ℓ, which is important for weakly anisotropic systems such as Ar-H_2, is not conserved for systems as anisotropic as Ar- or Xe-HCl. The K quantum number may be regarded either as the rotational angular momentum of a near-symmetric top about its axis, or as the vibrational angular momentum due to low frequency bending of a linear triatomic molecule; states with $K = 0$, 1, etc., are often designated Σ, Π, etc. In this paper, we will generally designate Rg-HCl energy levels by the three quantum numbers (b,K,n), with v taken to be 0 unless otherwise stated.

2. CALCULATIONS ON Xe-HCl

In order to assist in assigning these and future Xe-HCl spectra, close-coupling calculations are presented here for the lowest few bound states of ^{132}Xe-H^{35}Cl on the M5 potential of Ref. 1. Analogous results for Ar-HCl have already been published by Hutson.[2] Assignments to $K = 0$, 1, and 2, have been made on the basis of helicity decoupling calculations. The coupled equations were integrated from 3.0 to 15.0 Å using Manolopoulos's modified log-derivative propagator[17,18] with an interval size of 0.025 Å. The reduced mass was taken to be 28.266918 u, and the rotational constant used for HCl was 10.44019 cm^{-1}. The basis set used included all functions up to $j_{max}=8$. The resulting calculations are fully converged, and are essentially exact (for the M5 potential).

Table 1. Close-coupling calculation of J=0, K=0, states of Xe-HCl, using the M5 potential.

b	n	E/cm^{-1}	B/cm^{-1}	<P1>	<P2>
0	0	-180.909	0.03280	0.801	0.529
0	1	-147.187	0.03180	0.751	0.471
1	0	-134.214	0.03446	-0.539	0.284
0	2	-117.784	0.03067	0.693	0.438
1	1	-109.827	0.03352	-0.491	0.223
2	0	-96.399	0.03018	0.345	0.263
0	3	-88.864	0.03181	0.307	0.297
1	2	-86.454	0.03225	-0.478	0.238
0	4	-73.721	0.02908	0.401	0.267
1	3	-65.873	0.03092	-0.268	0.232
2	1	-62.946	0.03100	0.067	0.262
0	5	-53.563	0.02793	0.295	0.250
1	4	-47.785	0.02787	-0.174	0.233
2	2	-43.586	0.03216	0.033	0.225
There are other states in this gap					
3	0	-30.082	(0.0288)	0.104	0.215

Table 2. Close-coupling calculation of J=1, K=1, f parity states of Xe-HCl, using the M5 potential.

b	n	E/cm^{-1}	(B+q/2)/cm^{-1}	q/cm^{-1}	<P1>	<P2>
1	0	-129.408	0.03355	-0.00042	0.430	0.028
1	1	-101.740	0.03264	-0.00026	0.324	-0.010
2	0	-92.441	0.03355	-0.00035	-0.152	0.099
1	2	-76.441	0.03125	-0.00019	0.273	-0.044
2	1	-66.912	0.03240	-0.00139	-0.124	0.109
1	3	-54.513	0.02927	-0.00008	0.235	-0.066
2	2	-44.418	0.03130	-0.00328	-0.113	0.107
1	4	-36.581	0.02809	-0.00526	0.185	-0.055
3	0	-30.159	0.03029		-0.030	0.123
2	3	-24.084	0.02903	-0.00019	-0.038	0.043
1	5	-20.183	0.02684	-0.00154	0.099	-0.018
1	6	-9.512	0.02375	+0.00072	0.136	-0.109

Table 3. Close-coupling calculation of J=2, K=2, and J=3, K=3, f parity states of Xe-HCl, using the M5 potential.

b	n	E/cm^{-1}	B/cm^{-1}	<P1>	<P2>
J=2, K=2 States					
2	0	-77.452	0.03396	0.167	-0.216
2	1	-51.556	0.03279	0.140	-0.227
2	2	-29.384	0.03242	0.096	-0.189
J=3, K=3 State					
3	0	-10.922	(0.0338)	0.080	-0.305

451

The resulting energies, rotational constants and angular momentum expectation values are given in Tables 1 to 3, together with assignments of the quantum numbers b and n. It should of course be appreciated that these latter quantum numbers are only loosely defined since there is considerable mixing between the different modes in any representation.

The selection rules applying to far-infrared spectra are: $\Delta K = 0$, ± 1 and $\Delta J = 0$, ± 1 (or $\Delta J = \pm 1$ if K'=K"=0), together with weaker propensity rules for the less well-defined quantum numbers: $\Delta b = 0$, ± 1 and $\Delta n = 0$, ± 1. A rough estimate of the intensities of the different bands may be obtained by considering only the bending motion. If the well depth function $\epsilon(\theta)$ is taken to represent an effective bending potential for the complex, the bending wavefunctions may be calculated by diagonalising the (fixed-R) Hamiltonian in a basis set of HCl rotational functions. The intensities may then be estimated by calculating matrix elements of the dipole moment operator between these functions, assuming that the dipole moment vector lies along the HCl axis. The most intense fundamental band in the far-infrared spectrum of Xe-HCl will be that for the Π bending vibration, with a band origin around 51.5 cm^{-1}. The Σ bend at 46.7 cm^{-1} will be about a factor of 10 weaker. This is rather different from the situation for Ar-HCl, where the intensity ratio is smaller; the difference arises from the increased barrier to internal rotation in Xe-HCl, which prevents substantial overlap between the ground state and the Σ bend state. The ratio may be expected to be even greater in Xe-DCl. The intensity of the Σ stretching band, at 33.7 cm^{-1} for Xe-HCl, is not determined within the above approximation, but it will also be considerably weaker than the Π bend. The j = 2-0 "bending overtone" bands of Xe-HCl will be about a factor of 5 weaker than the Π bend.

Under the experimental conditions of the present paper, a number of vibrationally excited states will be populated and a rich hot band structure may be expected. All the allowed hot bands with $\Delta n = 0$, except (b,K) = (1,0)←(0,0) bands, are calculated to have reasonable intensity since the higher excited states are progressively more free-rotor like. The calculated band origins and relative intensities for the bending fundamentals and stronger hot bands are listed in Table 4.

3. EXPERIMENTAL RESULTS ON Ar-HCl

The present experiments were performed by recording FIR absorption spectra of equilibrium HCl + Ar or HCl + Xe mixtures at low temperatures in the range 135-150 K. We used a modified Bomem Model DA3.002 Fourier Transform spectrometer and a cooled, large aperture, multiple-traversal absorption cell. The cell had a base path length of 0.5 m and was used with 40 traversals for a total path of 20 m. The experimental conditions and apparatus are very similar to those used recently to study[19] (HCl)$_2$, and a more complete description is given there. Line widths in the spectra were limited by pressure broadening rather than by instrumental resolution, and reasonably high pressures were required in order to achieve significant absorption with the available path length.

Table 4. Calculated band origins and approximate relative intensities for Xe-HCl. The origins are taken from the close-coupling calculations of Tables 1-3, and the intensities from fixed-R calculations (see text). The intensities do not include Boltzmann factors. Only bands involving no change in n are listed.

$b' \leftarrow b''$		$K' \leftarrow K''$		n	Origin/cm^{-1}	Intensity
1	0	0	0	0	46.695	0.014
1	0	1	0	0	51.501	0.14
1	0	0	0	1	37.360	0.014
1	0	1	0	1	45.447	0.14
1	0	0	0	2	31.330	0.014
1	0	1	0	2	41.306	0.14
2	1	0	0	0	37.815	0.14
2	1	1	0	0	41.773	0.18
2	1	0	1	0	33.009	0.12
2	1	1	1	0	36.967	0.15
2	1	2	1	0	51.956	0.30
2	1	0	0	1	46.881	0.14
2	1	1	0	1	42.915	0.18
2	1	0	1	1	38.794	0.12
2	1	1	1	1	34.828	0.15
2	1	2	1	1	50.184	0.30
3	2	0	0	0	66.317	0.17
3	2	1	0	0	66.240	0.21
3	2	0	1	0	62.359	0.12
3	2	1	1	0	62.282	0.14
3	2	2	1	0	67.80	0.27
3	2	1	2	0	47.293	0.08
3	2	2	2	0	52.80	0.12
3	2	3	2	0	66.530	

The absorption spectrum of a low temperature Ar + HCl mixture in the 23-50 cm^{-1} region is shown in Fig. 1. The spectrum is, of course, dominated by the very intense pure rotational lines of the HCl monomer, of which the R(1) transition near 42 cm^{-1} is visible here. Also evident in the spectrum are lines of DCl and HBr, present as impurities in the gas sample. Most of the remaining weak but extensive features in Fig. 1 are due to Ar-HCl.

There are 3 experimentally known FIR bands of Ar-HCl, and each has been studied in detail: the Σ bend[14] at 23.7 cm^{-1}, the Σ stretch[13] at 32.4 cm^{-1}, and the Π bend[12] at 33.9 cm^{-1}. These bands correspond to the transitions from the ground vibrational state to the (b,K,n)=(100), (001), and (110) excited vibrations, respectively. Only the last of these is strong enough[3] to be clearly visible in our spectrum. A detailed look at the region near the origin of this Π bending fundamental is shown in Fig. 2. Our spectrum shows a series of lines between about 34.1 and 35.4 cm^{-1} which can clearly be associated with its Q-branch. The situation is closely analogous to that of the corresponding combination band, (v,b,K,n) = (1110)←(0000), as observed in the mid-

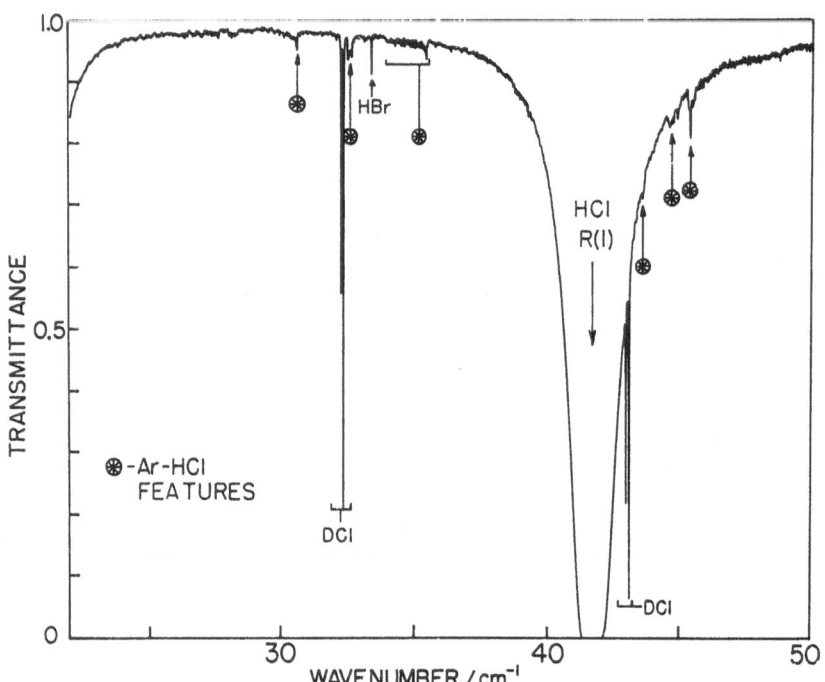

Fig. 1. Absorption spectrum of a mixture of HCl + Ar at a
pressure of 5 + 33 Torr and a temperature of 135 K. The
path was 20 m and the instrumental resolution was 0.010
cm^{-1}. DCl and HBr are present as impurities.

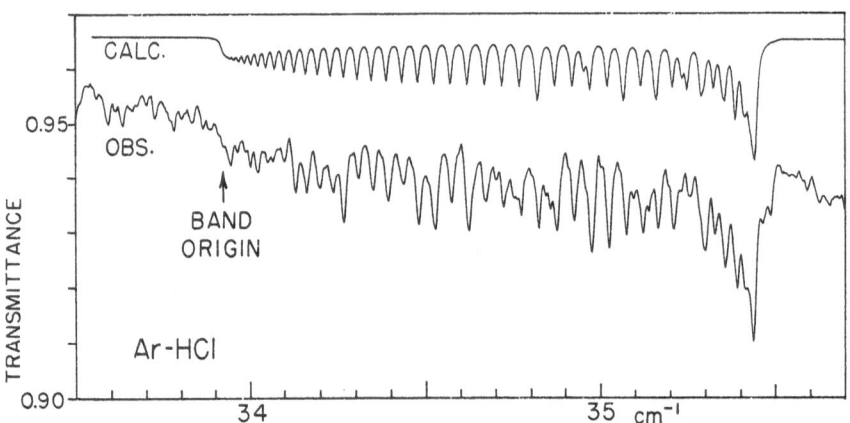

Fig. 2. A portion of the spectrum of Ar-HCl showing the Q-branch of
the Π bending fundamental band. Conditions are the same as
Fig. 1, except Ar pressure = 27 Torr, instrumental
resolution = 0.008 cm^{-1}. The calculated spectrum was
obtained using the parameters given in the text.

infrared spectrum of Ar-HCl by Howard and Pine[8] (see their Fig. 2) and by Lovejoy and Nesbitt.[15] In particular, both spectra show a prominent Q-branch head, located at 35.43 cm^{-1} in our spectrum and at 2919.49 cm^{-1} in Ref. 8. Using this analogy, along with the known[12] parameters of the Q-branch for low J-values, it was possible to arrive at a reasonable rotational assignment for the resolved lines in our spectrum and perform a least-squares fit. In the fit, the ground state parameters were fixed at their known values of B"=1678.51 MHz and D"=19.883 kHz. The fit was made using a power series in J(J+1), rather than J(J+1)-K^2, so that our value of ν_0 is lower than that of Ref. 12 by B', and our value of B' corresponds to their (B'+q/2). (Another definition of ν_0 is sometimes used[2] which corresponds to the position of the (unsplit) J=1 level and is 2B' greater than our ν_0.)

The parameters obtained in the fit were ν_0=34.922 cm^{-1} (fixed[12]), B'=1714.0(2) MHz, D'=24.1(6) kHz, H'=-0.9(6) Hz, and L'=-0.0001(2) Hz (3σ uncertainties are given in parentheses). This value of B' is in good agreement with the high resolution value[12] of 1714.33 MHz (which was obtained without including centrifugal distortion), especially considering the limited resolution of our results and the fact that we are analyzing only the more abundant isotope, Ar-H^{35}Cl, whereas the spectrum also contains contributions from Ar-H^{37}Cl. Our D' value is also very close to the value of 24.08 kHz obtained[15] for the analogous Π bend band in the mid-infrared spectrum. The results of the least squares fit are also shown in the form of a calculated spectrum in Fig. 2; the turning point corresponding to the Q-branch head was found to occur at about J=46. The existing high resolution laser measurements[9,10,12] of this band were limited to only a very few lines with values of J=3 and less. In order to make a more complete and unambiguous comparison with the present results, it would be very interesting to have an extended high resolution study of this band up to higher J-values at zero electric field using the tunable FIR laser technique of Busarow et al.[14]

Some other Ar-HCl features are indicated in Fig. 1 and shown in more detail in Fig. 3. There is a narrow (0.06 cm^{-1}) peak at 30.6 cm^{-1}, a wider (0.18 cm^{-1}) feature centered at 32.6 cm^{-1}, and extensive structure between 44.5 and 46 cm^{-1}, most notably a strong peak at 45.42 cm^{-1}. Much of the same structure was also observed at lower resolution by Boom and van der Elsken.[6,7] Calculated hot band origins for Ar-HCl have been tabulated by Hutson,[2] and a few of these are also shown in the form of a predicted spectrum by Clary and Nesbitt.[3] On the basis of these theoretical results, it is possible to give some tentative assignments for our observed spectral features. The strong absorption around 45.4 cm^{-1} is almost certainly due to Q-branch structure of the Coriolis-mixed hot bands (b,K,n)=(200)←(100) and (210)←(100), with predicted[2] origins at 45.2 and 45.6 cm^{-1}. The nearby structure in the 44.6 to 45.0 cm^{-1} region might be associated with the P-branches of these bands. The peak at 30.6 cm^{-1} resembles the observed Q-branch head at 35.43 cm^{-1}, and is probably due to the (111)←(001) hot band whose origin is predicted[2] to lie at 30.2 cm^{-1}.

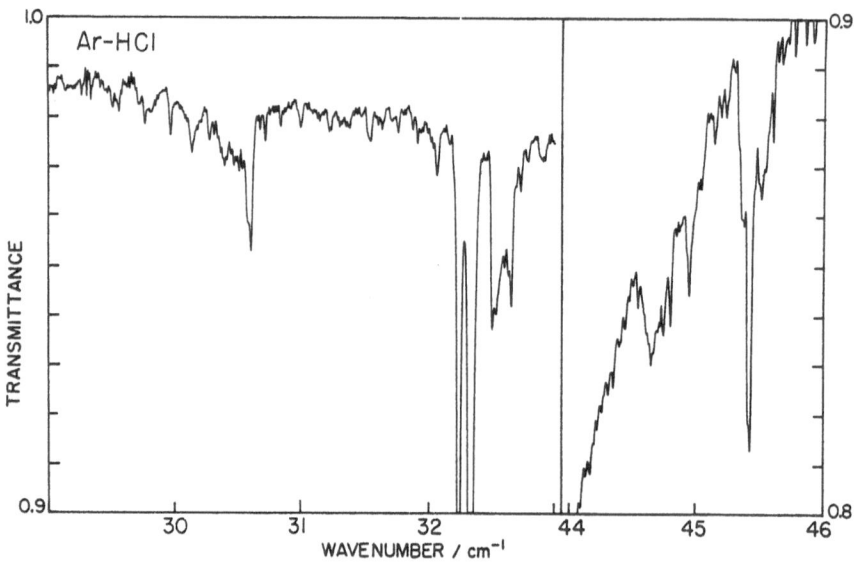

Fig. 3. Two portions of the spectrum of Ar-HCl. The two
strong lines at 32.2 and 32.3 cm^{-1} are due to DCl.
Conditions are the same as Fig. 1.

The 32.6 cm^{-1} feature (Fig. 3) does not coincide with any of
Hutson's[2] predicted hot band origins (the Σ stretch at 32.4 cm^{-1} is a
parallel band and does not have a Q-branch). However, we note from its
location that it may be analogous to the strong feature observed at
2916.7 cm^{-1} by Howard and Pine.[8] This association is supported by
Lovejoy and Nesbitt's[15] observation that the low frequency vibrations and
associated rotational constants for Ar-HCl complexes formed from v=1 HCl
are very similar to those formed from v=0 HCl. The same authors point
out that the 2916.7 cm^{-1} feature, which they did not observe due to low
source temperature, is very likely due simply to the P-branch head of the
Π bend band. The same assignment provides the most likely explanation
for our 32.6 cm^{-1} feature. In the present case, the feature has the
appearance of a double head, with the P-branch first turning around at
32.49 cm^{-1}, and then reversing again around 32.66 cm^{-1}.

4. EXPERIMENTAL RESULTS ON Xe-HCl

The spectrum of a low temperature Xe + HCl mixture in the 40-80
cm^{-1} region is shown in Fig. 4. Structure due to the Xe-HCl complex is
very extensive, and considerably stronger than that of Ar-HCl. Again,
much of this structure has been observed previously by Boom and van der
Elsken[6,7] at low resolution. The molecular beam electric resonance
spectrum of Xe-HCl has been studied by Chance et al.,[5] but there are no
reported high resolution mid- or far-infrared spectra of this complex
(though Howard and Pine[16] have some unpublished mid-infrared data). The
numerous isotopes of xenon and its relatively high cost may help to
explain the scarcity of experimental results.

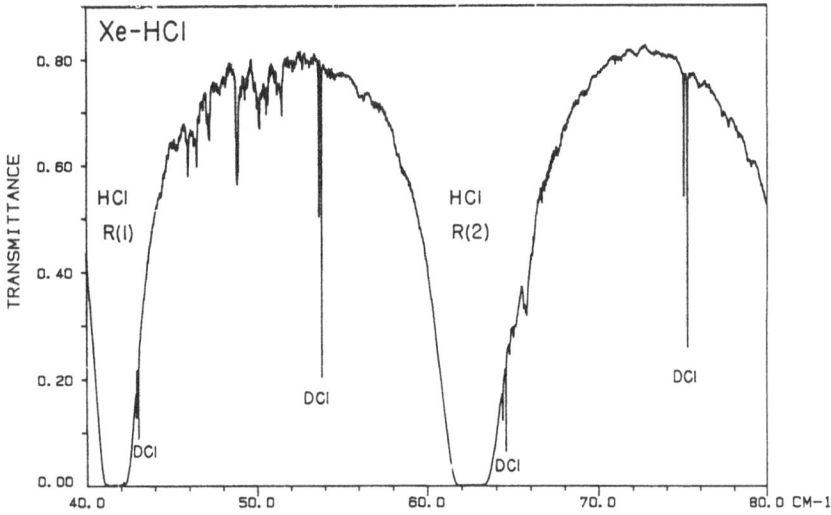

Fig. 4. Absorption spectrum of a mixture of HCl + Xe at a
pressure of 4 + 43 Torr and a temperature of 140 K.
The path was 20 m and the resolution was 0.010 cm^{-1}.

As indicated above in Sec. 2, the Σ stretch, Σ bend, and Π bend
fundamental bands of Xe-HCl are predicted to lie at about 33.8, 46.7, and
51.5 cm^{-1}, respectively, and the greater anisotropy in Xe-HCl means that
the intensity of the Π bend relative to the other fundamentals is even
greater than in Ar-HCl. Indeed, the region of strongest Xe-HCl
absorption in our spectrum is around 50 cm^{-1}, as shown in detail in Fig.
5 for three different sample pressures. At least some of the features
between about 50.0 and 51.5 cm^{-1} may correspond to the Q-branch of the Π
bend fundamental, perhaps with an origin at about 50 cm^{-1} and a Q-branch
head at 51.5 cm^{-1} in analogy with the Ar-HCl spectrum (Fig. 2). This
would suggest that the origin of the Xe-HCl Π bending band may be
slightly lower than the M5 calculated value of 51.50 cm^{-1}. In the
highest resolution Xe-HCl spectrum (the top trace of Fig. 5), there is
evidence for resolved Q transitions around 51 cm^{-1}, just like those
around 35 cm^{-1} for Ar-HCl. Some of the peaks observed in this region may
also be due to the hot bands (b,K,n) = (220)←(110) and (221)←(111), which
have calculated origins at 51.96 and 50.18 cm^{-1} and quite large
calculated intensities (see Table 4).

With this assignment of the 50-51.5 cm^{-1} structure to the Q-
branch of the Π bend fundamental, (b,K,n)=(110)←(000), analogy with Ar-
HCl then suggests a logical assignment for the structures observed at
46.0, 46.5 and perhaps 47.3 cm^{-1} (Fig. 5), namely to the Π bend hot band,
(111)←(001). The resemblence between the 50-51.5 cm^{-1} and 46-47.3 cm^{-1}
regions is especially evident in the highest pressure trace of Fig. 5.
The origin of the (111)←(001) hot band is predicted (Table 4) to be 45.45
cm^{-1}, lending further support to the assignment. We are then left with
the very strong peak at 48.9 cm^{-1}, which, by analogy with Ar-HCl, might
be due to a P-branch head of the Π bend fundamental.

457

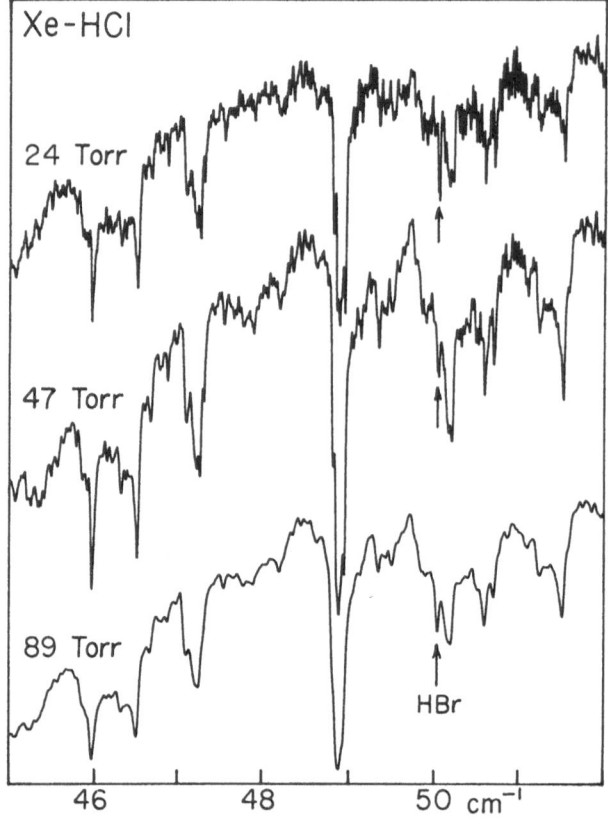

Fig. 5. A portion of the spectrum of Xe-HCl
recorded at three different pressures as
indicated. Partial pressures of HCl
ranged from 4 to 7 Torr. The 3 traces do
not share a common transmittance scale.

Another interesting region of the Xe-HCl spectrum, from 65 to 70 cm^{-1}, is shown in detail in Fig. 6. There is an unresolved feature located at 65.75 cm^{-1} with a width of about 0.25 cm^{-1} and a region of partially resolved lines around 67 cm^{-1}. (There is also a Q-branch of $(HCl)_2$ in this same region,[19] but the structure shown in Fig. 6 is mostly due to Xe-HCl.) Referring to Table 4, the most likely assignments for the structure in Fig. 6 are to the Coriolis-mixed (b,K,n)=(300)←(200) and (310)←(200) transitions, and the (320)←(210) transition, with predicted origins at 66.32, 66.24, and 67.80 cm^{-1}, respectively.

Some further Xe-HCl features, not illustrated here, are apparent in our spectra in the gap from 85 to 103 cm^{-1} between the HCl monomer R(3) and R(4) lines. In particular, there is a reasonably strong feature at 86.7 cm^{-1} with a width of about 0.2 cm^{-1} and a weaker, sharper feature at 89.0 cm^{-1}. These could be associated with branch heads of the (200)←(000) and (210)←(000) bending overtone bands with predicted origins at 84.5 and 88.5 cm^{-1}, respectively. The calculated spectrum of this region at very low temperature (10 K) in Fig. 17 of Clary and Nesbitt[3] lends some support to this assignment.

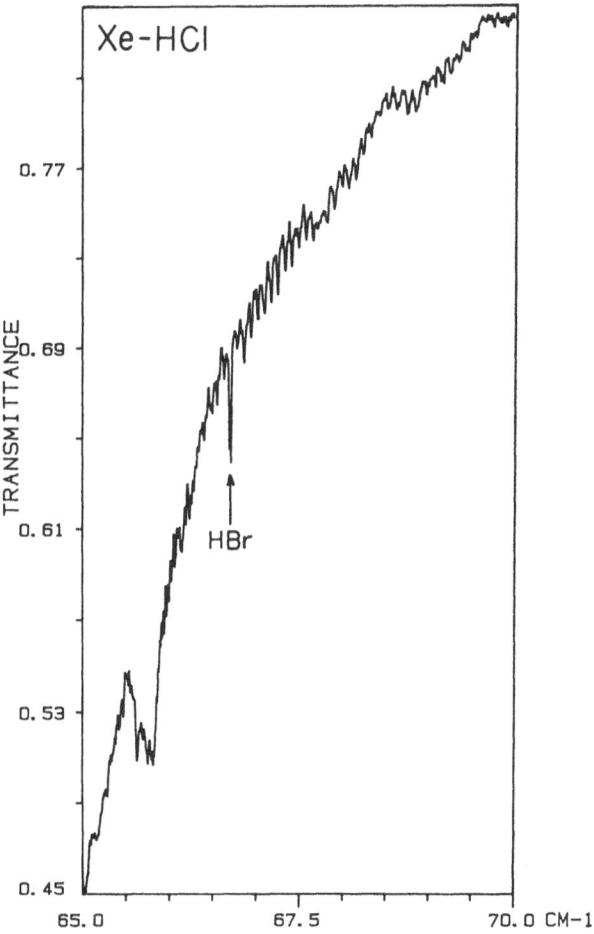

Fig. 6. A portion of the spectrum of Xe-HCl at a
total pressure of 24 Torr.

4. CONCLUSIONS

The experimental results presented in this paper do not have
sufficiently high resolution and sensitivity to be directly useful in
obtaining detailed new information on the Ar-HCl and Xe-HCl complexes.
From an apparatus point of view, what is required for improved
equilibrium spectra is a longer absorption path, so that lower gas
pressures can be used and higher spectral resolution will be obtained.
We are currently constructing a 2 m base path length cooled cell for the
FIR region which will give a fourfold increase in path compared to the
present results. With this improvement, there will be a reasonable
chance of obtaining rotationally resolved FIR spectra of the rare gas -
hydrogen halide complexes comparable to those obtained in the mid-
infrared by Howard and Pine.[8]

Our present theoretical and experimental data should be useful
for indicating promising regions of investigation for tunable FIR laser
studies, such as that of Busarow et al.[14] It is also possible that
comprehensive ab initio calculations of the "full" FIR spectra are now

within the realm of possibility, at least for Ar-HCl. Clary and Nesbitt[3] have made a good start in this direction. Given such a full calculation of line positions and intensities one could perform a band contour analysis on spectra like those shown here, and, by comparing theory with experiment, refine the intermolecular potential surface for regions far removed from equilibrium.

REFERENCES

1. J.M. Hutson and B.J. Howard, Mol. Phys. 43:493 (1981); 45:769 (1982); 45:791 (1982).
2. J.M. Hutson, J. Chem. Phys. 89:4550 (1988).
3. D.C. Clary and D.J. Nesbitt, J. Chem. Phys. 90:7000 (1989).
4. S.E. Novick, P. Davies, S.J. Harris, and W. Klemperer, J. Chem. Phys. 59:2273 (1973).
5. K.V. Chance, K.H. Bowen, J.S. Winn, and W. Klemperer, J. Chem. Phys. 70:5157 (1979).
6. E.W. Boom and J. van der Elsken, J. Chem. Phys. 73:15 (1980).
7. E.W. Boom and J. van der Elsken, J. Chem. Phys. 77:625 (1982).
8. B.J. Howard and A.S. Pine, Chem. Phys. Lett. 122:1 (1985).
9. M.D. Marshall, A. Charo, H.O. Leung, and W. Klemperer, J. Chem. Phys. 83:4924 (1985).
10. D. Ray, R.L. Robinson, D.-H. Gwo, and R.J. Saykally, J. Chem. Phys. 84:1171 (1985).
11. R.L. Robinson, D.-H. Gwo, D. Ray, and R.J. Saykally, J. Chem. Phys. 86:5211 (1987).
12. R.L. Robinson, D. Ray, D.-H. Gwo, and R.J. Saykally, J. Chem. Phys. 87:5149 (1987).
13. R.L. Robinson, D.-H. Gwo, and R.J. Saykally, J. Chem. Phys. 87:5156 (1987).
14. K.L. Busarow, G.A. Blake, K.B. Laughlin, R.C. Cohen, Y.T. Lee, and R.J. Saykally, Chem. Phys. Lett. 141:289 (1987).
15. C.M. Lovejoy and D.J. Nesbitt, Chem. Phys. Lett. 146:582 (1988).
16. Unpublished, see: B.J. Howard, High resolution infrared spectroscopy of van der Waals molecules, in: "Structure and Dynamics of Weakly Bound Molecular Complexes," A. Weber, ed., D. Reidel, Dordrecht (1987); and, G.T. Fraser and A.S. Pine, J. Chem. Phys. 85:2502 (1986).
17. D.E. Manolopoulos, J. Chem. Phys. 85:6425 (1986).
18. J.M. Hutson, BOUND computer code, distributed via Collaborative Computational Project No. 6 of the UK Science and Engineering Research Council, on Heavy Particle Dynamics.
19. N. Moazzen-Ahmadi, A.R.W. McKellar, and J.W.C. Johns, Chem. Phys. Lett. 151:318 (1988); The far infrared spectrum of the HCl dimer, J. Mol. Spectrosc., to be published.

PROGRESS ON THE DETERMINATION OF INTERMOLECULAR POTENTIAL ENERGY

SURFACES FROM HIGH RESOLUTION SPECTROSCOPY

David J. Nesbitt[*]

Joint Institute for Laboratory Astrophysics, National Institute
of Standards and Technology and Department of Chemistry and
Biochemistry, University of Colorado, Boulder, CO 80309-0440

INTRODUCTION

The determination of accurate intermolecular potentials has been a
key focus in the understanding of collision and half-collision dynamics,
but has been exceedingly difficult to obtain in quantitative detail for
even the simplest molecular systems. Traditional methods of obtaining
empirical intermolecular potential information have been from analysis
of nonideal gas behavior, second virial coefficients, viscosity data and
other transport phenomena.[1-3] However, these data sample highly averaged
collisional interactions over relative orientations, velocities, impact
parameters, initial and final state energies, etc. As a result intermolecular potential information from such methods is limited to estimates
of the molecular size and stickiness, i.e., essentially the depth and
position of the energy minimum for an isotropic well.

Molecular beam scattering methods can in many cases provide a "more
detailed" understanding of intermolecular potentials, in essence simply
by control of the collision parameters sampled. Cooling in supersonic
jets can be used in sufficiently small molecular systems to prepare a
single or a few rotational/vibrational levels, as well as to collimate
and tune collision velocities in the center of mass frame. Angle resolved, total differential scattering data can be used to observe rainbow features and diffraction oscillations which can determine accurate
isotropic well depths (ϵ) and potential minima (R_m).[4] With sufficient
angular resolution, information on anisotropy in the potential can be
inferred from damping of the diffraction oscillations with deflection
angle.[5] For suitably light collision systems with large rotational
spacings or collisionally accessible vibrational spacings in the target,
time of flight analysis of the final state velocities can be used to
infer state resolved, differential cross sections, and thereby eliminate
one additional layer of averaging between the intermolecular potentials
and experimental observation.[6,7] However, even infinitely resolved,
initial, and final state beam scattering data will always reflect an

---

[*]Staff Member, Quantum Physics Division, National Institute of Standards
and Technology.

Dynamics of Polyatomic Van der Waals Complexes
Edited by N. Halberstadt and K. C. Janda
Plenum Press, New York, 1990

461

average over all impact parameters of the collision plane. Finally, it should be mentioned that extraction of potential data from such experiments requires a parametrization of the potential form, forward convolution via quantum scattering theory, and least-squares comparison with the observed cross section data.

Many of these restrictions can in principle be overcome if we exploit the attractive wells between colliding species to orient the molecular species into a long lived complex. Since even excited internal states of the complex can be quite long lived, the quantum energy levels that develop from the "bound" region are well defined, sensitive to the exact topology of the potential surface, and can be determined to extreme precision via high resolution spectroscopic methods. If we are sufficiently clever, industrious, or dogged, the energy levels and eigenfunctions of these states might be used to determine the intermolecular potential directly, at least in the bound state region sampled by the eigenfunctions investigated. The hope, simply stated, is that the precision of high resolution spectroscopic methods (orders of magnitude more resolved than state-of-the-art time of flight energy loss measurements), can be used as leverage to map out intermolecular potentials in the bound region. It is the thrust of this paper to describe recent progress in our group toward that goal.

This paper will focus on the intermolecular vibrational data in complexes obtained via "combination band" methods in the near IR, and on the recent efforts we have made toward determining potential energy surfaces from such data. The organization of the presentation will be as follows. 1) We start with the simplest determination of 1-D (radial) potentials for a psuedodiatomic molecule (the "ball + ball" problem) based solely on extensive rotational data for the lowest vibrational state. 2) Extension of these 1-D rotational methods to 2-D (angular and radial) PES determination for rare gas-hydrogen halides (the "ball + stick" problem) is then described. 3) We then move to complexes of rare gases with asymmetric tops such as H_2O, where the internal rotor energy spacings provide insight into a potential energy surface for the "ball + top" problem.

EXPERIMENTAL

All of the near IR spectra discussed in this work are obtained via direct absorption laser spectroscopy in a slit supersonic jet expansion.[8] The slit jet geometry provides extremely long path lengths (4 cm × 12, with White cell optics) through high density regions (ρ proportional to $1/r$) of jet cooled (T_{rot} = 4-20 K) molecules. In addition, the high densities along the slit quench the velocity distributions such that intrinsic narrowing (down by 10 fold) of the Doppler widths occurs in an unskimmed free jet, which translates directly into both higher resolution and sensitivity per velocity group.

The source of cw IR light is either a tunable difference frequency laser (2.2-4.2μm) or a tunable diode laser (3.0-16μm, depending on diode). The difference frequency laser is formed from nonlinear subtraction of a frequency stabilized, single mode scanning dye laser and Ar^+ laser in a temperature phase matched, $LiNbO_3$ crystal.[9] Power outputs are on the order of 10 μW, with a frequency stability of a few MHz. The diode laser is a commercially available product, and tuned by combination of computer controlled current and temperature adjustment. Power outputs are very diode dependent, but range from 10 to 1000 μW on a single mode, with a frequency stability limited presently by current noise and acoustic vibration to 50 MHz. This can be easily reduced to <10 MHz with a simple servo loop locking to a reference cavity.

Scanning of the lasers, pulsing of the valve, and data acquisition, storage, and display are under microprocessor computer control. Transient absorption of the transmitted laser light through the jet on a molecular resonance is detected via dual beam imbalance on a pair of matched IR detectors with high common mode rejection transimpedance amplifiers. Data are digitized in real time during the 400-500 μsec gas pulse, integrated to improve S/N and compensate for small baseline fluctuations, and stored for later data analysis. For further information on the apparatus, the interested reader is referred to more complete descriptions already in the literature.[10,11]

ROTATIONAL RKR METHODS: "BALL + BALL"

Direct RKR inversion of vibrationally and rotationally resolved spectroscopic data for diatomics is now a fairly routine procedure. In normal RKR applications, however, the spectral data are exploited in a relatively limited fashion. One simply uses B(v) and E(v), the rotational constant and term value dependence on vibrational quantum v, respectively, to infer the inner and outer classical turning points at each v from a semiclassical analysis. In high resolution spectroscopy of van der Waals complexes, however, there is often far more <u>rotational</u> than <u>vibrational</u> data available. Consequently extensive information exists on very high order centrifugal effects on the radial coordinate, sometimes up to, and by virtue of centrifugal barriers, beyond the dissociation limit! The hope is that a simple extension of RKR ideas might be able to extract a 1-D potential directly from rotational data alone.

The key idea to this scheme is presented in Fig. 1. In weakly bound complexes, end-over-end rotation introduces a significant angular momentum barrier, which for a 1-D radial model modifies the potential in a simple, analytic fashion

$$V_{eff}(R;J) = V(R) + \frac{\hbar^2 J(J+1)}{2\mu R^2} \qquad (1)$$

This modification of the potential shifts the classical turning points for the lowest radial vibrational (v=0) level systematically outward in R as a function of J. These turning points for v=0 can be obtained from normal RKR theory, based on integration of the semiclassical equations only the small extent from v=-1/2 to v=0. The essence of the rotational RKR approach is to approximate this small integral harmonically, derive inner and outer turning points and energies for each J, use Eq. (1) to subtract off the centrifugal correction exactly, and thereby obtain the R dependence of the true (i.e. J=0) potential. This can then be used in an iterative fashion to provide successively better approximations to the semiclassical integral from harmonic and anharmonic corrections at each stage. Generally, however, even the initial harmonic determination of the potential is quantitatively acceptable, but certainly improves by iteration.[12]

A test of this method for 1-D radial potentials is shown in Fig. 2. Here we have taken an assumed intermolecular potential (shown in solid line) for a weakly bound diatomic molecule (which for realism had well depths and radial minima similar to a psuedo diatomic treatment of Ar-HF), and solved exactly for the J = 0, 5, 10, 15, 20, 25, and 30 eigenvalues for v=0 by exact close coupling quantum calculations. These eigenvalues are then taken as input into the rotational RKR method, which attempts to regenerate the initial potential.

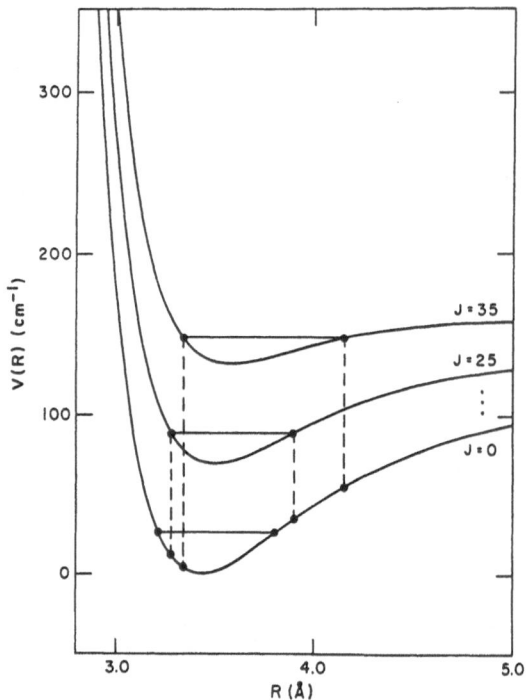

Fig. 1. Centrifugally modified 1-D potentials illustrating
the rotational RKR procedure for obtaining the J=0
potential from J dependent inner and outer turning
points.

The black and white circles represent the inner and outer rotational
RKR determined turning points, respectively. Agreement with the true potential is seen to be excellent, as highlighted by fractional residuals
(<0.1% of D_0) shown in Fig. 2a. In absolute terms, the rms deviation is
roughly 0.2 cm^{-1} (out of a well depth of 114 cm^{-1}!), in extremely good
agreement with the input potential. Even the small remaining deviations
are systematic (and negative), and are probably due to a failure of this
first order semiclassical analysis at low v. Hence, even more quantitative agreement might be obtained with next order WKB semiclassical corrections. The key point is that these calculations are extremely simple,
and require only the computing power of a hand calculator.

ROTATIONAL RKR IN HIGHER DIMENSIONS: "BALL + STICK"

In all fairness, however, the 1-D problem is sufficiently simple to
permit more numerically intensive approaches to achieve similar success.
The challenge is whether these ideas might be extendable into higher dimensions, such as a "ball + stick" described by two internal coordinates
(radial and angular). Our interest in this application arises out of our
work in rare gas-HX complexes, for which although there are strictly three
internal degrees of freedom, the high frequency HX stretch mode can be
treated as adiabatically averaged over the much slower intermolecular coordinates. One can at least see the superficial possibility of extension
to M + HX van der Waals species by drawing an analogy between electronic
motion in rotating diatomics and the (relatively) rapid bending and
stretching HX motion in a (relatively) slowly tumbling M-HX complex.

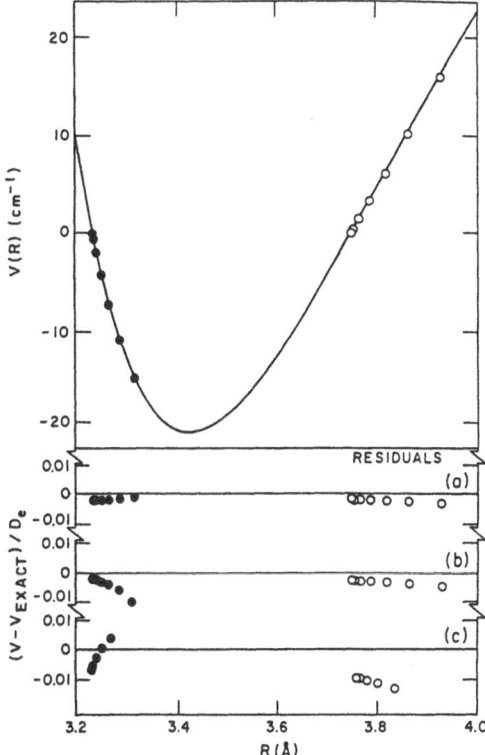

Fig. 2. 1-D fits to sample van der Waals data, illustrating the excellent reproduction of original potential (solid line) by the inner (•) and outer (o)
turning points determined via rotational RKR
methods.

Just as diatomic spectra consist of band systems involving transitions
between different electronic states, so are the vibrational bands in
van der Waals spectra associated with changes in the HX wave function,
coupled with end-over-end rotation of the complex. Hence one might hope
to build up diatomic-like potential curves in the van der Waals stretch
coordinate R for each type of HX motion.

    To capture the flavor of the method, consider the limit of a M-HX
complex in which the intermolecular potential is mostly isotropic, but
with sufficient anisotropy so the nearly free internal HX rotor motion
is oriented in the body fixed frame. The ground state of such a complex
would look like a $j=0$ HX rotor (i.e. an "s" orbital) bound to M, whereas
the three lowest excited HX bending states would approximate the three
$j=1$ rotor wave functions (i.e. three "p" orbitals) oriented with respect
to the end-over-end plane of rotation of the M-HX centers of mass, one in
a Σ and two in a Π configuration. The $j=0$ HX rotor state probes predominantly the <u>isotropic</u> part of the intermolecular radial potential, whereas
$j=1$ HX rotor states (the Σ and either one of the Π configurations) begin
to sample in addition the lowest order <u>anisotropic</u> parts of the potential.
The <u>radial</u> dependence of the intermolecular potential for each of these
three states can be determined from rotational RKR method. In principle,
these curves contain sufficient information to determine the three lowest

moments (i.e. coefficients of the P_0, P_1 and P_2 Legendre polynomials) of the angular dependence of the potential as a function of R.[13]

How such a method survives for realistic, appreciably anisotropic potentials is the focus of this discussion. As a test system, we consider the intermolecular potential between Ar and HF(v=1). This is best described as a strongly hindered internal rotor complex, and on which we have data for the (10^00), (11^10) and (12^00) from the slit jet spectrometer. These three states, respectively, correlate in the limit of weak anisotropy with precisely the three states (the "s," and the Π and Σ oriented "p" orbitals) in the above paragraph that sample the full range of angular coordinates.

Consistent with our assumption of weak (but finite) coupling between angular and radial degrees of freedom, the radially dependent differences in these three potentials can be converted into radially dependent coefficients, $V_i(R)$, of the three lowest Legendre polynomials, i.e.

$$V(R,\theta) = \Sigma_i V_i(R) P_\ell(\cos\theta) \tag{2}$$

This permits a reconstruction of full 2-D potential energy surface for Ar + HF(v=1), which is shown in Fig. 3, and exhibits the double minimum behavior (ArHF and ArFH configurations) predicted from multiproperty fits on rare gas-HCℓ complexes by Hutson and Howard.[14] The barrier between the two minima is considerably above the j=1 HF rotor energy, and thus the Ar-HF complex is an example of a strongly hindered internal rotor. It is worth noting that the radial minimum in the potential shifts with the angle θ, i.e. our inversion process does predict bend-stretch coupling in the potential that can be checked with experiment.

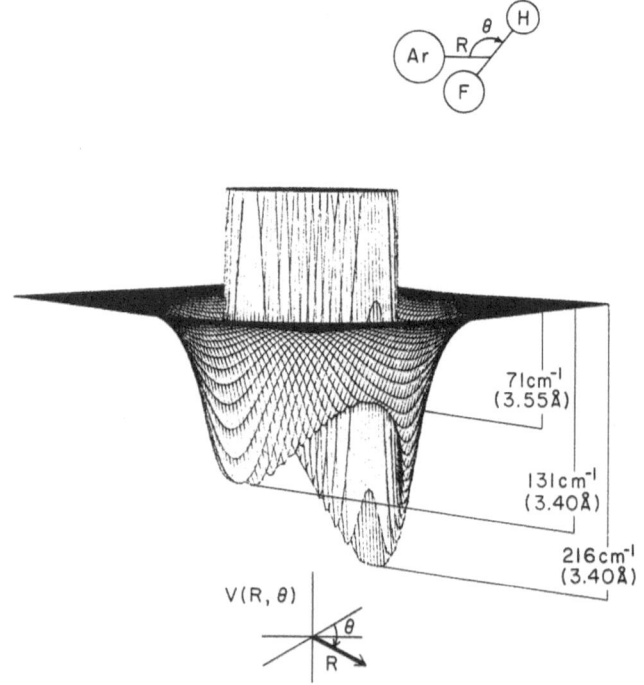

Fig. 3. 2D potential energy surface, $V(R,\theta)$, for
Ar+HF(ν=1) determined from the infrared data.

As a result of the high sensitivity of the slit jet apparatus, as well as the large amplitude nature of the bending and stretching motions, we have been able to observe an additional six of the remaining eight bound vibrational levels in ArHF.[15] These states include an excitation of one, two and four quanta in the van der Waals stretch [$(10^0 1)$, $(10^0 2)$ and $(10^0 4)$], as well as multiply excited bend and stretch states [$(11^1 e 0)$, $(11^1 e 1)$ and $(11^1 f 1)$]. Most importantly, two avoided crossings between van der Waals stretch and bend states are observed, which provides direct measure of the bend-stretch coupling in the potential.

An even more rigorous test of the accuracy of the bound state region of the Ar-HF(v=1) PES, therefore, would be its ability to predict the experimentally observable properties of vibrational levels not used in the rotational RKR inversion process. With this in mind, we have performed exact close coupling calculations to determine all the bound vibrational levels in ArHF(v=1) from J=0 to 20, using an efficient angular and radial basis set algorithm developed by Clary and Nesbitt.[16] Spectroscopic fits to the exact quantum values are made; vibrational origins, rotational constants, Coriolis coupling, ℓ-doubling, and even rotational levels of the avoided crossings are predicted and compared to experiment in Table I. Agreement between prediction and experiment is extremely good, at the fraction of a percent level for vibrational frequencies, and only a percent for rotational constants. Quite impressively, even the avoided crossings are predicted to within one or two J, and the ℓ-doubling magnitude of the Coriolis coupling between Σ and Π levels is predicted to roughly 5-20%. This latter observation is particularly noteworthy, since these interactions are occurring between states that differ by Δν=4, and hence would only be predicted to be nonzero at a high level of perturbation theory. Despite this quantitative agreement, it is important to note that our inversion process only utilizes information on the <u>bound</u> region of this potential, and thus one should not expect the repulsive inner wall to be as well characterized.

POTENTIAL INTERACTIONS IN WATER COMPLEXES: "BALL + TOP"

The next step in molecular complexity beyond "ball + stick" quantum mechanical motion involves interaction between an inert gas with a monomer with internal rotational structure, e.g. Ar + H_2O. For this system we have assistance from both the near and far IR. Cohen et al.[17] in the Saykally laboratories have utilized direct absorption far IR spectroscopy to detect two bands (Σ-Π and Π-Σ) in the Ar-H_2O complex. These were ascribed initially to rotation-tunneling transitions between K=0 and K=1 manifolds of a quasirigid complex, but more recently have been reinterpreted in terms of near free internal rotor motion of the H_2O in the presence of the Ar.[18] There have been recent efforts by Hutson to fit the two bands to an angular intermolecular potential, but there proved to be more important terms in the expansion than data and thus a family of possible curves could be inferred.[19]

In conjunction with data on Ar-H_2O from the near IR, however, these ambiguities can be resolved. For H_2O in a cold jet, complex formation only of the lowest para Ar-H_2O (correlating with para 0_{00} water) and ortho Ar-H_2O states (correlating to ortho 1_{01} water) will be appreciable. The (2j+1) degenerate ortho 1_{01} H_2O states will be split by interaction with the Ar induced anisotropy to yield states which can be characterized approximately by the projection of the internal rotor angular momentum onto the body fixed axis of the complex (i.e. Σ and Π), provided that the splittings are small with respect to the rotational spacings in free water. Note that this K is not the same as the projection of j onto the body fixed axis (k_a) of the water monomer, which is a very poor quantum

Table I. Comparison of experimental and calculated rovibrational data (in cm^{-1}) for Ar + HF ($\nu = 1$) potential energy surface

	exp	calc	error	exp	calc	error	exp	calc	error
	$(10^0 0)$[a]			$(12^0 0)$[a]			$(11^1 f_0)$[a]		
ν_0	--	--	--	57.3347	57.2925	-0.074%	70.3366	70.3130	-0.034%
B	0.102610	0.102970	+0.35%	0.100121	0.100522	-0.40%	0.100325	0.100395	+0.070%
$D/10^{-6}$	2.07	2.15	+4.0%	3.22	3.14	-2.5%	3.32	3.22	-3.0%
	$(10^0 1)$[b]			$(10^0 2)$[b]			$(11^1 e_0)$[b]		
ν_0	41.3349	41.3455	+0.026%	71.6109	72.7959	+1.7%	70.3403	70.3143	-0.037%
B	0.093504	0.094179	+1.3%	0.082509	0.083353	+1.0%	0.102609	0.102553	-0.055%
$D/10^{-6}$	3.25	2.94	-9.6%	5.93	6.11	+3.1%	3.64	3.70	+1.6%
β_{cor}	--	--	--	0.0168[c]	0.0131	-22%	--	--	--
J_{cross}	--	--	--	8[c]	10	--	--	--	--
	$(11^1 f_1)$[b]			$(11^1 e_1)$[b]			$(10^0 4)$[b]		
ν_0	101.8226	102.1316	+0.30%	101.8266	102.1335	+0.30%	--	102.9463	--
B	0.89666	0.89338	-0.37%	0.09218	0.09137	-0.88%	--	0.05473	--
$D/10^{-6}$	5.34	5.19	-2.8%	13.1	5.71	--	--	24.3	--
β_{cor}	--	--	--	--	--	--	0.0030-0.0036[f]	0.00609	--
J_{cross}	--	--	--	--	--	--	6[f]	5	--

[a]Vibrational states used in the determination of the potential.
[b]Vibrational states not used in the determination of the potential.
[c]The $(10^0 2)$ and $(11^1 e_0)$ states cross experimentally near $J\sim 8$, with a Coriolis interaction $\Delta\ell = \pm 1$ matrix element of $\beta_{cor}\sqrt{J(J+1)}$.
[d]ℓ-doubling of the $(11^1 e, f_0)$ levels arising from interacting with $(12^0 0)$ and $(10^0 0)$ at lower energy.
[e]ℓ-doubling of the $(11^1 e, f_1)$ levels from nearby Σ states, e.g., $(12^0 1)$ at lower energy.
[f]Coriolis interaction element (i.e., $\beta_{cor}\sqrt{J(J+1)}$) observed for $(10^0 4)$ crossing with $(11^1 1)$.

number for such an asymmetric top. In this jk_ak_c (K) notation for the states, we predict four strong vibrational bands in the asymmetric stretch region of water, which we label $1_{01}(0)\leftarrow 0_{00}(0)$, $1_{01}(1)\leftarrow 0_{00}(0)$ $0_{00}(0)\leftarrow 1_{01}(1)$ and $0_{00}(0)\leftarrow 1_{01}(0)$ and of which we have observed the latter three. Analysis of the last two bands indicates that the lower states are identical to the lower levels of the two bands observed by Saykally and coworkers,[17],[18] and hence the energy difference between the 1_{01} and 1_{10} levels in Ar-H_2O can be determined absolutely. Sample data from one of these bands, $1_{01}(0)$-$0_{00}(0)$, are shown in Fig. 4. Of course, information on the ν_3 vibrationally excited state is also obtained, but for this discussion we are interested in the potential for the $\nu_3=0$, ground vibrational state.

The quantum mechanical solution for angular motion of a "ball + top" can be readily determined from matrix diagonalization methods in an appropriate basis. Hutson[19] recently has described a treatment in body fixed basis functions, which permits helicity decoupling approximations to be implemented. As a result of previous experience in the group, we chose instead to perform exact quantum calculations in a space fixed frame, with a basis labeled by $|Jjk\ell>$ (ℓ is the end-over-end angular momentum of the complex) and incrementally increased until convergence in the eigenenergies to ± 0.001 cm^{-1} is observed. The angular potential can then be expanded in spherical harmonics, and the lowest three nonvanishing coefficients allowed by symmetry (V_{11} -V_{1-1}, V_{20} and $V_{22}+V_{2-2}$) least-squares adjusted to match the experimental data. The results are listed in Table II.

Interestingly, the potential is in general rather isotropic; terms of only 10-20 cm^{-1} in anisotropy are evident, in contrast with an approximately 130 cm^{-1} dissociation energy determined from psuedodiatomic calculations for a one-dimensional radial well. What anisotropy exists in-

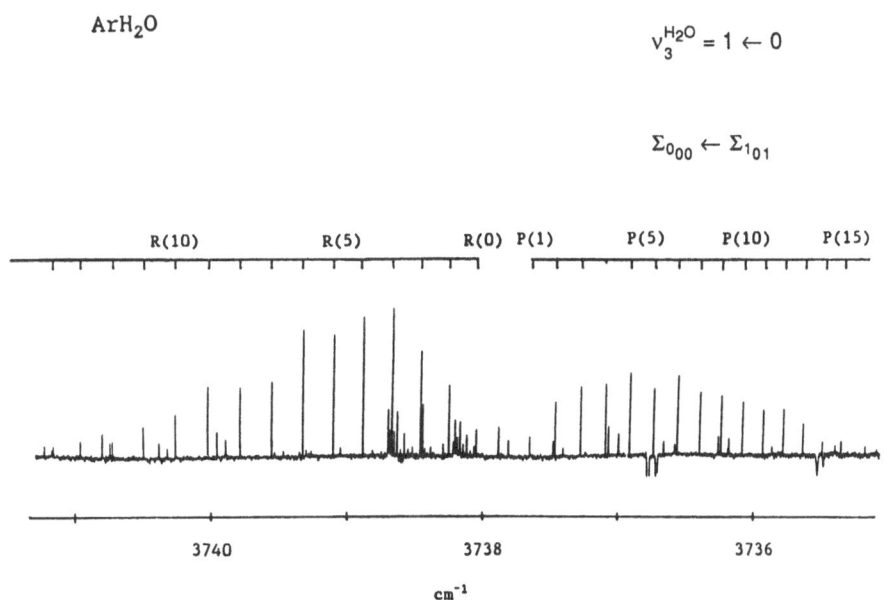

Fig. 4. Sample data on ArH$_2$O excited in the $\nu_3=1$ asymmetric stretch mode of H$_2$O. The unlabeled transitions are from P, Q, and R branch structure in the (H$_2$O)$_2$ complex also present in the expansion.

Table II. Spherical harmonic coefficients for ArH_2O internal rotor potential.

$V_{10} \leq 0.1$ cm^{-1}	$V_{21} = 0^*$
$V_{11} = 0^*$	$V_{2-1} = 0^*$
$V_{1-1} = 0^*$	$V_{22} = -11.6(0.2)$ cm^{-1}
$V_{20} = -19.7$ (1.2) cm^{-1}	$V_{2-2} = -11.6(0.2)$ cm^{-1}

*Zero by symmetry of the potential.

dicates relatively small barriers (10 cm^{-1}) to internal rotation of H_2O in the H_2O plane, with relatively large (40 cm^{-1}) barriers for H_2O internal rotation perpendicular to the H_2O plane. The inferred height of the barriers to rotation in the H_2O plane may change with the size of the Legendre expansion, which in turn will require observation of higher internal rotor excited levels. We hope this will be possible via either near or far IR spectral searches, whose progress should be considerably stimulated by predictions from this preliminary potential surface.

Acknowledgment

This work was supported by National Science Foundation (CHE86-05970 and PHY86-04504), the Petroleum Research Fund and Research Corporation. Further support from the Henry and Camille Dreyfus Foundation and Alfred P. Sloan Foundation is gratefully acknowledged. I would like to thank Mark Child for the many stimulating discussions we had during his stay in Boulder, and our subsequent collaboration which is the origin of many of the ideas discussed in this paper.

References

1. B. Schramm and V. Leuchs, Ber Bunsenges. Phys. Chem. 83, 847 (1979).
2. G. J. Q. Van der Peyl, D. Frenkel, and J. Van der Elsken, Chem. Phys. lett. 56, 602 (1978).
3. J. G. Kercy, G. J. Q. Van der Peyl, J. Van der Elsken, and D. Frenkel, J. Chem. Phys. 69, 4606 (1978).
4. L. Beneventi, P. Cassavecchia, F. Vecchiocattiori, G. G. Volpi, U. Buck, Ch. Lauenstein, and R. Schinke, J. Chem. Phys. 89, 4671 (1988).
5. C. V. Boughton, R. E. MIller, P. F. Vohralik, and R. O. Watts, Mol. Phys. 58, 827 (1986).
6. W. D. Held, E. Piper, G. Ringer, and J. P. Toennies, Chem. Phys. Lett. 75, 260 (1980).
7. G. Niedner, M. Noll, J. P. Toennies, and Ch. Schlier, J. Chem. Phys. 87, 2685 (1987).
8. C. M. Lovejoy and D. J. Nesbitt, Rev. Sci. Instrum. 58, 807 (1987).
9. A. S. Pine, J. Opt. Soc. Am. 64, 1683 (1974).
10. C. M. Lovejoy and D. J. Nesbitt, J. Chem. Phys. 86, 3151 (1987).
11. A. McIlroy and D. J. Nesbitt, J. Chem. Phys. 91, 104 (1989).
12. M. S. Child and D. J. Nesbitt, Chem. Phys. Lett. 149, 404 (1988).
13. D. J. Nesbitt, M. S. Child, and D. C. Clary, J. Chem. Phys. 90, 4855 (1989).
14. J. M. Hutson and B. J. Howard, Mol. Phys. 45, 791 (1982).
15. C. M. Lovejoy and D. J. Nesbitt, J. Chem. Phys. 91, 2790 (1989).
16. D. C. Clary and D. J. Nesbitt, J. Chem. Phys. 90, 7000 (1989).
17. R. C. Cohen, K. L. Busarow, K. B. Laughlin, G. A. Blake, M. Haventh, Y. T. Lee, and R. J. Saykally, J. Chem. Phys. 89, 4494 (1988).
18. R. C. Cohen, K. L. Busarow, T. Y. Lee, and R. J. Saykally, J. Chem. Phys. (in press).
19. J. M. Hutson, J. Chem. Phys. (in press).

A THEORETICAL STUDY OF Hg···Ar$_n$ (n=1, 2, 3) CLUSTERS

EXCITED IN THE Hg(^{3}P←^{1}S) SPECTRAL REGION

Octavio Roncero[a], J. Alberto Beswick

LURE[b], Université de Paris Sud, 91405 Orsay, France

Nadine Halberstadt   and   Benoît Soep

Laboratoire de Photophysique Moléculaire[c]
Université Paris-Sud, 91405 Orsay, France

ABSTRACT

A quantum mechanical theoretical study of Hg...Ar$_n$ ($n = 1, 2, 3$) Van der Waals complexes under Hg(^{3}P←^{1}S) electronic excitation is presented. The potential energy surfaces are calculated assuming additivity of the atom-atom pairwise interactions. The Hg···Ar potential is obtained from inversion of electronic spectra using a previously proposed model. For Ar$_2$ the best available empirical potential is used. In the excited electronic manifold, the nonspherical character of the Hg(^{3}P) state is taken into account by appropriate rotations of the wavefunctions in the molecular frame. This allows the determination of electronically excited diabatic potential energy surfaces and couplings. Diagonalization of the most strongly coupled states provides adiabatic potential energy surfaces. Electronic spectral shifts are estimated by computing vertical energy differences.

I. INTRODUCTION

Van der Waals hetero-clusters are ideal model systems for the detailed study of a variety of photophysical and photochemical processes such as energy transfer, vibrational predissociation, fine-structure relaxation, and half-collision chemical reactions. In addition, the spectroscopy of these species provides central information on intermolecular forces and their additivity properties.

Recently[1-15], several spectroscopic studies have been carried out on complexes of rare gas atoms with mercury atoms formed in supersonic expansions. From the analysis of laser induced spectra, interatomic potentials for the ground and several electronically excited states

---

[a] Perm. address: Inst. Estructura de la Materia, CSIC, Serrano 123, 28006 Madrid, Spain.
[b] Laboratoire du CNRS, CEA et MEN.
[c] Laboratoire du CNRS.

Dynamics of Polyatomic Van der Waals Complexes
Edited by N. Halberstadt and K. C. Janda
Plenum Press, New York, 1990

of Hg···X (X = Ne, Ar, Kr) complexes have been determined[1,3,4,6,7,10,13,14]. In the region of the $^3P_1$ state of Hg for instance, two bound states, A and B, have been identified and assigned to the two projections $\Omega_A = 0$ and 1 of the electronic angular momentum $J_A = 1$ of the mercury atom onto the interatomic axis. The A ($\Omega_A = 0$) state has a short bond length ($R_e = 3.34$ Å) and a rather large binding energy ($D = 380$ cm^{-1}) as compared with the ground state ($R_e = 3.98$ Å and $D = 130$ cm^{-1}). As a consequence, the spectrum associated to this state is red-shifted with respect to the atomic Hg($^3P_1 \leftarrow {}^1S_0$) line. On the other hand, the B state has a large bond length ($R_e = 4.66$ Å) and a rather weak binding energy ($D = 50$ cm^{-1}). Therefore it is blue-shifted. These results can be rationalized in a very simple Hund's case (c) model for the interatomic interaction. According to this scheme, the A($\Omega_A = 0$) state is a linear combination of Π orbitals which have electronic probability density outside the Van der Waals bond. On the contrary, the B ($\Omega_A = \pm 1$) state is a linear combination of Π and Σ orbitals with a substantial fraction of the electronic density localized between the two atoms. Hence in the A state the rare gas atom can approach the mercury atom closer and be more strongly polarized.

One very interesting and natural extension of this work is the study of clusters involving a mercury atom bound to more than one rare gas atom. Since the individual interatomic interactions are well known, potential energy surfaces for the ground and excited states of these clusters can be easily constructed by a sum of atom-atom interactions with appropriate rotations applied to the non-spherical states. A comparison between calculated and experimental spectra will lead to a direct check of the potential energy surfaces, in particular to study the effect of three-body terms. In addition, several very interesting problems can be addressed in these systems: evaporation, isomerization and phase transitions.

Very recently[14,15], experiments have been conducted on Hg···Ar$_n$ ($n = 2, 3$) clusters in the region of the Hg($^3P_1 \leftarrow {}^1S_0$) electronic transition. It is the purpose of this paper to present the calculation of potential energy surfaces for the ground $^1S_0$ and excited ^3P electronic states, and to provide some crude estimates of spectral shifts for Hg···Ar$_2$ and Hg···Ar$_3$ molecules. We begin by the calculation of the Hg···Ar spectra using a simple form for the potential curves. A comparison between theory and experiments provides the parameters of these curves. Then, using the best empirically determined potential for the Ar···Ar interaction in the ground state, we construct the potential energy surfaces for Hg···Ar$_2$ and Hg···Ar$_3$. Finally, the spectral shifts are estimated by calculating the vertical energy differences between excited and ground state potential energy surfaces in the region of the minimum of the ground state.

The organization of the paper is the following. In section II we present the general theory for the calculation of Hg···Ar$_n$ ($n = 1, 2, ...$). Although the theory is developped for Hg···Ar, it can apply to any cluster involving an atom with non-zero orbital electronic angular momentum and Hund's case (c) coupling scheme. The application to Hg···Ar$_2$ and Hg···Ar$_3$ complexes is presented in section III. Finally, section IV is devoted to the conclusions.

II. METHOD

We are interested in the calculation of the electronic absorption spectrum of Hg···Ar$_n$ complexes. For $n = 1$, we simulate the experimental spectrum by calculating the bound levels in the ground Ar···Hg($^1S_0$) and electronically excited state Ar···Hg(^3P), using the same assumptions as Duval et al.[7]. We use the atom-atom potential derived from Hg···Ar to construct potential energy surfaces and couplings for higher clusters, and we predict the region of maximum absorption of the spectrum by calculating the vertical energy difference from the ground electronic state to the excited energy surfaces.

A. Generalities

In the absence of *ab initio* calculations, we have built up a reasonable potential model for the Hg⋯Ar$_n$ complexes. Three-body and higher terms are neglected, and the interaction is supposed to be a sum of pairwise (atom-atom) potentials. The Ar⋯Ar potential is taken from the litterature[16−18] and is presented in Appendix I. The Hg⋯Ar potential is deduced from experiment, using the following assumptions:

- We assume the mercury atom to be described by a total orbital angular momentum **L** and a total electronic spin angular momentum **S**, with projections on the quantification axis equal to Λ and Σ respectively. The electronic manifold defined by **L** and **S**, consisting of the $(2L+1)(2S+1)$ states $|L, \Lambda\rangle |S, \Sigma\rangle$, is considered to be well isolated from other electronic states.

- We assume that the spin-orbit interaction H_{so} in mercury can be written as $H_{so} = g\,\mathbf{L \cdot S}$. We define the total electronic angular momentum $\mathbf{J}_A = \mathbf{L + S}$ and its projection $\Omega_A = \Lambda + \Sigma$ onto the quantification axis. Thus, the electronic basis set $|J_A, \Omega_A, L, S\rangle$ diagonalizes H_{so}, splitting up the $(2L+1)(2S+1)$ levels $|L, \Lambda\rangle |S, \Sigma\rangle$ in $2\min(L, S) + 1$ energy levels, each one being $(2J_A + 1)$ degenerated.

For the Hg⋯Ar interaction, the quantization axis is along the internuclear vector **R** from Hg to the argon atom. For more than one argon, the quantization axis is chosen along the direction **R** between the mercury atom and the center of mass of the argon atoms. A given $|J_A, \Omega_A, L, S\rangle$ state is then a linear combination of different $|J_A, \Omega'_A, L, S\rangle_{R_k}$, $\Omega'_A = -J_A, \cdots, J_A$, states, where subscript R_k indicates that $\mathbf{R}_k$ is the quantization axis for the individual Hg⋯Ar(k) interaction. Knowing the pair interactions in the $|J_A, \Omega_A, L, S\rangle_{R_k}$ basis set, *i.e.* the (diagonal) matrix elements $\langle J_A \Omega_A LS | H'_{el} | J_A \Omega_A LS \rangle_{R_k}$, we get the matrix elements of H'_{el} in the $|J_A, \Omega_A, L, S\rangle$ basis set. There are both diagonal $\langle J_A, \Omega_A, L, S | H'_{el} | J_A, \Omega_A, L, S \rangle$ (diabatic surfaces) and off-diagonal terms $\langle J'_A, \Omega'_A, L, S | H'_{el} | J_A, \Omega_A, L, S \rangle$ (couplings between the diabatic surfaces). Adiabatic potential energy surfaces have been calculated by diagonalizing the matrix of the electronic interaction hamiltonian for different atomic configurations.

B. Wave functions of Hg

As stated above, we assume the mercury atom to be described by its total electronic orbital angular momentum **L** and total electronic spin angular momentum **S**, with projections on the quantization axis equal to Λ and Σ, respectively. The spin-orbit interaction H_{so} is written as

$$H_{so} = g\,\mathbf{L.S} \tag{1}$$

H_{so} is then diagonal in the $|J_A, \Omega_A, L, S\rangle$ representation (Hund's case (c)), where $\mathbf{J}_A = \mathbf{L + S}$ is the total electronic angular momentum and $\Omega_A = \Lambda + \Sigma$ its projection on the quantization axis. Thus we have

$$|J_A, \Omega_A, L, S\rangle = \sum_{\substack{\Lambda, \Sigma \\ (\Lambda + \Sigma = \Omega_A)}} C(L, \Lambda, S, \Sigma; J_A, \Omega_A) |L, \Lambda\rangle |S, \Sigma\rangle \tag{2}$$

where C is a Clebsch-Gordan coefficient. The degeneracy between J_A levels is lifted by the diagonal matrix elements of H_{so}: if $E_{(J_A = 0)}$ is taken as the reference, $E_{(J_A = 1)} = 1767.220$ cm^{-1} and $E_{(J_A = 2)} = 6397.898$ cm^{-1} according to Mies *et al.*[19]. In writing Eq. (1), we have neglected the terms which couple different multiplets (*i.e.* states with different values of S). In particular, we neglect the interaction between ^{3}P and ^{1}P, which is responsible for the oscillator strength of the Hg(^{3}P) ← Hg(^{1}S$_0$) transition. There is a 2.49% mixture between these states[19].

473

C. Hg···Ar potentials

The electronic hamiltonian (without spin) is

$$H_{el} = H_{el}^{(0)} + H_{el}' \tag{3}$$

where $H_{el}^{(0)}$ is the asymptotic $(R \to \infty)$ non-relativistic electronic hamiltonian, while H_{el}' is the intermolecular interaction Hamiltonian which goes to zero as $R \to \infty$.

For finite values of the interatomic distance R, the $|J_A, \Omega_A, L, S\rangle$ states are coupled by H_{el}'. This is often referred to as the "radial" coupling[20]. Using a body-fixed system in which the z axis lies along the $\mathbf{R}$ vector (from Hg to Ar), the matrix elements of H_{el}' in the $|L, \Lambda, S, \Sigma\rangle$ basis set are given by

$$\langle L, \Lambda', S, \Sigma'|H_{el}'|L, \Lambda, S, \Sigma\rangle = \delta_{\Sigma'\Sigma}\,\delta_{\Lambda'\Lambda}\,V_\Lambda(R), \qquad \text{with} \qquad V_\Lambda = V_{-\Lambda}. \tag{4}$$

Using Eqs. (2) and (4), we get

$$\begin{aligned}
\langle J_A', &\Omega_A', L, S|H_{el}'|J_A, \Omega_A, L, S\rangle \\
&= \sum_{\Lambda',\Sigma'} \sum_{\Lambda,\Sigma} C(L, \Lambda', S, \Sigma'; J_A', \Omega_A')\, C(L, \Lambda, S, \Sigma; J_A, \Omega_A)\, \delta_{\Sigma'\Sigma}\,\delta_{\Lambda'\Lambda}\,V_\Lambda(R) \\
&= \delta_{\Omega_A'\Omega_A} \sum_{\Lambda} C(L, \Lambda, S, \Omega_A - \Lambda; J_A', \Omega_A)\, C(L, \Lambda, S, \Omega_A - \Lambda; J_A, \Omega_A)\, V_\Lambda(R).
\end{aligned} \tag{5}$$

Thus H_{el}' is diagonal in Ω_A. We will denote

$$\langle J_A', \Omega_A', L, S|H_{el}'|J_A, \Omega_A, L, S\rangle = V_{J_A', J_A}^{\Omega_A}(R)\, \delta_{\Omega_A', \Omega_A}, \tag{6}$$

with

$$V_{J_A', J_A}^{\Omega_A}(R) = \sum_{\Lambda} C(L, \Lambda, S, \Omega_A - \Lambda; J_A', \Omega_A)\, C(L, \Lambda, S, \Omega_A - \Lambda; J_A, \Omega_A)\, V_\Lambda(R). \tag{7}$$

We note that since $V_\Lambda = V_{-\Lambda}$ (Eq. (4)),

$$V_{J_A', J_A}^{\Omega_A} = (-1)^{J_A' + J_A}\, V_{J_A', J_A}^{-\Omega_A}. \tag{8}$$

For Hg···Ar, detailed spectroscopic data are available[1,7,10,13,14] concerning the ground $(\mathrm{X}, J_A = 0)$ state correlating to $\mathrm{Hg}(^1\mathrm{S}_0) + \mathrm{Ar}$ and the electronic excited states A $(\Omega_A = 0)$ and B $(\Omega_A = \pm 1)$ correlating to $\mathrm{Hg}(^3\mathrm{P}_1) + \mathrm{Ar}$. The corresponding potentials are denoted by

$$V_{J_A, J_A}^{\Omega_A} = \begin{cases} V_X & \text{for } \mathrm{Hg}(^1\mathrm{S}_0)\cdots\mathrm{Ar}; \\ \begin{cases} V_A & \text{for } J_A = 1,\ \Omega_A = 0 \\ V_B & \text{for } J_A = 1,\ \Omega_A = \pm 1 \end{cases} & \text{for } \mathrm{Hg}(^3\mathrm{P}_1)\cdots\mathrm{Ar}. \end{cases} \tag{9}$$

where V_X, V_A and V_B potentials have been fit to the following functional form

$$V(R) = \begin{cases} D\left\{ e^{-2\alpha(R - R_e)} - 2e^{-\alpha(R - R_e)} \right\} + (D - D_{exp}) & \text{if } R \leq R_0; \\ C_6\,R^{-6} + C_8\,R^{-8} & \text{if } R \geq R_0. \end{cases} \tag{10}$$

The parameters D, D_{exp}, α, R_e, R_0, C_6 and C_8 are given in Table 1.

In Fig. 1 we show the potential curves as well as the simulated absorption spectra corresponding to the transitions

$$\begin{aligned}
\mathrm{Ar}\cdots\mathrm{Hg}(\mathrm{X}) + \hbar\omega &\longrightarrow \mathrm{Ar}\cdots\mathrm{Hg}^*(^3\mathrm{P}_1, \mathrm{A}) \\
\mathrm{Ar}\cdots\mathrm{Hg}(\mathrm{X}) + \hbar\omega &\longrightarrow \mathrm{Ar}\cdots\mathrm{Hg}^*(^3\mathrm{P}_1, \mathrm{B}).
\end{aligned} \tag{11}$$

Table 1. Hg···Ar potential parameters in the X, A and B electronic states, calculated by fitting the experimental data of refs. 1, 7, 10, 13 and 14.

	Ar···Hg($^1S_0$)	Ar···Hg($^3P_1$)	
	X	A: $\Omega_A = 0$	B: $\Omega_A = \pm 1$
$D(\text{cm}^{-1})$	130.25	353.63	51.57
$D_{exp}(\text{cm}^{-1})$	136.75	382.6	67.7
$R_e(\text{Å})$	3.98	3.34	4.66
$\alpha(\text{Å})$	1.45	1.54	1.116
$R_0(\text{Å})$	6.	6.	7.
$C_6(\text{cm}^{-1}\text{Å}^6)^*$	-1.067×10^6	-5.090×10^6	-7.799×10^6
$C_8(\text{cm}^{-1}\text{Å}^8)^*$	4.743×10^6	1.150×10^8	2.471×10^8

* (calculated by imposing the continuity of $V(R)$ and its derivative at $R = R_0$)

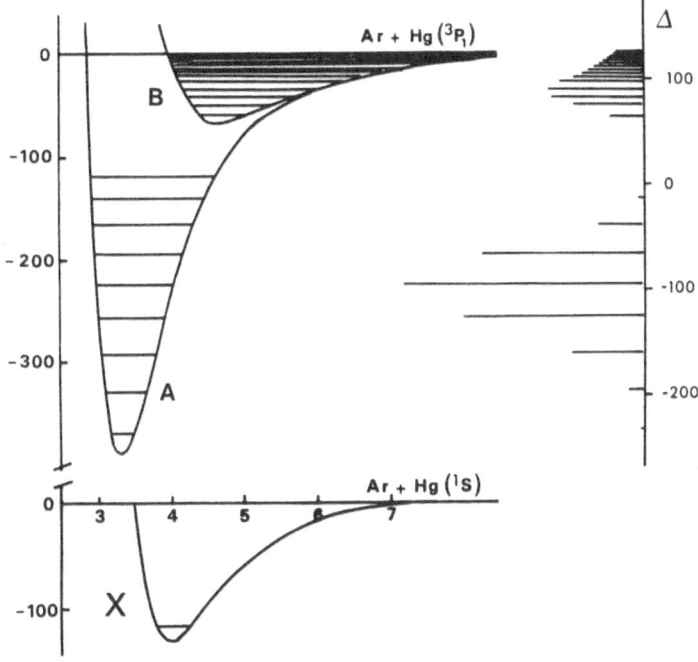

Figure 1. Hg···Ar atom-atom potential curves in the X, A and B electronic states (see Table 1) and simulation of the absorption spectra. Energies are in cm^{-1} and distances in Å. In the absorption spectra, Δ corresponds to the shift (in cm^{-1}) with respect to the Hg($^3P_1 \leftarrow {}^1S_0$) line.

The Franck-Condon approximation is used to calculate the intensities. The bound state energies and wave functions are obtained by numerically solving the Schrödinger equation:

$$H = -\frac{\hbar^2}{2m}\frac{\partial^2}{\partial R^2} + \frac{\ell^2}{2mR^2} + H_{el} + H_{so} \tag{12}$$

where $m = m_{Hg}m_{Ar}/(m_{Hg} + m_{Ar})$ is the reduced mass for the system, ℓ is the angular momentum operator associated to $\mathbf{R}$ (the orbital angular momentum of the fragments about the center of mass of the system), and H_{el} is the electronic hamiltonian (Eq. (3)). The numerical procedure is based on Numerov's integrator. We used $\hbar = 5.8064843$ (energies in cm^{-1}, lengths in Å and masses in g mol^{-1}), $m_{Hg} = 200.45183$ g mol^{-1} and $m_{Ar} = 39.9666387$ g mol^{-1}, giving $m = 33.322672$ g mol^{-1}.

Having determined V_X, V_A and V_B, we can then go back to the potentials (the matrix elements of H'_{el}) in the $|L, \Lambda\rangle|S, \Sigma\rangle$ representation. We define

for Hg($^1S_0$)···Ar: $\quad V_{\Sigma}^g(R) = \langle L = 0, \Lambda = 0|H'_{el}|L = 0, \Lambda = 0\rangle$

for Hg($^3P_1$)···Ar: $\begin{cases} V_{\Sigma}^e(R) = \langle L = 1, \Lambda = 0|H'_{el}|L = 1, \Lambda = 0\rangle \\ V_{\Pi}^e(R) = \langle L = 1, \Lambda = 1|H'_{el}|L = 1, \Lambda = 1\rangle \\ \qquad\quad = \langle L = 1, \Lambda = -1|H'_{el}|L = 1, \Lambda = -1\rangle \end{cases}$ $\tag{13}$

Then using Eq. (7) together with the definitions given in Eq. (9) we get

$$\begin{cases} V_X = V_{\Sigma}^g \\ V_A = V_{\Pi}^e \\ V_B = \frac{1}{2}(V_{\Sigma}^e + V_{\Pi}^e) \end{cases} \quad \text{from which} \quad \begin{cases} V_{\Sigma}^g = V_X \\ V_{\Sigma}^e = 2V_B - V_A \\ V_{\Pi}^e = V_A \end{cases} \tag{14}$$

Within this framework, it is possible to calculate the potential curves for all the $|J_A \Omega_A LS\rangle$ states of the ^3P manifold ($L = 1, S = 1$). Using Eqs. (7) and (14), we get

$$V_{J_A=0, J_A=0}^{\Omega_A=0} = (V_{\Sigma}^e + 2V_{\Pi}^e)/3 = (V_A + 2V_B)/3$$

$$V_{J_A=2, J_A=2}^{\Omega_A=0} = (2V_{\Sigma}^e + V_{\Pi}^e)/3 = (-V_A + 4V_B)/3$$

$$V_{J_A=2, J_A=2}^{\Omega_A=1} = (V_{\Sigma}^e + V_{\Pi}^e)/2 = V_B$$

$$V_{J_A=2, J_A=2}^{\Omega_A=2} = \qquad V_{\Pi}^e \qquad = V_A \tag{15}$$

In principle, the couplings between states with same Ω_A's but different J_A's can be calculated using the same procedure. However, these couplings are not considered here since they are weak as compared to the energy difference.

D. Comparison with experiment

The observables calculated with these potentials are in very good agreement with the experimental ones, as can be seen in Table 2. It can be seen from Fig. 1 that the continuum of the B state can be directely excited. This allows to get the D_0 values for the three states X, A and B from the shifts. The uncertainty in determining the beginning of the continuum spectrum (hence D_0) has been dramatically reduced by Zwier and coworkers[14] using a filter of gaseous mercury in the LIF experiment to eliminate the fluorescence of the Hg fragment: comparing the spectra with and without the filter, they could get the onset of the continuum very accurately.

Table 2. Comparison between experimental[7] and theoretical (this work) values of some Hg···Ar observables in the X (Ar···Hg($^1S_0$)), A (Ar···Hg($^3P_1, \Omega_A = 0$)) and B (Ar···Hg($^3P_1, \Omega_A = \pm 1$)) states.

	X		A		B	
	expt.	theory	expt.	theory	expt.	theory
D_0 (cm^{-1})	125.*	125.24	362.*	362.30	62.*	62.16
B_0 (cm^{-1})	0.0311	0.0313		0.0447		0.0227
$r_{v=0}$ (Å)	4.03	4.028		3.369	4.76	4.738
$r_{v=1}$ (Å)			3.43	3.430	4.92	4.908
$r_{v=2}$ (Å)			3.48	3.496	5.08	5.108
$r_{v=3}$ (Å)			3.54	3.567	5.24	5.346
$r_{v=4}$ (Å)				3.644		5.639
$r_{v=5}$ (Å)			3.67	3.729		6.001
$r_{v=6}$ (Å)			3.75	3.822		6.352

* (ref. 11)

The states corresponding to the potentials deduced in Eqs. (14): $^3P_0$ for ($J_A = 0, \Omega_A = 0$), $^3P_2$ for ($J_A = 2, \Omega_A = 0, \pm 1$ and ± 2), cannot be directly accessed experimentally. However, they can be detected in the fluorescence spectrum from the C state, which correlates to Ar + Hg(^3S)[7]. This emission corresponds mainly to a transition to the $\tilde{b}(J_A = 2, \Omega_A = \pm 2)$ state, but also to the $\tilde{d}(J_A = 2, \Omega_A = 0)$ state. Unfortunately, this is not possible in the case of the $\tilde{c}(J_A = 2, \Omega_A = \pm 1)$ state. Duval et al.[7] have simulated these emission spectra from the C state with the potentials inferred from the B←X and A←X transitions using the same assumptions that lead to Eqs. (14). The agreement between the simulated and the experimental spectra is good. Thus we can conclude that the $\tilde{b}, \tilde{c}, \tilde{d}$ and $\tilde{a}$ states are well represented using Eqs. (15). The diatomic potentials obtained here for the A and B states slightly differ from those of Duval et al.[7] because additional information from Tsuchiya et al.[10,13] and from Zwier et al.[14] is utilized in our fit.

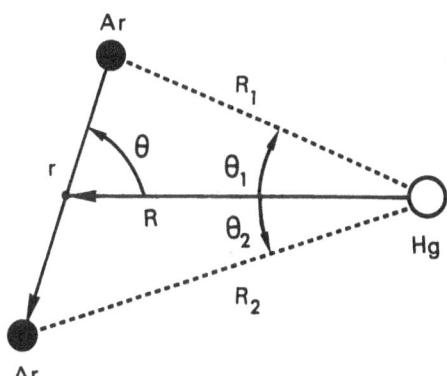

Figure 2. The body-fixed system of axes for Hg···Ar$_2$.

E. The Hg···Ar interaction in Hg···Ar$_n$, $n > 1$

In a complex with more than one argon atom, we use the body-fixed system of axes illustrated in Fig. 2 for Hg···Ar$_2$. The z axis is chosen along the $\mathbf{R}$ vector joining the mercury atom to the center of mass of the argons; the x axis lies in the plane defined by the z axis and the first argon atom Ar(1). The quantization axis is the body-fixed z axis. The corresponding atomic basis set is denoted by $|L, \Lambda, S, \Sigma\rangle$ or $|J_A, \Omega_A, L, S\rangle$ for Hund's case (c). For the Hg···Ar(k) interaction on the other hand, the quantization axis is along the $\mathbf{R}_k$ vector going from Hg to the Ar(k) atom. If we denote by $|L, \Lambda, S, \Sigma\rangle_{R_k}$ or $|J_A, \Omega_A, L, S\rangle_{R_k}$ the corresponding basis set, and by $\mathcal{R}(\phi_k, \theta_k, 0)$ the rotation with Euler angles $(\phi_k, \theta_k, 0)$ that brings $\mathbf{R}$ into $\mathbf{R}_k$, then $|J_A, \Omega_A, L, S\rangle$ is the transformed of $|J_A, \Omega_A, L, S\rangle_{R_k}$ under the inverse rotation:

$$
\begin{aligned}
|J_A, \Omega_A, L, S\rangle &= \mathcal{R}^{-1}(\phi_k, \theta_k, 0)\, |J_A, \Omega_A, L, S\rangle_{R_k} \\
&= \sum_{\Omega'_A} D^{J_A}_{\Omega'_A \Omega_A}(0, -\theta_k, -\phi_k)\, |J_A, \Omega'_A, L, S\rangle_{R_k} \\
&= \sum_{\Omega'_A} D^{J_A *}_{\Omega_A \Omega'_A}(\phi_k, \theta_k, 0)\, |J_A, \Omega'_A, L, S\rangle_{R_k}.
\end{aligned}
\tag{16}
$$

where $D^{J_A}_{\Omega_A \Omega'_A}$ is the Wigner rotation matrix[21]. The matrix elements of the Hg···Ar(k) interaction $H_{el}^{\prime\, \mathrm{Ar}(k)}$ in the $|J_A, \Omega_A, L, S\rangle$ basis set are given by

$$
\begin{aligned}
&\langle J'_A, \Omega'_A, L, S | H_{el}^{\prime\, \mathrm{Ar}(k)} | J_A, \Omega_A, L, S\rangle \\
&\qquad = \sum_{\Omega_{A_k}} D^{J'_A}_{\Omega'_A, \Omega_{A_k}}(\phi_k, \theta_k, 0)\, D^{J_A *}_{\Omega_A, \Omega_{A_k}}(\phi_k, \theta_k, 0)\, V^{\Omega_{A_k}}_{J'_A, J_A}(R_k)
\end{aligned}
\tag{17}
$$

where we have used Eq. (16) together with Eq. (6). The final expression (using Eq. (7)) is

$$
\begin{aligned}
&\langle J'_A, \Omega'_A, L, S | H_{el}^{\prime\, \mathrm{Ar}(k)} | J_A, \Omega_A, L, S\rangle \\
&\qquad = \sum_{\Lambda} \sum_{\Omega_{A_k}} C(L, \Lambda, S, \Omega_{A_k} - \Lambda; J'_A, \Omega_{A_k})\, C(L, \Lambda, S, \Omega_{A_k} - \Lambda; J_A, \Omega_{A_k}) \\
&\qquad\quad \times D^{J'_A}_{\Omega'_A, \Omega_{A_k}}(\phi_k, \theta_k, 0)\, D^{J_A *}_{\Omega_A, \Omega_{A_k}}(\phi_k, \theta_k, 0)\, V_\Lambda(R_k),
\end{aligned}
\tag{18}
$$

where V_Λ has been defined in Eqs. (4) and (13), and is given as a function of only V_A and V_B in Eq. (14). In the calculations presented here, we can make directly use of the expression given in Eq. (17) together with the expressions for $V^{\Omega_A}_{J'_A, J_A}$ obtained in Eqs. (9) and (15) as a function of only V_A and V_B for the ^{3}P state. Hence the total interaction $H_{el}^{\prime\, \mathrm{Ar}_n}$, which is taken as the sum of the individual interactions Hg···Ar(k), is determined by the knowledge of only two potentials that can be deduced from relatively simple spectroscopic experiments on Hg···Ar. We have

$$
H_{el}^{\prime\, \mathrm{Ar}_n} = \sum_k H_{el}^{\prime\, \mathrm{Ar}(k)}
\tag{19}
$$

and, using Eq. (17), we get

$$
\begin{aligned}
&\langle J'_A, \Omega'_A, L, S | H_{el}^{\prime\, \mathrm{Ar}_n} | J_A, \Omega_A, L, S\rangle \\
&\qquad = \sum_k \sum_{\Omega_{A_k}} D^{J'_A}_{\Omega'_A, \Omega_{A_k}}(\phi_k, \theta_k, 0)\, D^{J_A *}_{\Omega_A, \Omega_{A_k}}(\phi_k, \theta_k, 0)\, V^{\Omega_{A_k}}_{J'_A, J_A}(R_k)
\end{aligned}
\tag{20}
$$

The total interaction is therefore not diagonal in Ω_A, nor in J_A. However, it has been argued above that the couplings between different J_A's can be neglected in the calculation of the Hg···Ar spectrum (§C, D). Therefore a fairly good approximation can be obtained by taking into account the effect of the coupling between different Ω_A's only. Using Eq. (8), we get

$$\langle J'_A, -\Omega'_A, L, S | H'^{Arn}_{el} | J_A, -\Omega_A, L, S \rangle = (-1)^{J'_A + J_A + \Omega'_A + \Omega_A} \langle J'_A, \Omega'_A, L, S | H'^{Arn}_{el} | J_A, \Omega_A, L, S \rangle^* \tag{21}$$

F. Parity with respect to the plane $\phi = 0$

In the case of Hg$\cdots$Ar$_2$, the plane of the molecule is a plane of symmetry: if we choose wave functions that are either symmetric or antisymmetric for an inversion with respect to that plane, then the symmetric and the antisymmetric functions are decoupled from each other. In the case of Hg$\cdots$Ar$_n$, $n > 2$, the (z, x) plane of the body-fixed system is no longer a symmetry plane for the electronic interaction. However, keeping the same linear combinations is convenient since the matrix elements of H'_{el} are then real. Thus we define the following wave functions

$$|J_A, \Omega_A, L, S, p\rangle = \frac{i^p}{\sqrt{2(1 + \delta_{\Omega_A, 0})}} \left\{ |J_A, \Omega_A, L, S\rangle + (-1)^{p + J_A - \Omega_A} |J_A, -\Omega_A, L, S\rangle \right\} \tag{22}$$

where $p = 0$ or 1, and $\Omega_A \geq 0$.

The matrix elements of H'^{Arn}_{el} in this new basis set are obtained from Eqs. (20-22)

$$\langle J'_A, \Omega'_A, L, S, p' | H'^{Arn}_{el} | J_A, \Omega_A, L, S, p\rangle$$

$$= \frac{i^{p-p'}}{2\sqrt{(1 + \delta_{\Omega_A})(1 + \delta_{\Omega'_A})}} \sum_k \sum_{\Omega_{Ak}=0}^{J^{min}_A} \frac{1}{1 + \delta_{\Omega_{Ak}, 0}} V^{\Omega_{Ak}}_{J'_A, J_A}(R_k)$$

$$\times \left\{ \left[e^{i(\Omega_A - \Omega'_A)\phi_k} + (-1)^{p+p'} e^{-i(\Omega_A - \Omega'_A)\phi_k} \right] \right. \tag{23}$$

$$\times \left[d^{J'_A}_{\Omega'_A, \Omega_{Ak}}(\theta_k) d^{J_A}_{\Omega_A, \Omega_{Ak}}(\theta_k) + (-1)^{J'_A + J_A} d^{J'_A}_{\Omega'_A, -\Omega_{Ak}}(\theta_k) d^{J_A}_{\Omega_A, -\Omega_{Ak}}(\theta_k) \right]$$

$$+ (-1)^{p' + J'_A} \left[e^{i(\Omega_A + \Omega'_A)\phi_k} + (-1)^{p+p'} e^{-i(\Omega_A + \Omega'_A)\phi_k} \right]$$

$$\times (-1)^{\Omega_{Ak}} \left. \left[d^{J'_A}_{\Omega'_A, -\Omega_{Ak}}(\theta_k) d^{J_A}_{\Omega_A, \Omega_{Ak}}(\theta_k) + (-1)^{J'_A + J_A} d^{J'_A}_{\Omega'_A, \Omega_{Ak}}(\theta_k) d^{J_A}_{\Omega_A, -\Omega_{Ak}}(\theta_k) \right] \right\}$$

where $J^{min}_A = \min(J'_A, J_A)$. Hence for $p = p'$ we have

$$\langle J'_A, \Omega'_A, L, S, p' | H'^{Arn}_{el} | J_A, \Omega_A, L, S, p\rangle$$

$$= \frac{1}{\sqrt{(1 + \delta_{\Omega_A})(1 + \delta_{\Omega'_A})}} \sum_k \sum_{\Omega_{Ak}=0}^{J^{min}_A} \frac{1}{1 + \delta_{\Omega_{Ak}, 0}} V^{\Omega_{Ak}}_{J'_A, J_A}(R_k)$$

$$\times \left\{ \cos((\Omega_A - \Omega'_A)\phi_k) \left[d^{J'_A}_{\Omega'_A, \Omega_{Ak}}(\theta_k) d^{J_A}_{\Omega_A, \Omega_{Ak}}(\theta_k) + (-1)^{J'_A + J_A} d^{J'_A}_{\Omega'_A, -\Omega_{Ak}}(\theta_k) d^{J_A}_{\Omega_A, -\Omega_{Ak}}(\theta_k) \right] \right.$$

$$+ (-1)^{p' + J'_A} \cos((\Omega_A + \Omega'_A)\phi_k)$$

$$\times (-1)^{\Omega_{Ak}} \left. \left[d^{J'_A}_{\Omega'_A, -\Omega_{Ak}}(\theta_k) d^{J_A}_{\Omega_A, \Omega_{Ak}}(\theta_k) + (-1)^{J'_A + J_A} d^{J'_A}_{\Omega'_A, \Omega_{Ak}}(\theta_k) d^{J_A}_{\Omega_A, -\Omega_{Ak}}(\theta_k) \right] \right\}$$

$$\tag{24}$$

and for $p \neq p'$:

$$\langle J'_A, \Omega'_A, L, S, p' | H'^{Arrow}_{el} | J_A, \Omega_A, L, S, p \rangle$$

$$= \frac{(-1)^{p'}}{\sqrt{(1 + \delta_{\Omega_A})(1 + \delta_{\Omega'_A})}} \sum_k \sum_{\Omega_{A_k}=0}^{J_A^{min}} \frac{1}{1 + \delta_{\Omega_{A_k},0}} V_{J'_A, J_A}^{\Omega_{A_k}}(R_k)$$

$$\times \left\{ \sin\left((\Omega_A - \Omega'_A)\phi_k\right) \left[d_{\Omega'_A, \Omega_{A_k}}^{J'_A}(\theta_k) d_{\Omega_A, \Omega_{A_k}}^{J_A}(\theta_k) + (-1)^{J'_A + J_A} d_{\Omega'_A, -\Omega_{A_k}}^{J'_A}(\theta_k) d_{\Omega_A, -\Omega_{A_k}}^{J_A}(\theta_k) \right] \right.$$

$$+ (-1)^{p' + J'_A} \sin\left((\Omega_A + \Omega'_A)\phi_k\right)$$

$$\left. \times (-1)^{\Omega_{A_k}} \left[d_{\Omega'_A, -\Omega_{A_k}}^{J'_A}(\theta_k) d_{\Omega_A, \Omega_{A_k}}^{J_A}(\theta_k) + (-1)^{J'_A + J_A} d_{\Omega'_A, \Omega_{A_k}}^{J'_A}(\theta_k) d_{\Omega_A, -\Omega_{A_k}}^{J_A}(\theta_k) \right] \right\}$$

(25)

G. Adiabatic potential energy surfaces

In the excited states considered here, the electronic angular momentum of Hg is different from zero. The atom has no longer spherical symmetry and its interaction with an Ar atom depends on the relative orientation of the interatomic vector and the Hg orbital. Since the quantization axis in Hg$\cdots$Ar$_n$ is not the interatomic axis for $n > 1$, this produces a mixing between electronic states with the same J_A but different values of Ω_A. The system is well described in the diabatic representation $|J_A, \Omega_A, L, S\rangle$ only at large distances.

In order to get a better description at shorter distances, the non-diagonal matrix elements of the interaction are taken into account inside each of the J_A subspaces. This is done by defining a new basis set, the adiabatic basis set $|J_A, N_A, L, S\rangle$, obtained by diagonalization of the diabatic electronic hamiltonian for a given value of J_A. Thus

$$|J_A, N_A, L, S\rangle = \sum_{\Omega_A} c_{\Omega_A}^{J_A, N_A} |J_A, \Omega_A, L, S\rangle \tag{26}$$

where the $c_{\Omega_A}^{J_A, N_A}$'s are obtained by diagonalizing $\langle J_A, \Omega'_A, L, S | H'^{Ar_n}_{el} | J_A, \Omega_A, L, S\rangle$ for a given J_A and a given configuration of the nuclei (*i.e.* the $c_{\Omega_A}^{J_A, N_A}$ coefficients depend on the interatomic distances and angles).

III. RESULTS

A. Potential surfaces for Hg$\cdots$Ar$_2$ Hg in its ground electronic state is spherical. The interactions are only function of the distance between atoms, independently of the quantization axis considered. Hence the equilibrium configuration is T-shaped, with the individual atom-atom distances given by the minimum of the corresponding two body-potentials, the well depth being the sum of the atom-atom well depths: $D_e = 373.045$ cm^{-1}, $r_e = 3.76$ Å and $R_e = 3.51$ Å. Potential contours for this state are presented in Fig. 3. The calculated ground state energy is $E_g = -335$ cm^{-1}. The amplitudes of the stretching and bending motions at that energy determine the Franck-Condon region.

For mercury in the excited state $^3P_1$, the situation is more complicated. Fig. 4 displays the potential contour plots for the diabatic A state (Ar$_2\cdots$Hg($^3P_1, \Omega_A = 0$)). The surface has two minima of about the same energy ($\simeq -500$ cm^{-1}) corresponding to the configurations shown in Fig. 5a) and 5b). There is an interesting configuration where the mercury atom sits in between the two argons (Fig. 5c). This correspond to a saddle point (with energy

Figure 3.
Potential contour plots for
the Hg⋯Ar_2 system in the
ground electronic state. All
distances are in Å.
Contour key:
    1)-300 cm^{-1}
    2)-250 cm^{-1}
    3)-200 cm^{-1}
    4)-150 cm^{-1}
    5)-100 cm^{-1}
    6)-50 cm^{-1}.
a) Ar_2 is frozen at its equilibrium distance $r_e = 3.76$ Å

(the two argons are located
on the horizontal axis).
b) T-shaped configuration
($\theta = \pi/2$, see Fig. 2): the
coordinates are r (the Ar-
Ar distance) and R (from
Hg to the center of mass of
Ar_2).

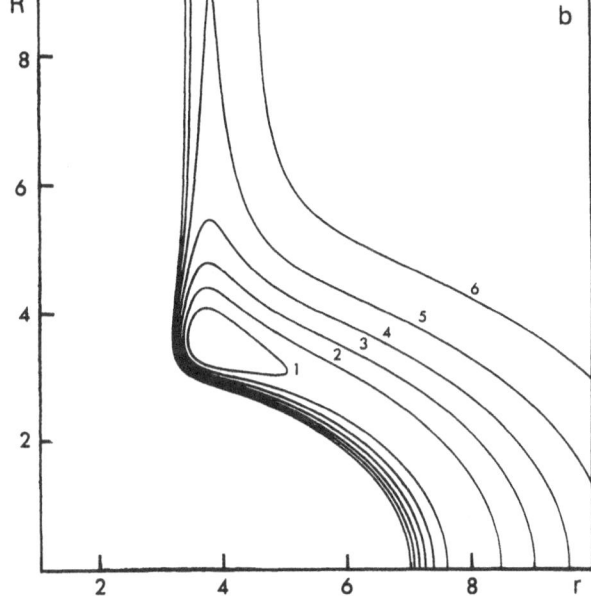

481

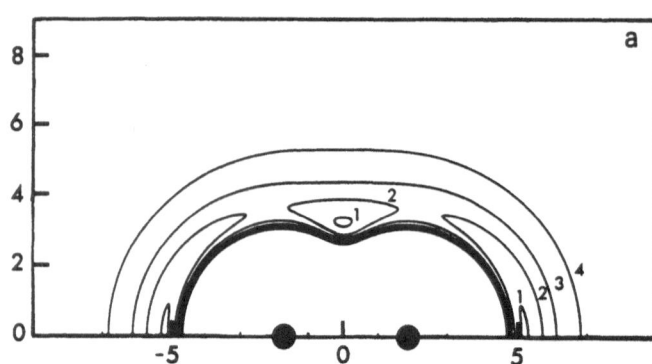

Figure 4. Same as Fig. 3, for the A diabatic electronic state.

Contour key:
1) -500 cm^{-1}
2) -400 cm^{-1}
3) -300 cm^{-1}
4) -200 cm^{-1}
5) -100 cm^{-1}.

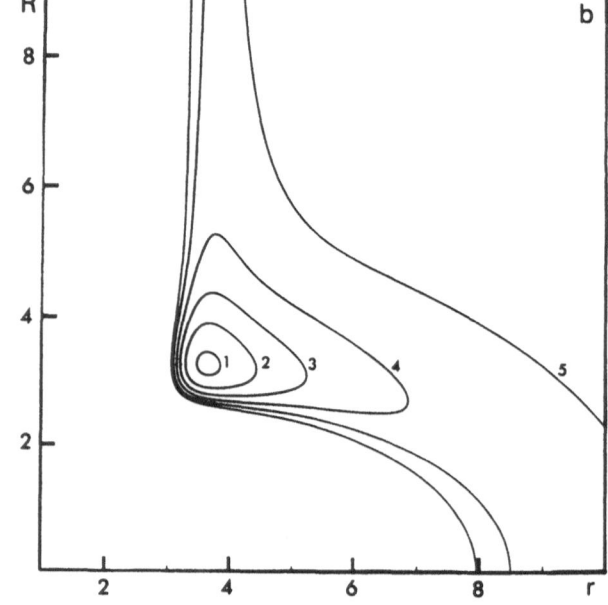

Figure 5.
Some particular configurations of the Ar$_2$···Hg ($J_A =$ 1, A) surface.

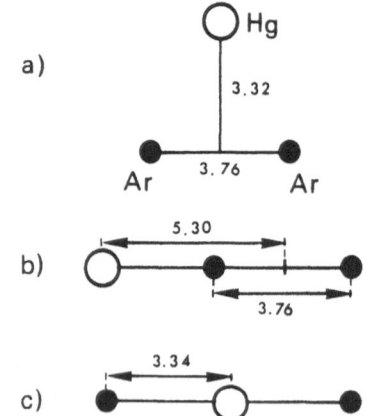

$\simeq -136$ cm^{-1}) which can be seen in Fig. 4b at $R = 0$ and $r = 9.32$ Å, which corresponds to two Ar$\cdots$Hg fragments in the B state. This is because in the diabatic picture, the A state becomes the B state in the in that particular configuration. By diagonalizing the coupled diabatic potentials (A and B$^+$), two adiabatic surfaces denoted $\tilde{A}$ and $\tilde{B}^+$ are obtained. Fig. 6 shows contour plots for the $\tilde{A}$ surface. In this case, as the mercury atom approaches Ar$_2$ in the perpendicular configuration, the p-orbital of Hg "rotates" and Ar$\cdots$Hg$\cdots$Ar has twice the energy of the A state ($V \simeq -760$ cm^{-1}) at $R = 0$ and $r = 6.70$ Å (see Fig. 6b).

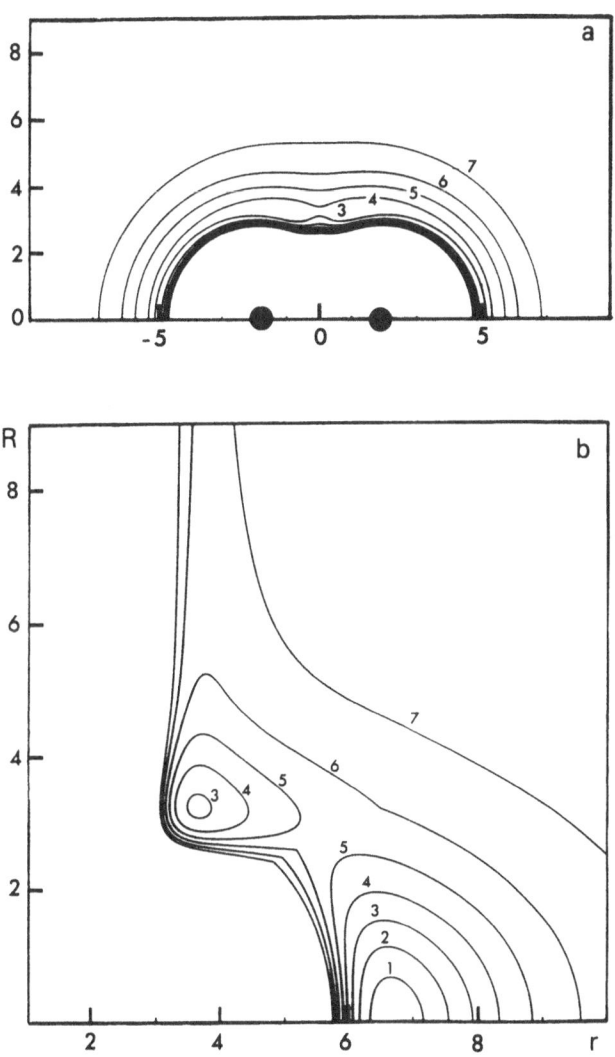

Figure 6. Same as Fig. 3, for the $\tilde{A}$ adiabatic electronic state.
Contour key: 1)-700 cm^{-1}; 2)-600 cm^{-1}; 3)-500 cm^{-1}; 4)-400 cm^{-1}; 5)-300 cm^{-1}; 6)-200 cm^{-1}; 7)-100 cm^{-1}.

In Figs. 7 and 8, the B^+ and $\tilde{B}^+$ states are presented. Again, the effect of the diabatic coupling between the $\tilde{B}^+$ and the A state produces an adiabatic $\tilde{B}^+$ potential energy surface wich differs substantially from the B^+.

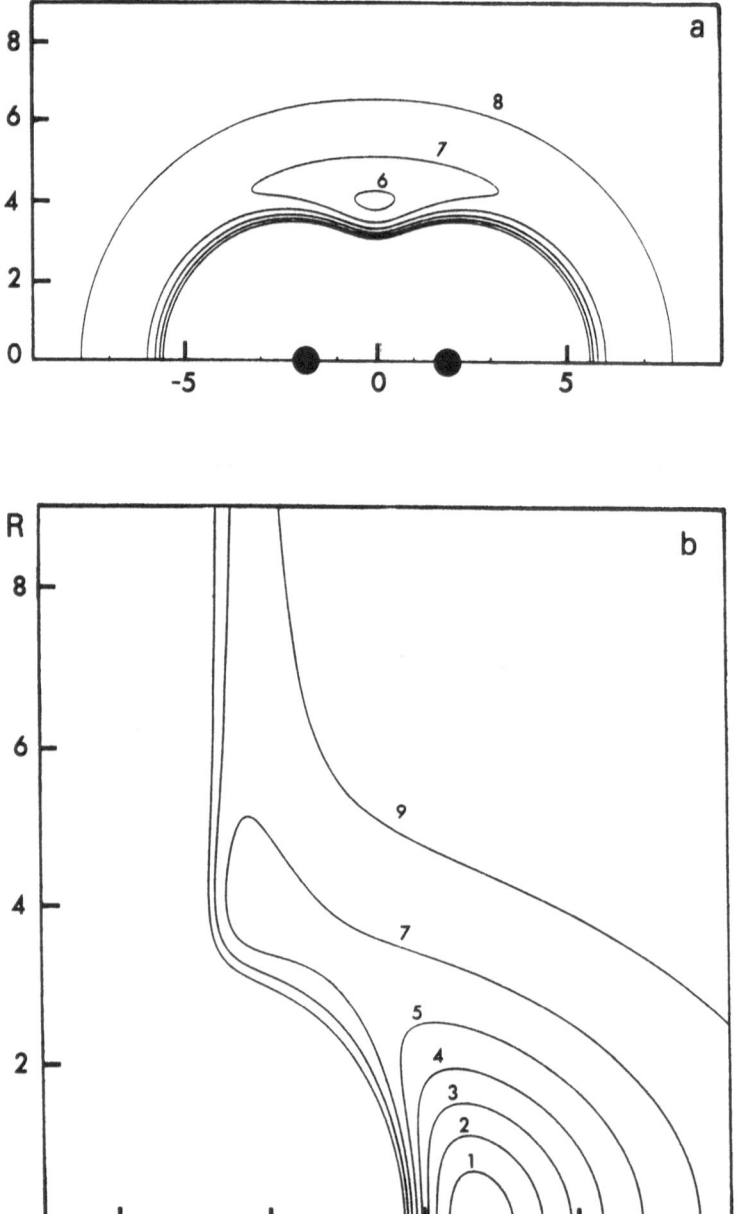

Figure 7. Same as Fig. 3, for the B^+ diabatic electronic state.
Contour key: 1)-700 cm^{-1}; 2)-600 cm^{-1}; 3)-500 cm^{-1}; 4)-400 cm^{-1}; 5)-300 cm^{-1}; 6)-250 cm^{-1}; 7)-200 cm^{-1}; 8)-150 cm^{-1}; 9)-100 cm^{-1}.

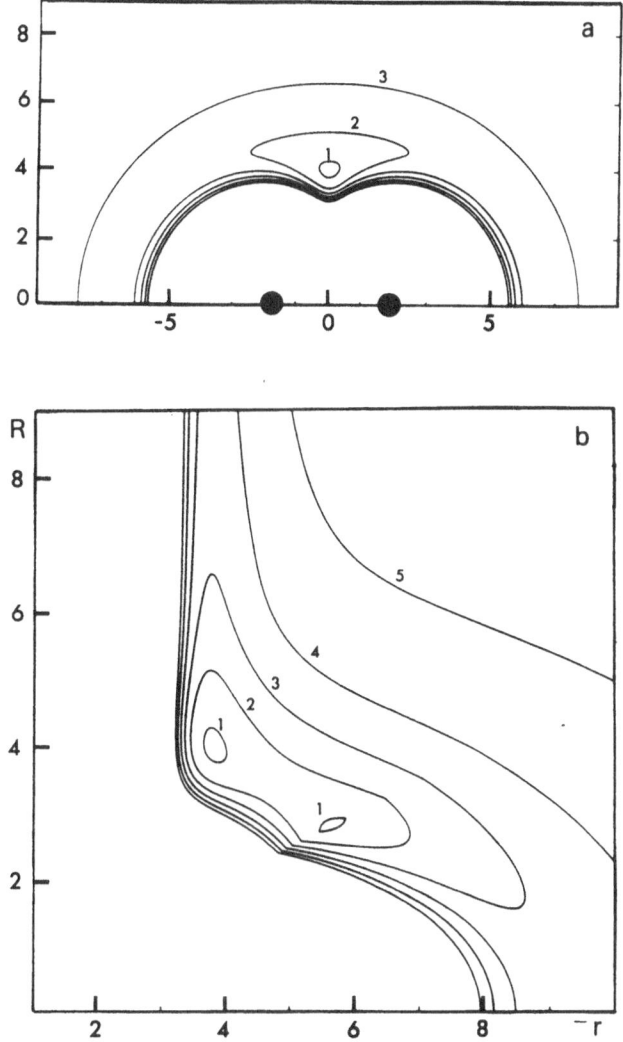

Figure 8. Same as Fig. 3, for the B$^+$ adiabatic electronic state.
Contour key: 1)-250 cm^{-1}; 2)-200 cm^{-1}; 3)-150 cm^{-1}; 4)-100 cm^{-1}; 5)-50 cm^{-1}.

Finally, in Fig. 9 we present the potential energy surface for the B$^-$ state. Since this state is uncoupled from A and B$^+$, the diabatic and adiabatic surfaces coincide.

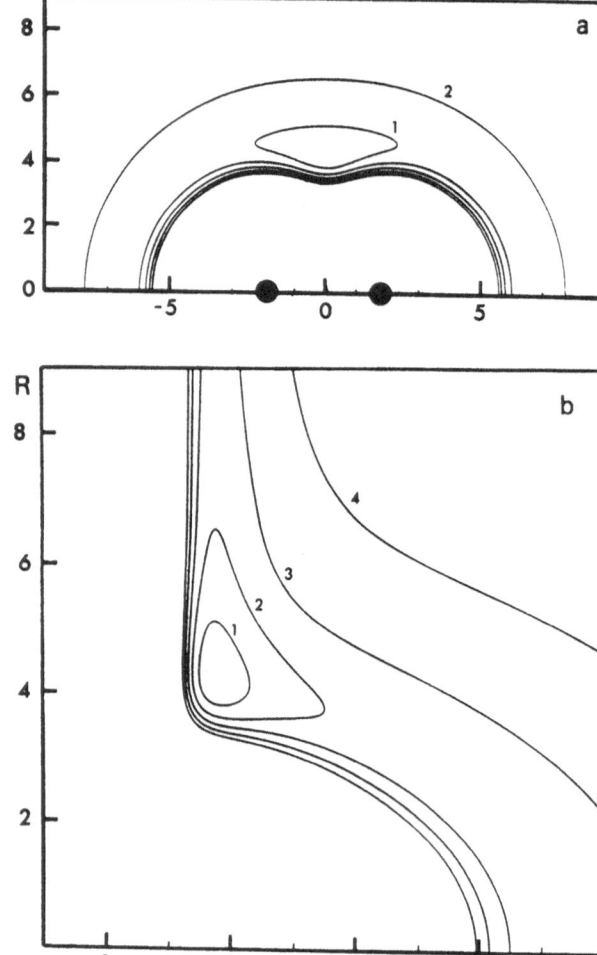

Figure 9.
Same as Fig. 3, for the
B$^-$ electronic state.
Contour key:
  1)-200 cm^{-1}
  2)-150 cm^{-1}
  3)-100 cm^{-1}
  4)-50 cm^{-1}.

B. Spectral shifts for Hg$\cdots$Ar$_n$

We have estimated the electronic spectral shifts for Hg$\cdots$Ar$_n$ with respect to the corresponding Hg line, using the following procedure. We first compute the values of the potential energy of the electronically excited states of Hg$\cdots$Ar which correspond to the maximum of the calculated absorption spectra. We obtain $V_A = -222$ cm^{-1} and $V_B = -36$ cm^{-1} for the A and B states, respectively. These values correspond to the vertical energy differences from the zero-point vibrational level of the ground electronic state for an internuclear distance of about 4.1 Å. Notice that this is larger than the average bond length for that level. We then use this value of the Hg$\cdots$Ar distance to compute vertical energies differences for higher clusters correcting for the difference in zero-point energies. For Hg$\cdots$Ar$_2$ have calculated a zero-point energy of 38 cm^{-1}. For Hg$\cdots$Ar$_3$ we estimate the zero-point energy to be of about 75 cm^{-1}. The final results are collected in Table 3.

Table 3. Electronic spectral shifts for $Hg\cdots Ar_n$, $n = 1$, 2 and 3. Shifts are measured in cm^{-1} with respect to the Hg line (a negative number indicating a red shift).

n	$\tilde{A}$	$\tilde{B}^+$	$\tilde{B}^-$
1	-96.	90.	90.
2	-114.	95.	177.
3	-129.	186.	186.

Recent experiments[14,15] have recorded spectra for these clusters. Zwier *et al.*[14] for instance, have detected fluorescence spectra from the B state of $Hg\cdots Ar_2$, shifted by about 120 cm^{-1} to the blue of the Hg line. They have also observed A-state fluorescence for this cluster shifted to the red by about the same amount than for the Hg$\cdots$Ar molecule. This is consistent with our calculation of -114 cm^{-1} red shift of the $Hg\cdots Ar_2$ Ã state. On the other hand, we predict two blue-shifted progressions with maxima at +95 and +177 cm^{-1}. In the experiment there is no evidence for any fluorescence beyond +150 cm^{-1}. One simple reason for this discrepancy can be the inaccuracy of the vertical approximation in calculating the shifts. We have checked this approximation by changing slightly the internuclear distance R_0 at which the calculation was made. It became clear that since we are in the region of the repulsive branch of the B potential (see Fig. 1), a small change in R_0 can affect the shifts by as much as 20 or 30 cm^{-1}. It will be important in the future to compute exactly the absorption spectrum of this complex to assess the validity of our approximate calculations.

For the $Hg\cdots Ar_3$ cluster the situation is quite different. The experiments[15] observe fluorescence shifted by about 200 cm^{-1} to the blue (as predicted by our estimates) but none to the red. This is not so easy to understand. All our calculations predict a red-shifted state in the region of -100 to -150 cm^{-1}. One possible explanation is that for higher clusters the A state predissociates electronically to produce Hg ($^3P_0$) atoms.

IV. CONCLUSION

We have presented a semiempirical method to calculate potential energy surfaces for clusters involving one atom with non-zero electronic angular momentum and several rare gas atoms. It starts by determining the potential energy curves for the molecule containing only one rare gas atom through a fitting of the available experimental spectra. Then appropriate rotations are performed on the wave functions to calculate the potential energy surface of higher clusters.

We have applied this method to $Hg\cdots Ar_n$ ($n = 1, 2, 3$) for which experimental information is available. We have estimated the shifts of the electronic spectra of these clusters in the region of the $^3P_1 \leftarrow {}^1S_0$ transition. Although semiquantitative agreement between theory and experiments was obtained for the $Hg\cdots Ar_2$ molecule, it seems important to perform more accurate calculations in the future to check some of the assumptions made in the estimates of the shifts.

For the $Hg\cdots Ar_3$ cluster our calculations predict a red-shifted band similar to the ones observed for the Hg$\cdots$Ar and $Hg\cdots Ar_2$ molecules. This has not been found in the experiments. One possible explanation could be that the strongly bound A state for the $Hg\cdots Ar_3$ cluster undergoes electronic predissociation producing Hg ($^3P_0$) atomic fragments.

Acknowledgements

We would like to thank C. Jouvet, D. Solgadi and C. Lardeux-Dedonder for very useful discussions and for communication of their results prior to publication.

APPENDIX I: The Ar_2 potential

To describe the Ar···Ar interaction in the $^1\Sigma$ state we have used the analytical potential of Aziz et al.[16] of the form:

$$V^{Ar\cdots Ar}(r) = \varepsilon V(x)$$

with

$$V(x) = Ae^{(-\alpha x + \beta x^2)} - F(x)\sum_{j=0}^{2} \frac{C_{2j+6}}{x^{2j+6}}$$

and

$$F(x) = \begin{cases} e^{-(D/x - 1)^2} & x < D; \\ 1 & x \geq D. \end{cases}$$

where $x = r/r_m$ and all the parameters are given in Table 4. The minimum for this potential is $\varepsilon = 99.545$ cm^{-1} at $r_m = 3.7565$ Å.

Table 4. Parameters of the Ar_2 potential (from Aziz et al.[16]).

ε(cm^{-1})	99.545
r_m(Å)	3.7565
$A \times 10^{-5}$	2.26210716
α	10.77874743
β	-1.8122004
D	1.36
C_6	1.10785136
C_8	0.56072459
C_{10}	0.34602794

Table 5. Ar_2 diatomic data calculated with the potential from Aziz et al.[16]: energy E_v, average interatomic distance r_v and rotational constant B_v for each bound level v

v	E_v(cm^{-1})	r_v(Å)	B_v(cm^{-1})
0	0	3.8342	0.05775
1	25.6616	4.0153	0.05336
2	46.1560	4.2465	0.04843
3	61.7121	4.5523	0.04288
4	73.0362	4.8828	0.03791
5	83.3888	4.9327	0.03775

This potential has been extracted from the Ar$_2$ absorption spectrum of Colbourn and Douglas[17] and is in very good agreement with recent spectroscopic data[18]. With this potential six bound levels of Ar$_2$ are found (see Table 5). Their energies are in good agreement with the values previously obtained by Aziz et al.[16].

It should be stressed that as v increases the equilibrium distance r_v changes from 3.8 Å for $v = 0$ to 4.9 Å for $v = 5$. Thus if we describe the triatomic complex Hg$\cdots$Ar$_2$ as an atom (Hg) bound to a diatomic molecule (Ar$_2$), the triatomic potential strongly depends on the vibrational state of the diatomic molecule.

APPENDIX II: $H_{el}'^{\text{Ar}_2}$ potentials in the excited state

In this appendix the explicit forms of the triatomic potentials and couplings used in the calculations

$$\langle J_A', \Omega_A', L, S, p' | H_{el}'^{\text{Ar}_2} | J_A, \Omega_A, L, S, p \rangle$$

defined in Eq. (23) are given in terms of the A and B diatomic energy curves (Eqs. (9,10) and Table 1).

■ Coordinates
In the case of Hg$\cdots$Ar$_2$, the coordinates of the two argon atoms in the body-fixed system are (see Fig.2):

$$\text{Ar}(1) : (R_1, \phi_1 = 0, \theta_1); \qquad \text{Ar}(2) : (R_2, \phi_2 = \pi, \theta_2). \qquad \text{(AII} - 1)$$

■ Diagonal terms
With these values, Eq. (24) becomes for the diagonal terms

$$\langle J_A' = 1, \Omega_A' = \Omega_A, L, S, p' | H_{el}'^{\text{Ar}_2} | J_A = 1, \Omega_A, L, S, p \rangle = \delta_{p',p} \, W_{1,\Omega_A;1,\Omega_A}^{p} \qquad \text{(AII} - 2)$$

with

$$W_{1,\Omega_A;1,\Omega_A}^{p} = \frac{1}{(1+\delta_{\Omega_A,0})} \sum_{0 \le \Omega_{A_k} \le 1} \frac{1}{(1+\delta_{\Omega_{A_k},0})}$$

$$\left\{ \left(d_{\Omega_A,\Omega_{A_k}}^{1}(\theta_1) + (-1)^{p+1-\Omega_{A_k}} d_{\Omega_A,-\Omega_{A_k}}^{1}(\theta_1) \right)^2 V_{1,1}^{\Omega_{A_k}}(R_1) \right.$$

$$\left. + \left(d_{\Omega_A,\Omega_{A_k}}^{1}(\theta_2) + (-1)^{p+1-\Omega_{A_k}} d_{\Omega_A,-\Omega_{A_k}}^{1}(\theta_2) \right)^2 V_{1,1}^{\Omega_{A_k}}(R_2) \right\}$$

$$\text{(AII} - 3)$$

The diagonal terms used in the calculations are then

$$W_{1,1;1,1}^{0} = V_B(R_1) + V_B(R_2)$$

$$W_{1,0;1,0}^{1} = \cos^2\theta_1 \, V_A(R_1) + \sin^2\theta_1 \, V_B(R_1) + \cos^2\theta_2 \, V_A(R_2) + \sin^2\theta_2 \, V_B(R_2) \qquad \text{(AII} - 4)$$

$$W_{1,1;1,1}^{1} = \sin^2\theta_1 \, V_A(R_1) + \cos^2\theta_1 \, V_B(R_1) + \sin^2\theta_2 \, V_A(R_2) + \cos^2\theta_2 \, V_B(R_2)$$

■ Off-diagonal terms

Similarly, using Eq. (AII-1) in Eq. (24), the off-diagonal terms for $J_A = 1$ are

$$W^p_{1,\Omega'_A;1,\Omega_A} = \frac{1}{\sqrt{(1+\delta_{\Omega'_A,0})(1+\delta_{\Omega_A,0})}} \sum_{0 \le \Omega_{A_k} \le J_A} \frac{1}{(1+\delta_{\Omega_{A_k},0})}$$

$$\left\{ \left(d^1_{\Omega_A,\Omega_{A_k}}(\theta_1) + (-1)^{p+1-\Omega_{A_k}} d^1_{\Omega_A,-\Omega_{A_k}}(\theta_1)\right) \right.$$

$$\left(d^1_{\Omega'_A,\Omega_{A_k}}(\theta_1) + (-1)^{p+1-\Omega_{A_k}} d^1_{\Omega'_A,-\Omega_{A_k}}(\theta_1)\right) V^{\Omega_{A_k}}_{1,1}(R_1)$$

$$+(-1)^{\Omega'_A-\Omega_A} \left(d^1_{\Omega_A,\Omega_{A_k}}(\theta_2) + (-1)^{p+1-\Omega_{A_k}} d^1_{\Omega_A,-\Omega_{A_k}}(\theta_2)\right)$$

$$\left. \left(d^1_{\Omega'_A,\Omega_{A_k}}(\theta_2) + (-1)^{p+1-\Omega_{A_k}} d^1_{\Omega'_A,-\Omega_{A_k}}(\theta_2)\right) V^{\Omega_{A_k}}_{1,1}(R_2) \right\}$$

$$(AII-5)$$

Since $\Omega_A = 1$ is the only state with $p = 0$, it is decoupled from the two other states. For $J_A = 1$, the only non-zero coupling element is between $\Omega_A = 0$ and $\Omega_A = 1$, with $p = 1$:

$$W^1_{1,0;1,1} = -\sin\theta_1 \cos\theta_1 \left[V_A(R_1) - V_B(R_1)\right] + \sin\theta_2 \cos\theta_2 \left[V_A(R_2) - V_B(R_2)\right] \quad (AII-6)$$

REFERENCES

1 - K. Fuke, T. Saito and K. Kaya, J. Chem. Phys. **79**, 2487 (1983); *ibid.*, **81**, 2591 (1984).

2 - a) C. Jouvet and B. Soep, J. Chem. Phys. **80**, 2229 (1984); b) C. Jouvet, Thesis, Université de Paris-Sud (1985); c) C. Jouvet and J.A. Beswick, J. Chem. Phys. **86**, 5500 (1987).

3 - M.C. Duval, C. Jouvet and B. Soep, Chem. Phys. Lett. **119**, 317 (1985).

4 - W.H. Breckenridge, M.C. Duval, C. Jouvet and B. Soep, Chem. Phys. Lett. **122**, 181 (1985).

5 - K. Fuke, S. Nonose and K. Kaya, J. Chem. Phys. **85**, 1696 (1986).

6 - K. Yamanouchi, J. Fukuyama, H. Horiguchi, S. Tsuchiya, K. Fuke, T. Saito and K. Kaya, J. Chem. Phys. **85**, 1806 (1986).

7 - a) M.C. Duval, O. Benoist D'Azy, W.H. Breckenridge, C. Jouvet and B. Soep, J. Chem. Phys. **85**, 6324 (1986); b) M.C. Duval, Thesis, Université de Paris-Sud (1989).

8 - M.C. Duval and B. Soep, Chem. Phys. Lett. **141**, 225 (1987).

9 - K. Fuke, T. Saito, S. Nonose and K. Kaya, J. Chem. Phys. **86**, 4745 (1987).

10 - K. Yamanouchi, S. Isogai, M. Okunishi and S. Tsuchiya, J. Chem. Phys. **88**, 205 (1988).

11 - M.C. Duval, B. Soep, R.D. van Zee, W.B. Bosme and T.S. Zwier, J. Chem. Phys. **88**, 2148 (1988).

12 - K. Yamanouchi, S. Isogai, S. Tsuchiya, M.C. Duval, C. Jouvet, O. Benoist d'Azy and B. Soep, J. Chem. Phys. **89**, 2975 (1988).

13 - T. Tsuchizawa, K. Yamanouchi and S. Tsuchiya, J. Chem. Phys. **89**, 4646 (1988).

14 - a) R.D. van Zee, S.C. Blankespoor and T.S. Zwier, Chem. Phys. Lett. **158**, 306 (1989); b) T.S. Zwier, private communication.

15 - C. Jouvet, C. Lardeux-Dedonder, M. Richard-Viard and D. Solgadi, to be published.

16 - R.A. Aziz and M.J. Slaman, Mol. Phys. **58**, 679 (1986).

17 - E.A. Colbourn and A.E. Douglas, J. Chem. Phys. **65**, 1741 (1976).

18 - P.R. Herman, P.E. La Rocque and B.P. Stoicheff, J. Chem. Phys. **89**, 4535 (1988).

19 - F.H. Mies, W.J. Stevens and M. Krauss, J. Mol. Spect. **72**, 303 (1978).

20 - E.E. Nikitin, J. Chem. Phys. **43**, 744 (1965); Adv. Chem. Phys. **28**, 317 (1975); E.E. Nikitin and B.M. Smirnov, Sov. Phys. Ups. **21**, 95 (1978).

21 - R.N. Zare, *in* Angular Momentum, A Wiley-Interscience Publication, New York, 1988.

NONADIABATIC EFFECTS ON THE DYNAMICS OF THE NeICl VAN DER WAALS

COMPLEX

Thomas A. Stephenson,* Yujian Hong and Marsha I. Lester +

Department of Chemistry
University of Pennsylvania
Philadelphia, PA 19104-6323
USA

ABSTRACT

Electronic state changing processes have been directly observed in the dissociation of NeICl complexes excited to a high-lying ion-pair state of ICl ($\sim 39,000$ cm^{-1}). The complex is prepared in the ion-pair state by an optical-optical double resonance technique through a long-lived intermediate level in the A electronic state. Following excitation of the complex, dispersed fluorescence has been detected on the $E \rightarrow X$ and $D' \rightarrow A'$ transitions, indicating the presence of nonadiabatic interactions in the complex that are absent in uncomplexed ICl. Vibrational predissociation with a strong propensity for $\Delta v_{ICl} = -1$ relaxation has also been observed. The electronic state changing pathways are found to be sensitive to both the initial ICl vibrational level of the NeICl complex as well as the degree of excitation of the van der Waals vibrational modes.

INTRODUCTION

The electronically excited van der Waals complexes of halogen and interhalogen diatomic molecules with rare gas atoms have proven to be important and convenient systems for the examination of the structure and vibrational predissociation dynamics of weakly bound molecular complexes.[1-8] To date, however, these studies have focused on the behavior of the van der Waals complexes confined to single adiabatic potential energy curves. The influence of additional electronic states and nonadiabatic couplings has been inferred from the behavior of certain vibrational levels in electronically excited ArI$_2$,[9] HeCl$_2$[4] and HeICl[10] and has been suggested as a possible mechanism for the relaxation of rare gas - iodine complexes upon excitation above the $B^3\Pi_{0^+u}$ dissociation limit.[11] Direct observation of nonadiabatic electronic effects in polyatomic van der Waals complexes has been limited, however, to fine structure relaxation in HgN$_2$.[12] In the latter case, excitation of the complex to a molecular electronic state correlating with a $^3P_1$ mercury atom leads to dissociation and formation of a $^3P_0$ mercury atom. In this manuscript, we report on the direct observation of such electronic state changing processes in the dissociation of a diatom-rare gas complex, namely NeICl in the β ($\Omega = 1$) ion-pair state.

---

* On sabbatical leave (1988-89) from the Department of Chemistry, Swarthmore College, Swarthmore, Pennsylvania, 19081, USA.
+ Alfred P. Sloan Research Fellow and Camille and Henry Dreyfus Foundation Teacher-Scholar.

Dynamics of Polyatomic Van der Waals Complexes
Edited by N. Halberstadt and K. C. Janda
Plenum Press, New York, 1990

493

Previous reports from this laboratory have described in great detail the vibrational predissociation dynamics of the HeICl [7] and NeICl [6] complexes excited to the A(1) and B(0⁺) valence electronic states. In these investigations a pump laser promotes complexes from the ground electronic state to either the A or B state; a probe laser interrogates the resulting photofragments by excitation to an ICl ion-pair state. Emission from the ion-pair state is monitored as a function of probe laser wavelength, thus providing a fluorescence excitation spectrum of the nascent A or B state ICl products. For several A [13] and B [14] state vibrational levels in NeICl, the predissociation lifetime of the van der Waals complex is comparable to the pulse duration of the pump and probe lasers (10 ns). By temporally and spatially overlapping the lasers, it proves possible to promote the NeICl complex to an ion-pair electronic state. In our most recent experiments, we have obtained wavelength resolved emission spectra following optical-optical double resonance excitation of NeICl to the ion-pair state. These spectra demonstrate the presence of both vibrational predissociation and nonadiabatic electronic processes.[15]

EXPERIMENTAL

For the investigations discussed in this report, our existing apparatus[8] has been modified to incorporate a scanning monochromator for dispersed emission studies. Briefly, a continuous expansion of ICl (Sigma) seeded in 75 psig of first-run grade Ne (Airco) is excited by two independently tunable XeCl excimer-pumped dye lasers. The first dye laser (616-622 nm; "laser I") selectively promotes either ground state NeICl or uncomplexed ICl to the v = 14 or 15 levels of the A(1) electronic state. The second dye laser (430-437 nm; "laser II") excites NeICl (or ICl) from the A(1) valence state to the ion-pair state. The resulting laser-induced fluorescence is collected and collimated by a 2 inch diameter, f/1 lens and is focused onto the entrance slit of a 0.25 meter f/4 scanning monochromator with a 2 inch diameter f/4 lens. A blue sensitive phototube (Thorn/EMI 9535QA) is mounted directly on the exit slit body to detect emission from the ion-pair state. A red sensitive phototube (Thorn/EMI 9658B) detects the A → X emission collected by a second set of f/1 collection optics. The resulting emission signals are processed by gated integrators (PAR) and transferred to a laboratory computer for signal averaging and graphics output.

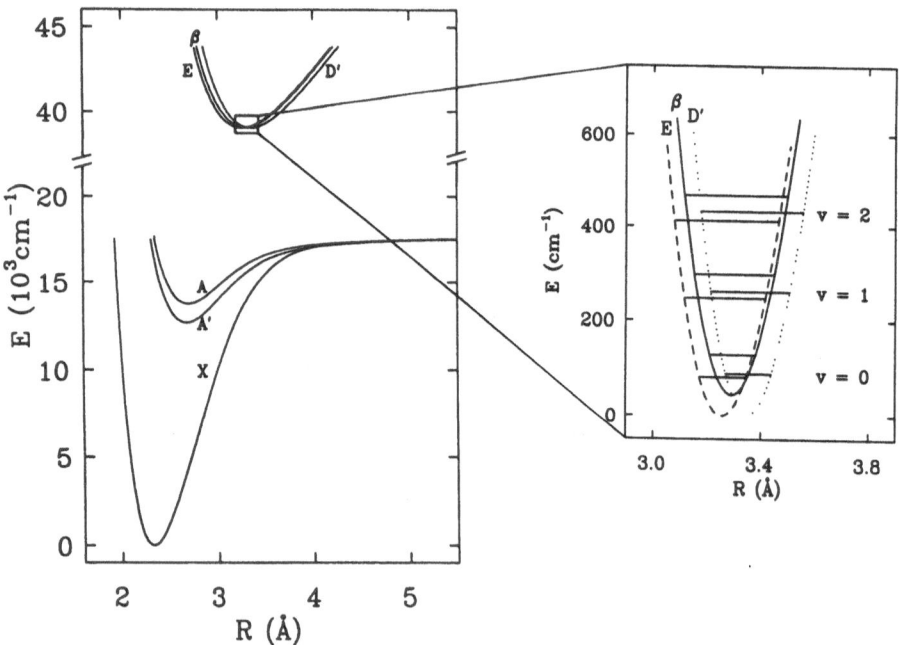

Fig. 1. RKR potential energy curves for the lowest lying valence and ion-pair electronic states in ICl. The RKR data for the X, A, A' and ion-pair states are taken from References 16, 17, 18 and 19, respectively.

RESULTS AND DISCUSSION

Three strongly bound electronic states correlating with the ionic atoms $I^+(^3P_2) + Cl^-(^1S_0)$ form the lowest energy tier of ICl ion-pair states (Figure 1). Conventionally labeled (using Hund's case c notation) as $E(0^+)$, $D'(2)$ and $\beta(1)$, these states are closely spaced ($T_e = 39059.5$ cm^{-1}, 39061.8 cm^{-1}, and 39103.7 cm^{-1}, respectively) and have similar, though slightly displaced, potential energy curves.[19] The ion-pair states can be excited from the lower lying valence states of ICl with a strong propensity for $\Delta\Omega = 0$ electronic transitions. Both the $D'(2)$ and $E(0^+)$ states are heterogeneously mixed with the $\beta(1)$ state resulting in absorption to all three ion-pair states from the $A(1)$ valence state.[19] The $E \leftarrow A$ and $D' \leftarrow A$ excitation features are, however, at least a factor of 200 weaker than the $\beta \leftarrow A$ features.

In figures 2(a) - 2(f), we present the dispersed emission spectra from the $v = 0$ and 1 levels of the E, β and D' states of uncomplexed ICl. Two features of these spectra are worthy of note:

1) Emission from the E state lies in a clearly distinguishable wavelength region from the D' and β state emissions, which are significantly overlapped. The strong $\Delta\Omega = 0$ propensity rule for ion-pair - to - valence state electronic transitions dictates that the $E(0^+)$ emission spectrum is dominated by transitions to the $X(0^+)$ ground state, while $\beta(1) \to A(1)$ and $D'(2) \to A'(2)$ transitions are prominent in emission from the β and D' states, respectively.

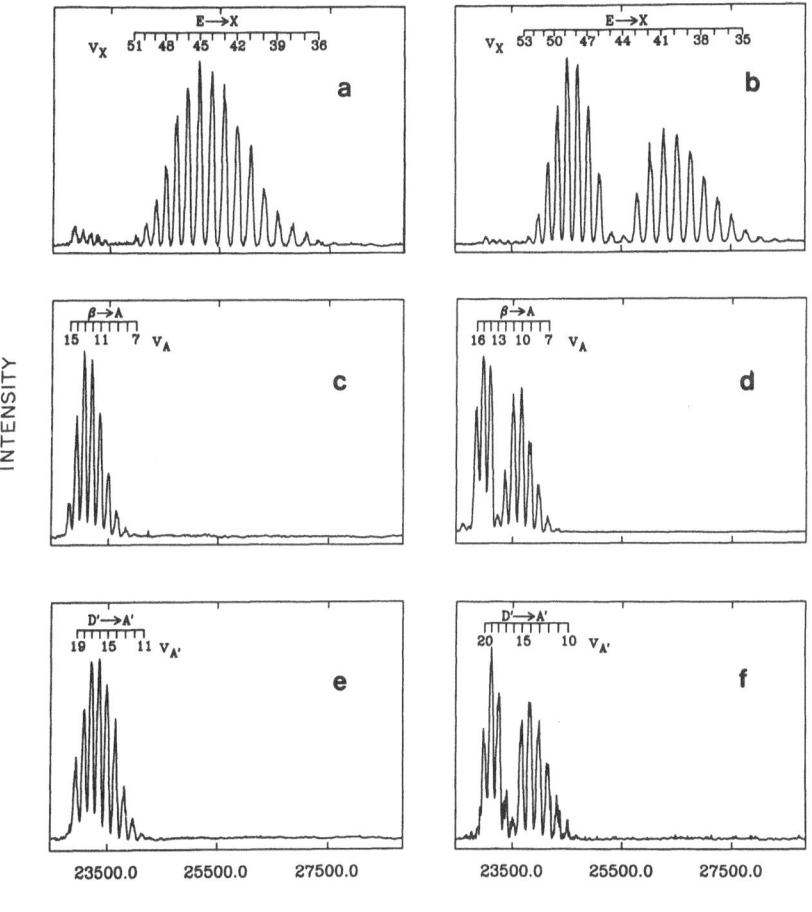

Fig. 2. Dispersed emission spectra following excitation of the ion-pair states in ICl. Emission from: (a) E(v = 0); (b) E(v = 1); (c)β(v = 0); (d) β(v = 1); (e) D'(v = 0); (f) D'(v = 1). X, A and A' state vibrational assignments are derived from References 16, 17 and 18, respectively.

2) The vibrational quantum number of the initially excited ion-pair state is reflected in the shape of the envelope (no nodes for v = 0, one node for v = 1, etc.) of the dispersed emission spectrum. The equilibrium internuclear separations of the ion-pair states are shifted significantly to larger values relative to the A(1), A'(2) and X(0⁺) states. Since the lower state vibrational wavefunctions are highly oscillatory, the only internuclear separations that contribute to the Franck-Condon overlap with the ion-pair states are those near the lower states' outer turning points. Emission is observed, roughly, to lower state levels whose outer turning points fall within the range of internuclear separations which have significant ion-pair state wavefunction amplitude. The intensity of each emission transition is largely determined by the amplitude of the ion-pair state wavefunction in the vicinity of the lower state (A, A' or X, for β, D' or E state emission, respectively) outer turning point. The similarity of the A and A' potential energy curves in the Franck-Condon accessible region means that there are only small differences in the bound-bound emission profiles for the D' and β states.

To distinguish between emission from the β and D' states, we have recently extended our dispersed emission scans to lower frequencies than those displayed in Figure 2. In the vicinity of 21100 cm⁻¹, weak continuum features have been observed that allow us to clearly identify emission as arising from either the β or D' ion-pair states. These continuum features, which appear at various emission frequencies in the spectra of all three ion-pair states, arise from bound-free transitions to the multitude of repulsive curves correlating with the ground and low-lying excited states of I + Cl. Our analysis of these features will be presented elsewhere. For the purposes of our discussion of the dynamics of the NeICl complex, it suffices to note that the dispersed emission spectra, including the bound-free transitions, are an unambiguous signature of the vibrational and electronic character of the emitting level.

In Figure 3, we present the fluorescence excitation spectrum that results from fixing laser I in the bandhead of the (14,0) A ← X transition in NeICl while scanning laser II in the vicinity of the (1,14) β ← A transition. In this experiment, the emission from the ion-pair state is

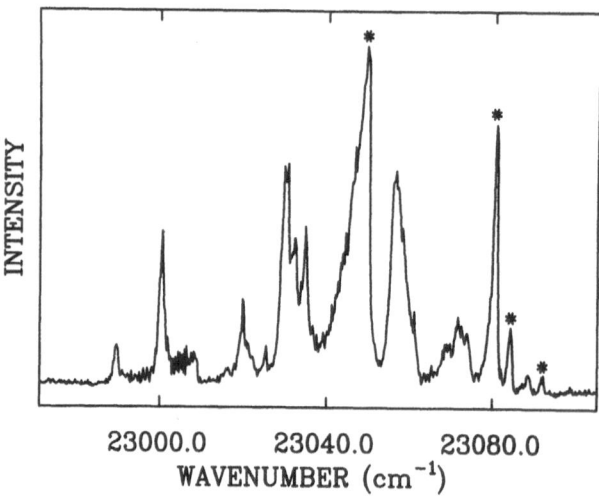

Fig. 3. Fluorescence excitation spectrum of NeICl, detecting emission from the β and D' states while scanning laser II in the vicinity of the (1,14) β ← A transition. Laser I is positioned in the bandhead of the (14,0) A ← X transition in NeICl. All features, except those marked with a '*', are assigned to NeICl. Features marked with a '*' arise from 1) excitation of A state ICl photofragments from the vibrational predissociation of NeICl (A, v = 14) and 2) non-resonant two photon excitation of ICl.

detected after passing through the 0.25 meter monochromator. Because of the large difference in the frequency of emission from the E state as compared to the β and D' states, the monochromator (with wide open slits) has been used as a filter to selectively detect only emission from the E state or only emission from the β and/or D' states as a function of excitation wavelength. To record the spectrum shown in Figure 3, the monochromator transmits emission with a bandwidth of ≈ 700 cm^{-1} while the grating position is fixed so that only emission from the β and/or D' states is detected. The number of features observed in Figure 3 and the overall width of the progression(s) suggests that excitation of NeICl to the ion-pair states is accompanied by a large change in the geometry and/or binding energy relative to the A electronic state. Based on the strong propensity for β ← A transitions in uncomplexed ICl, the excitation features shown in Figure 3 are attributed to the bound states of a NeICl potential that correlates with the β ion-pair state of ICl. Our analysis of this spectrum in terms of the stretching and bending vibrational coordinates of the Ne atom relative to ICl is in progress.

In Figure 4, the dispersed emission spectrum that arises following excitation of the NeICl feature at 23057 cm^{-1} is displayed. By comparing this spectrum with the emission profiles from uncomplexed ICl (Figure 2), two conclusions can be drawn.[15] 1) Emission from more than one electronic state is observed. Emission is detected at frequencies corresponding to E → X vibronic transitions along with emission that could be assigned to either the β → A or the D' → A' transition. Examination of the continuum features (not shown) following excitation of this NeICl feature demonstrates conclusively that most of the emission observed in the vicinity of 23500 cm^{-1} is due to the D' → A' transition. 2) The Franck-Condon profiles observed in Figure 4 are characteristic of emission from the v = 0 levels of the ion-pair states. We attribute this emission as arising from ICl fragments following vibrational predissociation of the complex. (We note, however, that our experimental resolution is insufficient to distinguish between emission from ICl and NeICl.) The absence of any emission from the v = 1 level indicates that vibrational predissociation is rapid compared to the timescale for emission from the ion-pair state ($\tau_{fluor} \approx 10$ ns[20]).

The presence of emission from the E and D' states upon excitation of NeICl to the β state is a direct indication of nonadiabatic interactions in the ion-pair electronic states that are absent in uncomplexed ICl. To examine the dependence of these interactions on the degree of excitation of van der Waals stretching and/or bending coordinates at the $v_{ICl} = 1$ level, we have recorded

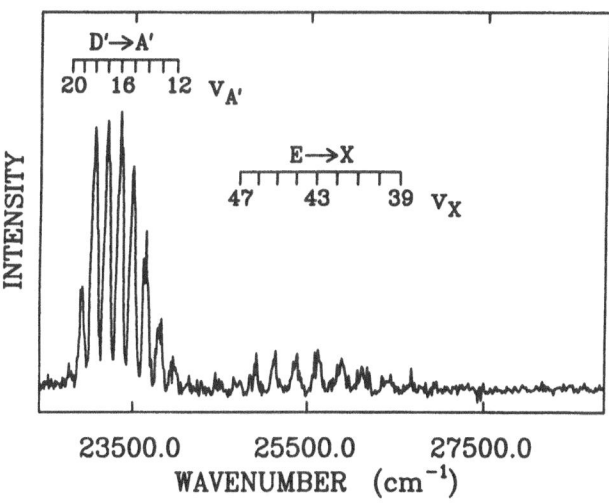

Fig. 4. Dispersed emission spectrum following double resonance excitation of NeICl to the $v_{ICl} = 1$ level. NeICl is excited from the A(v = 14) level using the 23057 cm^{-1} valence - to - ion-pair transition (see Figure 3).

the "selective" excitation spectrum, analogous to Figure 3, in which the monochromator grating is positioned to detect only emission from the E ion-pair state. Within experimental uncertainty, both the line positions and relative intensities of the NeICl excitation features are independent of the emission channel detected. Thus, the branching ratio for populating the E state is insensitive to the degree of excitation of the van der Waals stretching and/or bending coordinates at the $v_{ICl} = 1$ level. In addition, we have also recorded the dispersed emission spectrum following excitation of the NeICl feature at 23030 cm^{-1}. Within experimental uncertainty, this spectrum is identical to that shown in Figure 4, confirming the trend observed in the "selective" excitation spectra with respect to emission from the E state and suggesting that the predominance of emission from the D' state may be a general phenomenon.

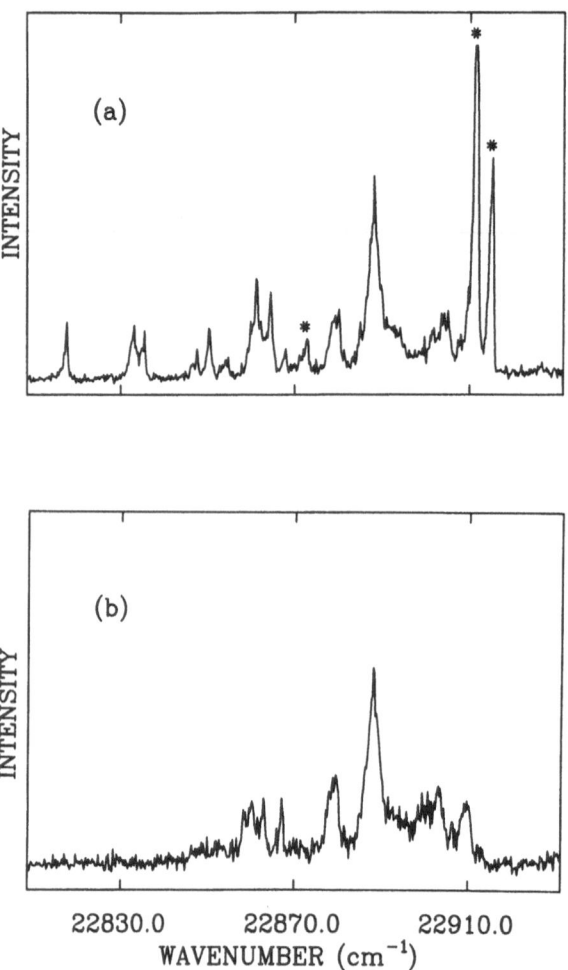

Fig. 5. Fluorescence excitation spectra arising from scanning laser II in the vicinity of the (0,14) β ← A transition. Laser I is positioned in the bandhead of the (14,0) A ← X transition in NeICl. (a): Emission from the β and D' states is selectively detected. (b): Emission from the E state is selectively detected. The '*' notation is the same as that used in Figure 3.

The fluorescence excitation spectra that result from scanning laser II in the vicinity of the (0,14) β ← A transition are shown in Figure 5. Spectrum 5(a) was recorded by selectively detecting only emission from the β and/or D' states, while spectrum 5(b) results when only E state emission is detected. The relative positions of the peaks assigned to excitation of NeICl in Figure 5(a) match those shown in Figure 3. This suggests that the features shown in Figures 3 and 5(a) represent progressions in the same set of NeICl transitions, but correlating with two different vibrational levels of the ICl molecule. A comparison of Figures 5(a) and (b) illustrates that the appearance of emission from the E state is dependent on the degree of excitation of the van der Waals vibrational coordinates. This behavior at the $v_{ICl} = 0$ level of excitation is very different from that found at the $v_{ICl} = 1$ level. Excitation of the lower energy levels does not result in emission from the E state; at excitation energies higher than 22859 cm^{-1}, however, E state emission is again observed.

The dispersed emission spectra that result from excitation of the NeICl features at 22861 and 22888 cm^{-1} (see Figure 5) are quite similar to that shown in Figure 4. Examination of the continuum emission features demonstrates that for these two transitions, the bulk of the emission detected in Figure 5(a) is from the D' and not the β state. Limitations on the signal-to-noise ratio in dispersed emission spectra make it impossible at the present time to wavelength resolve the emission that results from excitation of the van der Waals levels that are not detected in the E state "selective" excitation spectrum. Thus, we cannot comment on whether the D'/β branching ratio changes for these lower energy van der Waals vibrational levels.

From the frequency shift of the NeICl excitation features observed relative to those of ICl and an estimate of the van der Waals dissociation energy in the A electronic state,[6] we can determine the energetic position of the NeICl energy levels relative to the asymptotic energies of the E, β, and D' states in uncomplexed ICl. We find that all of the NeICl levels accessed in our experiments with $v_{ICl} = 0$ lie below the energy of the v = 0 level of the E ion-pair state in ICl. Thus, all of the emission detected in the excitation scans shown in Figure 5 must arise from the NeICl van der Waals complex, not from ICl, as is presumed to be the case for excitation of $v_{ICl} \geq 1$. The mechanism by which the ion-pair states at $v_{ICl} = 0$ are populated may, therefore, be quite different from higher levels and must reflect interactions between the NeICl bound states correlating with the E, β and D' states in ICl.

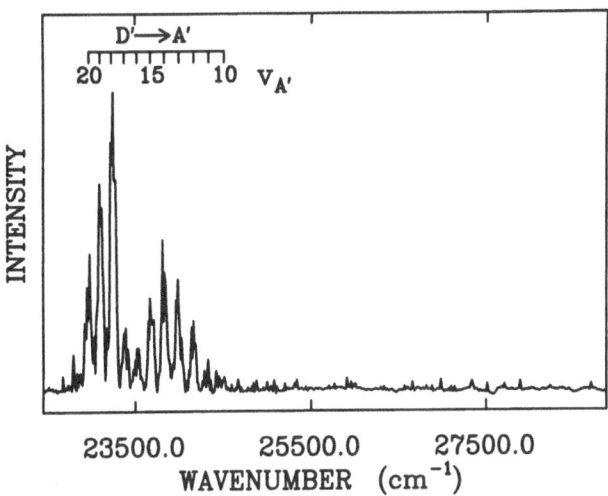

Fig. 6. Dispersed emission spectrum following double resonance excitation of NeICl to the $v_{ICl} = 2$ level.

We have also examined the behavior of NeICl upon excitation to the $v_{ICl} = 2$, 3 and 4 vibrational energy levels. The dispersed emission spectrum that results from excitation of NeICl to the $v_{ICl} = 2$ level is shown in Figure 6. The emission at lower frequency is clearly due to the $v = 1$ level of the β and/or D' states, suggesting that a $\Delta v_{ICl} = -1$ propensity rule is followed for vibrational predissociation in the ion-pair electronic states. No emission from the $v_{ICl} = 0$ level is evident in this spectrum. An examination of continuum features not shown in Figure 6 indicates that the emission observed is predominately due to the D' state. Very surprising is the complete absence of any emission from the E ion-pair state. "Selective" excitation scans of the region about the $(2,14)$ $\beta \leftarrow$ A transition demonstrate that the absence of E state emission is a general characteristic of all of the NeICl resonances at this level of I—Cl vibrational excitation. All three of these observations - the $\Delta v_{ICl} = -1$ propensity rule, the dominance of emission from the D' state, and the absence of emission from the E state - are also characteristic of our studies at the $v_{ICl} = 3$ and 4 levels of excitation.

The presence of the emission from the D' state in these spectra is consistent with our findings at $v_{ICl} = 0$ and 1. The most surprising result of this set of experiments is that the appearance of emission from the E ion-pair state is strongly dependent on the value of the ICl vibrational quantum number, v_{ICl}. Because of the relatively small changes with I—Cl vibrational excitation in both the vibrational spacings and the matrix elements that weakly couple the E, β and D' states in ICl, it is difficult to reconcile such dramatic changes in the electronic state changing dynamics. We hope to construct model NeICl potential energy surfaces in the near future to aid in our understanding of the origin of these processes. We anticipate that these calculations, when coupled with our experimental studies, will provide crucial tests for the emerging theory of nonadiabatic electronic effects on the dynamics of small van der Waals complexes.[11,21]

ACKNOWLEDGEMENTS

Acknowledgement is made to the Donors of the Petroleum Research Fund, administered by the American Chemical Society, for partial support of this work. Partial equipment support for this research was provided by the National Science Foundation through the Physical Chemistry Program. M.I.L. thanks the Natural Science Association at the University of Pennsylvania for a Young Faculty Award. T.A.S. thanks Swarthmore College for a Eugene M. Lang Faculty Fellowship (1988-89).

REFERENCES

1. D.H. Levy, Adv. Chem. Phys. 47:323 (1981).
2. K.C. Janda, Adv. Chem. Phys. 60:201 (1985).
3. D.D. Evard, C. R. Bieler, J.I. Cline, N. Sivakumar, and K.C. Janda, J. Chem. Phys. 89:2829 (1988) and references cited therein.
4. J.I. Cline, B.P. Reid, D.D. Evard, N. Sivakumar, N. Halberstadt, and K.C. Janda, J. Chem. Phys. 89:3535 (1988).
5. N. Sivakumar, J.I. Cline, C.R. Bieler, and K.C. Janda, Chem. Phys. Lett. 147:561 (1988) and references cited therein.
6. J.C. Drobits and M.I. Lester, J. Chem. Phys. 89:4716 (1988) amd references cited therein.
7. R.L. Waterland, J.M. Skene, and M.I. Lester, J. Chem. Phys. 89:7277 (1988) and references cited therein.
8. J.C. Drobits and M.I. Lester, J. Chem. Phys. 86:1662 (1987).
9. G. Kubiak, P.S.H. Fitch, L. Wharton, and D.H. Levy, J. Chem. Phys. 68:4477 (1978).
10. J.M. Skene and M.I. Lester, Chem. Phys. Lett. 116:93 (1985).
11. J.A. Beswick, R. Monot, J.-M. Philippoz, and H. van den Bergh, J. Chem. Phys. 86:3965 (1987).
12. C. Jouvet and B. Soep, J. Chem. Phys. 80:2229 (1984).
13. J.C. Drobits, J.M. Skene, and M.I. Lester, J. Chem. Phys. 84:2896 (1986).
14. J.M. Skene, Ph.D. thesis, University of Pennsylvania, 1988.
15. T.A. Stephenson, Y. Hong and M.I. Lester, Chem. Phys. Lett. 159:549 (1989).

16. J.C.D. Brand and A. R. Hoy, <u>J. Mol. Spectrosc.</u> 114:197 (1985).
17. J.A. Coxon and M.A. Wickramaaratchi, <u>J. Mol. Spectrosc.</u> 79:380 (1980).
18. J.C.D. Brand, D. Bussières, and A.R. Hoy, <u>J. Mol. Spectrosc.</u> 113:388 (1985).
19. D. Bussières and A.R. Hoy, <u>Can J. Phys.</u> 62:1941 (1984).
20. J.G. Eden, M.L. Dlabal, and S.B. Hutchison, <u>IEEE J. Quant. Elec.</u> 17:1085 (1981).
21. C. Jouvet and J.A. Beswick, <u>J. Chem. Phys.</u> 86:5500 (1987).

THE INFRARED PHOTODISSOCIATION SPECTRA AND THE INTERNAL MOBILITY OF SF$_6$-, SiF$_4$- AND SiH$_4$-DIMERS

J.W.I. van Bladel and A. van der Avoird

Institute of Theoretical Chemistry, University of Nijmegen
Toernooiveld, 6525 ED Nijmegen, The Netherlands

ABSTRACT

We present an analysis of the couplings originating from different intermolecular interactions (electrostatic, exchange, dispersion, induction) which split and shift the frequencies of the vibrational transitions in Van der Waals dimers, and determine their intensities. Model potential calculations illustrate the importance of the various contributions in $(SF_6)_2$, $(SiF_4)_2$ and $(SiH_4)_2$ and their dependence on the monomer orientations. The results, in conjunction with calculated equilibrium structures, barriers to internal rotation and (harmonic) Van der Waals vibrational frequencies, lead to several observations which are relevant for the interpretation of the infrared photodissociation spectra of these complexes. We confirm that in $(SF_6)_2$ and $(SiF_4)_2$ (orientation-independent) resonant dipole-dipole coupling dominates the appearance of the spectra. For $(SiH_4)_2$ we conclude, however, that other than electrostatic terms are not negligible and, moreover, that the electrostatic coupling leads to orientation-dependent vibrational frequencies and intensities. This orientational dependence is related to the large displacements of the hydrogen atoms in the ν_4 mode of SiH_4. We also find that the internal rotations in $(SF_6)_2$ and $(SiF_4)_2$ are more strongly locked than those in $(SiH_4)_2$. Especially the geared internal rotations in the latter dimer could easily occur at the experimental molecular beam temperatures.

1. INTRODUCTION

The photodissociation of $(SF_6)_2$, $(SiF_4)_2$ and $(SiH_4)_2$ by means of CO_2 laser light in the frequency range from 880 to 1100 cm^{-1} has been the subject of detailed experimental studies[1-7]. In all these complexes the monomers have an infrared-active vibrational mode in this range: the ν_3 modes of SF_6 and SiF_4 at 948 cm^{-1} and 1031 cm^{-1}, respectively, and the ν_4 mode of SiH_4 at 913 cm^{-1}. In the free monomers these modes are threefold degenerate; in the dimers their degeneracy is lifted. Photodissociation spectra have been obtained by irradiating a beam of dimers with a CO_2 laser and monitoring, by a mass spectrometer[1-3] or by a bolometer[4-7], the changes in dimer concentration. These spectra show the frequency splittings of the monomer modes in the dimers, the intensities of the infrared allowed dimer transitions and the line widths of these transitions, which are related to the dimer predissociation life times. Additional information has been extracted from elegant two- and three- pump/probe laser experiments and from the resolution of the peaks originating from different S and Si isotopes[4,6,7].

Dynamics of Polyatomic Van der Waals Complexes
Edited by N. Halberstadt and K. C. Janda
Plenum Press, New York, 1990

Before we summarize the results of these studies, let us note that for none of these Van der Waals complexes the structure has been determined. In particular, it is not known whether the monomer rotations are (strongly) hindered in the dimers and, if they are, what the orientations of the monomers are. The following results emerge from the spectra. In isotopically homogeneous $(SF_6)_2$ and $(SiF_4)_2$ there are two peaks with an intensity ratio of nearly $1:2$ which are shifted from the monomer ν_3 frequency by amounts $\Delta\nu$ which have a ratio close to $(-2):1$. This can be explained by a splitting of the threefold degenerate monomer ν_3 mode due to resonant dipole-dipole coupling in the dimer. The two peaks correspond with the transition-dipole components of the coupled ν_3 mode that are parallel and perpendicular to the dimer bond axis, respectively. Also the further splitting of these two peaks in isotopically mixed SF_6– and SiF_4–dimers can be explained by the resonant dipole-dipole coupling mechanism. As shown below, the dipole-dipole coupling mechanism is equally effective for all orientations of the monomers and, so, the splitting of the monomer vibrational peaks in $(SF_6)_2$ and $(SiF_4)_2$ cannot be used to determine the monomer orientations.

In $(SiH_4)_2$ the threefold degenerate ν_4 mode splits into two peaks as well, but the intensity ratio and the shifts of these peaks from the monomer frequency are qualitatively different from the other dimers and cannot be explained by the resonant dipole-dipole coupling mechanism[4]. An attempt to add to this mechanism the effect of dipole-induced dipole interactions[5] gave a slight improvement of the calculated results, but the experimental observations could still not be explained in a satisfactory manner. So in $(SiH_4)_2$ the mechanism which dominates the coupling between the resonant ν_4 transitions is yet uncertain.

Another question regards the line widths of the observed dimer peaks. The broadening of these lines is partly homogeneous, related to the vibrational predissociation life time, and partly inhomogeneous. Snels and Fantoni[4] have tried to explain the inhomogeneous broadening by including rotational transitions of the monomers, which were assumed to rotate freely in the dimer, as well as dimer end-over-end rotational transitions. Hole burning experiments by Heymen et al.[6,7] have separated the homogeneous and inhomogeneous line broadening effects and disproved the model of Snels and Fantoni[4]. The model invoked by Heymen et al. to explain these more recent experiments[7] includes the centrifugal distortion associated with the dimer end-over-end rotation, as well as Coriolis interactions associated with internal monomer rotations. So, also in this model free internal rotations of the monomers are assumed. The conclusions do not critically depend on this assumption, however, since the most relevant Coriolis interactions actually arise from the simultaneous rotation of both monomers about the dimer bond axis. We think, therefore, that the question to what extent the monomer rotations are hindered in the dimer is still open.

The present study concerns the two main questions which remain after the analysis of the experimental data. First, we have derived which mechanisms, in addition to the resonant dipole-dipole interaction, may couple the infrared-active monomer vibrations in $(SF_6)_2$, $(SiF_4)_2$ and $(SiH_4)_2$ and lift the degeneracy of these vibrations. In this derivation we have started from an intermolecular potential which includes the electrostatic, exchange, dispersion and induction interactions. In order to estimate the relative importance of each coupling contribution we have replaced the potential by an empirical atom-atom potential, supplemented with induction terms. From this analysis it follows also how the splitting between the dimer vibrational peaks depends on the monomer orientations.

Secondly, we have calculated from the same potential the equilibrium structure of each of the Van der Waals dimers and estimated the barriers to internal rotation of the monomers. We have also estimated the frequencies of the Van der Waals vibrations by means of a harmonic analysis. These Van der Waals vibrations are not directly observed in the photodissociation spectra, but since they affect the monomer orientations and the distance between the monomers, they influence the positions and the widths of the observed dimer peaks.

2. VIBRATIONAL COUPLING

In order to demonstrate which mechanisms couple the monomer vibrations in $(SF_6)_2$, $(SiF_4)_2$ and $(SiH_4)_2$ and, thereby, determine the photodissociation spectra of these dimers, it is convenient to model the intermolecular interactions by an atom-atom potential. Such a potential depends on the orientations of the monomers and, thus, it gives an idea to what extent the monomers are free to rotate in these Van der Waals complexes. It also depends (implicitly) on the monomer internal coordinates and, therefore, it yields all the terms that couple the monomer vibrations. An atom-atom potential between two molecules A and B is usually written as

$$V_{AB} = \sum_{i \in A} \sum_{j \in B} \left[A_{ij} \exp(-B_{ij} r_{ij}) - C_{ij} r_{ij}^{-6} + q_i q_j r_{ij}^{-1} \right], \tag{1}$$

where the first term represents the exchange interactions, the second term the dispersion interactions and the third term the electrostatic interactions. The parameters in this potential are A_{ij}, B_{ij}, C_{ij} and the fractional charges q_i and q_j on the atoms i belonging to molecule A and the atoms j belonging to B. For all these interactions it is common to assume pairwise additivity, as reflected by Eq. (1). The induction interactions, however, contain three-body terms which are equally important as the pairwise terms and we model them by

$$\begin{aligned} V_{AB}^{\text{ind}} = &-\frac{1}{2} \sum_{i,i' \in A} \sum_{j \in B} q_i q_{i'} \alpha_j r_{ij}^{-2} r_{i'j}^{-2} \left(\hat{\boldsymbol{r}}_{ij} \cdot \hat{\boldsymbol{r}}_{i'j} \right) \\ &-\frac{1}{2} \sum_{i \in A} \sum_{j,j' \in B} q_j q_{j'} \alpha_i r_{ij}^{-2} r_{ij'}^{-2} \left(\hat{\boldsymbol{r}}_{ij} \cdot \hat{\boldsymbol{r}}_{ij'} \right). \end{aligned} \tag{2}$$

We have assumed here that the atomic polarizabilities α_i and α_j are isotropic and we have neglected the (higher order) interactions between the induced moments. The interatomic vectors $\boldsymbol{r}_{ij}$ can be written as $\boldsymbol{r}_{ij} = \boldsymbol{R} + \boldsymbol{r}_j - \boldsymbol{r}_i$, where $\boldsymbol{r}_i$ and $\boldsymbol{r}_j$ denote the position vectors of the atoms i and j relative to the centers of mass of the molecules A and B, respectively, and $\boldsymbol{R}$ is the vector that points from the center of mass of molecule A to that of molecule B.

The dependence of the intermolecular potential V_{AB} on the intramolecular vibrational coordinates can be made explicit by writing it as a Taylor expansion in the atomic displacement coordinates

$$\begin{aligned} \boldsymbol{u}_i &= \boldsymbol{r}_i - \boldsymbol{r}_i^{(0)} \\ \boldsymbol{u}_j &= \boldsymbol{r}_j - \boldsymbol{r}_j^{(0)}, \end{aligned} \tag{3}$$

505

which we shall arrange in column vectors u_A and u_B with $3N_A$ and $3N_B$ components. N_A and N_B are the numbers of atoms in the molecules A and B, $r_i^{(0)}$ and $r_j^{(0)}$ the equilibrium positions of these atoms. Up to second order this expansion yields

$$V_{AB} = V^{(0)} + V_A^{(1)} u_A + V_B^{(1)} u_B + \tfrac{1}{2} u_A^T V_{AA}^{(2)} u_A + u_A^T V_{AB}^{(2)} u_B + \tfrac{1}{2} u_B^T V_{BB}^{(2)} u_B \quad (4)$$

The "static potential" $V^{(0)}$ has the same form as Eqs. (1) and (2) with all the (instantaneous) interatomic vectors r_{ij} replaced by the equilibrium vectors $r_{ij}^{(0)} = R + r_j^{(0)} - r_i^{(0)}$. Explicit expressions for the first and second derivatives $V_X^{(1)}$ and $V_{XY}^{(2)}$, where X,Y = A or B, are given in a forthcoming paper[8].

The splitting and shifts of the monomer vibrational frequencies are easily calculated if we first transform Eq. (4) to the normal coordinates Q_A and Q_B of (specific) intramolecular vibrations. The connection between the atomic displacements u_A and u_B and these normal coordinates is simply given by

$$\begin{aligned} u_A &= R_A L_A Q_A \\ u_B &= R_B L_B Q_B. \end{aligned} \quad (5)$$

The matrices L_A and L_B can be calculated from the (harmonic) force fields of the free monomers by the standard GF-matrix method[9]. The rotation matrices R_A and R_B depend on the orientations of the monomers in the dimer; they rotate the cartesian components of the atomic displacements with respect to the monomer frames to the corresponding components relative to the dimer frame. After this transformation Eq. (4) becomes:

$$V_{AB} = V^{(0)} + W_A^{(1)} Q_A + W_B^{(1)} Q_B + \tfrac{1}{2} Q_A^T W_{AA}^{(2)} Q_A + Q_A^T W_{AB}^{(2)} Q_B + \tfrac{1}{2} Q_B^T W_{BB}^{(2)} Q_B \quad (6)$$

with

$$\begin{aligned} W_X^{(1)} &= V_X^{(1)} R_X L_X \\ W_{XY}^{(2)} &= L_X^T R_X^T V_{XY}^{(2)} R_Y L_Y \end{aligned} \quad (7)$$

for X, Y = A or B.

If the monomer vibrations are assumed to be harmonic, it becomes fairly obvious how the coupling terms in Eq. (6) will affect their frequencies. Each of the vibrations we are interested in, ν_3 in $(SF_6)_2$ and $(SiF_4)_2$, ν_4 in $(SiH_4)_2$, is threefold degenerate in each monomer. Their components, which we denote by x, y and z, span the irreducible representations T_{1u} and T_2 of the point groups O_h and T_d, respectively. So in the non-interacting monomers we have six equivalent harmonic modes, which define the unperturbed states

$$| n_{x_A}, n_{y_A}, n_{z_A}, n_{x_B}, n_{y_B}, n_{z_B} \rangle.$$

In the ground state $| 0 \rangle$ all these quantum numbers n are equal to zero. In the sixfold degenerate first excited state with components $| x_A \rangle$, $| y_A \rangle$, $| z_A \rangle$, $| x_B \rangle$, $| y_B \rangle$, and $| z_B \rangle$ the quantum number n that corresponds to the mode indicated has been raised to $n = 1$. The splitting and shifts of the monomer fundamental vibrational frequencies can be calculated by taking Eq. (6) as the perturbation and using first order perturbation theory for the ground state and the degenerate first excited state of the dimer. The normal mode coordinates Q_A in Eq. (6) refer to the modes x_A, y_A and z_A and the coordinates Q_B are x_B, y_B and z_B. By the use of standard harmonic oscillator algebra it follows that the fundamental vibrational excitation frequencies of the dimers are the eigenvalues of the 6×6 matrix

$$\begin{pmatrix} \hbar\omega_A \mathbf{1} + \frac{1}{2}\hbar\omega_A^{-1}\mathbf{W}_{AA}^{(2)} & \frac{1}{2}\hbar(\omega_A\omega_B)^{-1/2}\mathbf{W}_{AB}^{(2)} \\ \frac{1}{2}\hbar(\omega_A\omega_B)^{-1/2}\mathbf{W}_{AB}^{(2)T} & \hbar\omega_B \mathbf{1} + \frac{1}{2}\hbar\omega_B^{-1}\mathbf{W}_{BB}^{(2)} \end{pmatrix} \qquad (8)$$

where ω_A and ω_B are the monomer fundamental excitation frequencies.

It is important to note that all the interactions represented by Eqs. (1) and (2), (exchange, dispersion, electrostatic and induction) contribute to the second derivatives $V_{XY}^{(2)}$ in Eq. (4) and, therefore, to the matrix elements $\mathbf{W}_{XY}^{(2)}$. The blocks $\mathbf{W}_{AA}^{(2)}$ and $\mathbf{W}_{BB}^{(2)}$ on the diagonal lead to (first order) shifts and splitting of the monomer frequencies. The off-diagonal block $\mathbf{W}_{AB}^{(2)}$ couples the monomer vibrations and leads to a further splitting of the frequencies. The dimer eigenstates will be mixed monomer states. The mixing is determined by the coupling matrix elements $V_{AB}^{(2)}$ in Eq. (4) which depend on the interatomic vectors $\boldsymbol{r}_{ij}^{(0)}$ and, therefore, on the orientations of the monomers in the dimer. Only for specific orientations, when the x, y and z axes become symmetry axes of the dimer and the monomers are identical, the matrix in Eq. (8) can be diagonalized by symmetry projection and the excited eigenstates are simply

$$| x_A \rangle \pm | x_B \rangle , \; | y_A \rangle \pm | y_B \rangle , \; | z_A \rangle \pm | z_B \rangle . \qquad (9)$$

In general, the mixing of the monomer states will be more complicated, however.

The first derivatives $V_A^{(1)}$ and $V_B^{(1)}$ in Eq. (4) vanish at the exact equilibrium geometry of the dimer. But even if we determine first the monomer equilibrium structures, in the intramolecular force fields, and next the dimer equilibrium geometry from the intermolecular potential $V^{(0)}$ in Eq. (4), they are very small. Moreover, they have no effect on the vibrational frequencies in first order perturbation theory. In second order they will lead to further, but small, shifts of the monomer frequencies, which we have not calculated.

The dipole-dipole resonance mechanism, which has been held responsible for the vibrational splittings in the literature,[1-7] is implicitly contained in Eqs. (4) to (8). In order to derive this mechanism explicitly, we have to expand the matrices $V_{XY}^{(2)}$ in Eq. (4), which depend on the interatomic vectors $\boldsymbol{r}_{ij}^{(0)}$, about the molecular centers of mass. For the electrostatic and induction interactions this expansion is equivalent to the molecular multipole expansion. The leading electrostatic term in the coupling matrix $V_{AB}^{(2)}$ is the (vibrational) dipole-dipole interaction. As is known from the earlier papers[1-7], this interaction splits the vibrational frequencies by amounts $\pm\Delta$, $\pm\Delta$ and $\mp 2\Delta$, corresponding to the eigenstates of Eq. (9) (with z along the vector $\boldsymbol{R}$). The splitting parameter $\Delta = \mu_{01}^A \mu_{01}^B R^{-3}$ is determined by the transition-dipole moments

$$\mu_{01}^A = \langle \, 0_{x_A} \, | \mu_{x_A} | \, 1_{x_A} \, \rangle \quad , \quad \mu_{01}^B = \langle \, 0_{x_B} \, | \mu_{x_B} | \, 1_{x_B} \, \rangle . \qquad (10)$$

The leading multipole contribution to the matrices $V_{AA}^{(2)}$ and $V_{BB}^{(2)}$ is the dipole-induced dipole term. If we assume that the atomic polarizabilities α_i and α_j are related to the molecular polarizabilities α_A and α_B by $\alpha_A = \sum_{i \in A} \alpha_i$ and $\alpha_B = \sum_{i \in B} \alpha_j$, this leading induction term adopts the form given in Ref. 5 and it contributes to the shifts of the vibrational frequencies as indicated there.

The splitting and shifts of the vibrational frequencies caused by the leading electrostatic (dipole-dipole) and induction (dipole-induced dipole) interactions appear to be independent of the monomer orientations. At first sight this seems surprising, since the matrices $W_{XY}^{(2)}$ that determine the splitting and the shifts of the frequencies depend on the monomer orientations through the rotation matrices R_A and R_B, see Eq. (7). It can be proved, however, that the rotations R_A and R_B just amount to a similarity transformation of the dynamical matrix in Eq. (8), when this matrix contains only the dipole-dipole and dipole-induced dipole couplings.

We can make a similar expansion of the exchange and dispersion contributions to the coupling matrices $V_{XY}^{(2)}$. They give rise to splitting and shifts of the vibrational frequencies which display a different pattern than obtained from the electrostatic interactions only. In general, the splitting and shifts caused by the leading exchange and dispersion terms depend on the monomer orientations. Also the effects of the higher electrostatic and induction terms on the vibrational frequencies depend on the monomer orientations.

The contributions of the various intermolecular interactions to the vibrational coupling in Van der Waals complexes have been calculated explicitly for $(SF_6)_2$, $(SiF_4)_2$ and $(SiH_4)_2$. To try and simulate the dimer vibrational spectra (see Section 4) in the frequency range from 880 to 1100 cm^{-1}, we have also calculated the infrared intensities of the dipole allowed transitions. We concentrate, in particular, on the dependence of the calculated spectra on the monomer orientations. In line with the atom-atom model used for the intermolecular potential, we write the following expression for the vibrational dipole moment operator of a dimer

$$\mu = \sum_{i \in A} q_i r_i + \sum_{j \in B} q_j r_j + \sum_{i,j} q_i \alpha_j r_{ij}^{-3} r_{ij} - \sum_{i,j} q_j \alpha_i r_{ij}^{-3} r_{ij}. \qquad (11)$$

The first two terms are the vibrational dipole moments of the monomers A and B, as represented by a model with fractional charges q_i and q_j assigned to the vibrating atoms. Actually we have chosen these charges such that the known transition strengths of the ν_3 vibrations in SF_6 and SiF_4 and the ν_4 vibration in SiH_4 are exactly reproduced. We have added to each monomer term the induced dipole moment caused by the (vibrating) charges on the other monomer. Substituting Eq. (3) into Eq. (11) and making a Taylor expansion, we find μ explicitly as a function of the vibrational atomic displacements u_A and u_B. For calculating the transition strengths in the harmonic approximation we need only the first order expansion

$$\mu = \mu^{(0)} + \mu_A^{(1)} u_A + \mu_B^{(1)} u_B \qquad (12)$$

where $\mu^{(0)}$ is obtained from Eq. (11) by replacing r_{ij} by $r_{ij}^{(0)}$ and the first order coefficients are given in Ref. 8. Using Eq. (5) we obtain μ in terms of the normal coordinates Q_A and Q_B of the monomer vibrations

$$\mu = \mu^{(0)} + M_A^{(1)} Q_A + M_B^{(1)} Q_B \qquad (13)$$

with

$$M_X^{(1)} = \mu_X^{(1)} R_X L_X \quad \text{for} \quad X = A, B.$$

The transition strengths of the dimer excitations are given by

$$T_{0 \to 1} = | \langle 0 |\boldsymbol{\mu}| 1 \rangle |^2 \tag{14}$$

where $| 1 \rangle$ is an eigenstate of the matrix given by Eq. (8). Substituting Eq. (13) for μ and the appropriate linear combinations of monomer excited states for $| 1 \rangle$ and using the standard harmonic oscillator algebra, we can easily calculate the transition strengths of all the vibrational transitions in the dimer.

If the excited states $| 1 \rangle$ would simply be given by Eq. (9) and μ is restricted to the first two terms in Eq. (11), then we obtain the well known result that the transitions to the minus-states in Eq. (9) are forbidden and those to the plus-states are allowed with equal intensities. This yields the characteristic spectrum[1-7] with one peak at position $\omega_0 + \Delta$ and another peak at position $\omega_0 - 2\Delta$, and an intensity ratio of 2 : 1, that is obtained when only the electrostatic dipole-dipole interactions are included ($\omega_0 = \omega_A = \omega_B$). In general, the excited states $| 1 \rangle$ from Eq. (8) are more complex, however. For arbitrary orientations of the monomers all dimer transitions become allowed, in principle, and their intensities become orientation-dependent, just as their frequencies.

3. VAN DER WAALS VIBRATIONS

Given the potential expansion in Eq. (4) it is relatively easy to calculate the Van der Waals vibrations of the dimers, in the harmonic approximation. Although we realize that the harmonic model will not be appropriate for the larger amplitude Van der Waals vibrations, it is still interesting to consider the harmonic frequencies and the corresponding normal modes, since these give already a clear indication of the extent to which specific monomer rotations will be hindered (see Section 4). First we find, by direct minimisation of $V^{(0)}$, the equilibrium structure of the dimer, i.e. the equilibrium distance R and the equilibrium orientations of the monomers. Next, we use the second derivatives $V_{AB}^{(2)}$, $V_{AA}^{(2)}$ and $V_{BB}^{(2)}$ and we transform these as in Eq. (7). Instead of the matrices L_A and L_B that correspond to the normal modes of vibration Q_A and Q_B of the monomers, we use the matrices $\tilde{L}_A$ and $\tilde{L}_B$ that correspond to the Eckart coordinates[9] for the center of mass translations and the overall rotations of the monomers. This yields a 12 dimensional force constant matrix with blocks $\widetilde{W}_{AA}^{(2)}$, $\widetilde{W}_{AB}^{(2)}$, $\widetilde{W}_{AB}^{(2)T}$ and $\widetilde{W}_{BB}^{(2)}$ which, together with the appropriate inertia matrix that contains the molecular masses and moments of inertia, gives the 6 harmonic frequencies of the Van der Waals vibrations (and 6 frequencies zero that correspond with the overall translations and rotations of the dimer) and the corresponding normal modes.

One complication has yet been overlooked in this simple description. The space spanned by the linearized Eckart coordinates for the monomer rotations is not invariant under the overall rotations of the dimer. For rotational invariance of the dynamical problem[10] it is necessary, and in the harmonic model sufficient, to include additional terms in the atomic displacements, Eq. (5), which are quadratic in the Eckart coordinates $\tilde{Q}_A$ and $\tilde{Q}_B$ of the monomer rotations. The transformation matrices occurring in these terms, which are derived in Ref. 11, must be multiplied with the first derivatives

$V_A^{(1)}$ and $V_B^{(1)}$ from Eq. (4), and be added to the diagonal blocks $\widetilde{W}_{AA}^{(2)}$ and $\widetilde{W}_{BB}^{(2)}$, as described in Ref. 12, for instance. Thus, the 3 eigenfrequencies of the 12-dimensional harmonic eigenvalue problem that correspond with the overall rotations of the dimer will indeed be zero. This correction affects the frequencies of the five rotational Van der Waals vibrations.

4. RESULTS AND CONCLUSIONS

In order to make more quantitative estimates of the various coupling terms derived in Section 2, which may affect the photodissociation spectra of $(SF_6)_2$, $(SiF_4)_2$ and $(SiH_4)_2$, we have substituted literature values for the parameters occurring in the atom-atom interaction potential, Eqs. (1) and (2), and in the vibrational dipole-moment operator, Eq. (11). Using these values, given in Ref. 8, we have first determined the equilibrium structure of each dimer by minimizing the "static" intermolecular energy $V^{(0)}$. The dimer structures are shown in Fig. 1. The effect of the electrostatic and induction interactions on the binding energies are small; we are dealing with real Van der Waals dimers which are bound mainly by the dispersion attraction. In general, the calculated dimer equilibrium structures depend rather sensitively on the atom-atom parameters chosen for the exchange and dispersion interactions. So, we have avoided drawing conclusions that depend too specifically on the equilibrium structure calculated.

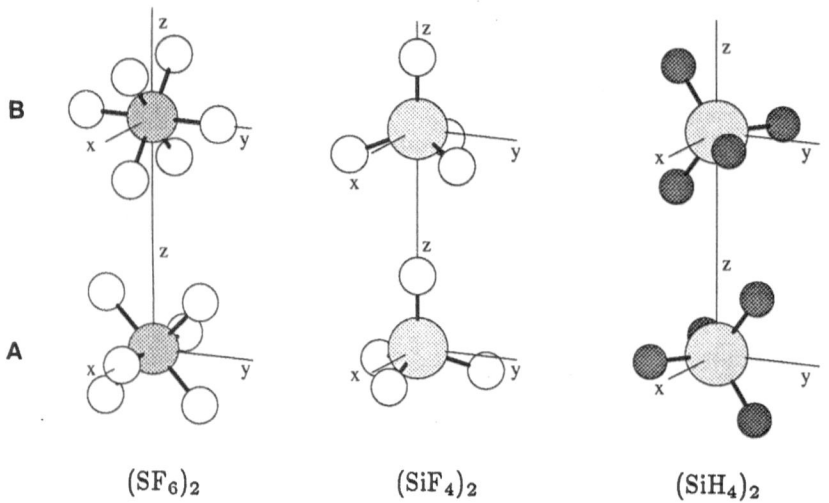

$(SF_6)_2$ $(SiF_4)_2$ $(SiH_4)_2$

Fig. 1 Equilibrium structures of the dimers calculated with the atom-atom potential.

In order to get some insight into the internal mobility of these Van der Waals dimers we have calculated the harmonic frequencies of the Van der Waals vibrations, as described in Section 3. The results shown in Table 1 indicate that in all dimers the geared internal rotations are relatively soft and the anti-geared rotations are considerably stiffer. The torsional frequencies are intermediate between these extremes. The Van der Waals stretch mode has typical frequencies of 30 to 50 cm^{-1}, substantially higher than the 11 to 13 cm^{-1} estimated in Ref. 7 for $(SF_6)_2$.

Table 1. Van der Waals vibrations (harmonic)

		ω (cm^{-1})	mode A	mode B	symmetry	
$(SF_6)_2$	geared rotation	5.4	$0.6R_x - 0.8R_y$	$-0.8R_x + 0.6R_y$	E	(D_{2d})
	geared rotation	5.4	$0.6R_x + 0.8R_y$	$-0.8R_x - 0.6R_y$		
	torsion	21.9	R_z	$-R_z$	B_1	
	stretch	30.5	z	$-z$	A_1	
	antigeared rotation	38.2	$0.8R_x - 0.6R_y$	$0.6R_x - 0.8R_y$	E	
	antigeared rotation	38.2	$0.8R_x + 0.6R_y$	$0.6R_x + 0.8R_y$		
$(SiF_4)_2$	geared rotation	7.7	R_y	$-R_y$	E	(C_{3v})
	geared rotation	7.7	R_x	$-R_x$		
	torsion	10.4	R_z	$-R_z$	A_2	
	stretch	49.0	z	$-z$	A_1	
	antigeared rotation	60.2	R_x	R_x	E	
	antigeared rotation	60.2	R_y	R_y		
$(SiH_4)_2$	geared rotation	16.3	R_y	$-R_y$	B_u	(C_{2h})
	geared rotation	27.4	R_x	$-R_x$	A_u	
	stretch	35.0	z	$-z$	A_g	
	torsion	36.3	R_z	$-R_z$	A_u	
	antigeared rotation	50.6	R_y	R_y	A_g	
	antigeared rotation	112.0	R_x	R_x	B_g	

A further exploration of the potential energy surface was made by calculating the barriers to internal rotation along the paths indicated by the Van der Waals normal modes. It is confirmed by these calculations that the geared internal rotations are much easier than the anti-geared ones. Still, in $(SF_6)_2$ and $(SiF_4)_2$ the rotational barriers associated with the geared rotations are higher than 125 cm^{-1}. In $(SiH_4)_2$, however, the barriers for the geared internal rotations are only 17 cm^{-1}, when the distance R is relaxed. This value is comparable with the zero-point energy for the geared rotation-vibrations, see Table 1. So we can expect large amplitude hindered rotations for this degree of freedom in $(SiH_4)_2$. Another mode which seems relatively soft and has a low barrier to rotation is the torsional mode in $(SiF_4)_2$, but this depends specifically on the calculated equilibrium structure. At the experimental molecular beam temperatures[1-7] of about 20 K several Van der Waals vibrational states will be populated in all the dimers considered (cf. the frequencies in Table 1). Most of the orientational modes, especially in $(SF_6)_2$ and $(SiF_4)_2$, are "locked"; the Van der Waals vibrations will be librations about the equilibrium angles. For the geared internal rotations in $(SiH_4)_2$ they will be hindered rotations with large amplitudes, just below and above the rotation barriers. The amplitudes for the rotational vibrations in the latter dimer will be larger anyway, because of the large rotational constant of SiH_4. This contradicts the lesser orientational mobility assumed for $(SiH_4)_2$ in Ref. 4. The interpretation of the photodissociation spectra of these dimers will have to be consistent with this picture.

Table 2. Vibrational frequency shifts[a]

	mode	electrostatic[b]	exchange[b]	dispersion[b]	induction[b]	total[b]
(SF$_6$)$_2$	1	−11.20 (−12.36)	0.23 (0.02)	−0.13 (−0.07)	−1.44 (−1.13)	−12.55 (−13.54)
	2,3[c]	−6.44 (−6.18)	0.01 (0.00)	0.01 (0.01)	−0.46 (−0.28)	−6.89 (−6.46)
	4,5[c]	6.20 (6.18)	0.00 (0.00)	0.02 (0.01)	−0.43 (−0.28)	5.79 (5.91)
	6	11.48 (12.36)	0.24 (0.06)	−0.12 (−0.09)	−1.35 (−1.13)	10.25 (11.21)
(SiF$_4$)$_2$	1	−8.84 (−7.31)	3.51 (0.08)	−0.79 (−0.12)	−1.61 (−0.63)	−7.74 (−7.98)
	2,3[c]	−3.25 (−3.65)	−0.09 (−0.03)	0.05 (0.03)	−0.22 (−0.16)	−3.51 (−3.81)
	4,5[c]	3.35 (3.65)	−0.11 (0.00)	0.05 (0.01)	−0.23 (−0.16)	3.06 (3.51)
	6	7.29 (7.31)	0.93 (0.31)	−0.28 (−0.19)	−0.79 (−0.63)	7.15 (6.79)
(SiH$_4$)$_2$	1	−5.10 (−5.51)	1.50 (0.27)	−0.82 (−0.40)	−0.91 (−0.49)	−5.33 (−6.13)
	2	−3.06 (−2.76)	−0.65 (0.04)	0.37 (−0.04)	−0.26 (−0.12)	−3.61 (−2.88)
	3	−0.82 (−2.76)	1.74 (−0.02)	−0.78 (0.08)	−0.06 (−0.12)	0.09 (−2.82)
	4	0.32 (2.76)	1.01 (0.04)	−0.36 (−0.04)	−0.27 (−0.12)	0.70 (2.63)
	5	3.92 (2.76)	−0.36 (−0.02)	0.25 (0.07)	−0.22 (−0.12)	3.59 (2.68)
	6	6.72 (5.51)	4.60 (0.27)	−1.58 (−0.45)	−0.64 (−0.49)	9.11 (4.84)

a) In cm^{-1}, calculated for the dimer equilibrium structures.

b) The numbers in parentheses are the leading terms in the expansion about the molecular centers of mass; for the electrostatic contribution this is the resonant dipole-dipole splitting, for the induction contribution this is the dipole-induced dipole shift.

c) These modes are degenerate.

Next we have calculated the frequencies and the intensities of the dimer vibrational transitions in the range of the ν_3 vibrations of SF$_6$ and SiF$_4$ and the ν_4 vibration of SiH$_4$. The force fields used for the monomers are given in Ref. 8. We have analyzed the various contributions to the splitting and shifts of the monomer frequencies, and we have investigated the orientational dependence of each contribution, as well as the orientational dependence of the dimer transition strengths. From the results in Table 2 we conclude that the electrostatic contributions are dominant in all cases. As explained in Section 2, the electrostatic dipole-dipole shifts (indicated in parentheses) are $-2\Delta, -\Delta, \Delta$ and 2Δ, for all dimers, independent of the monomer orientations. Also the dipole-induced dipole shifts have constant ratios $-4 : -1 : -1 : -4$, independent of the monomer orientations.

In (SF$_6$)$_2$ the effects of the exchange and dispersion couplings on the dimer frequencies are completely negligible. Therefore, the (SF$_6$)$_2$ photodissociation spectrum shows the characteristic two peaks with shifts -2Δ and Δ, and intensities $1 : 2$, from the excitation of the modes 1 and 4, 5 (the other excitations are forbidden in this case). In (SiF$_4$)$_2$ the exchange and dispersion effects are somewhat larger, but they are partly cancelled by the non-dipolar electrostatic and induction effects and, so, the (SiF$_4$)$_2$ spectrum still shows the characteristic two peaks, see Fig. 2. As illustrated in Fig. 2, the calculated (SiF$_4$)$_2$ spectrum is nearly independent of the monomer orientations, just as the (SF$_6$)$_2$ spectrum.

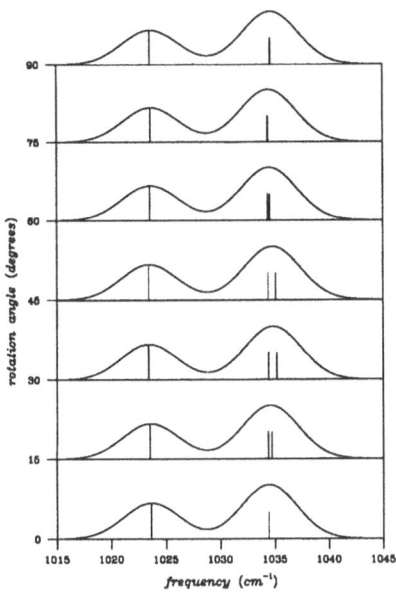

Fig. 2

Effect of the geared monomer rotation R_y on the dimer vibrational spectrum of $(SiF_4)_2$. The stick spectrum indicates the calculated frequencies and transition strengths, the envelope is obtained by adopting Gaussian line profiles, fwhm 6 cm^{-1}. The effect of the relaxation of R is included in the spectra, but it is hardly visible.

For $(SiH_4)_2$ the situation is markedly different. Here, we find (see Table 2) that the vibrational frequency splitting and shifts caused by the exchange and dispersion interactions are not negligible. Still, they are dominated by the electrostatic effects. It is most remarkable, however, that the electrostatic coupling terms in this case do not lead to the usual two peaks with the familiar 1 : 2 intensity relation. A more irregular pattern is obtained, both for the peak positions and their intensities, which, moreover, depends strongly on the monomer orientations. For the two directions of easy (geared) rotation in $(SiH_4)_2$ this is illustrated in Fig. 3. The effect of the relaxation of the distance R, which accompanies the geared rotations, is included in Fig. 3. This effect is very small, however, in comparison with the changes in the spectrum that are caused directly by the monomer rotations. From an analysis of the calculated results which have yielded Fig. 3 we conclude that in $(SiH_4)_2$ not only the leading (dipole-dipole) term in the multipole expansion of the electrostatic coupling $V_{AB}^{(2)}$ is important, but also the higher terms. These higher terms lead to resonance effects between the monomer vibrations in $(SiH_4)_2$ which are orientationally dependent. The mixing of the monomer excitations is more complicated than given by Eq. (9), which is reflected by the frequencies of the dimer vibrations, as well as by their intensities. Excitations of all dimer modes become allowed, in this case.

Given this conclusion we can at least understand why the photodissociation spectrum of $(SiH_4)_2$ is qualitatively different from the spectra of $(SF_6)_2$ and $(SiF_4)_2$. For all these complexes, several Van der Waals vibrational states will be populated at the experimental beam temperatures. The structure of the vibrational spectra in the range from 880 to 1100 cm^{-1} must be explained by summation of the spectra of all these Van der Waals states. Together with the effects from Coriolis couplings[7], this will lead to inhomogeneous line broadening, because of the variations in the average intermolecular bond length $\langle R \rangle$ for the different Van der Waals vibrational states. In all dimers we have orientational vibrations also, but this will only lead to additional broadening (and shifting) of the lines when the line positions depend on the monomer orientations.

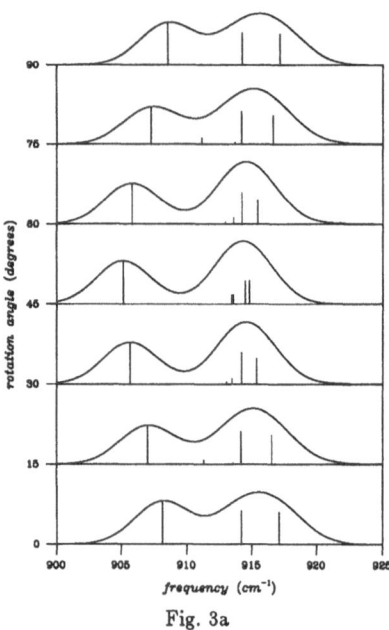

Fig. 3a

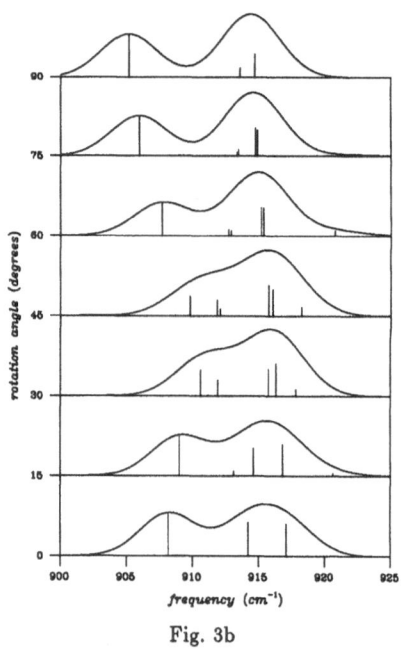

Fig. 3b

As Fig. 2, for the geared monomer rotation R_y in $(SiH_4)_2$.

As Fig. 2, for the geared monomer rotation R_x in $(SiH_4)_2$.

This occurs for $(SiH_4)_2$ only, cf. Figs. 2 and 3, and it is more important than the line shifts caused by the variations of $\langle R \rangle$. Moreover, the amplitudes of the orientational vibrations, especially of the geared modes, are larger in $(SiH_4)_2$. Although we may expect that the origin of the qualitatively different $(SiH_4)_2$ spectrum is related to these effects, we cannot yet reproduce the experimental spectrum from our calculations. We would need to calculate explicitly the large amplitude Van der Waals vibrational states and to take a thermal average over those states. This would be an enormous task which is not practically possible without drastic approximations. The intermolecular model potential is not sufficiently accurate to justify such an effort. Moreover, it would be preferable to first obtain more detailed experimental information on the $(SiH_4)_2$ spectrum. In particular, it would be very informative to study $(SiH_4)_2$ by the same double-resonance techniques[6,7] that have been applied to $(SF_6)_2$ and $(SiF_4)_2$. For the latter dimers these techniques have confirmed the resonant dipole-dipole mechanism of the vibrational coupling; for $(SiH_4)_2$ they will yield new information that verifies the effects found in our model calculations.

We have checked whether our conclusions do not depend specifically on the parameters chosen for the atom-atom potential. We have replaced the values for the F-parameters in $(SiF_4)_2$ and the values for the H-parameters in $(SiH_4)_2$ by values from another reference (see Ref. 8). Although this leads to somewhat different equilibrium structures of the dimers and, therefore, slightly different sizes of the vibrational frequency shifts, it does not change the overall conclusions. The size of our calculated splitting is not accurate, anyway, because we have used the values of R_e calculated with the atom-atom potential, rather than extracted R_0 from the spectra as in Ref. 4.

Finally, we have addressed the question why the higher (orientationally dependent) electrostatic resonance couplings are important in $(SiH_4)_2$ but not in $(SiF_4)_2$ and $(SF_6)_2$. The ratio between the intramolecular bond length $r_i^{(0)}$ and the Van der Waals bond length R_e, which determines the convergence of the molecular multipole expansion of the couplings, is practically the same for all dimers. However, the amplitude of vibration of the light H-atoms in SiH_4 is much larger than the amplitude of the F-atoms in SiF_4 and SF_6, in both the ν_3 and ν_4 modes. Through the transformation of the coupling matrices to monomer normal coordinates Q_A and Q_B, see Eqs. (5) to (7), this enters the calculated shifts of the vibrational frequencies in the dimers, see Eq. (8). So, the contribution of the ligand displacements to these shifts is much larger in $(SiH_4)_2$ than in $(SF_6)_2$ and $(SiF_4)_2$ and, therefore, the orientationally dependent (higher) coupling terms are considerably more important.

ACKNOWLEDGEMENT

We thank prof. Jörg Reuss for stimulating discussions.

REFERENCES

1. J. Geraedts, S. Setiadi, S. Stolte, and J. Reuss, Laser induced predissociation of SF₆ clusters, *Chem. Phys. Lett.* 78:277 (1981).

2. J. Geraedts, S. Stolte, and J. Reuss, Vibrational predissociation of SF₆ dimers and trimers, *Z. Phys.* A304:167 (1982).

3. J. Geraedts, M. Waayer, S. Stolte, and J. Reuss, Dimer spectroscopy, *Faraday Discuss. Chem. Soc.* 73:375 (1982).

4. M. Snels and R. Fantoni, IR dissociation of dimers of high symmetry molecules: SF₆, SiF₄ and SiH₄, *Chem. Phys.* 109:67 (1986).

5. M. Snels and J. Reuss, Induction effects on IR-predissociation spectra of (SF₆)₂, (SiF₄)₂ and (SiH₄)₂, *Chem. Phys. Lett.* 140:543 (1987).

6. B. Heymen, C. Liedenbaum, S. Stolte, and J. Reuss, Hole burning in the IR predissociation spectrum of SF₆-dimers, *Laser Chem.* 8:275 (1988).

7. B. Heymen, A. Bizzarri, S. Stolte, and J. Reuss, IR–IR double resonance experiments on SF₆ and SiF₄ clusters, *Chem. Phys.* 132:331 (1989).

8. J.W.I. van Bladel and A. van der Avoird, to be published.

9. E.B. Wilson, J.C. Decius, and P.C. Cross, "Molecular Vibrations", McGraw-Hill, New York (1955).

10. S. Califano, V. Schettino, and N. Neto, "Lattice Dynamics of Molecular Crystals", Springer, Berlin (1981).

11. N. Neto, G. Taddei, S. Califano, and S.H. Walmsley, Lattice dynamics of molecular crystals using a molecular force field and an intermolecular potential function with application to the atom-atom model, *Mol. Phys.* 31:457 (1976).

12. A.P.J.M. Jongenelis, T.H.M. van den Berg, A.P.J. Jansen, J. Schmidt, and A. van der Avoird, Vibron band structure in chlorinated benzene crystals; lattice dynamics calculations and Raman spectra of 1,2,4,5-tetrachlorobenzene, *J. Chem. Phys.* 89:4023 (1988).

VIBRATIONAL PREDISSOCIATION OF THE $He - I_2(B, v = 34) - Ne$ COMPLEX: SEQUENTIAL MECHANISM AND DIRECT DISSOCIATION

P. Villarreal, S. Miret-Artés, J. Campos-Martínez and G. Delgado-Barrio
Instituto de Estructura de la Materia, C.S.I.C.
Serrano 123, 28006 Madrid, SPAIN.

Abstract

The competition between direct dissociation of $He - I_2 - Ne$, producing either a diatomic fragment $He - Ne$ or separated rare atoms plus I_2, and sequential processes in which a rare atom and a triatomic complex emerge in a first step, is estimated by means of quantal approximations resting on the Fermi's Golden Rule. Assuming a non rotating I_2 molecule and restricting the rare atoms to move on a perpendicular plane to its axis, the four-body problem is reduced to an effective three-body one after a diabatic separation of the iodine vibration is performed. Bond coordinates are used to describe sequential mechanisms as well as double continuum fragmentation, while Jacobi coordinates appear as more convenient to analyse the dissociation into two diatomic, I_2 and $He - Ne$, fragments.

1   Introduction

Janda and co.[1] have recently reported the presence of all the possible channels of fragmentation in the vibrational predissociation of Ne_2Cl_2, that is, dissociation either in two separated Ne atoms plus Cl_2 or in two diatomic Ne_2 and Cl_2 molecules or sequentially in a Ne atom plus a $NeCl_2$ complex and further dissociation of this last one. With the exception of energy constraints, the two first cases may proceed via the lost of one vibrational quantum by Cl_2 while the third one needs two quanta to take place. Due to the wealth of mechanisms of fragmentation that may appear in this type of complexes with one chemical and two weak bonds, it deserves to estimate, at least in an approximate way, their relative contribution to the dissociation process.

However, as a first attempt, we study the vibrational predissociation of the $He - I_2 - Ne$ complex presenting some advantages as regards to the dynamical

Dynamics of Polyatomic Van der Waals Complexes
Edited by N. Halberstadt and K. C. Janda
Plenum Press, New York, 1990

approximations that may be applied in the theoretical treatment. Mainly the very small rotational constant induced by the large mass of iodine suggests the possibility of neglecting its rotation without losing of physical significance. In addition, the kinetic couplings appearing in bond coordinates[2] may also be neglected owing to the presence of the inverse of the iodine mass as a common factor to all of them[3].

Assuming a near equilibrium configuration of the complex, as it was done to study the sequential fragmentation of $He-I_2-Ne$[3] and also the double continuum problem in the $He-I_2-He$[4], we estimate in this work the incidence of the channel leading to two diatomic fragments. Briefly, in the particular configuration already mentioned a diabatic approximation in describing the iodine vibration faces us to a three-body problem with a heavy I_2 pseudo-atom and two rare atoms, for which a total angular momentum $J = 0$ is prefixed. To study sequential and double continuum fragmentations it is quite natural to use bond coordinates, i.e., the distances between each rare atom and the center of mass of I_2 and the angle formed by the corresponding vectors, that may be adiabatically treated. However, to analyse the fragmentation in two diatomic molecules, it seems more suitable to work in Jacobi coordinates, that means, the distance from the center of mass of I_2 to that of $He - Ne$, the distance between the rare atoms and the angle defined by the corresponding vectors that, again, may be considered as an adiabatic variable. We are aware of the troubling introduced by the mixed using of different coordinates. But, as it will be shown, the initial metastable ground level, as regards at least to its energy, becomes equivalently described by both treatments. Hence, the estimation of the different contributions to the fragmentation rate may be representative.

This paper is organized as follows. Section II is devoted to present the theoretical background needed to calculate the different rates within a Golden Rule frame. Finally, in Section III, we present the results obtained for an initial excitation $v = 34$, for which the $\Delta v = -1$ channel only allows the fragmentation into He plus I_2Ne, all the other processes taking place on the $\Delta v = -2$ channel.

2 Energy Levels and Fragmentation Rates

2.1 Bond coordinates

In the particular configuration mentioned above, and denoting by r the iodine bondlenght, $\underline{R}_1$ and $\underline{R}_2$ the vectors going from the I_2 center of mass to the Ne and He atoms, respectively, that form an angle γ, we write the total wavefunction as [3]

$$\Psi(r, \underline{R}_1, \underline{R}_2) \simeq \chi_v(r)\phi^{(v)}(R_1, R_2; \gamma)F^{(v)}(\hat{R}_1, \hat{R}_2) \qquad (1)$$

where χ is an eigenfunction of the I_2 vibrator with associated eigenvalue $E_{I_2}(v)$

$$\left[-\frac{\hbar^2}{m_I} + V_{I_2}(r) - E_{I_2}(v) \right] \chi_v(r) = 0 \qquad (2)$$

and $\phi^{(v)}$ is solution, at each γ value, of the equation

$$\left[-\frac{\hbar^2}{2\mu_1} \frac{\partial^2}{\partial R_1^2} - \frac{\hbar^2}{2\mu_2} \frac{\partial^2}{\partial R_2^2} + <\chi_v\,|[V_{Ne-I_2}(R_1,r) + V_{He-I_2}(R_2,r)]|\,\chi_v> + \right.$$

$$\left. +V_{He-Ne}(R_1,R_2,\gamma) - \epsilon\right]\phi^{(v)}(R_1,R_2;\gamma) = 0 \qquad (3)$$

with $\mu_1 = 2m_I m_{Ne}/(2m_I + m_{Ne})$ and $\mu_2 = 2m_I m_{He}/(2m_I + m_{He})$ being the relevant reduced masses. Eq.(3) has discrete as well as continuum and also double continuum solutions. In the first case, ϕ depends on an additional quantum number, k, necessary to specify the two-dimensional stretching problem, while its eigenvalue ϵ becomes an angular dependent function $W_k^{(v)}(\gamma)$ constituting an effective potential for the bending motion described by the F function,

$$\left[\frac{\hbar^2}{4\mu_1} \{\underline{j}_1^2, <\phi_k^{(v)}\left|R_1^{-2}\right|\phi_k^{(v)}>\} + \frac{\hbar^2}{4\mu_2} \{\underline{j}_2^2, <\phi_k^{(v)}\left|R_2^{-2}\right|\phi_k^{(v)}>\} + \right.$$

$$\left. +W_k^{(v)}(\gamma) - E_{vkl}\right]F_{kl}^{(v)}(\hat{R}_1, \hat{R}^2) = 0 \qquad (4)$$

Here, $\underline{j}_1$ and $\underline{j}_2$ are angular momentum operators associated to $\underline{R}_1$ and $\underline{R}_2$, respectively, and $\{\,,\,\}$ denotes an anticommutator introduced to overcome the noncommutativity between angular operators and angular dependent functions[5]. The symbols $<\,>$ stand for quadratures on the radial variables, and all the dynamical couplings[2] have been neglected since they are of the order of $1/(2m_I)$. The energy of a level labelled by the set of quantum numbers (v,k,l) becomes E_{vkl} measured from the iodine energy $E_{I_2}(v)$.

Eqs. (1) to (4) involve a diabatic separation of the I_2 vibration and an angular adiabatic approximation[2] in which the "transparency" of the stretching function $\phi_k^{(v)}$ to the action of the angular operators has been implicitly assumed. Such a function may be obtained by expansion in a numerical basis of products of triatomic stretching functions

$$\phi_k^{(v)}(R_1,R_2;\gamma) = \sum_{mn} a_{mn}^{(v,k)}(\gamma)\xi_m^{(v)}(R_1)\eta_n^{(v)}(R_2) \qquad (5)$$

verifying

$$\left[-\frac{\hbar^2}{2\mu_1} \frac{\partial^2}{\partial R_1^2} + <\chi_v\,|V_{Ne-I_2}(R_1,r)|\,\chi_v> - E_{Ne-I_2}^{(v,m)}\right]\xi_m^{(v)}(R_1) = 0 \qquad (6)$$

$$\left[-\frac{\hbar^2}{2\mu_2} \frac{\partial^2}{\partial R_2^2} + <\chi_v\,|V_{He-I_2}(R_2,r)|\,\chi_v> - E_{He-I_2}^{(v,n)}\right]\eta_n^{(v)}(R_2) = 0 \qquad (7)$$

The coefficients $a_{m,n}^{(v,k)}(\gamma)$ together with the eigenvalues $W_k^{(v)}(\gamma)$ are being obtained by simple diagonalization of the following matrix,

$$H_{mn;m'n'}^{(v)}(\gamma) = \left[E_{Ne-I_2}^{(v,m)} + E_{He-I_2}^{(v,n)}\right]\delta_{mm'}\delta_{nn'} +$$

$$+ < \xi_m^{(v)}(R_1)\eta_n^{(v)}(R_2) \, |V_{He-Ne}(R_1, R_2, \gamma)| \, \xi_{m'}^{(v)}(R_1)\eta_{n'}^{(v)}(R_2) > \qquad (8)$$

Similarly, eq.(4) may be solved by expanding the F function in an angular basis set and diagonalizing the representation of the hamiltonian appearing in that equation in such a basis[3].

In order to estimate rates for sequential fragmentation, we consider for each value of γ the continuum states describing a Ne atom plus a triatomic $He - I_2$ molecule in a given n'' state, where I_2 is in a lower $v' < v'$ vibrational level

$$\varphi_{v'\epsilon_1 n''}^{cont, Ne}(r, R_1, R_2; \gamma) = \xi_{\epsilon_1}^{(v', n'')}(R_1; \gamma)\eta_{n''}^{(v')}(R_2)\chi_{v'}(r) \qquad (9)$$

where χ and η are already defined by Eqs. (2) and (7), respectively, and ξ is now a continuum wavefunction normalized in energy to a δ distribution fulfilling

$$\left[-\frac{\hbar^2}{2\mu_1}\frac{\partial^2}{\partial R_1^2} + < \chi_{v'} \, |V_{Ne-I_2}(r, R_1)| \, \chi_{v'} > + \right.$$

$$\left. + < \eta_{n''}^{(v')} \, |V_{He-Ne}(R_1, R_2, \gamma)| \, \eta_{n''}^{(v')} > -\epsilon_1 \right] \xi_{\epsilon_1}^{(v', n'')}(R_2; \gamma) = 0 \qquad (10)$$

with $\epsilon_1 = E_{I_2}(v) - E_{I_2}(v') + W_k^{(v')}(\gamma) - E_{He-I_2}^{(v', n'')}$ being the kinetic energy between the fragments. Thus, in the Golden Rule frame, the rate for escaping Ne becomes

$$\Gamma_{vk}^{Ne}(\gamma) = \pi \sum_{v'<v} \sum_{n''} \left| \sum_{mn} a_{mn}^{(v,k)}(\gamma) \left[< \eta_n^{(v)} \, | \, \eta_{n''}^{(v')} > < \xi_m^{(v)} \, | \, \chi_v \, | \, V_{Ne-I_2} \, | \, \chi_{v'} > + \right. \right.$$

$$\left. \left. + < \xi_m^{(v)} \, | \, \xi_{\epsilon_1}^{(v', n'')} > < \eta_n^{(v)} \, |< \chi_v \, | \, V_{He-I_2} \, | \, \chi_{v'} >| \, \eta_{n''}^{(v')} > \right] \right|^2 \qquad (11)$$

A similar treatment can be obviously done to calculate the rate for fragmentation of the complex in He plus $I_2 - Ne$. In this way, the total rate for dissociation in a rare atom and the corresponding triatomic rest is estimated by averaging on the angular functions[3]

$$\Gamma_{vkl}^{SD} = < F_{kl}^{(v)}(\hat{R}_1, \hat{R}_2)| \left[\Gamma_{vk}^{Ne}(\gamma) + \Gamma_{vk}^{He}(\gamma) \right] |F_{kl}^{(v)}(\hat{R}_1, \hat{R}_2) > \qquad (12)$$

where "SD" denotes single dissociation.

To calculate double continuum rates corresponding to the fragmentation of the complex in iodine plus separated rare atoms we need to obtain double continuum wavefunctions. To this end, we apply one step of a SCF procedure[6] and write down for each γ value such a function as a product[4]

$$\varphi_{v', \epsilon_1, E-\epsilon_1}^{DC}(r, R_1, R_2; \gamma) = \chi_{v'}(r)\xi_{\epsilon_1}^{(v')}(R_1)\eta_{\epsilon_1, E-\epsilon_1}^{(v')}(R_2; \gamma) \qquad (13)$$

where ξ describes the escaping of Ne on a potential $< \chi_{v'}|V_{Ne-I_2}|\chi_{v'} >$ with kinetic energy ϵ_1 relative to I_2,

$$\left[-\frac{\hbar^2}{2\mu_1}\frac{\partial^2}{\partial R_1^2} + <\chi_{v'}|V_{Ne-I_2}(r,R_1)|\chi_{v'}> -\epsilon_1 \right] \xi_{\epsilon_1}^{(v')}(R_1) = 0 \qquad (14)$$

while η, describing the escaping of He, feels a potential $He - I_2$ distorted by the presence of the Ne atom

$$\left[-\frac{\hbar^2}{2\mu_2}\frac{\partial^2}{\partial R_2^2} + <\chi_{v'} \mid V_{He-I_2}(r,R_2) \mid \chi_{v'}> + \right.$$

$$\left. + \int_0^E d\epsilon_1' <\xi_{\epsilon_1}^{(v')} \mid V_{He-Ne}(R_1,R_2,\gamma) \mid \xi_{\epsilon_1'}^{(v')}> +\epsilon_1 - E \right] \eta_{\epsilon_1,E-\epsilon_1}^{(v')}(R_2;\gamma) = 0 \quad (15)$$

where $E = E_{I_2}(v) - E_{I_2}(v') + W_k^{(v)}(\gamma)$ is the angular dependent available energy that is being shared, in all the possible ways, by the escaping rare atoms. The corresponding rate may be obtained by quadrature[4]

$$\Gamma_{vk}^{DC}(\gamma) = \pi \sum_{v'<v} \int_0^E d\epsilon_1' \left| \sum_{mn} a_{mn}^{(v,k)}(\gamma) \left[<\eta_n^{(v)} \mid \eta_{\epsilon_1,E-\epsilon_1}^{(v')}> \langle \xi_m^{(v)} \mid <\chi_v \mid V_{Ne-I_2} \mid \chi_v >\mid \right.\right.$$

$$\left.\left. \mid \xi_{\epsilon_1}^{(v')}\rangle + <\xi_m^{(v)} \mid \xi_{\epsilon_1}^{(v')}> \langle \eta_n^{(v)} \mid <\chi_v \mid V_{He-I_2} \mid \chi_{v'} > \eta_{\epsilon_1,E-\epsilon_1}^{(v')}\rangle \right] \right|^2 \qquad (16)$$

Again, the total rate for double continuum fragmentation is obtained through an average

$$\Gamma_{vkl}^{DC} = <F_{vl}^{(v)} \mid \Gamma_{vk}^{DC}(\gamma) \mid F_{kl}^{(v)}> \qquad (17)$$

Let us analyse briefly the expressions obtained till now. In Eq. (11), discrete-discrete $<\eta_n^{(v)}|\eta_{n''}^{(v')}>$ and discrete continuum $<\xi_m^{(v)}|\xi_{\epsilon_1}^{(v',n'')}>$ overlaps do appear. With the exception of large variations of the corresponding atom-iodine diabatic potentials with the vibrational level of I_2, that would be reflected in large variations of the blue-shift in the absorption spectrum of the complex, those become almost 1 for $n'' = n$ and 0 otherwise while these are expected to be really small if the interaction $He - Ne$ is very weak, as it has to be. Using the same argument, the presence of discrete-continuum overlaps in Eq.(16) suggest a very small contribution of the double continuum mechanism to the vibrational predissociation of the complex. On the other hand, it seems reasonable to expect the main contribution to averaged rates coming from the region of equilibrium of the complex, i.e., the angular region of the well in the potentials $W_k^{(v)}(\gamma)$.

2.2 Jacobi coordinates

Dissociation of the complex in two diatomic fragments is described, as it was mentioned above, in Jacobi coordinates. We shall call ρ to the distance $He - Ne$, R to the distance between the iodine center of mass and that of $He - Ne$, and θ

to the angle defined by the two corresponding vectors. To consider the vibrational predissociation of the ground van der Waals level of the complex, that is actually our interest, we propose a very crude treatment. It consists in applying a diabatic approximation not only in the I_2 vibration but also in the $He-Ne$ one. Moreover, this diatomic van der Waals molecule is assumed to remain in its ground level ever and ever, let us say $q = 0$. Finally, an angular adiabatic approximation is applied[7] and the discrete wavefunction is written down as

$$\Psi^{dis}(r, \rho, R; \theta) \simeq \chi_v(r) f_{q=0}(\rho) \varphi_{v_s}(R; \theta) \mathcal{F}_{v_s, v_b}(\hat{\rho}, \hat{R}) \qquad (18)$$

where f is a vibrational wavefunction of the isolated $He-Ne$ molecule

$$\left[-\frac{\hbar^2 (m_{Ne} + m_{He})}{2 m_{Ne} \cdot m_{He}} \frac{\partial^2}{\partial \rho^2} + V_{He-Ne}(\rho) - E_{q=0} \right] f_{q=0}(\rho) = 0 \qquad (19)$$

while ρ, for each θ value, fulfills

$$\left[-\frac{\hbar^2}{2\mu} \frac{\partial^2}{\partial R^2} + < \chi_v \left\| [V_{Ne-I_2}(r, \bar{\rho}, R; \theta) + V_{He-I_2}(r, \bar{\rho}, R; \theta)] \right\| \chi_v > - \right.$$

$$\left. -U_{v_s}(\theta) \right] \varphi_{v_s}(R; \theta) = 0 \qquad (20)$$

where the $He-Ne$ bondlength has been fixed at its equilibrium value $\rho = \bar{\rho}$ and $\mu = 2m_I \cdot (m_{Ne} + m_{He}) / (2m_I + m_{Ne} + m_{He})$ is the relevant reduced mass. In Eq.(20), v_s denotes the stretching quantum number and the eigenvalues U, varying with θ, constitute effective potentials for the bending motion

$$\left[Be \underline{j}^2 + \frac{\hbar^2}{2\mu \bar{R}^2} \underline{l}^2 + U_{v_s}(\theta) - E_{v v_s, v_b} \right] \mathcal{F}_{v_s, v_b}(\hat{\rho}, \hat{R}) = 0 \qquad (21)$$

with Be being the rotational constant of $He-Ne$ and $\underline{j}$ and $\underline{l}$ angular momentum operators associated to $\underline{\rho}$ and $\underline{R}$, respectively. Here, v_b stands for the bending quantum number. Note that the R distance has been fixed at its equilibrium value to describe the end-over-end rotational energy.

At each θ value we also consider the continuum states representing two diatomic fragments with relative kinetic energy $\epsilon = E_{I_2}(v) - E_{I_2}(v') + U_{v_s}(\theta)$

$$\psi^{cont} \simeq \chi_{v'}(r) f_{q=0}(\rho) \varphi_\epsilon(R; \theta) \qquad (22)$$

where φ satisfies

$$\left[-\frac{\hbar^2}{2\mu} \frac{\partial^2}{\partial R^2} + < \chi_{v'} \left\| [V_{Ne-I_2} + V_{He-I_2}] \right\| \chi_{v'} > - \epsilon \right] \varphi_\epsilon(R; \theta) = 0 \qquad (23)$$

Using again the Golden Rule expression, we have the rate for fragmentation into two diatoms (TD),

$$\Gamma_{vv_s}^{TD}(\theta) = \pi \left| \langle \varphi_{v_s}(R;\theta) \left| < \chi_v \right| [V_{NeI_2}(r,\bar{\rho},R,\theta) + V_{He-I_2}(r,\bar{\rho},R,\theta)] \left| \chi_{v'} > \right| \varphi_\epsilon^{(R;\theta)} \rangle \right|^2$$

$$(24)$$

and the averaged rate

$$\Gamma_{vv_s v_b}^{TD} = < \mathcal{F}_{v_s v_b}(\hat{\rho},\hat{R}) | \Gamma_{vv_s}^{TD}(\theta) | \mathcal{F}_{v_s v_b}(\hat{\rho},\hat{R}) > \qquad (25)$$

3 Calculations and Results

The simple addition of pairwise atom-atom interactions was used to described the potential energy surface. Each one of them was represented by means of Morse functions, where the relevant parameters were the same already employed[8] (see Table I there).

All the integrals needed were performed by Gauss-Chebyshev quadratures. Vibrational iodine wavefunctions where calculated in a grid of 3000 points in the range $[2,8]$Å through a mixed Truhlar-Numerov procedure[9], from which 500 points of the gaussian quadrature were numerically interpolated. The same way was followed to get 100 points from a grid of 2000 points in the range $[2.5,20]$Å for triatomic discrete wavefunctions. A Fox-Numerov propagation method and matching of the function to a sinus at long distances[10] was used to get the initial 2000 points of continuum functions in the same range. Also, the energy integrals appearing in Eqs.(15) and (16) where performed through 20 points Gauss-Chebyshev quadratures. 20 values of γ, as well as of θ, equally spaced in the range $[0,\pi]$ where taken into account, and the corresponding angular functions were expanded in series of Legendre polynomials by a collocation procedure. A basis of 25 space-fixed angular functions[11] were used to represent the bending hamiltonians for a total angular momentun $J = 0$. Finally, all the diagonalizations were performed by a standard Jacobi method because the small size of the corresponding matrices.

We have studied the ground van der Waals state for an initial vibrational excitation of the iodine $v = 34$. Its energy is $-93.24cm^{-1}$ measured from the I_2 level when bond coordinates are used, while a value of $-93.47cm^{-1}$ is obtained through Jacobi coordinates if the only one level of $He - Ne$, $-2.32cm^{-1}$, is added to the lowest energy of Eq.(21), $-91.15cm^{-1}$. Taking into account the number of approximations included, in particular when Jacobi coordinates were used, we consider as satisfactory the agreement obtained by both treatments using different coordinates.

Concerning with the different mechanisms of fragmentation, the averaged rate for single dissociation gets a value $\Gamma_{v=34,0,0}^{SD} = 0.392cm^{-1}$, the main contribution corresponding to a $\Delta v = -1$ channel producing He and $I_2 - Ne$, $0.377cm^{-1}$, and the rest to $\Delta v = -2$ giving rise to the same products and also Ne plus $I_2 - He$. As it was expected, the double continuum fragmentation is almost negligible, with a rate $\Gamma_{34,0,0}^{DC} = 8 \times 10^{-6}cm^{-1}$. Despite we are aware that the simple approach to get double continuum wave-functions used here is not completely satisfactory (for instance, the result from a much more expensive perturbative treatment[4] differs by a factor of 3 from this approximation for $He - I_2 - He)$, we belive that this estimation is representative enough. Finally, the rate for fragmentation in two diatoms is, from Eq.(25), $\Gamma_{34,0,0}^{TD} = 0.004 cm^{-1}$. From this results we conclude that for $He - I_2(B, v = 34) - Ne$, the sequential mechanism dominates and contributes to its vibrational predissociation as high as 99%, the rest corresponding to dissociation into two diatomic fragments since the complete, double continuum, fragmentation mechanism becomes almost forbidden. Of course, these contributions may dramatically change for a complex like $Ne_2 Cl_2$, where Ne_2 shows several discrete levels and a larger coupling between Cl_2 vibrations and Ne_2 ones through the van der Waals $Ne_2 - Cl_2$ bond may be expected. In addition, the Cl_2 rotation may play an important role and a more sophisticated treatment than the used here needs to be carried out.

References

[1] S.R. Hair, J.I. Cline, C.R. Bieler, and K.C. Janda, J. Chem. Phys. 90, 2935(1989).

[2] a) G.A. Natanson, G.S. Ezra, G. Delgado-Barrio, and R.S. Berry, J. Chem. Phys. 81, 3400 (1984).

b) Ibid 84, 2035 (1986).

[3] P. Villarreal, A. Varadé, and G.Delgado-Barrio, J. Chem. Phys. 90, 2684 (1989).

[4] P. Villarreal, S. Miret-Artés, O. Roncero, S. Serna, J. Campos-Martínez, and G.Delgado-Barrio, submitted to J. Chem. Phys.

[5] a) M. Aguado, P. Villarreal, G.Delgado-Barrio, P. Mareca, and J.A. Beswick, Chem. Phys. Lett. 102, 227 (1983).

b) F.A. Gianturco, A. Palma, P. Villarreal, G.Delgado-Barrio, and O.Roncero, J. Chem. Phys. 87, 1054 (1987).

c) Y.S. Kim, M. Hutchinson, and T.F. George, J. Chem. Phys. 86, 5515 (1987).

[6] R.B. Gerber, and M.A. Ratner, Advan. Chem. Phys. LXX(1), 97 (1988).

[7] J.A. Beswick, and G. Delgado-Barrio, J. Chem. Phys. 73, 3653 (1980).

[8] G. Delgado-Barrio, P. Villarreal, A. Vardé, N. Martín, and A. García-Vela in *Structure and Dynamics of Weakly Bound Molecular Complexes*, NATO ASI Series C 212, p. 573, A.Weber Ed. (Reidel, Dordrecht, 1987).

[9] G. Delgado-Barrio, A.M. Cortina, A.Varadé, P. Mareca, P. Villarreal, and S. Miret-Artés, J. Compt. Chem. 7, 208 (1986).

[10] O. Roncero, S. Miret-Artés, G. Delgado-Barrio, and P. Villarreal, J. Chem. Phys. 85, 2084 (1986).

[11] A.M. Arthurs, and A. Dalgarno, Proc. Roy. Soc. London, A 256, 540 (1960).

ORGANIZING COMMITTEE

Directors

N. HALBERSTADT
Laboratoire de Photophysique
  Moléculaire CNRS,
Université de Paris-Sud,
91405 - ORSAY CEDEX, France
UPPMØ52 at FRORS12

K.C. JANDA
Department of Chemistry,
University of Pittsburgh,
PITTSBURGH PA 15260, U.S.A.
JANDA at PITTVMS

Organisational Committee

A. BESWICK
LURE
Université de Paris-Sud,
91405 - ORSAY CEDEX, France
ULURØØ3 at FRORS31

O. DUBOST
Laboratoire de Photophysique
  Moléculaire CNRS,
Université de Paris-Sud,
91405 - ORSAY CEDEX, France
UPPMØ63 at FRORS12

A. KELLER
Service des Atomes et des Surfaces,
CEN Saclay,
91191 - GIF/YVETTE CEDEX, France
and

Laboratoire de Photophysique
  Moléculaire CNRS
Université de Paris-Sud,
91405 - ORSAY CEDEX, France
UPPMØ59 at FRORS31

O. RONCERO
LURE, Université de Paris-Sud,
91405 - ORSAY CEDEX, France
ULURØ2Ø at FRORS31
and
Instituto de Estructura de la Materia,
CSIC,
Serrano 123
28006 - MADRID, Spain
IMTORØ1 at EMDCSIC1

G.G. BALINT-KURTI
School of Chemistry,
University of Bristol,
BRISTOL BS8 1TS, U.K.
GGB at UKACRL

E.R. BERNSTEIN
Department of Chemistry,
Colorado State University,
FORT COLLINS CO 80523, U.S.A.

P. BRECHIGNAC
Laboratoire de Photophysique
    Moléculaire CNRS,
Université de Paris-Sud,
91405 - ORSAY CEDEX, France,
UPPMØ52 at FRORS31

U. BUCK
Max-Planck-Institut für
    Strömungsforschung
Bunsenstrasse 10,
D3400 - GOTTINGEN, F.R.G.
UBUCK at DGOGWDG1

P.R. BUNKER
Herzberg Institute of Astrophysics,
National Research Council
    of Canada,
OTTAWA, Ontario K1A 0R6,
    Canada
PRB at NRCVMØ1

P. CASAVECCHIA
Dipartimento di Chimica,
Università degli Studi di Perugia,
Via Elce di Sotto 8,
06100 - PERUGIA, Italy
PGSCAT at IPGUNIV

M.S. CHILD
Theoretical Chemistry Department,
Oxford University,
5 South Parks Road,
OXFORD OX1 3UB, U.K.
CHILDMS at VAX.OX.AC.UK (Janet)

D.C. CLARY
University Chemical
    Laboratory
Lensfield Road,
CAMBRIDGE CB2 1EW, U.K.
DCC4 at PHX.CAM.AC.UK (Janet)

G. DELGADO BARRIO
Instituto de Estructura de la Materia,
CSIC,
Serrano 123,
28006 - MADRID, Spain
IMTV1Ø at EMDCSIC1

G.E. EWING
Department of Chemistry,
Indiana University,
BLOOMINGTON IN 47005, U.S.A.
EWINGG at IUBACS

R.B. GERBER
Department of Physical
    Chemistry and
The Fritz Haber Research Center,
The Hebrew University,
JERUSALEM, 91904, Israel
BENNY at batata.Huji.AC.IL
and
Department of Chemistry,
University of California,
IRVINE CA 92717, U.S.A.
BGERBER at UCIVMSA

S.K. GRAY
Department of Chemistry,
Northern Illinois University,
DELKAB IL 60115, U.S.A.
T4ØSKG1 at NIU

J.M. HUTSON
Department of Chemistry,
University of Durham,
Science Laboratories,
South Road
DURHAM DH1 3LE, U.K.
J.M.HUTSON at
    DURHAM.AC.UK (Janet)

M. ITO,
Quantum Chemistry Laboratory,
Department of Chemistry,
Faculty of Science
Tohoku University,
SENDAI 980, Japan

C. JOUVET
Laboratoire de Photophysique
   Moléculaire CNRS,
Université de Paris-Sud,
91405 - ORSAY CEDEX, France

W.A. KLEMPERER
Department of Chemistry,
Harvard University,
12 Oxford Street,
CAMBRIDGE MA 02138, U.S.A.
KLEMPERER at HUCHE1

R.J. LE ROY
Guelph-Waterloo Centre for
   Graduate Work in Chemistry,
University of Waterloo,
Department of Chemistry,
WATERLOO, Ontario N2L 3G1,
   Canada
LEROY at WATDCS,
LEROY at WATDCS.UWARLOO.CA

M.I. LESTER
Chemistry Department,
University of Pennsylvania,
231 South 34th St.
PHILADELPHIA PA 19104-6323,
   U.S.A.
LESTER at C.CHEM.UPENN.EDU

S. LEUTWYLER
Institut für Anorganische,
   Analytische und Physikalische
   Chemie,
Universität Bern,
Freiestr. 3,
CH-3000 BERN 9, Switzerland
U30K at CBEBDA3T

R.E. MILLER
Department of Chemistry,
University of North Carolina,
Venable Hall 045A,
CHAPEL HILL NC 27514, U.S.A.
MILLER at UNCVX1

D.J. NESBITT
Joint Institute Laboratory
   for Astrophysics,
University of Colorado,
Box 440,
BOULDER CO 80309, U.S.A.
DJN at JILA

J. REUSS,
Fysich Laboratorium,
Katholieke Universiteit Nijmegen,
Toernooiveld,
6525 ED-NIJMEGEN,
   The Netherlands
U630012 at HNYKUN11

R.P.H. RETTSCHNICK
Laboratory of Physical Chemistry,
University of Amsterdam,
Nieuwe Achtergracht 127
1018 WS AMSTERDAM,
   The Netherlands

S.A. RICE
The James Franck Institute,
The University of Chicago,
5640 South Ellis Avenue,
CHICAGO IL 60637, U.S.A.

B. SOEP
Laboratoire de Photophysique
   Moléculaire CNRS,
Université de Paris-Sud,
91405 - ORSAY CEDEX, France
UPPM040 at FRORS31

A.J. STONE
University Chemical Laboratory,
Lensfield Road
CAMBRIDGE CB2 1EW, U.K.
AJS1 at PHX.CAM.AC.UK (Janet)

J. TENNYSON
Department of Physics and
   Astronomy,
University College London,
Gower Stret,
LONDON WC1E 6BT, U.K.
TENNYSON at EUCLID.UCL.AC.UK
   (Janet)

A. TRAMER
Laboratoire de Photophysique
   Moléculaire CNRS,
Université de Paris-Sud
91405 - ORSAY CEDEX, France

D.G. TRUHLAR
139 Smith Hall,
Department of Chemistry,
University of Minnesota,
207 Pleasant St. SE,
MINNEAPOLIS MN 55455, U.S.A.
TRUHLAR at UMNACVX

PARTICIPANTS

L. BENEVENTI,
Dipartimento di Chimica,
Università degli Studi di Perugia,
Via Elce di Sotto, 8,
06100 PERUGIA, Italy
PGSCAT at IPGUNIV

J. CAMPOS-MARTINEZ,
Instituto de Estructura de la Materia,
CSIC
Serrano 123,
28006 MADRID, Spain
IMTJC15 at EMDCSIC1

C.E. CHUAQUI
Guelph Waterloo Centre for
   Graduate Work in Chemistry,
University of Waterloo,
Department of Chemistry,
WATERLOO, Ontario N2L 3G1,
   Canada
CHUAQUI at GWCHEM
CHUAQUI at GWCHEM.UWATERLOO.CA

J.P. DAUDEY
Laboratoire de Physique Quantique,
Université Paul Sabatier,
118, Route de Narbonne,
31062 - TOULOUSE CEDEX, France
DAUDEY at FRPQT51

I. DIMICOLI
Département d'Etude des Lasers
   et de la Physico-Chimie,
IRDI, DESICP,DLPC,
CEN Saclay, Bât. 522,
BP2,
91191 - GIF/YVETTE CEDEX, France

M.C. DUVAL
Laboratoire de Photophysique
   Moléculaire CNRS,
Université de Paris-Sud,
91405 - ORSAY CEDEX, France

A. GULDBERG
Chemistry Laboratory IV,
H.C. Ørsted Institute,
Universitetsparken 5,
2100 - COPENHAGEN Ø,
   Denmark
GULDBERG at DKUCCC11

P. HALVICK
Laboratoire de Physico-Chimie
   Théorique,
Université de Bordeaux 1,
351 Cours de la Libération
33405 - TALENCE CEDEX, France
DUGUAY at FRBDX11

A. KARPFEN
Institut für Theoretische Chemie
   und Strahlenchemie
   der Universität Wien,
Waehringerstr. 17,
A1090 WIEN, Austria
A8441DAB at AWIUNI11

E. R. KERSTEL
Princeton University,
Department of Chemistry,
PRINCETON NJ 08544, U.S.A.
KERSTELE at PUCC

F. LAHMANI
Laboratoire de Photophysique
   Moléculaire CNRS,
Université de Paris-Sud,
91405 - ORSAY CEDEX, France

C. LARDEUX
Laboratoire de Photophysique
   Moléculaire CNRS,
Université de Paris-Sud,
91405 - ORSAY CEDEX, France

N. LIPKIN,
Department of Physical Chemistry,
Technion,
HAIFA 32000, Israel
CHR31NM at TECHNION

A.R.W. Mc KELLAR
Herzberg Institute of Astrophysics,
National Research Council of
  Canada,
OTTAWA, Ontario K1A 0R6,
  Canada
ARM at NRCVM01

M.D. MARSHALL,
Department of Chemistry,
Amherst College,
AMHERST MA 01002, U.S.A.
MDMARSHALL at AMHERST

W.L. MEERTS
Fysich Laboratorium,
Katholieke Universiteit Nijmegen,
Toernooiveld,
6525 ED-NIJMEGEN,
  The Netherlands
U630013 at HNYKUN11

B. SCHMIDT
Max-Planck Institut für
  Strömungsforschung,
Bunsenstrasse 10,
D-3400 GOTTINGEN, F.R.G.
BSCHMID1 at DGOGWDG1

D. SOLGADI
Laboratoire de Photophysique
  Moléculaire CNRS,
Université de Paris-Sud,
91405 - ORSAY CEDEX, France

T.A. STEPHENSON
Department of Chemistry,
Swarthmore College,
SWARTHMORE PA 19081, U.S.A.
STEPHENSON at SWARTHMR

J.W.I. VAN BLADEL
Faculteit Natuurwetenschappen,
Vakgroep Theoretische Chemie,
Toernooiveld,
6525 ED NIJMEGEN,
  The Netherlands

A. VAN DER AVOIRD
Institute of Theoretical Chemistry,
Faculty of Mathematics and
  Natural Sciences,
Katholieke Universiteit Nijmegen,
Toernooiveld,
6525 ED NIJMEGEN,
  The Netherlands
U644101 at HNYKUN11

P. VILLARREAL
Instituto de Estructura de la Materia,
CSIC,
Serano 123,
28006 MADRID, Spain
IMTVI01 at EMDCSIC1

J.P. VISTICOT,
Service de Physique des Atomes
  et des Surfaces,
CEN Saclay,
91191 - GIF/YVETTE CEDEX, France

INDEX

542